To JIMMY
Best Wishes
Michael Milken

Richard Perle

PROSTATE CANCER

CURRENT CLINICAL ONCOLOGY

Maurie Markman, MD, Series Editor

PROSTATE CANCER

Signaling Networks, Genetics, and New Treatment Strategies

Edited by

RICHARD G. PESTELL, MD, PhD

MARJA T. NEVALAINEN, MD, PhD

Kimmel Cancer Center, Thomas Jefferson University, Philadelphia, PA

Foreword by

MICHAEL MILKEN, MBA

Prostate Cancer Foundation, Santa Monica, CA

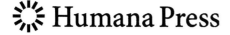 Humana Press

Editors
Richard G. Pestell
Department of Cancer Biology
Kimmel Cancer Center
Thomas Jefferson University
Philadelphia, PA

Marja T. Nevalainen
Department of Cancer Biology
Kimmel Cancer Center
Thomas Jefferson University
Philadelphia, PA

Series Editor
Maurie Markman
MD Anderson Cancer Center
University of Texas
Houston, TX

ISBN 978-1-58829-741-9   e-ISBN 978-1-60327-079-3

Library of Congress Control Number: 2007941659

Cover illustration: Fig. 3, Chapter 17, "Advances in Surgical Intervention of Prostate Cancer: Comparison of the Benefits and Pitfalls of Retropubic, Perineal, and Laparoscopic Radical Prostatectomy," by Jay B. Basillote, Thomas E. Ahlering, and Douglas W. Skarecky.

Printed on acid-free paper

9 8 7 6 5 4 3 2 1

springer.com

Foreword

The minute I walked in the door, I knew something was wrong. My wife waited until I took off my coat and sat down at the kitchen table. "I had some bad news today," Lori began. "Mom has cancer." We talked for hours about what it meant and what we could do to help. It was 1972, only a year after President Nixon had declared "war on cancer," and many of us thought that a concerted national effort could lead to a cure in the same way that President Roosevelt's targeting of polio and President Kennedy's quest to put men on the moon had succeeded.

Over the next 20 years, I acquired an extensive layman's knowledge of cancer in the process of supporting research and seeking effective treatments for my mother-in-law's breast cancer and, later, my father's malignant melanoma, and the cancers of several other close relatives. By 1982, my brother and I had assembled a professional staff to formalize our philanthropy through our family charity, the Milken Family Foundation. Working closely with the foundation's medical and scientific advisors, we became very familiar with the leading-edge work of our grant recipients. We were inspired by the progress of pioneers such as Dennis Slamon in breast cancer, Bert Vogelstein in cancer genetics, Owen Witte in leukemia, Lawrence Einhorn in testicular cancer, and many other recipients of the foundation's cancer research awards.

By the time I was diagnosed with advanced prostate cancer in 1993, I thought I knew a lot about cancer. So the shock of my diagnosis was compounded by the realization that I knew almost nothing about prostate cancer. How could I have spent two decades working with cancer researchers and possess so little knowledge of this disease that had already spread from my prostate to my abdominal lymph nodes? How could I not know that this disease affects one in six American men or that a man is more likely to develop prostate cancer than a woman is to develop breast cancer?

It turned out that I was not alone. The public knew next to nothing about prostate cancer. Articles in the popular press, which so often chronicled the importance of pap smears, mammograms, and smoking cessation, rarely mentioned that little walnut-sized organ surrounding men's urethras. As far as most men were concerned, this disease was something they didn't want to think about. Men seem to be more fatalistic than women and believe they're either living or dying so there's no point in getting tested.

Even more surprising was the lack of interest in the medical community. The National Cancer Institute didn't fund much research on prostate cancer because they received few grant applications. Physician–scientists weren't submitting the applications because there appeared to be little funding available. It was a vicious circle. The field was so moribund that one young investigator was told by his mentor to avoid the "career suicide" of prostate cancer research.

Meanwhile, the pharmaceutical and biotechnology industries weren't allocating enough research funds to cancer drug development because they didn't think the return on the investment would justify the risk. And as I traveled around the country to major academic research centers, I felt a growing sense of frustration, and even anger, when I realized that each of these elite institutions considered the others to be competitors rather than collaborators in cancer research.

After extensive discussions with the heads of these centers and other advisors, I concluded that a new organization was needed to bring focus and a sense of urgency to the field of prostate cancer research. This organization, which would become the Prostate Cancer Foundation (PCF), would need to:

- Identify promising research not being funded by the National Cancer Institute;
- Recruit the best and brightest investigators to energize the field;
- Reduce paperwork requirements and fund projects quickly;
- Require awardees to share the results of their work;
- Help build centers of excellence in prostate cancer and link them digitally;
- Encourage public–private partnerships;
- Pursue a venture-funding model;
- Act with urgency;
- Build public awareness.

The Milken Family Foundation jump-started the process with early funding, but since then, the majority of funds for more than 1400 competitive research awards have been contributed by the public. Awardees gather each year at the PCF's Scientific Retreat to present their findings. Many of them are affiliated with member institutions of the PCF Therapy Consortium comprising eight leading cancer centers that now collaborate on prostate cancer programs.

Over the past 15 years, we've reached several important milestones. The Department of Defense Prostate Cancer Research Program exceeds three-quarters of a billion dollars in cumulative spending. Hundreds of bright young investigators are launching careers in prostate cancer research. Prostate cancer SPORE grants have increased fivefold. Articles about prostate cancer in popular periodicals increased from 2500 in 1993 to nearly 36,000 in 2006. Federal and state government funding of prostate cancer research is 20 times the 1993 level. Major states like New York and California allow taxpayers to check off a donation to prostate cancer research on their tax returns. Institutions in dozens of countries around the world now participate in PCF collabora-tions. The latest treatment options are described in a patient guide, and in a separate professional guide, published by the PCF. Millions more men now know about prostate-specific antigen (PSA) tests and DREs. Far more drugs targeting prostate cancer are available or in development than even a few years ago.

None of these achievements would mean much, however, if we weren't keeping more men alive and if they weren't able to enjoy a good quality of life. Fortunately, there's progress here, too. Close to 40,000 prostate cancer patients were dying each year in the early 1990s. With the aging of the baby-boom cohort, that number was expected to increase to as many as 60,000 deaths today. Instead, the number has fallen to below 28,000. The reasons are complex and no one initiative should receive the credit, nor can we be complacent, because without major breakthroughs, death totals could rise again as the first baby boomers move through their seventh and eighth decades.

The work described in *Prostate Cancer: Signaling Networks, Genetics, and New Treatment Strategies*, however, is cause for optimism. The authors of the chapters included here—16 of whom are PCF awardees—are moving quickly on the frontiers of science. They're providing hope that the two million Americans currently living with prostate cancer and the three million projected to join them in the next decade will live long, fulfilling lives.

Building on the lessons of the PCF, we've established a separate organization dedicated to removing the barriers to progress that so often frustrate the efforts of the best researchers in all fields of medicine. This new group, known as FasterCures and headquartered in Washington, D.C., does not fund medical research. Rather, it figures out how we can improve the *process* of research by creating more effective incentives, eliminating unnecessary bureaucracy, improving professional training, linking biobanks, and other steps that shorten the time from idea to bench to bedside. All readers of this book are invited to join us in our effort to make cancer something that our grandchildren will know only by reading history books.

Michael Milken, MBA

Michael Milken, called "The Man Who Changed Medicine" in a 2004 Fortune magazine cover story, is the founder and chairman of the Prostate Cancer Foundation and of FasterCures/The Center for Accelerating Medical Solutions. He has supported medical research for 35 years.

Preface

Prostate cancer remains a major healthcare challenge in the United States. Currently, prostate cancer is the most commonly diagnosed malignancy and the second leading cause of cancer-related deaths in men in the United States. Alternate therapy approaches based on a deeper understanding of prostate cancer are of vital importance. At this time, more than 218,000 new cases of prostate cancer will be diagnosed per year in the United States, and more than 27,000 men will die annually from this disease. We now know that the economic, physical, and psychological burden will be significantly greater for certain groups, including African American men. At this time, an African American man is approximately 2.5 or more times likely to die from prostate cancer than a Caucasian American man. *Prostate Cancer: Signaling Networks, Genetics, and New Treatment Strategies* describes the most current understanding of the molecular mechanisms underlying the onset and progression of prostate cancer. In an attempt to identify new molecular targets for therapy development of prostate cancer, current concepts of steroid receptor and protein kinase signaling pathways are reviewed. In addition, new perspectives in radiation therapy, prediction of therapeutic response, new directions in hormonal treatment, surgical intervention, and targeted therapies are described.

In the opening chapter, new information of histological changes in the prostate associated with cellular atrophy and inflammation provides insight into the pathogenesis of prostate cancer. Chapters 2 through 5 are focused on the key genetic changes involved in prostate carcinogenesis and progression and specific epigenetic abnormalities that accompany prostate cancer progression to advanced disease. The molecular mutations, both low and high penetrant variants, which predispose to and/or modify the response to treatment of prostate cancer, are described. In Chapter 4, Dr. Gelmann focuses on the role of cell cycle control, DNA repair, and oncogenic and tumor suppressor drivers in prostate cancer. In Chapter 5, Drs. Helenius, Waltering, and Visakorpi introduce the role of the somatic genetic changes and the important role of the androgen receptor. Chapters 6 through 9 articulate the role of nuclear hormone receptors in the onset and progression of prostate cancer. We find that the androgen receptor is post-translationally modified not only by phosphorylation but also by acetylation, and these specific post-translational modifications provide new avenues for intervention. In Chapters 8 (Drs. Imamov, Lopatkin, and Gustafsson) and 9 (Drs. Prins and Korach), the authors present an important and balanced view on the role of both estrogen receptors α and β in prostate tumorigenesis.

Chapters 10 through 13 summarize recent advances in intracellular signaling pathways, including the importance of hypoxia-inducible factor 1, the Ras–MAP kinase pathway, the transcription factors STAT5 and STAT3, and the role of Akt and PI3K kinase signaling in prostate cancer progression. Transcription factor Stat5 as a therapeutic target and prognostic factor of poor clinical outcome is described.

With the goal of identifying key molecular targets for therapeutic stratification and prognostication, Chapters 14 through 20 focus on predictors of clinical outcome and the values of specific molecular targets in the management of prostate cancer. Advances in radiation therapy, hormonal therapy, and surgical intervention are highlighted in Chapters 16 and 17. In Chapter 17, Drs. Basillote, Ahlering, and Skarecky highlight recent data using the Da Vinci Surgical System. In Chapter 18, Drs. Heath and Carducci outline key opportunities given by more than 200 novel agents currently under evaluation in the treatment of prostate cancer. Chapter 19 describes new perspectives in chemotherapy of prostate cancer. In closing, opportunities for early detection and treatment of prostate cancer are outlined by Dr. Gomella and Dr. Valicenti.

As a collective medical community, our responsibility lies with engaging all individuals participating in early detection and valuable preventative measures. We are most grateful for the participation of our colleagues in creating this book for the improvement of the treatment of patients with prostate cancer. We acknowledge and share our gratitude to our patients and families who inspire and guide us on a daily basis.

Richard G. Pestell, MD, PhD
Marja T. Nevalainen, MD, PhD

Contents

Contributors

THOMAS E. AHLERING, MD, FACS • *Division of Urologic Oncology, Department of Urology, University of California Irvine, Orange, CA*

SALEH ALTUWAIJRI, DVM, PhD • *Department of Pathology and Laboratory Medicine, University of Rochester Medical Center, Rochester, NY; King Faisal Specialist Hospital and Research Center, Riyadh, Saudi Arabia*

JAY B. BASILLOTE, MD • *Department of Urology, Richmond University Medical Center, Staten Island, NY*

MICHAEL A. CARDUCCI, MD • *Prostate Cancer Program, Chemical Therapeutics, The Sydney Kimmel Comprehensive Cancer Center at Johns Hopkins, Baltimore, MD*

JOHN D. CARPTEN, PhD • *Division of Integrated Cancer Genomics, Translational Genomics Research Institute (TGen), Phoenix, AZ*

CHAWNSHANG CHANG, PhD • *George Whipple Laboratory for Cancer Research, Departments of Pathology and Urology, University of Rochester Medical Center, Rochester, NY*

SEAN COLLINS, MD, PhD • *Department of Radiation Medicine, Lombardi Comprehensive Cancer Center, Georgetown University School of Medicine, Washington, DC*

ZORAN CULIG, MD • *Department of Urology, Medical University of Innsbruck, Innsbruck, Austria*

ANGELO M. DE MARZO, MD, PhD • *Department of Pathology, Johns Hopkins University School of Medicine, Baltimore, MD*

ANATOLY DRITSCHILO, MD • *Departments of Radiation Medicine and Oncology, Lombardi Comprehensive Cancer Center, Georgetown University School of Medicine, Washington, DC*

MARIO A. EISENBERGER, MD • *Division of Medical Oncology, Bunting-Blaustein Cancer Research Building, Johns Hopkins University, Baltimore, MD*

GREGORY GAGNON, MD • *Department of Radiation Medicine, Lombardi Comprehensive Cancer Center, Georgetown University School of Medicine, Washington, DC*

EDWARD P. GELMANN, MD • *Departments of Oncology and Medicine, Lombardi Comprehensive Cancer Center, Georgetown University, Washington, DC*

DANIEL GIOELI, PhD • *Department of Microbiology and Cancer Center, University of Virginia Health System, Charlottesville, VA*

LEONARD G. GOMELLA, MD • *Department of Urology, Kimmel Cancer Center, Thomas Jefferson University, Philadelphia, PA*

JAN-ÅKE GUSTAFSSON, MD, PhD • *Department of Biosciences and Nutrition, Director of Center for Biotechnology, Karolinska Institute, NOVUM, Stockholm, Sweden*

JULIA H. HAYES, MD • *Department of Medical Oncology, Dana Farber Cancer Institute, Boston, MA*

ELISABETH I. HEATH, MD • *Department of Medicine and Oncology, Karmanos Cancer Institute, Wayne State University, Detroit, MI*

MERJA A. HELENIUS, PhD • *Institute of Medical Technology, University of Tampere and Tampere University Hospital, Tampere, Finland*

HAOJIE HUANG, PhD • *Departments of Laboratory Medicine and Pathology, The Cancer Center, University of Minnesota, Minneapolis, MN*

OTABEK IMAMOV, MD • *Department of Biosciences and Nutrition, Karolinska Institute, NOVUM, Stockholm, Sweden; Reconstructive Urology Unit, Moscow Institute of Urology, Moscow, Russia*

ROBERT B. JENKINS, MD, PhD • *Department of Laboratory Medicine and Pathology, The Mayo Clinic, Rochester, MN*

PHILIP KANTOFF, MD • *Division of Solid Tumor Oncology, Lank Center for Genitourinary Oncology, Dana Farber Cancer Institute, Prostate Cancer Program and Prostate Cancer SPORE at Dana Farber Harvard Cancer Center, Harvard Medical School, Dana, Boston, MA*

KENNETH S. KORACH, PhD • *Environmental Disease and Medicine Program, Laboratory of Reproductive and Developmental Toxicology, NIEHS/NIH, Research Triangle Park, NC*

SARAH KRAUS, PhD • *Department of Microbiology and Cancer Center, University of Virginia Health System, Charlottesville, VA*

FUSHENG LAN, MD, PhD • *Department of Laboratory Medicine and Pathology, The Mayo Clinic, Rochester, MN*

SHENGWEN LI, PhD • *Department of Cancer Biology, Kimmel Cancer Center, Thomas Jefferson University, Philadelphia, PA*

MICHAEL P. LISANTI, MD, PhD • *Department of Cancer Biology, Kimmel Cancer Center, Thomas Jefferson University, Philadelphia, PA*

NIKOLAI A. LOPATKIN, MD, PhD • *Academician of Russian Academy of Medical Sciences, Director, Moscow Institute of Urology, Moscow, Russia*

DONALD MCRAE, PhD • *Department of Radiation Medicine, Lombardi Comprehensive Cancer Center, Georgetown University School of Medicine, Washington, DC*

MICHAEL MILKEN, MBA • *Prostate Cancer Foundation, Santa Monica, CA*

HIROSHI MIYAMOTO, MD, PhD • *Department of Pathology and Laboratory Medicine, University of Rochester Medical Center, Rochester, NY*

WILLIAM G. NELSON, MD, PhD • *Department of Oncology, Sidney Kimmel Comprehensive Cancer Center at Johns Hopkins, Brady Urological Institute, Bunting-Blaustein Cancer Research Building, Baltimore, MD*

MARJA T. NEVALAINEN, MD, PhD • *Department of Cancer Biology, Kimmel Cancer Center, Thomas Jefferson University, Philadelphia, PA*

RICHARD G. PESTELL, MD, PhD • *Kimmel Cancer Center, Department of Cancer Biology, Thomas Jefferson University, Philadelphia, PA*

MICHAEL POWELL, BS • *Department of Cancer Biology, Kimmel Cancer Center, Thomas Jefferson University, Philadelphia, PA*

GAIL S. PRINS, PhD • *Department of Urology, University of Illinois at Chicago, Chicago, IL*

MARK A. RUBIN, MD • *Department of Urologic Pathology, Brigham and Women's Hospital, Dana Farber Cancer Institute, Harvard Medical School, Boston, MA*

JONATHAN W. SIMONS, MD • *Prostate Cancer Foundation, Santa Monica, CA*

DOUGLAS W. SKARECKY, BS • *Department of Urology, University of California Irvine, Orange, CA*

DONALD J. TINDALL, PhD • *Departments of Urology and Biochemistry and Molecular Biology, Mayo Clinic College of Medicine, Rochester, MN*

JEFFREY M. TRENT, PhD • *Translational Genomics Research Institute, Phoenix, AZ*

ALEXANDER VALDMAN, MD, PhD • *Department of Pathology, Johns Hopkins University School of Medicine, Baltimore, MD*

RICHARD K. VALICENTI, MD • *Vice Chairman of Clinical Affairs, Department of Radiation Oncology, Kimmel Cancer Center, Thomas Jefferson University, Philadelphia, PA*

TAPIO VISAKORPI, MD, PhD • *Institute of Medical Technology, University of Tampere and Tampere University Hospital, Tampere, Finland*

KATI K. WALTERING, MSc • *Institute of Medical Technology, University of Tampere and Tampere University Hospital, Tampere, Finland*

CHENGUANG WANG, PhD • *Department of Cancer Biology, Kimmel Cancer Center, Thomas Jefferson University, Philadelphia, PA*

MICHAEL J. WEBER, PhD • *Department of Microbiology and Cancer Center, University of Virginia Health System, Charlottesville, VA*

SRINIVASAN YEGNASUBRAMANIAN, MD, PhD • *Sidney Kimmel Comprehensive Cancer Center at Johns Hopkins, Brady Urological Institute, Bunting-Blaustein Cancer Research Building, Baltimore, MD*

1

Histopathology and Molecular Biology of Prostate Atrophy

A Lesion Associated with Inflammation, Prostate Intraepithelial Neoplasia, and Prostate Cancer

Alexander Valdman, MD, PhD,
Robert B. Jenkins, MD, PhD,
Fusheng Lan, MD, PhD,
and Angelo M. De Marzo, MD, PhD

CONTENTS

1. INTRODUCTION

Prostate atrophy has been considered a possible precursor of prostate cancer since the 1920s *(1–3)*. Prostate inflammation has come into focus over the last several years as a potential stimulus for the development of prostate cancer. Because many prostate atrophy lesions are associated with inflammation, and most inflammatory lesions are associated with atrophic epithelium,

From: *Current Clinical Oncology: Prostate Cancer:*
Signaling Networks, Genetics, and New Treatment Strategies
Edited by: R. G. Pestell and M. T. Nevalainen © Humana Press, Totowa, NJ

it is plausible that prostate atrophy represents a critical link between inflammation and prostate cancer. There have been recent reviews focused on inflammation and prostate cancer *(4–9)* and the histomorphology of prostate atrophy *(10)*. Because there are no systematic reviews of studies that have examined histopathological mapping of prostate atrophy lesions in relation to prostate cancer, or of somatic genomic alterations in prostate atrophy, these topics will be the subject of this chapter.

2. CLASSIFICATION OF FOCAL ATROPHY IN THE PROSTATE

The early systematic description and classification of epithelial atrophy of the prostate can be attributed to Franks *(11)*. He divided morphological variants of prostate atrophy into five patterns:

1. Simple atrophy
2. Sclerotic atrophy
3. Post-atrophic hyperplasia (PAH), of which there were two subtypes:

   a. Lobular hyperplasia
   b. Sclerotic atrophy with hyperplasia

4. Secondary hyperplasia

Although it was not explicitly stated, secondary hyperplasia would appear to best correspond to what we would refer to today as high-grade prostatic intraepithelial neoplasia (HGPIN). In order to standardize terminology regarding focal atrophy of the prostate, De Marzo et al. *(12)* developed a working group classification of focal atrophy of the prostate, and the outcome of this is a new classification system for the various morphological patterns of focal atrophy lesions. According to the working group, focal prostate atrophy can be classified into one of four subtypes as indicated in Table 1. In addition to the classification above, it was proposed to refer to most simple atrophy and PAH lesions as "proliferative inflammatory atrophy" (PIA) *(13)*. This grouping term emphasizes the fact that many atrophy lesions are in fact associated with inflammation *(14)*. In those atrophic lesions that are not associated with inflammation, the term proliferative atrophy (PA) may be used (see Table 1 for usage details).

3. HISTOLOGICAL MAPPING STUDIES

Several studies addressed the frequency and topographical relationship of focal atrophy and prostate cancer. In the classic article by Franks *(15)*, a case–control study was performed in autopsy material. He reported the incidence

Table 1

Working Group Classification of Focal Atrophy of the Prostate

Histology on H & E Sections

	Normal	*SA*	*PAH*	*SACF*	*Partial*
Cytoplasm amount	Abundant	Little, variable	Little, variable	Usually little or very little	More than other atrophy, less than normal
Cytoplasm color	Clear[a]	Variable, often dark	Variable, often dark	Variable, often clear	Clear
Papillae	Abundant	None, but some undulations	None, few undulations	None with vascular cores, but some projections remain	None, more subtle undulations
Gland size	Medium to large	Similar to normal, but more variable	Small	Medium to very large	Small to medium
Gland shape	Compound tubuloalveolar	Similar to normal, less complex	Mostly round	Round	Variable
Gland packing	Well spaced	Similar to normal	Close	Close	Variable
Inflammation	Absent	Usually present	Usually present	Usually absent	Usually absent

(*Continued*)

3

Table 1
(Continued)

	Histology on H & E Sections				
	Normal	*SA*	*PAH*	*SACF*	*Partial*
Fibrosis of stroma	Absent	Variable	Variable	No	No
May be considered PIA or PA	No	Yes	Yes	No	No
Relation to Franks system (13)	SA	SA	PAH or "sclerotic atrophy with hyperplasia"	SA	Similar lesions not discussed

SA, simple atrophy; PAH, post-atrophic hyperplasia; SACF, simple atrophy with cyst formation; partial, partial atrophy; PIA, proliferative inflammatory atrophy; PA, proliferative atrophy.

Mixed lesions may be classified by dominant (primary) pattern with mention of secondary pattern. Atrophic lesions not conforming to these criteria are classified as "focal prostate atrophy not otherwise specified."

[a]Except in central zone which may be somewhat dark.

Reproduced with permission (12).

4

Table 2
Data Summarized from Franks *(15)*

Cancer	Post-sclerotic hyperplasia		
	No	Yes	Total
No	68 (58.6%)	48 (41.3%)	116 (62.7%)
Yes	29 (42%)	40 (58%)	69 (37.3%)
Total	97 (52.3%)	88 (47.7%)	185 (100%)

of sclerotic atrophy and post-sclerotic hyperplasia[1] (Table 2). Of 69 cases in which prostate cancer was found, which he regarded as "latent carcinoma," 21% contained sclerotic atrophy and 56% contained post-sclerotic hyperplasia. In comparison, in 116 cases from men over 50 years of age without prostate carcinoma, sclerotic atrophy alone was found in 17% cases and post-sclerotic hyperplasia in 41.5% cases. Although he did not report statistical analysis, a Pearson χ^2 test of his results indicates that there was a statically significant increase in the fraction of cancer cases with post-sclerotic hyperplasia ($\chi^2 = 4.7754$, $p = 0.029$) than in cases without carcinoma.

In the group of 32 men under 50 years of age, sclerotic atrophy alone was found in 15.75% of cases and post-sclerotic hyperplasia in 9.25%. These latter results show a clear positive correlation of the extent of some types of focal atrophy and age. Franks also reported direct merging of post-sclerotic hyperplasia atrophy lesions at times with small invasive carcinoma lesions *(15)*.

Liavåg *(16)* reported the incidence of atrophy in 324 autopsies of males at the age of 40 and above in which 87[2] (26.8%) were found to have incidental cancer. In this study, which is the only reported one that attempted to measure the extent of prostate atrophy, atrophy was categorized as slight, moderate, or marked. Both atrophy and carcinoma increased with age. Atrophy was found in 100% of the prostates with carcinoma and in 90.3% of the prostates without carcinoma, and this difference was significant ($p < 0.01$). There was a highly significant difference in the extent of atrophy in cases versus controls ($p < 0.00001$), such that atrophy was more "severe" in prostate specimens with cancer (85% severe atrophy in cancer cases and 49% in non-cancer patients; the difference was indicated to be statistically significant, but the p value was not given). In the same study, Liavåg also reported direct merging of atrophy lesions with small invasive carcinoma lesions.

---

[1]We assume this corresponds to the Franks description of the pattern of PAH referred to as sclerotic atrophy with hyperplasia.

[2]Note that in this study, only three sections of the prostate were examined.

Billis *(17)* assessed the presence of prostate atrophy in the peripheral zone of prostates from 100 consecutively autopsied men older than 40 years of age. He found focal atrophy in 85 prostates and subtyped it into simple, hyperplastic (PAH), and sclerotic. In 65 (76.47%) of 85 cases, the histologic subtypes were combined. In 33 (50.76%) of these 65 cases, the three subtypes were seen concomitantly. Cancer was found in 29 of 100 cases. Atrophy was reported in 24 of 29 cases with carcinoma (83%) and in 61 of 71 cases without carcinoma (86%). The author concluded that there was no relation of atrophy to "latent carcinoma." In a more recent study by the same authors *(18)*, neither a topographic relation nor a morphologic transition was seen between prostatic atrophy and carcinoma or HGPIN. One potential limitation of these studies, however, was that the authors only reported on whether atrophy was present or absent, and not on the extent of atrophy. Given that almost all cases harbored some atrophy, it follows that without examining large numbers of specimens (i.e., >300 cases as in Liavåg *(16)*) in order to uncover a potential link between prostate atrophy and carcinoma in these types of studies, it is likely that merely indicating its presence is not adequate; rather the quantitative extent of atrophy should be measured and correlated with cancer.

Putzi and De Marzo *(19)* examined the two-dimensional topographic relationship between simple atrophy and PAH to HGPIN lesions and prostate carcinoma lesions in 14 radical prostatectomy (RP) specimens. The prostate-ctomy specimens chosen were those in which only minimal carcinoma was present (<0.5 cc total tumor volume) in order to reduce the possibility that HGPIN lesions represented intraprostatic dissemination of carcinoma and to examine the topology of carcinoma lesions that were very likely to be close to their point of inception. In these 14 specimens, the total number of spatially separate HGPIN lesions was 629. The topographic relation was described as "merging" when HGPIN or carcinoma merged directly with focal atrophy within a given acinus or duct. Lesions were considered "adjacent" when the neoplastic-appearing epithelium was in very close proximity (<100 μm), but did not merge with atrophy. Lesions were described as "near" when the individual duct/acinus of HGPIN or carcinoma was separated from a distinct acinus/duct containing PIA by less than 1 mm but were further away than "adjacent." Finally, lesions were described as distant when the HGPIN or carcinoma lesion was more than 1 mm from any atrophic area. The results showed that HGPIN merged with atrophy in 267 (42.5%) of 629 lesions. It was adjacent in 57 lesions (9%), was near in 233 lesions (37%), and was distant from atrophy in 72 lesions (11.5%). Thus, focal atrophy was within 1 mm of HGPIN in 88.5% of such PIN lesions. In the same study, carcinoma was not found to merge directly with focal atrophy; it was adjacent in 24 (30.4%) of 79 lesions, was near in 46 lesions (58.2%), and was distant from focal atrophy in nine lesions (11.4%). Thus, for all of these "microcarcinoma" lesions, focal atrophy was within 1 mm 88.6% of the time. This was higher than the fraction of carcinoma lesions that were within 1 mm of HGPIN (70%). In many cases, there were regions in which atrophic areas appeared to demonstrate gradations of increasing nuclear atypia from no atypia in the purely atrophic cells to significant atypia in the HGPIN regions (Fig. 1). In summary, the study revealed frequent morphologic transitions between HGPIN and focal atrophy. In subsequent studies, we have noted direct merging at times between focal atrophy and small carcinoma lesions occurs *(20)*, as did Montironi et al. *(21)*. The primary limitation of this study was the relatively small number of patients, albeit very high-resolution mapping was performed such that every PIN lesion and every cancer lesion was examined in detail. Other limitations of the study were that the pattern of prostate atrophy was not noted, just that it was either PAH or simple atrophy, and that no morphometric measurements were used to quantify the gradations in nuclear atypia between atrophy and PIN lesions.

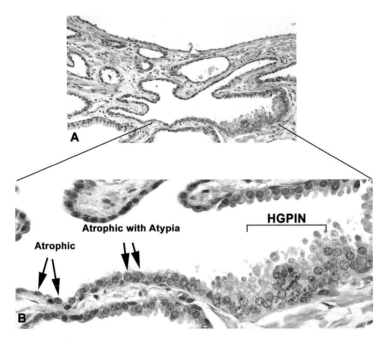

Fig. 1. High-grade prostatic intraepithelial neoplasia (HGPIN) merging with focal atrophy lesion. **(A)** Medium power view of a focus of proliferative inflammatory atrophy (PIA) showing a single prostatic acinus (center) illustrating merging of atrophic epithelium with epithelium containing HGPIN (original magnification ×100, hematoxylin and eosin). **(B)** High-power view of area outlined in **(A)**. Arrows indicate atrophic acini and atrophic acini with atypia. Note the progression from atrophic acini with atypia to PIN. The characteristics of atypical atrophic cells are slightly enlarged nuclei and occasional nucleoli, whereas cell with HGPIN show enlarged nuclei, numerous prominent nucleoli, and chromatin clumping (original magnification ×600, hematoxylin and eosin). Reprinted with permission *(19)*.

Anton *(22)* assessed whole mount sections from 272 randomly selected RP and 44 cystoprostatectomy specimens for the presence, location, and number of foci of PAH, and then correlated these with the presence and location of carcinoma foci. PAH was identified in 86 (32%) prostatectomy and in 12 (27%) cystoprostatectomy specimens. The distribution of PAH foci was the following: peripheral zone (91%), transition zone (8%), central zone (1%), and apex (49%), mid (39%), and base (12%). For prostatectomy specimens, 183 foci of PAH showed no atrophy in a mirror image area of the prostate opposite the focus of PAH. Of the foci, 33% showed carcinoma either within or within 2 mm of the focus of PAH. For the mirror image area without PAH, carcinoma was identified either within or within 2 mm of the area in 40% ($p = 0.19$). The frequency of PAH in cystoprostatectomy specimens and its relationship to incidental carcinoma were not significantly different from those of RP specimens ($p = 0.60$, χ^2). Therefore, PAH was found to be a relatively common lesion, most often seen in the peripheral zone of the apical third of the gland. PAH did not, however, appear to have any association with carcinoma. A limitation of this study is that the authors did not consider other patterns of prostate atrophy, which are more common than PAH.

Tsujimoto et al. *(23)* examined 28 RP specimens containing carcinoma and reported on PAH lesions. They found that 7 of 28 (25%) of cases harbored PAH and these lesions were multifocal in 85.7% of cases. They also reported on details of histological findings including the presence and size of nucleoli and presented a review table of the literature up until that point. They found that PAH lesions were "near" HGPIN and carcinoma lesions in 43% of cases.

In a recent study by Tomas et al. *(24)*, the authors evaluated the extent and type of atrophy lesions in 50 patients with prostate carcinoma and 31 patients with benign prostatic hyperplasia (BPH). Atrophy lesions were classified according to the working group *(12)*. The study revealed that atrophy foci were present in 100% cases with and without carcinoma. However, in cases with carcinoma, atrophy associated with inflammation (PIA) was a significantly more frequent finding than atrophy without inflammation (PA) (1.63 and 0.76 foci per slide, respectively; $p < 0.001$). In BPH patients, the opposite was seen; PA was more frequent than PIA (2.28 and 1.27, respectively; $p < 0.001$).

If prostate atrophy is a precursor to prostate cancer, then it is plausible that having focal atrophy on prostate needle biopsy that does not contain carcinoma may be associated with an increased risk of prostate cancer on follow-up biopsies. In a study by Postma et al. *(25)* in the group of 202 randomly selected benign sextant biopsies with a follow-up time of at least 8 years and total incidence of atrophy of 94%, there was no association between atrophy and incidence of prostate cancer or HGPIN. As in the other studies indicated above, none of these studies examining the association between prostate atrophy on needle biopsy specimens and subsequent cancer found on needle biopsy used quantitative methods to assess the degree of atrophy.

Nevertheless, because our work supports the concept that some forms of prostate atrophy may be risk factor lesions and at times precursors to HGPIN and/or microcarcinomas, it is not expected that the mere presence or absence of focal atrophy on a prostate biopsy would be predictive of cancer on a repeat biopsy. We expect that many years, likely a decade or more, would need to elapse before atrophy lesions might progress to a carcinoma that would be large enough to sample on random sextant or even 10–12 core biopsies.

4. MOLECULAR BIOLOGY OF PROSTATE ATROPHY

A number of molecular changes have been reported in prostate atrophy. Chromosome 8 abnormalities represent a frequent molecular change in prostate cancer. Several studies have been reported in which fluorescent in situ hybridization (FISH) was used to examine focal atrophy lesions. In the study by Macoska et al., (26) FISH was performed simultaneously with two probes, one for a locus on chromosome 8p [cosmid containing the lipoprotein lipase (LPL)] and another with sequences that recognize the centromeric region from chromosome 8 (8c). *Disomy* was defined in individual cells when the FISH signals for the 8p and 8c probes were each 2 (2,2). Chromosomal alterations were defined in individual cells as *8p loss* when the number of FISH signals for 8p,8c in an individual cell was 0,2 or 1,2; *concomitant 8p loss and 8c gain* was determined by 8p,8c counts of 0,3, 1,3, or 2,3. Then, after counting 200 individual cells for each of these chromosome-alteration categories for each tissue type in each patient examined, the percentage of nuclei in each category was obtained. Then for each lesion or tissue type (atrophy, benign, cancer, and hyperplastic), the means of each of these percentages were computed and compared between groups. Although atrophy tended to have a mean percentage of nuclei with *8p loss* that was greater than normal (8p loss nuclei = 34.7% in atrophy and 28.6% in benign and 35.7% in hyperplastic tissues), and 8c signals that were greater than normal (data not shown), these differences were not statistically significant, albeit only seven cases of atrophy were examined. However, the percentage of nuclei with *concomitant 8p loss and 8c gain* was statistically significantly higher in atrophy (1.57%) compared with "benign" (0.88%) and hyperplasic tissues (0.6%) ($p < 0.05$). Although this analysis provides some information regarding the population of cells within lesions as a whole, it does not attempt to determine whether an individual case of prostate atrophy contains true *clonal* alterations. Therefore, Macoska et al. developed a cut point approach in which a given tissue type in a given case was judged to show loss or gain (26). Because they found that in all benign tissues the mean percentage of disomic nuclei was 41.4% and the range was 35.2–47.6%, they considered that an individual case was disomic if it fell within that range. For 8p loss, they determined that mean percentage of nuclei with *8p loss* in all benign tissues was 33.1% with an SD of 11.2 Therefore, *8p loss* was defined when the percentage of nuclei with 0,2 or 1,2 8p,8c signal counts was at least 44.3% (mean + 1 SD). When defined in this manner, none of the benign tissues (normal, atrophy, and hyperplastic) showed *8p loss*, yet 84% of cancer lesions showed 8p loss (10/25 with *8p loss* only, and 11/25 with *concomitant 8p loss and 8c gain*).

Shah et al. *(27)* also examined prostate atrophy lesions (simple atrophy and PAH) compared with normal, HGPIN, and carcinoma lesions, using FISH in which the number of cells with three or more signals for chromosome 8 centromere was counted. They found that the fraction of cells with three signals increased going from normal to all other histological lesions, although this was not statistically significant. They did find, however, when examining the percentage of cells with three or more chromosome 8 centromere signals, that there was a significant increase in the percentage of cells in atrophy—the mean percentage was 1.3% in benign, 2.1% in simple atrophy, 2.8% in HGPIN, 4% in PAH, and 6% in carcinoma, respectively *(27)*. In this study, the authors did not attempt to categorize individual cases of atrophy as having gain of 8c, just the overall mean and SD for the percentage of cells with 3 or more 8c signals was enumerated. Note that Macoska et al. *(26)* did not comment on 8c gain only.

In our recent study *(28)*, we examined prostate tissues using tissue microarrays (TMAs) with FISH simultaneously using a cocktail of three different probes, which included 8p22 (LPL), 8 centromere, and 8q24. For each tissue type in each patient (normal-appearing epithelium, atrophy, PIN, and carcinoma) from RP specimens, we recorded the percentage of cells with 1, 2, 3, or more signals for each of the three differently labeled probes. At least 30 cells were counted in each Tissue microarray (TMA) spot. In order to determine whether *clonal* genetic alterations were present, we used a stringent cut point based method. For example, to categorize a lesion as having 8p loss, 60% or more cells should have 0 or 1 8p22 signal and the overall 8p22/8c signal is less than 0.8. The gain category for any of the three probes was determined if 30% or more epithelial nuclei contained three or more signals for the probe. Although there was a significant increase in the percentage of cells in atrophy that harbored three or more signals for 8c compared with normal (2.4% for atrophy vs. 1.2% for normal; $p = 0.024$), no TMA cores containing prostate atrophy were considered to have gained chromosome 8 centromere in a *clonal* fashion. Using these strict cut points, we also found that (i) no cases of normal lesions contained 8p22 loss or 8q24 gain; (ii) 12% of HGPIN lesions contained 8p22 loss; (iii) 0% of HGPIN contained 8q24 gain; and (iv) 52% of carcinoma lesions contained 8p22 loss, and this correlated with Gleason grade (74% 8p22 loss for Gleason patterns 4–5 and 33% 8p22 loss for Gleason pattern 3). It was noted that the fraction of cases of HGPIN with 8p22 loss and 8q24 gain was lower than in any other previously published study, although certainly (for 8p22 loss) higher than in atrophy lesions.

Yildiz-Sezer et al. *(29)* also examined prostate atrophy using FISH with the same 8p22 (LPL), 8 centromere, and 8q24 probes used by Bethel et al. using prostate TMAs from RP specimens. For each case, they also counted the number of signals for each probe in each nucleus and determined the average number of signals of each per case. Like Macoska et al., they placed their nuclei into the following categories of disomy, monosomy 8, 8p22 loss, 8q24 gain, and concomitant 8p22 loss and 8q24 gain. About 3.6% of normal nuclei from non-cancer patients had 8p22 loss, and 10.26% of normal-appearing epithelial nuclei from cancer patients, 14.17% of atrophy nuclei, 17.08% of PIN nuclei, and 21.2% of cancer nuclei were considered to harbor 8p22 loss. It is not clear, however, how many cases of each tissue type harbored what would be considered clonal changes, yet we presume it was quite low, and perhaps even zero in atrophy, given the overall means that were reported.

In another recent study, Yildiz-Sezer et al. *(30)* examined focal prostate atrophy lesions for genomic abnormalities on the X chromosome. Strikingly, using comparative genomic hybridization (CGH) analysis, they reported that 90% of prostate cancer lesions and 70% of focal atrophy lesions contained gains of the entire X chromosome. The results were further verified using FISH with an X chromosome-specific centromeric probe. Although these results are certainly novel and intriguing, as in the other studies reported above, it seems that clonal alterations were not found. For example, if an entire chromosome X was gained in the majority of cells in the population of interest, than the ratio of signals in the lesional tissue compared with the normal tissue would be approximately 2. However, in this study the authors used a cut point for gain of 1.2. Again, although statistically significantly different than the normal tissues in this study, the results suggest that only a subset of cells harbor this change. Similarly, by FISH analysis, the mean number of cells with gain of X chromosome centromeric signals was 4.28% in normal epithelium, 18.4% in atrophy, and 23.9% in carcinoma. Using a cutoff of 10% of cells showing gain in which to consider a given lesion to harbor gain, 13/20 (65%) of atrophy lesions showed gain and 18/20 (90%) of carcinoma lesions showed gain. Consistent with the idea that these changes do not represent true clonal alterations that were selected for, there have been no previous reports of gain of the entire X chromosome in prostate cancer (see Sun et al. *(31)*, for recent meta-analysis).

In summary, chromosomal abnormalities similar to those found in PIN and carcinoma occur in a subset of atrophic lesions. In these studies, however, it appears that there were no cases in which clonal alterations were identified in atrophy. Assuming that these changes are due to biological differences and not some non-biologically based systematic experimental artifact, such as difficulties with overlapping nuclei and counting FISH signals, these changes are likely indicative of genomic damage and/or the emergence of genomic instability in PIA/PA. Because some of the changes reported (8c gain, 8p loss, and 8q gain) are similar to changes that are found as clonal alterations in the truly neoplastic appearing cells in invasive carcinoma, and at times in PIN, the findings are consistent with a hypothesis in which non-clonal DNA alterations begin to arise in atrophy lesions at an increased frequency as compared with normal-appearing epithelium from the same patients, and that these could later be selected for during the process of neoplastic transformation. This is consistent with the findings that atrophy lesions tend to show an apparent stress response suggestive of oxidative DNA damage that could lead to chromosomal alterations.

4.1. Mutational Analysis in Prostate Atrophy Lesions

There have been very few studies in which specific genes have undergone sequence analysis using DNA isolated from focal atrophy lesions of the prostate. Tsujimoto et al. (23) used laser capture microdissection of PAH (N = 7 patients with 89 lesions) (a form of focal atrophy), BPH, HGPIN, and adenocarcinoma lesions to isolate genomic DNA and perform mutational analysis for exons 5–8 of p53. p53 mutations were found in 2 of 38 PAH lesions (5.3%), 4 of 16 carcinoma lesions (25%), and 1 of 24 PIN lesions (4.2%), yet benign glands never showed mutations. In this study, cases were first screened by single-stranded conformation polymorphism (SSCP) analysis in order to uncover potential mutations. Interestingly, in the figure shown, the SSCP band appeared to be a minor component of the total, suggesting that the mutations detected were likely not clonal.

Tsujimoto et al. (32) also examined normal, PAH, PIN, and carcinoma lesions by microdissection for somatic mutations in the AR gene (encoding the androgen receptor), in which they used nested PCR to determine the repeat length in the CAG repeat in exon 1, the GGC repeat in exon 1, and the BAT-25 and the BAT-26 repeats in AR alleles. They reported that although there were no somatic alterations in the GGC, BAT-25, or BAT-26 repeats in any of the lesions, there were somatic decreases in the CAG repeat length in 3 of 89 (3.4%) cancer lesions, 6 of 75 (8%) PIN lesions, 4 of 24 (16.7%) PAH lesions, and 0 of 56 (0%) benign areas. In the cases shown, there was a prominent band at the shorter repeat location along with a weak but present band at the wild-type location. It is not clear whether this represents contamination with normal cells or a heterogeneous population in the lesional tissues. It was of interest here too that the presumed precursor lesions harbored a higher frequency of changes than the cancer lesions. Thus, it appears that there is some evidence of genetic instability in PAH lesions, but that AR mutations do not appear to have been selected for in most cases of localized carcinoma lesions.

4.2. Epigenetic Somatic Genome Alterations

GSTP1 CpG island hypermethylation is a very common somatic genome alteration described for human prostate cancer *(33)*. *GSTP1* CpG island hypermethylation is associated with acquired defects in defense mechanisms against oxidant and electrophilic DNA damage *(33)*. As a consequence, cells lacking GSTP1 activity are more sensitive to oxidative stress caused by inflammation. In a study from our group, we compared CpG island methylation status in normal epithelium ($n = 48$), BPH ($n = 22$), and PIA lesions ($n = 64$). We found hypermethylation of the *GSTP1* promoter region in a small subset of atrophy cases. *GSTP1* promoter methylation was found in 0 of 48 regions form normal-appearing epithelium (0%), 0 of 22 regions of BPH (0%), 4 of 64 PIA lesions (6.3%), 22 of 32 HGPIN lesions (68.8%), and 30 of 33 of carcinoma lesions (90.9%) *(20)*. The results regarding PIN and carcinoma were similar to result obtained previously by Nelson et al. *(34,35)*.

5. SUMMARY AND CONCLUSIONS

In summary, focal prostate atrophy is known to be associated with chronic inflammation in the majority of cases, and we have put forth the terms proliferative inflammatory atrophy and proliferative atrophy for most of these lesions. We reviewed studies in which morphological and molecular evidence have been explored to relate these lesions to prostate cancer. From the data we have so far, it seems that these extremely common lesions may at times represent precursors to PIN and or adenocarcinoma. Because they are so common and often extensive, most are not going to directly evolve into carcinoma. Rather, the molecular genetic data so far accumulated appear to indicate that a fraction of cells in some atrophy lesions have developed somatic DNA alterations consistent with those found as clonal changes in carcinoma lesions, and less frequently in PIN lesions. Thus, the term "risk factor lesion" may be somewhat more appropriate than "precursor lesion" for focal atrophy. Although PIN lesions have morphological features of neoplastic cells, whereas atrophy lesions do not, the likelihood of individual PIN lesions progressing to cancer is also presumably quite low. Much work needs to be done before focal atrophy is considered a true risk factor for prostate cancer. Nevertheless, because it is so tightly linked to inflammation, strategies to prevent prostate cancer may eventually involve the suppression of prostate inflammation.

REFERENCES

1. Oerteil, H. Involutionary changes in prostate and female breast in relation to cancer development. *Can Med Assoc J* 16, 237–241 (1926).
2. Moore, R. A. The evolution and involution of the prostate gland. *Am J Pathol* 12, 599–624 (1936).
3. Rich, A. R. On the frequency of occurrence of occult carcinoma of the prostate. *J Urol* 33, 215–223 (1934).
4. De Marzo, A. M. et al. Pathological and molecular mechanisms of prostate carcinogenesis: implications for diagnosis, detection, prevention, and treatment. *J Cell Biochem* 91, 459–477 (2004).
5. Nelson, W. G., De Marzo, A. M. and Isaacs, W. B. Prostate cancer. *N Engl J Med* 349, 366–381 (2003).
6. Platz, E. A. and De Marzo, A. M. Epidemiology of inflammation and prostate cancer. *J Urol* 171, S36–40 (2004).
7. Palapattu, G. S. et al. Prostate carcinogenesis and inflammation: emerging insights. *Carcinogenesis* 26, 1170–1181 (2005).
8. Lucia, M. S. and Torkko, K. C. Inflammation as a target for prostate cancer chemoprevention: pathological and laboratory rationale. *J Urol* 171, S30–34; discussion S35 (2004).
9. De Marzo, A. M. et al. Inflammation in prostate carcinogenesis. *Nat Rev Cancer* 7, 256–269 (2007).
10. De Marzo, A. M. in *Prostate Cancer: Biology, Genetics and the New Therapeutics* (Chung, L. W. K., Isaacs, W. B. and Simons, J. W., eds.) Humana Press, Totawa, NJ. (2007).
11. Franks, L. M. Atrophy and hyperplasia in the prostate proper. *J Pathol Bacteriol* 68, 617–621 (1954).
12. De Marzo, A. M. et al. A working group classification of focal prostate atrophy lesions. *Am J Surg Pathol* 30, 1281–1291 (2006).
13. De Marzo, A. M., Marchi, V. L., Epstein, J. I. and Nelson, W. G. Proliferative inflammatory atrophy of the prostate: implications for prostatic carcinogenesis. *Am J Pathol* 155, 1985–1992 (1999).
14. McNeal, J. E. Normal histology of the prostate. *Am J Surg Pathol* 12, 619–633 (1988).
15. Franks, L. M. Latent carcinoma of the prostate. *J Pathol Bacteriol* 603–616 (1954).
16. Liavag, I. Atrophy and regeneration in the pathogenesis of prostatic carcinoma. *Acta Pathol Microbiol Scand* 73, 338–350 (1968).
17. Billis, A. Prostatic atrophy: an autopsy study of a histologic mimic of adenocarcinoma. *Mod Pathol* 11, 47–54 (1998).
18. Billis, A. and Magna, L. A. Inflammatory atrophy of the prostate. Prevalence and significance. *Arch Pathol Lab Med* 127, 840–844 (2003).
19. Putzi, M. J. and De Marzo, A. M. Morphologic transitions between proliferative inflammatory atrophy and high-grade prostatic intraepithelial neoplasia. *Urology* 56, 828–832 (2000).
20. Nakayama, M. et al. Hypermethylation of the human glutathione S-transferase-pi gene (GSTP1) CpG island is present in a subset of proliferative inflammatory atrophy lesions but not in normal or hyperplastic epithelium of the prostate: a detailed study using laser-capture microdissection. *Am J Pathol* 163, 923–933 (2003).
21. Montironi, R., Mazzucchelli, R. and Scarpelli, M. Precancerous lesions and conditions of the prostate: from morphological and biological characterization to chemoprevention. *Ann N Y Acad Sci* 963, 169–184 (2002).
22. Anton, R. C., Kattan, M. W., Chakraborty, S. and Wheeler, T. M. Postatrophic hyperplasia of the prostate: lack of association with prostate cancer. *Am J Surg Pathol* 23, 932–936 (1999).

23. Tsujimoto, Y., Takayama, H., Nonomura, N., Okuyama, A. and Aozasa, K. Postatrophic hyperplasia of the prostate in Japan: histologic and immunohistochemical features and p53 gene mutation analysis. *Prostate* 52, 279–287 (2002).
24. Tomas, D. et al. Different types of atrophy in the prostate with and without adenocarcinoma. *Eur Urol* 51, 98–103; discussion 103–104 (2007).
25. Postma, R., Schroder, F. H. and van der Kwast, T. H. Atrophy in prostate needle biopsy cores and its relationship to prostate cancer incidence in screened men. *Urology* 65, 745–9 (2005).
26. Macoska, J. A., Trybus, T. M. and Wojno, K. J. 8p22 loss concurrent with 8c gain is associated with poor outcome in prostate cancer. *Urology* 55, 776–782 (2000).
27. Shah, R., Mucci, N. R., Amin, A., Macoska, J. A. and Rubin, M. A. Postatrophic hyperplasia of the prostate gland: neoplastic precursor or innocent bystander? *Am J Pathol* 158, 1767–1773 (2001).
28. Bethel, C. R. et al. Decreased NKX3.1 protein expression in focal prostatic atrophy, prostatic intraepithelial neoplasia and adenocarcinoma: association with Gleason score and chromosome 8p deletion. *Cancer Res* 66, 10683–10690 (2006).
29. Yildiz-Sezer, S. et al. Assessment of aberrations on chromosome 8 in prostatic atrophy. *BJU Int* 98, 184–188 (2006).
30. Yildiz-Sezer, S. et al. Gain of chromosome X in prostatic atrophy detected by CGH and FISH analyses. *Prostate* 67, 433–8 (2007).
31. Sun, J. et al. DNA copy number alterations in prostate cancers: a combined analysis of published CGH studies. *Prostate* (2007).
32. Tsujimoto, Y. et al. In situ shortening of CAG repeat length within the androgen receptor gene in prostatic cancer and its possible precursors. *Prostate* 58, 283–290 (2004).
33. Lin, X. et al. GSTP1 CpG island hypermethylation is responsible for the absence of GSTP1 expression in human prostate cancer cells. *Am J Pathol* 159, 1815–1826 (2001).
34. Lee, W. H. et al. Cytidine methylation of regulatory sequences near the pi-class glutathione S-transferase gene accompanies human prostatic carcinogenesis. *Proc Natl Acad Sci USA* 91, 11733–11737 (1994).
35. Brooks, J. D. et al. CG island methylation changes near the GSTP1 gene in prostatic intraepithelial neoplasia. *Cancer Epidemiol Biomarkers Prev* 7, 531–536 (1998).

2 Epigenetic Gene Silencing in Prostate Cancer

Srinivasan Yegnasubramanian, MD, PhD, and William G. Nelson, MD, PhD

CONTENTS

1. INTRODUCTION

Alterations in gene expression are caused by a number of epigenetic processes including DNA methylation and chromatin remodeling. Aberrations in these processes, leading to abnormal gene expression patterns, are nearly ubiquitous in human cancers and can carry the same importance as mutations in the initiation and progression of human cancers, including prostate cancer. In this chapter, we will first provide an overview of these epigenetic processes in normal physiology and in carcinogenesis. Then we will describe some of the

From: *Current Clinical Oncology: Prostate Cancer:*
Signaling Networks, Genetics, and New Treatment Strategies
Edited by: R. G. Pestell and M. T. Nevalainen © Humana Press, Totowa, NJ

specific epigenetic abnormalities that accompany prostate cancer progression. Finally, we will discuss how these epigenetic abnormalities can be targeted to enhance prostate cancer detection, risk stratification, prevention, and therapy.

2. EPIGENETIC PROCESSES IN PHYSIOLOGY AND CANCER PATHOPHYSIOLOGY

Virtually all somatic cells within any individual contain identical primary genomic DNA sequence information. Yet cells of different lineages, organs, and even different microenvironments within the same organs have vastly differing phenotypes and gene expression profiles. The heritable processes by which cells establish unique gene expression patterns without changing their primary gene sequence are referred to as epigenetic processes. These processes constitute an entire level of coding beyond the primary gene sequence and are likely responsible for establishing the vast spectrum of gene expression changes observed during development and differentiation. Dysregulation of these processes appears to be one of the earliest and most frequent somatic changes in human cancers, contributing to the initiation of malignant transformation and progression to advanced disease.

2.1. DNA Methylation

Among the most widely studied of these epigenetic processes is DNA methylation. In vertebrate genomes, DNA methylation occurs predominantly at the 5-position of cytosine (C) in self-complementary CpG dinucleotides by the action of DNA methyltransferase (DNMT) enzymes. This process is known to be central to several physiological processes including development, imprinting (1–3), X chromosome inactivation (4), suppression of parasitic and repetitive DNA elements (5–8), and transcriptional regulation (9–11).

One of the most striking illustrations of the importance of DNA methylation occurs early in development, just after fertilization. In certain mammals, after fertilization, the male pronucleus, even before any DNA replication and before fusing with the egg pronucleus, undergoes a rapid process of active DNA demethylation, reaching peak demethylation within a few hours (12). The egg genome also undergoes widespread demethylation, but in a slower process dependent on DNA replication (12). Although the cause and consequence of these DNA demethylation events are largely unknown, it has been conjectured that demethylation of the parental genomes is essential in order to begin reprogramming of epigenetic processes in the developing embryo by erasing the epigenetic programming that marked the parental genomes. These demethylation steps may be a necessary component for establishing undifferentiated, toti- and pluri-potent stem cells. Likewise, it is possible that such demethylation events occurring abnormally in somatic cells might lead to formation of cancer stem cells and carcinogenesis. These possibilities are currently under intense investigation.

Another important aspect of this epigenetic reprogramming during development is the establishment of imprinting. Imprinting is the process by which a given gene's expression is limited to either the maternal or paternal copy, but not both. CpG methylation modifications are often found to mark either the active or inactive allele in distinct regions called differentially methylated domains in imprinted genes. Loss of imprinting (LOI) often occurs in cancer cells, including prostate cancer cells, leading to inappropriate expression or repression from both the maternal and paternal alleles of a normally imprinted gene (13,14). This LOI is often associated with changes in the DNA methylation patterns at these genes (15,16). Whether these DNA methylation changes are a cause or effect of the LOI process is still debated.

The process of X chromosome inactivation is presumably used for gene dosage limitation in female cells and is mediated by the expression and binding of the *Xist* RNA to the target X chromosome, widespread CpG methylation throughout the target X chromosome, and recruitment of chromatin remodeling complexes that tightly package the target X chromosome into transcriptionally inactive heterochromatin [reviewed in (17)]. These inactive X chromosomes are referred to morphologically as Barr bodies, appearing as highly condensed heterochromatin-like regions (18–20). The precise role of DNA methylation changes in silencing the inactive X chromosome, and whether this process is dysregulated in cancer cells, are still largely unknown.

CpG methylation also appears to be involved in the transcriptional repression of parasitic transposable elements and repetitive elements by maintaining them in a closed chromatin state that suppresses transcription and genomic rearrangements (5–8). Indeed, the majority of CpG dinucleotides contained in repetitive elements such as *LINE1* retrotransposon sequences are normally methylated in adult somatic cells [reviewed in (21)]. Cancer cells have a tendency to develop undermethylation at these repetitive elements (22), and therefore may be more prone to genomic rearrangements (23,24), possibly by homologous recombination or by expression of intact retrotransposons.

The role of DNA methylation in transcriptional regulation has been the subject of much recent research. The self-complementary CpG dinucleotide is usually methylated in the normal somatic cell genome and is highly under-represented compared with all other dinucleotides (25). This under-representation presumably occurs because spontaneous hydrolytic deamination of 5-methyl-cytosine (5mC) to thymine in germ cell genomes has led to depletion of CpG dinucleotides during evolution (25). Despite this overall under-representation of CpG dinucleotides, dense clusters of CpG dinucleotides, termed CpG islands (CGIs), which are usually unmethylated in normal somatic cell genomes, are found at the transcriptional regulatory regions of ~60% of mammalian genes *(26)*. In the unmethylated state, these CGIs can be housed in chromatin structures that take on active conformations.

DNA methylation at these CGIs is associated with recruitment of chromatin-remodeling complexes that condense the local chromatin in a manner that resembles the facultative heterochromatin seen in the inactive X chromosome in female somatic cells *(25)*. This condensed local chromatin structure is highly resistant to loading of RNA polymerase II at the transcriptional start site of the associated gene and therefore leads to its transcriptional inactivation *(25)*. Such DNA methylation-induced gene-silencing events have long been supposed to mediate tissue- and developmental/differentiation stage-specific gene expression profiles. However, only recently, with the use of unbiased genome methylation detection technologies, have such tissue differentially methylated and expressed genes been identified in a systematic fashion in a mammalian genome *(27)*.

Much work has focused on the derangement of these physiological DNA methylation processes in the initiation and propagation of human malignancies. Early studies examining aberrations in DNA methylation in human cancers showed that cancer genomes have decreased genomic 5mC content compared with normal genomes and also become undermethylated at CpG dinucleotides within the coding sequences of known genes *(28–33)*. Although the exact consequences of these changes were unknown, subsequent work has suggested that this undermethylation of DNA sequences may result in genomic instability due to increased rearrangements *(23,24,34,35)*.

Cancer genomes have also been identified to often harbor abnormal DNA hypermethylation at CGI sequences resulting in an inappropriate silencing of the associated gene *(36,37)*. Like gene deletions and mutational silencing, DNA hypermethylation and the resulting epigenetic transcriptional repression have been postulated to be an important means by which cancer cells acquire and maintain their malignant phenotype *(38)*. These findings underline a funda-mental enigma in the generation of abnormal DNA methylation patterns in cancer cells, in which there may be a decrease in overall genomic 5mC content with a paradoxical increase in CpG methylation at certain CGIs *(39)*.

Independent of the mechanism(s) by which abnormal epigenetic gene silencing arises during cancer initiation, abnormal DNA methylation changes, resulting in phenotypic gene expression changes, appear subject to selection for cell growth and/or survival. In one experiment, using Luria Delbruck fluctu-ation analysis, Holst et al. showed that a small minority of normal human mammary epithelial cells (HMECs) develop hypermethylation and silencing of the *p16/INK4a* gene, which encodes a cell cycle regulatory protein, and that these cells are highly selected for during the passaging of HMECs in culture *(40)*. Indeed, almost all of the HMECs that escaped senescence harbored hypermethylated and silenced *p16/INK4a* alleles *(40)*. This somatic epigenetic alteration therefore permits the cells to continue proliferating while unaffected HMECs undergo cell senescence *(40)*. The equivalence of epigenetic and genetic alterations during cancer development is further demonstrated by experiments with human colorectal carcinoma 116 (HCT-116) colorectal cancer cells, which, like some of the HMEC cells described above, lack *p16/INK4a* function. HCT-116 cells contain one mutant gene encoding *p16/INK4a*, with a frameshift mutation in the coding sequence, and one wild-type gene, showing marked hypermethylation at the CGI region and repression of expression from this normal allele *(41)*. Remarkably, *p16/INK4a* CGI hypermethylation changes are only present at the wild-type and not the mutant gene *(41)*, presumably because hypermethylation of the CGI at the mutant allele would not have provided a growth advantage. CGI hypomethylation changes may also be subject to selection. For example, by long-term exposure to doxorubicin and other antineoplastic drugs that are substrates for P-glycoprotein-mediated efflux pumping, it is possible to select for rare variants of MCF7 breast cancer cells that stably express high levels of the P-glycoprotein and GSTP1 due primarily to loss of hypermethylation at the corresponding regulatory CGIs *(42,43)*. Nonetheless, the detailed mechanisms by which cancer cells first acquire *de novo* methylation changes and then maintain them through the subsequent growth expansion and progression of the transformed cells remain largely unknown.

2.2. DNA Methyltransferase Enzymes

The mammalian DNMTs, which include DNMT1, DNMT3a, and DNMT3b, are central to the establishment and maintenance of methylation changes during physiological processes as well as during carcinogenesis. These enzymes catalyze the transfer of a methyl group from *S*-adenosyl-methionine to the 5-position of cytosine bases in CpG dinucleotides. On the basis of their propensity to modify C to 5mC in unmethylated versus hemimethylated double-stranded DNA oligonucleotides *in vitro*, the mammalian DNMTs have been classified as primarily "*de novo*" (DNMT3a and DNMT3b) or "maintenance" (DNMT1) methyltransferases *(44–47)*. Under this classification, the *de novo* methyltransferases would initiate new CpG methylation patterns, whereas the maintenance methyltransferase would maintain established CpG methylation patterns during replication and mitosis. The idea that DNMT3a and DNMT3b are *de novo* methyltransferases was further supported by data showing that targeted disruption of these genes in mice results in a blockage of *de novo* methylation in embryonic stem cells and early embryos without abrogating maintenance of pre-existing imprinted methylation patterns *(48)*. While it is likely that DNMT1, which is targeted to the advancing replication fork, is most responsible for maintaining methylation patterns during genome duplication, it is clear that the DNMTs may cooperate and/or complement each other to establish and maintain methylation patterns. For example, HCT-116 human colon cancer cells carrying targeted homozygous disruptions of the *DNMT1* or *DNMT3b* genes lose only 20 and 3% of their genomic methylation levels, respectively *(49,50)*. However, HCT-116 colon cancer cells carrying targeted homozygous disruptions of both *DNMT1* and *DNMT3b* lose approximately 95% of their genomic methylation levels, suggesting that DNMT3b may cooperate with and/or complement DNMT1 in maintaining genomic methylation patterns during genome duplication and mitosis *(50)*.

Although it is clear that the fidelity of CpG methylation pattern maintenance must be somehow corrupted in cancer cells, the means by which *de novo* increases and decreases in CpG dinucleotide methylation appear during carcinogenesis have not been fully established. Aberrant DNMT expression and function may contribute both to DNA hypermethylation and DNA hypomethylation during cancer development. In this regard, of all the DNMTs, DNMT1 appears most likely to play a major role in cancer development. Both too much and too little DNMT1 function has been implicated in the generation of the abnormal DNA methylation patterns typical of cancer cells. Forced overexpression of DNMT1 in normal cells directly causes increased DNA methylation and epigenetic gene silencing *(51–53)*. Additionally, DNMT1 is required for *c-fos* transformation of rodent fibroblasts *in vitro*, for intestinal polyp development in $Apc^{Min/+}$ mice, and for tobacco carcinogen-induced murine lung cancer development *in vivo (54–57)*. By contrast, under-production of the enzyme also results in carcinogenesis; mice carrying one disrupted *Dnmt1* allele and one hypomorphic *Dnmt1* allele and exhibiting only 10% of normal DNMT activity develop genomic instability and T-cell lymphomas *(58,59)*. Whether such manipulations to increase and decrease DNMT1 activity are appropriate models for true endogenous DNMT1 function is still unknown, but is an area of active investigation. Some evidence for increased DNMT1 function in cancer cells has recently emerged, however. For instance, increases in the endogenous expression of DNMT1 with accompanying abnormalities in DNA methylation have been reported for mouse prostate cells carrying disrupted *Rb* genes, linking the pRb–E2F pathway to regulation of DNA methylation *(60)*. However, levels of mRNA encoding *DNMT1*, when normalized to proliferation, does not appear to be commonly over- or under-expressed in cancer cells compared with normal cells *(61)*. Nonetheless, DNMT1 protein levels appear to be extensively regulated via targeted ubiquitin-mediated proteasome degradation pathways, and many cancer cells display marked defects in this DNMT1 degradation pathway. As a result, DNMT1 protein over-expression appears to occur even in the absence of increases in *DNMT1* mRNA levels *(62)*. The degree to which increased DNMT1 protein levels contribute to aberrant DNA methylation patterns, epigenetic gene silencing, and carcinogenesis is still unknown.

2.3. Selective Methylated-DNA-Binding Proteins, Chromatin Structure, and Histone Modifications

The mechanisms by which DNA methylation changes are translated to epigenetic gene silencing are under intense investigation. In one model, CpG methylation occurring in transcription-factor-binding sites directly prevents binding of the transcription factor and thereby prevents gene expression. In a second model, chromatin-remodeling complexes that include selective methylated-DNA-binding (mDB) proteins, histone deacetylases (HDACs), and histone methyltransferases (HMTs) are recruited to sites of DNA methylation and condense the local chromatin into a transcriptionally non-permissive conformation. It is likely that both of these models occur, each in different genomic contexts [reviewed in *(63,64)*].

Currently, two major classes of mDB proteins have been recognized based on the structural domains that allow these proteins to selectively bind methylated, but not unmethylated, DNA (Fig. 1). The first of these to be identified were the so-called methylated-DNA-binding domain (MBD) containing proteins [reviewed in *(64)*]. There are five well-characterized members of this family (MBD1, MBD2, MBD3, MBD4, and MECP2), though a recent bioinformatics search of genome and protein databases suggested that there may be up to 11 of these proteins *(65)*. However, of the five well-characterized members, only MBD1, MBD2, and MECP2 are known to bind with high affinity and specificity to methylated DNA. Members of the more recently identified second class of mDB proteins, called ZBTB proteins, contain kruppel-like C2H2 zinc fingers, rather than MBD, to facilitate selective binding to methylated DNA *(66,67)*. These members include Kaiso, ZBTB4, and ZBTB38, of which Kaiso is the best characterized *(66,67)*. Kaiso is a bimodal DNA-binding protein capable of binding a specific unmethylated consensus sequence as well as methylated DNA containing two consecutive methylated CpG dinucleotides *(66,68)*. Nonetheless, Kaiso, ZBTB4, and ZBTB38 have all been shown to bind methylated DNA and facilitate transcriptional repression *(66,67)*.

In addition to their characteristic methylated-DNA-binding domains, the mDB proteins contain transcriptional repression domains (TRDs) that recruit other components of the repressive chromatin-remodeling complex, which ultimately allow transcriptional silencing [Fig. 1; reviewed in *(64)*]. When we speak of chromatin structure, we refer to the structural conformation of local genomic DNA and all bound proteins. Typically, genomic DNA is not free-floating but rather intricately wrapped around histone octamers, forming nucleosomes, and bound by numerous protein complexes to form complex quaternary structures. Modifications on histone octamer subunits lead to structural changes that affect the spacing of these nucleosomes as well as the accessibility of the surrounding DNA to other proteins, including

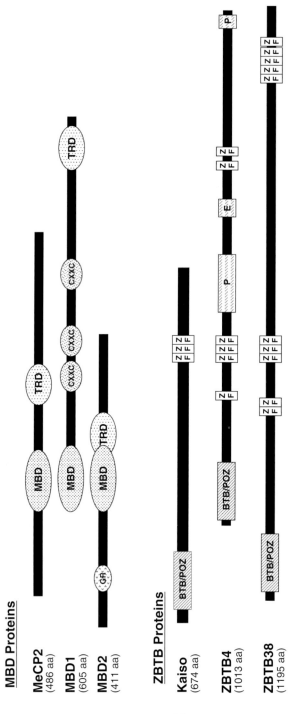

Fig. 1. Schematic representation of the primary structure of methylated-DNA-binding (mDB) proteins. MBD, methyl-binding domain; TRD, transcriptional repression domain; CXXC, cysteine-rich zinc finger domain; GR, glycine, arginine repeat; ZBTB, zinc finger/BTB domain; BTB/POZ, BR-C/ttk/bab/Pox virus and zinc finger transcriptional repression domain; ZF, zinc finger domain; P, proline-rich region; E, glutamate-rich region; aa, amino acids

24

RNA polymerases [reviewed in *(69)*]. Among the best characterized of these modifications are methylation and acetylation of histone subunits. Acetylation of H3 subunits, catalyzed by histone acetyltransferases (HATs), leads to open chromatin structures, whereas deacetylation of H3 subunits, catalyzed by HDACs, leads to condensed, transcriptionally repressed, chromatin structures [reviewed in *(69)*]. Methylation and demethylation at histone subunits are catalyzed by the actions of HMTs and histone demethylases (HDMs), respectively *(70–73)* [reviewed in *(74)*]. These histone methylation modifications make up a complex code that directs the local chromatin to be either permissive or non-permissive for transcription. The precise nature of this code is only now being unraveled. Nonetheless, the mDB proteins provide a clear link between DNA methylation changes, which they bind through their methylated-DNA-binding domains, and these chromatin structural changes, through the action of their TRDs. For instance, MECP2's TRD allows recruitment of Sin3 and Sin3-bound HDACs that facilitate condensation of the local chromatin structure and transcriptional repression *(75)*. The MBD2 protein, a member of the large 1 Mega-Dalton MeCP1 transcriptional repression complex, contains a TRD that recruits other members of this complex, including Mi-2/NuRD chromatin-remodeling complex components such as MBD3, HDAC1, and HDAC2, histone-binding proteins RbAp46 and RbAp48, the SWI/SNF helicase/ATPase domain-containing protein Mi-2, MTA2, and other proteins *(76)*. MBD1 can recruit the SETDB1 histone H3-K9 methyltransferase, coupling DNA methylation with histone methylation *(77)*. SETDB1 can in turn recruit DNMT3a to establish *de novo* methylation patterns, suggesting that there is a cyclical interplay between chromatin modification and DNA methylation *(78)*. Kaiso facilitates DNA methylation-dependent transcriptional repression by recruitment of the HDAC-containing N–CoR corepressor complex *(79)*. Therefore, these mDB proteins can transduce DNA methylation changes to transcriptional repression changes via the recruitment of chromatin altering co-repressor complexes. This transduction is crucial to DNA methylation-mediated epigenetic gene silencing. For instance, cells from MBD2-deficient mice, as well as human cancer cells treated with siRNA-targeting *MBD2* mRNA, are unable to repress transcription from exogenously hypermethylated promoters in transient transfection assays *(80,81)*.

However, the degree of functional redundancy/complementarity between the mDB proteins still needs clarification. Although complete disruption of DNMT1 or DNMT3b in murine models results in severe developmental deficits and embryonic lethality, *Mbd2*$^{-/-}$ and *Mecp2*$^{-/-}$ mice are completely viable *(48,80,82,83)*. Indeed, their only phenotypes are that the *Mbd2*$^{-/-}$ mice exhibit an abnormality in maternal nurturing behavior and the *Mecp2*$^{-/-}$ mice develop characteristics highly similar to those of patients with Rett syndrome, which results from mutations in the *MECP2* gene *(80,83)*. The viability of these mice is perhaps due to a certain amount of redundancy between the numerous different mDB proteins. However, it is also evident that many of the mDB proteins have unique properties both *in vitro* and *in vivo* that cannot be complemented by other mDB proteins. Methylated CGIs in cells are often occupied by either MECP2 or MBD2, but not both *(84)*. Some of this specificity may be due to their biochemical properties. For instance, whereas MBD2 appears to bind methylated CpGs in any genomic context, MECP2 seems to prefer binding to methylated CpG sites that are adjacent to a run of four or more A/T bases *(84)*, while Kaiso prefers binding of at least two consecutive methylated CpG dinucleotides *(68)*. Nonetheless, it is likely that mDB proteins may have some capacity to complement each other, but still maintain non-redundant, unique functions.

Additionally, chromatin structure and gene expression can be modulated even in the absence of DNA methylation changes. A multitude of histone modification enzymes, including HMTs, HDMs, HATs, and HDACs, have been described that alter chromatin structure and gene expression without the requirement for recruitment to methylated DNA sites. Among these, the polycomb group complexes, which contain the EZH2 HMT, are crucial for establishing selective gene repression during development *(85)*. Several studies have even suggested that silencing by these enzymes can occur independently of DNA methylation changes and in some cases can even instruct downstream DNA methylation changes, in order to set up long-term repression, as evidenced by the recent observation that EZH2 can recruit DNMTs to genomic sites targeted for transcriptional repression *(85–87)*.

3. ABNORMALITIES IN EPIGENETIC GENE SILENCING DURING PROSTATE CANCER INITIATION AND PROGRESSION

The first gene found to be silenced via somatic CGI hypermethylation in prostate cancer was *GSTP1*, which encodes the π-class glutathione *S*-transferase (GST) enzyme *(36)*. This genome change remains the most common somatic genome abnormality of any kind (>90% of cases) reported thus far for prostate cancer, appearing earlier and more frequently than other gene defects, including the recently described fusions between *TMPRSS2* and ETS family genes, that arise during prostate cancer development *(88,89)*. The GST enzymes catalyze the detoxification of carcinogens and reactive chemical species via conjugation to glutathione. Loss of π-class GST function by CGI hypermethylation and silencing likely sensitizes prostatic epithelial cells to damage from dietary carcinogens and inflammatory oxidants, perhaps explaining the well-documented contribution of diet and lifestyle factors to prostatic carcinogenesis *(88,90)*. In support of this hypothesis, mice carrying disrupted

Gstp1/2 genes are more susceptible to developing skin cancers after treatment with the carcinogen 7,12-dimethylbenz[a]anthracene (DMBA) than wild-type mice *(91)*. Furthermore, π-class GST-deficient human prostate cancer cells in culture accumulated high levels of genome damage when exposed to 2-hydroxyamino-1-methyl-6-phenylimidazo[4,5-b]pyridine (N–OH–PhIP), a dietary agent known to have toxicity in the prostate *(92–94)*.

Provocatively, *GSTP1* CGI hypermethylation, which is not present in normal prostatic cells (or any other normal cells), seems to arise first in a fraction of proliferative inflammatory atrophy (PIA) lesions, the earliest prostate cancer precursors. The blunted, dysfunctional luminal epithelia in these PIA lesions, which are often surrounded by an inflammatory infiltrate, typically express high levels of stress response genes, including *GSTP1*, *GSTA1*, and *COX2* *(95–97)*. Induction of these genes is likely a reaction to the electrophilic and oxidative stress in their inflammatory milieu *(95,98)*. However, a fraction of PIA lesions exhibit loss of *GSTP1* expression *(99)*. Recent reports suggest that this loss is likely due to *GSTP1* promoter CGI hypermethylation in these PIA lesions, as evidenced by the observation that a similar fraction of PIA lesions exhibit *GSTP1* CGI hypermethylation *(99)*. As the prostate lesions progress to prostatic intraepithelial neoplasia (PIN) lesions, which are known prostate cancer precursors, and to prostatic carcinomas, there is a progressive accumulation of cells with *GSTP1* CGI hypermethylation and loss of *GSTP1* expression, suggesting that these characteristics are selected for during the earliest stages of prostate cancer progression *(90,99–101)*. The recognition that DNA hypermethylation changes characteristic of prostate cancer cells first appear in PIA lesions suggests that chronic or recurrent inflammation may be involved in the *de novo* acquisition of abnormal DNA methylation patterns. In support of this hypothesis, a recent report suggested that interleukin 1B-triggered nitric oxide generation led to silencing of the *FMR1* and *HPRT* genes by hypermethylation of their regulatory CGIs *(102)*. Activated macrophages, expressing high levels of the inducible form of nitric oxide synthetase (iNOS), have been detected around PIA lesions in human prostate tissues (Lee B.H. et al., personal communication, 2006).

Since the recognition by Lee et al. in 1994 that the *GSTP1* CGI was frequently hypermethylated in prostate cancers, more than 40 other genes have been reported to be targets of CGI hypermethylation-associated epigenetic gene silencing in prostate cancers [reviewed in *(103)*]. Synthesizing evidence from existing reports, we found that CGI hypermethylation changes occur in at least two waves: first in prostate cancer precursor lesions, possibly driving the initiation of neoplastic transformation, and second in malignant prostate carcinoma cells, possibly driving malignant progression to advanced disease. For example, in one case series, hypermethylation of CGIs at *GSTP1*, *APC*, *RASSF1a*, *COX2*, and *MDR1* was present in the >90% of localized prostate cancer lesions and faithfully maintained in the majority of advanced metastatic cancers. These changes likely occurred in a highly coordinated wave very early during prostate carcinogenesis. On the contrary, hypermethylation at the *ERα*, *hMLH1*, and *p16/INK4a* CGIs was rare in primary cancers, but more common in metastatic cancer deposits *(104)*. Additionally, in a study of metastatic prostate cancers obtained from rapid autopsies of men succumbing to their advanced prostate cancer, CGI hypermethylation profiles appeared to be maintained in a nearly clonal pattern during the process of metastatic dissemination to multiple anatomically distinct sites. Indeed, almost all metastatic deposits from the same patient harbored nearly identical patterns of methylation. This observation provides evidence that abnormal DNA methylation patterns may arise before prostate cancer cell growth and expansion at metastatic sites *(104)*.

As discussed earlier, paradoxically, cancer genomes are thought to have a decreased 5mC content despite hypermethylation at certain CGIs. Although somatic DNA hypomethylation has also been described in prostate cancer, it has not been studied in as great a detail thus far as somatic hypermethylation. In an early analysis, reduction of total 5mC levels was found to be rare in primary prostate cancers, but more common in prostate cancer metastases *(32)*. In a more recent study, methylation of *LINE-1* sequence promoter CGIs, which are repeated thousands of times in the human genome, was found to be decreased in 53% of all the prostate cancer cases analyzed. Interestingly, *LINE-1* hypomethylation changes occurred in 67% of cases with lymph node metastases but only 8% of cases without lymph node metastases *(105)*. Additionally, CGI hypermethylation changes at *GSTP1*, *RARβ2*, *RASSF1a*, and *APC* appeared to precede these *LINE-1* hypomethylation changes, which were generally detected in cancers of higher stage and histologic grade *(106)*. These hypomethylation changes may be associated with genetic instability, as evidenced by reports describing a correlation between DNA hypomethylation and losses or gains of sequences on chromosome 8 *(58,107)*.

Chromatin structural changes, marked by histone modifications, and alterations in the protein complexes binding the regulatory regions of genes and modulating their expression, have also been described in prostate cancer, but have not yet been catalogued in a detailed fashion. One change that has been found in microarray studies is the pronounced over-expression in prostate cancer metastases of the enhancer of zeste homolog 2 (*EZH2*), a histone H3-K27 methyltransferase component of the polycomb group complexes *(108)*. This over-expression likely contributes to prostate cancer progression by the dysregulated repression of specific genes through H3-K27 methylation. Indeed, siRNA-mediated reduction of EZH2 in prostate cancer cells resulted in inhibition of cell proliferation whereas forced EZH2 over-expression triggered repression of a specific set of genes *(108)*. This repression was likely because of histone H3-K27 methylation by EZH2, facilitating assembly of repressive chromatin structures at these sites *(85,108)*. One candidate target of EZH2-mediated repression is the *DAP2IP* gene, encoding a GTPase-activating protein that can effect Ras signaling and tumor necrosis factor (TNF)-associated apoptosis *(109,110)*. Forced over-expression of EZH2 in normal prostate cells led to histone H3-K27 methylation at the *DAB2IP* promoter and associated *DAB2IP* suppression, whereas siRNA depletion of EZH2 in cancer cells increased *DAP2IP* expression *(110)*. However, whether EZH2 over-expression, which is likely regulated by the pRB–E2F pathway and required for cell replication *(111)*, simply reflects increased proliferation in prostate cancer metastases or indicates epigenetic gene dysregulation has not been firmly established.

4. CGI HYPERMETHYLATION AS A MOLECULAR BIOMARKER FOR PROSTATE CANCER

Molecular screening for prostate cancer by assaying for prostate-specific antigen (PSA) in serum has had dramatic consequences on the recognized natural history of prostate cancer, allowing detection of prostate cancer as localized disease amenable to definitive treatment with radical prostatectomy and/or radiation therapy *(112)*. This trend may be responsible, in part, for the recent decline in prostate cancer mortality. However, PSA screening is riddled with several limitations, including a relatively high false-negative rate. For instance, in the Prostate Cancer Prevention Trial (PCPT), 24.4% of men on the placebo treatment arm who entered the study with "normal" serum PSA values and underwent prostate biopsies at the end of the trial were found to have prostate cancer *(113,114)*. The current approach to prostate biopsy for prostate cancer detection and diagnosis, featuring ultrasound-guided random sampling of ˜0.3% of prostate tissue rather than targeted sampling of a radiographically imaged lesion (as is routine for other cancers), leaves much to be desired. With the current approach, several controversies remain concerning the optimal number of sampling tissue cores to be obtained and regarding which men should undergo repeat biopsies if cancer is not detected *(115,116)*. Furthermore, autopsy studies suggest that 29% of men between the age of 30 and 40 and 64% of men between the age 60 and 70 harbor small prostate cancers *(117)*. Clearly, only a minority of these men, approximately 5% of all men, will develop symptomatic or life-threatening disease. Consequently, the wisdom of prostate cancer screening and early detection has been questioned *(118)*. To confront these challenges, researchers have sought new molecular biomarkers that could be useful for prostate cancer screening and diagnosis and for directing treatment choices for men with prostate cancer.

Somatic epigenetic alterations, particularly DNA methylation changes, offer a great source of potential molecular biomarkers for prostate cancer for several reasons. First, somatic CGI hypermethylation changes have been nearly universally identified in all human cancers, including prostate cancer. Second, these somatic CGI hypermethylation changes appear to be more prevalently associated with prostate cancer and other cancers than other somatic genetic changes such as mutations, deletions, and translocations. Finally, a number of sensitive and specific strategies are being developed to detect CGI methylation from scant genomic DNA sources such as bodily fluids and biopsy specimens *(119–121)*.

There are currently three major strategies for distinguishing methylated DNA from unmethylated DNA (Table 1). The first of these takes advantage of the selectivity of so-called methylation-sensitive restriction enzymes (MSRE) to digest only unmethylated DNA while leaving methylated DNA intact. The undigested methylated DNA can then be detected by a variety of techniques including Southern blot *(30,122–124)*, PCR *(125,126)*, real-time PCR *(121)*, microarray hybridization *(127)*, and multiplex ligation-dependent amplification *(128)*. While this strategy can be extremely sensitive when coupled with PCR, capable of detecting single copies of methylated DNA, it also has a few important limitations, including a propensity for false-positive results arising from incomplete digestion of unmethylated DNA, and the inability to interrogate methylation at CpG dinucleotides outside of the recognition sequence of the MSRE. MSRE-based sequencing and microarray approaches have also been developed to characterize whole-genome methylation patterns in a relatively unbiased manner *(27,129,130)*, but are limited to interrogating CpG methylation only within the recognition sequence of the specific restriction enzyme. Utilization of the McrBC methylation-specific homing endonuclease, which digests at a very wide variety of sequence contexts, can circumvent this problem *(131)* but is still limited by the unpredictability of digestion.

A second strategy uses sodium bisulfite to deaminate cytosine to uracil while leaving 5mC intact *(132)*, creating DNA sequence differences at C versus 5mC after PCR amplification. These DNA methylation-based sequence differences can then be mapped at single-base resolution by a technique called bisulfite genomic sequencing *(133,134)*. In this technique, PCR primers are complementary to the bisulfite converted alleles but do not overlap with potentially methylated cytosines (those in CpG dinucleotides). PCR amplification, cloning of PCR products into plasmids, and subsequent sequencing will reveal the prevalence of each pattern of CpG methylation in the original sample at single-base resolution. Although this technique has become the gold standard for determination of DNA methylation patterns, it is limited by the fact that it is very labor intensive and not easily amenable to high-throughput analysis, and is not well-suited to sensitive identification of low-prevalence methylation patterns.

To circumvent these problems, one can use another bisulfite-based strategy, called methylation-specific PCR (MSP) *(135)*, which uses PCR primers

Table 1
Summary of DNA Methylation Detection Assays

Discrimination basis	Detection strategy	Name	Sample throughput	Target Throughput*	Quantitative	Sensitivity	References
MSRE	Southern Blot	MSRE-Southern	Low	Low	Yes	Low	30,122–124
	PCR	MSRE-PCR	High	Low	No	High	125,126
	Real-time PCR	MSRE-QPCR	High	Low	Yes	High	121
	Target Specific Microarray	MAD	Low	High	Semi	Mod	127
	Multiplex Amplification	MS-LPA	Low	Mod	Semi	High	128
	2-D Gel, Sequencing	MS-RLGS	Low	Mod-High	No	Low	27
	Unbiased sequencing	MSDK	Low	Massive	Semi	Low-Mod	129
	Whole Genome Microarrays	MSRE-WGMA, McrBC-WGMA	Low	Massive	Semi	Mod	130,131
Sodium Bisulfite	Sequencing	Bisulfite Genomic Sequencing	Low	Low	Semi	Mod	133,134
	PCR	MSP	Mod-High	Low	No	High	135

(continued)

31

Table 1
(Continued)

Discrimination basis	Detection strategy	Name	Sample throughput	Target Throughput*	Quantitative	Sensitivity	References
	Real-time PCR	RT-MSP, MethyLight, HeavyMethyl, MethylQuant	High	Low	Yes	High	104,136–138
	Restriction Digestion	COBRA	Low	Low	Yes	Mod	139
	Single Nucleotide Primer Extension	MS-SNuPE	Low	Low	Yes	Mod-High	140
	Pyrosequencing	PyrcMeth	High	Low	Yes	High	141
	Target Specific Microarray	DMH, MSO, Epigenomics Microarray	Low	Mod-High	Semi	Mod	142–144
	Golden Gate/Bead Arrays	Methylation Specific Golden Gate Universal Bead Array	Low	Mod-High	Semi	Mod-High	145

	Base-Specific Cleavage/Mass Spectrometry	Methylation-Specific BSC-MALDI Mass Spectrometry	High	Low-Mod	Yes	High	146
Methylated DNA Binding Proteins	MBD Affinity Enrichment	MECP2-MBD Column Chromatography, MIRA, ICEAMP, MBD-SPM, MB-PCR	Low-Mod	Low	No	Mod-High	26,147–151
	MBD/Anti-5mC-Ab Affinity Enrichment/Microarray	MIRA-AMA, MCIp-Chip, MeDIP-Chip, COMPARE-MA	Low	Massive	Semi	Low-Mod	150,153–155

(continued)

Table 1
(Continued)

Discrimination basis	Detection strategy	Name	Sample throughput	Target Throughput*	Quantitative	Sensitivity	References
Combination of Methylated DNA Binding Proteins and MSRE	Real-time PCR	COMPARE-MS	High	Low	Yes	High	152

* Degree to which multiple targets can be analyzed in a single assay; Mod. Moderate; MSRE, Methylation Sensitive Restriction Enzymes; PCR, Polymerase Chain Reaction; QPCR, Quantitative PCR; MAD, Methylation Amplification DNA-Chip; MS-LPA, Methylation Specific Ligation-Dependent Probe Amplification; MS-RLGS, Methylation Specific Restriction Landmark Genome Scanning; MSDK, Methylation Sensitive Digital Karyotyping; WGMA, Whole Genome Microarray; McrBC, methylation sensitive homing endonuclease; MSP, Methylation Specific PCR; RT-MSP, Real-time MSP; COBRA, Combined Bisulfite Restriction Assay; MS-SNuPE, Methylation Specific Single Nucleotide Primer Extension; DMH, Differential Methylation Hybridization; MSO, Methylation-specific oligonucleotide microarray; BSC-MALDI, Base Specific Cleavage/Matrix Assisted Laser Desorption Ionization Mass Spectrometry; MECP2-MBD, Methylated CpG-binding Protein 2 Methyl Binding Domain; MIRA, Methylated-CpG Island Recovery Assay; ICEAMP, Identification of CpG-Islands Altered Methylation Patterns; MBD-SPM, MBD Enrichment/Separation of Partially Melted Sequences; MB-PCR, Methyl-binding PCR; MIRA-AMA, MIRA-Assisted Microarray Analysis; MCIp-Chip, Methylated CpG-Island Immunoprecipitation DNA-Chip; MeDIP-Chip, Methylated-DNA Immunoprecipitation DNA-Chip; COMPARE-MA, Combination of Methylated-DNA Precipitation and Restriction Enzymes/Microarray Analysis; COMPARE-MS, Combination of Methylated-DNA Precipitation and Methylation Sensitive Restriction Enzymes.

targeting the bisulfite-induced sequence changes to specifically amplify either methylated or unmethylated alleles, and can be used to detect the presence of a single pattern of methylation in each reaction. Quantitative variations of this technique, such as MethyLight *(136)*, HeavyMethyl *(137)*, RT-MSP *(104)*, and MethylQuant *(138)*, employ methylation-specific oligonucleotide primers in conjunction with Taqman probes or SYBR Green-based real-time PCR amplification to quantitate alleles with a specific pattern of methylation. Other bisulfite-based strategies use restriction enzyme digestion of bisulfite-conversion-generated restriction sites (COBRA) *(139)*, single-nucleotide primer extension (MS-SNuPE) *(140)*, pyrosequencing *(141)*, microarray hybridization *(142–144)*, golden gate assay/universal bead arrays *(145)*, or base-specific cleavage/mass spectrometry *(146)* to provide quantitative information regarding the levels of methylation at individual or groups of CpG dinucleotides. A number of bisulfite-based techniques capable of interrogating methylation patterns at multiple sequences in parallel have also been developed. All of these techniques can be highly sensitive and specific for detection of DNA methylation. However, the bisulfite-based techniques are in general somewhat cumbersome, involving time- and labor-intensive chemical treatments that damage DNA, limiting sensitivity and throughput. Additionally, PCR primer design becomes difficult because of the reduction in genome complexity after bisulfite treatment, leading to an inability to interrogate the methylation pattern at some or all CpG dinucleotides in a genomic locus of interest.

A third strategy for detection of DNA methylation, first introduced in 1994 by Cross et al., uses column- or bead-immobilized recombinant methylated-CpG-binding domain (MBD) proteins or anti-5mC antibodies to enrich for methylated DNA fragments for subsequent detection by Southern blot, PCR, or microarray hybridization *(26,147–151)*. New assays featuring capture and enrichment of methylated DNA coupled with PCR appear highly sensitive and specific and easily adapted to high-throughput platforms. One limitation of these assays, however, might be a propensity for false-positive results due to non-specific capture of unmethylated DNA fragments. However, one recent approach, termed COMPARE-MS, uses MBD-mediated capture in conjunction with MSRE digestion to further enhance specificity and eliminate false-positive results while maintaining exquisite sensitivity down to five genomic equivalents of methylated alleles *(152)*. Finally, of all the techniques mentioned, the methylated DNA capture and enrichment strategies show the most promise for determination of whole-genome methylation patterns by utilization of whole-genome tiling microarrays and whole-genome promoter microarrays (Yegna-subramanian et al., unpublished data, 2006) *(150,153–155)*.

One of the most promising epigenetic gene-silencing-based biomarkers for prostate cancer is the CGI hypermethylation of the *GSTP1* gene regulatory region *(36,88,90)*. The CGI in the *GSTP1* promoter is devoid of 5mC in normal cells of the prostate and other tissues, but in almost all prostate cancers that have been carefully studied, the *GSTP1* CGI is densely methylated and the gene is transcriptionally silenced *(90,156,157)*. In a recent review of 24 published studies, it was found that more than 81% of the 1071 prostate cancer cases analyzed in these studies harbored *GSTP1* CGI hypermethylation *(103)*. The sensitivity of assays for *GSTP1* CGI hypermethylation varied considerably and depended on the specific assays used and the region of the *GSTP1* CGI targeted by these assays *(103)*. Using one of the new methylated DNA capture and enrichment strategies, called COMPARE-MS, we found that *GSTP1* CGI hypermethylation exhibited 99.2% sensitivity and 100% specificity for distinguishing prostate cancer DNA from normal prostate tissue DNA *(152)*. Because of this exquisite potential for screening/diagnostic sensitivity and specificity, several different *GSTP1* CGI hypermethylation assays are under clinical development for prostate cancer screening and diagnosis. Many of these assays have already demonstrated efficacy in detecting *GSTP1* hypermethylation and thus the presence of prostate cancer DNA, from multiple tissue sources including prostate tissue biopsies, prostate secretions, urine, and blood *(121,158–160)*. CGI hypermethylation changes at other loci, including *APC, RASSF1a, PTGS2*, and *MDR1*, may also serve as useful biomarkers in distinguishing prostate cancer from non-cancerous tissue with high sensitivity (97.3–100%) and specificity (92–100%) *(104)*. Undoubtedly, new epigenetically-silenced genes will be added to this list in the future.

CGI hypermethylation changes may also serve as useful biomarkers for risk stratification in prostate cancer. A recent study demonstrated that the quantity of hypermethylated *GSTP1* CGIs in the serum of patients with localized prostate cancer directly correlated with Gleason grade, pathologic stage, and PSA recurrence after radical prostatectomy *(121)*. Likewise, hypermethylation at the *EDNRB, RARβ, RASSF1a, ERβ*, and *TIG1* have been correlated with known prognostic indicators, such as Gleason score and tumor stage *(104,161–164)*. Additionally, the quantitative CGI hypermethylation levels at the *PTGS2* gene regulatory region predicted prostate cancer recurrence after radical prostatectomy, independently of tumor stage and Gleason grade *(104)*. These genes, and others yet to be discovered, may be targets for epigenetic silencing during progression to a more malignant phenotype. Identification of such genes may lead to a better understanding of the molecular pathophysiology of prostate cancer progression, help identify new targets for drug therapy of advanced prostate cancer, and define patient subpopulations that may benefit from existing therapies. For example, *EDNRB* encodes the endothelin-B receptor, a clearance receptor for endothelin-1 (ET1). ET1 is produced at high levels by metastatic prostate cancers as part of autocrine and paracrine signaling loops *(165,166)*. Loss of this clearance receptor by CGI hypermethylation

(167) could lead to unchecked activation of the endothelin-A receptor, which is thought to be involved in the pathogenesis of osteoblastic bony prostate cancer metastases *(168–170)*. Therefore, loss of the endothelin-B receptor may promote the formation of prostate cancer metastases to bone. Atrasentan, an endothelin-A receptor antagonist, has shown promise for treatment of refractory prostate cancer in randomized clinical trials *(171,172)*. It is possible that CGI hypermethylation-induced epigenetic silencing of *EDNRB* might define prostate cancer cases for which ET1-signaling loops contribute to disease progression, and may therefore help identify a subset of patients that might benefit most from atrasentan therapy.

So far, assays for detection of epigenetic changes as biomarkers for cancer detection and prognostication have focused on the identification of CGI hyper-methylation changes, with little attention to other epigenetic alterations, such as genomic hypomethylation, LOI, and chromatin structural changes at specific genes. As more robust assays for these other alterations become available, the correlations between these epigenetic changes and prostate cancer disease initiation and progression can be tested, perhaps providing new molecular biomarkers for the disease.

5. EPIGENETIC GENE SILENCING AS A TARGET FOR PROSTATE CANCER PREVENTION AND TREATMENT

Unlike mutations, deletions, translocations, and amplifications, somatic changes in DNA methylation and chromatin structure are potentially reversible, making epigenetic genome defects one of the most attractive rational thera-peutic targets in human cancer. Several therapeutic approaches have been undertaken so far to reactivate expression from epigenetically silenced genes in cancer cells. The two general strategies under most advanced development feature interference with the maintenance of abnormally hypermethylated CpG dinucleotides at the promoters of the silenced genes, and/or interruption of the action of histone- and chromatin-modifying enzymes responsible for the construction of repressive chromatin.

Several inhibitors of DNMTs, capable of reducing DNA methylation at the loci of epigenetically silenced genes, are under development for cancer treatment, including (i) the nucleoside analogs 5-aza-cytidine (Vidaza®) and 5-aza-deoxycytidine (decitabine or Dacogen®), both of which have been approved by the US Food and Drug Administration (FDA) for the treatment of myelodysplasia, (ii) zebularine, an orally bioavailable agent, and (iii) the non-nucleosides procainamide, procaine, and hydralazine *(173–178)*. Unfor-tunately, even though 5-aza-cytidine and decitabine clearly provide benefit when used to treat myelodysplasia, whether this benefit is attributable to the cytotoxic actions of the drugs against neoplastic cells or to the reactivation of silenced gene expression remains to be determined. Nucleoside analogs inhibit DNMTs only after incorporation into genomic DNA during replication, as DNMT catalytic attack on the modified cytosine base in DNA results in trapping of the enzyme in a covalent reaction intermediate, leading both to a reduction in genome-wide DNA methylation and to cell death *(179)*. Clearly, cell death, or arrest of cell growth, that is associated with nucleoside analog DNMT inhibitor treatment might then be either a direct consequence of drug-associated cytotoxicity or an indirect consequence of reactivation of silenced cancer genes. Stable reversal of epigenetic gene silencing of course can only be seen among cells that survive nucleoside analog treatment. Nonetheless, the drugs do reduce DNA methylation when used clinically, especially when administered at certain doses: in a study of myeloid leukemias treated with decitabine, the 5mC content in genomic DNA was found to fall by an average of 14% (from 4.3% of all cytosine bases to 3.7%), with a near-linear decrease in methylation when decitabine was used at low doses (between 5 and 20 mg/m^2 each day), but not when the drug was used at higher doses (between 100 and 180 mg/m^2 each day) *(180)*. Perhaps, at lower doses of nucleoside analog DNMT inhibitors, a reduction in DNA methylation is the predominant treatment response, whereas at higher doses, cytotoxicity overwhelms this epigenetic effect. Of note, a phase II clinical trial ($n = 14$) for men with androgen-independent metastatic prostate cancer has been conducted with decitabine, given at a fairly high dose of 75 mg/m^2 intravenously every 8 h for three doses, repeated every 5–8 weeks *(181)*. The treatment resulted in stable disease for two of the 12 men who could be assessed for response for as long as 10 weeks *(181)*. There has not been a completed study with 5-aza-cytidine or decitabine used at lower doses for prostate cancer treatment to truly test the potential efficacy of epigenetic gene "reactivation." This quandary even bedevils preclinical experiments for prostate cancer. As an example, in the transgenic adenocarcinoma of the mouse prostate (TRAMP) mouse model of prostatic carcinogenesis, prolonged treatment with decitabine prevented both the appearance of hypermethylation at the *Mgmt* promoter and the progression of PIN to invasive metastatic prostate cancer *(182)*. However, even in this controlled model system, whether these beneficial effects of decitabine reflect reversal of gene silencing has not been established. As for safety, nucleoside analogs are known to cause myelotoxicity as the major dose-limiting side effect, and the incorporation of an abnormal base into the DNA template clearly carries risks of mutations *(183)*.

Non-nucleoside DNMT inhibitors offer the possibility of fewer safety concerns and the prospect of a more direct "proof-of-concept" test of epigenetic gene reactivation therapy for prostate cancer and other neoplastic diseases. Procainamide, a drug approved by the FDA for the treatment of cardiac arrhythmias, procaine, an approved anesthetic agent, and hydralazine, a drug approved for the treatment of hypertension, all inhibit DNMTs without myelo-toxicity or mutations *(184–186)*. Detailed mechanistic studies of procainamide inhibition of DNA methylation have revealed several features of the drug that may make it an attractive candidate for further development as epigenetic therapy in certain clinical settings. The agent has been found to be a highly selective inhibitor of DNMT1 at concentrations that can be achieved clini-cally, with little or no activity toward DNMT3a or DNMT3b *(186)*. Also, the drug only antagonizes the activity of the enzyme on hemimethylated DNA substrates, but not on unmethylated DNA substrates, suggesting that maintenance methylation activity during DNA replication may be targeted selectively over *de novo* methylation activity *(186)*. Furthermore, even when inhibiting DNMT1 activity on hemimethylated DNA substrates, procainamide functions as a partial competitive inhibitor, never fully stopping methylation. For these reasons, although the drug may have somewhat less activity than nucleoside DNMT inhibitors in the treatment of some cancers, the agent may have a more favorable safety profile. Mice carrying one disrupted *Dnmt1* allele and one hypomorphic *Dnmt1* allele, with both reduced maintenance and *de novo* DNA methylation activity, exhibit genomic instability and develop T-cell lymphomas, hinting that excessive inhibition of DNMT1 activity might cause certain cancers (e.g., lymphomas) even while preventing or treating others [e.g., epithelial tumors; see *(58,59)*]. It is possible that procainamide, by partially inhibiting *only* the maintenance methylation activity of DNMT1, might maintain efficacy for prostate cancer treatment while avoiding the risk for lymphomagenesis. In preclinical studies using LNCaP cells propagated as xenograft tumors on immunodeficient mice, procainamide reactivated silenced *GSTP1* expression, with a trend toward greater anti-tumor activity than decitabine *(176)*. Long-term use of procainamide has not been associated with genetic instability or with an increased lymphoma risk, although prolonged procainamide treatment has been correlated with drug-induced lupus, more commonly arising in women than in men *(185,187)*. Procainamide is clearly ready for clinical "proof-of-concept" testing in men with prostate cancer.

The other general therapeutic approach to epigenetic gene silencing in cancer features the targeting of enzymes that contribute to the construction and/or maintenance of repressed chromatin complexes encompassing the transcrip-tional regulatory regions of key cancer genes. The most advanced drug discovery activity has targeted HDACs, the enzymes responsible for antago-nizing activation of transcription accompanying the activity of HATs, with a growing portfolio of small molecule HDAC inhibitors reaching early clinical and/or advanced preclinical development, including sodium phenylbutyrate, valproic acid, suberoylanilide hydroxamic acid (SAHA), pyroxamide, *N*-acetyl dinaline (CI-994), LAQ824, LBH-589, MS-275, depsipeptide (FR901228), and many others *(188–191)*. These inhibitors have generally displayed very promising efficacy in cancer models, including prostate cancer models *(192–200)*. The agents thus far tested in clinical trials have produced a number of side effects, such as nausea, vomiting, diarrhea, fatigue, and edema, but have not commonly caused severe adverse events *(188,189,191)*. With this favorable safety profile, combinations of HDAC inhibitors and other agents will likely be feasible. Already, there has emerged strong preclinical evidence that combinations of DNMT inhibitors and HDAC inhibitors more effectively reactivate silenced gene expression in cancer cells *(57,201)*. The clinical activity of HDAC inhibitors against prostate cancer is under active clinical study.

REFERENCES

1. Feinberg, A. P., Cui, H., and Ohlsson, R. DNA methylation and genomic imprinting: insights from cancer into epigenetic mechanisms. *Semin Cancer Biol, 12:* 389–398, 2002.
2. Onyango, P., Jiang, S., Uejima, H., Shamblott, M. J., Gearhart, J. D., Cui, H., and Feinberg, A. P. Monoallelic expression and methylation of imprinted genes in human and mouse embryonic germ cell lineages. *Proc Natl Acad Sci USA, 99:* 10599–10604, 2002.
3. Tilghman, S. M. The sins of the fathers and mothers: genomic imprinting in mammalian development. *Cell, 96:* 185–193, 1999.
4. Norris, D. P., Brockdorff, N., and Rastan, S. Methylation status of CpG-rich islands on active and inactive mouse X chromosomes. *Mamm Genome, 1:* 78–83, 1991.
5. Challita, P. M., Skelton, D., el-Khoueiry, A., Yu, X. J., Weinberg, K., and Kohn, D. B. Multiple modifications in cis elements of the long terminal repeat of retroviral vectors lead to increased expression and decreased DNA methylation in embryonic carcinoma cells. *J Virol, 69:* 748–755, 1995.
6. Shinar, D., Yoffe, O., Shani, M., and Yaffe, D. Regulated expression of muscle-specific genes introduced into mouse embryonal stem cells: inverse correlation with DNA methylation. *Differentiation, 41:* 116–126, 1989.
7. Chapman, V., Forrester, L., Sanford, J., Hastie, N., and Rossant, J. Cell lineage-specific undermethylation of mouse repetitive DNA. *Nature, 307:* 284–286, 1984.
8. Tolberg, M. E., Funderburk, S. J., Klisak, I., and Smith, S. S. Structural organization and DNA methylation patterning within the mouse L1 family. *J Biol Chem, 262:* 11167–11175, 1987.
9. Razin, A. and Riggs, A. D. DNA methylation and gene function. *Science, 210:* 604–610, 1980.
10. Siegfried, Z. and Cedar, H. DNA methylation: a molecular lock. *Curr Biol, 7:* R305–307, 1997.
11. Siegfried, Z., Eden, S., Mendelsohn, M., Feng, X., Tsuberi, B. Z., and Cedar, H. DNA methylation represses transcription in vivo. *Nat Genet, 22:* 203–206, 1999.
12. Santos, F., Hendrich, B., Reik, W., and Dean, W. Dynamic reprogramming of DNA methylation in the early mouse embryo. *Dev Biol, 241:* 172–182, 2002.
13. Rainier, S., Johnson, L. A., Dobry, C. J., Ping, A. J., Grundy, P. E., and Feinberg, A. P. Relaxation of imprinted genes in human cancer. *Nature, 362:* 747–749, 1993.
14. Jarrard, D. F., Bussemakers, M. J., Bova, G. S., and Isaacs, W. B. Regional loss of imprinting of the insulin-like growth factor II gene occurs in human prostate tissues. *Clin Cancer Res, 1:* 1471–1478, 1995.
15. Steenman, M. J., Rainier, S., Dobry, C. J., Grundy, P., Horon, I. L., and Feinberg, A. P. Loss of imprinting of IGF2 is linked to reduced expression and abnormal methylation of H19 in Wilms' tumour. *Nat Genet, 7:* 433–439, 1994.

16. Cui, H., Onyango, P., Brandenburg, S., Wu, Y., Hsieh, C. L., and Feinberg, A. P. Loss of imprinting in colorectal cancer linked to hypomethylation of H19 and IGF2. *Cancer Res, 62:* 6442–6446, 2002.

17. Li, E. Chromatin modification and epigenetic reprogramming in mammalian development. *Nat Rev Genet , 3:* 662–673, 2002.

18. Barr, M. L. and Bertram, E. G. A morphological distinction between neurons of the male and female. *Nature, 163:* 676–677, 1949.

19. Ohno, S., Kaplan, W. D., and Kinosita, R. Formation of the sex chromatin by a single X-chromosome in liver cells of Rattus norvegicus. *Exp Cell Res, 18:* 415–418, 1959.

20. Lyon, M. F. Gene action in the X-chromosome of the mouse (Mus musculus L.). *Nature, 190:* 372–373, 1961.

21. Bender, J. Cytosine methylation of repeated sequences in eukaryotes: the role of DNA pairing. *Trends Biochem Sci, 23:* 252–256, 1998.

22. Ushijima, T., Morimura, K., Hosoya, Y., Okonogi, H., Tatematsu, M., Sugimura, T., and Nagao, M. Establishment of methylation-sensitive-representational difference analysis and isolation of hypo- and hypermethylated genomic fragments in mouse liver tumors. *Proc Natl Acad Sci USA, 94:* 2284–2289, 1997.

23. Suzuki, K., Suzuki, I., Leodolter, A., Alonso, S., Horiuchi, S., Yamashita, K., and Perucho, M. Global DNA demethylation in gastrointestinal cancer is age dependent and precedes genomic damage. *Cancer Cell, 9:* 199–207, 2006.

24. Rodriguez, J., Frigola, J., Vendrell, E., Risques, R. A., Fraga, M. F., Morales, C., Moreno, V., Esteller, M., Capella, G., Ribas, M., and Peinado, M. A. Chromosomal instability correlates with genome-wide DNA demethylation in human primary colorectal cancers. *Cancer Res, 66:* 8462–9468, 2006.

25. Bird, A. P. CpG-rich islands and the function of DNA methylation. *Nature, 321:* 209–213, 1986.

26. Cross, S. H., Charlton, J. A., Nan, X., and Bird, A. P. Purification of CpG islands using a methylated DNA binding column. *Nat Genet, 6:* 236–244, 1994.

27. Song, F., Smith, J. F., Kimura, M. T., Morrow, A. D., Matsuyama, T., Nagase, H., and Held, W. A. Association of tissue-specific differentially methylated regions (TDMs) with differential gene expression. *Proc Natl Acad Sci USA, 102:* 3336–3341, 2005.

28. Feinberg, A. P., Gehrke, C. W., Kuo, K. C., and Ehrlich, M. Reduced genomic 5-methylcytosine content in human colonic neoplasia. *Cancer Res, 48:* 1159–1161, 1988.

29. Goelz, S. E., Vogelstein, B., Hamilton, S. R., and Feinberg, A. P. Hypomethylation of DNA from benign and malignant human colon neoplasms. *Science, 228:* 187–190, 1985.

30. Feinberg, A. P. and Vogelstein, B. Hypomethylation distinguishes genes of some human cancers from their normal counterparts. *Nature, 301:* 89–92, 1983.

31. Feinberg, A. P. and Vogelstein, B. Hypomethylation of ras oncogenes in primary human cancers. *Biochem Biophys Res Commun, 111:* 47–54, 1983.

32. Bedford, M. T. and van Helden, P. D. Hypomethylation of DNA in pathological conditions of the human prostate. *Cancer Res, 47:* 5274–5276, 1987.

33. Gama-Sosa, M. A., Slagel, V. A., Trewyn, R. W., Oxenhandler, R., Kuo, K. C., Gehrke, C. W., and Ehrlich, M. The 5-methylcytosine content of DNA from human tumors. *Nucleic Acids Res, 11:* 6883–6894, 1983.

34. Feinberg, A. P. and Tycko, B. The history of cancer epigenetics. *Nat Rev Cancer, 4:* 143–153, 2004.

35. Cadieux, B., Ching, T. T., Vandenberg, S. R., and Costello, J. F. Genome-wide hypomethylation in human glioblastomas associated with specific copy number alteration, methylenetetrahydrofolate reductase allele status, and increased proliferation. *Cancer Res, 66:* 8469–8476, 2006.

36. Lee, W. H., Morton, R. A., Epstein, J. I., Brooks, J. D., Campbell, P. A., Bova, G. S., Hsieh, W. S., Isaacs, W. B., and Nelson, W. G. Cytidine methylation of regulatory sequences near the pi-class glutathione S-transferase gene accompanies human prostatic carcinogenesis. *Proc Natl Acad Sci USA, 91:* 11733–11737, 1994.

37. Esteller, M., Corn, P. G., Baylin, S. B., and Herman, J. G. A gene hypermethylation profile of human cancer. *Cancer Res, 61:* 3225–3229, 2001.

38. Jones, P. A. and Laird, P. W. Cancer epigenetics comes of age. *Nat Genet, 21:* 163–167, 1999.

39. Ehrlich, M. DNA methylation in cancer: too much, but also too little. *Oncogene, 21:* 5400–5413, 2002.

40. Holst, C. R., Nuovo, G. J., Esteller, M., Chew, K., Baylin, S. B., Herman, J. G., and Tlsty, T. D. Methylation of p16(INK4a) promoters occurs in vivo in histologically normal human mammary epithelia. *Cancer Res, 63:* 1596–1601, 2003.

41. Myohanen, S. K., Baylin, S. B., and Herman, J. G. Hypermethylation can selectively silence individual p16ink4A alleles in neoplasia. *Cancer Res, 58:* 591–593, 1998.

42. Lin, X. and Nelson, W. G. Methyl-CpG-binding domain protein-2 mediates transcriptional repression associated with hypermethylated GSTP1 CpG islands in MCF-7 breast cancer cells. *Cancer Res, 63:* 498–504, 2003.

43. David, G. L., Yegnasubramanian, S., Kumar, A., Marchi, V. L., De Marzo, A. M., Lin, X., and Nelson, W. G. MDR1 promoter hypermethylation in MCF-7 human breast cancer cells: changes in chromatin structure induced by treatment with 5-aza-cytidine. *Cancer Biol Ther, 3:* 540–548, 2004.

44. Pradhan, S., Bacolla, A., Wells, R. D., and Roberts, R. J. Recombinant human DNA (cytosine-5) methyltransferase. I. Expression, purification, and comparison of *de novo* and maintenance methylation. *J Biol Chem, 274:* 33002–33010, 1999.

45. Gowher, H. and Jeltsch, A. Enzymatic properties of recombinant Dnmt3a DNA methyltransferase from mouse: the enzyme modifies DNA in a non-processive manner and also methylates non-CpG [correction of non-CpA] sites. *J Mol Biol, 309:* 1201–1208, 2001.

46. Okano, M., Xie, S., and Li, E. Cloning and characterization of a family of novel mammalian DNA (cytosine-5) methyltransferases. *Nat Genet, 19:* 219–220, 1998.

47. Jeltsch, A. Beyond Watson and Crick: DNA methylation and molecular enzymology of DNA methyltransferases. *Chembiochem, 3:* 274–293, 2002.

48. Okano, M., Bell, D. W., Haber, D. A., and Li, E. DNA methyltransferases Dnmt3a and Dnmt3b are essential for *de novo* methylation and mammalian development. *Cell, 99:* 247–257, 1999.

49. Rhee, I., Jair, K. W., Yen, R. W., Lengauer, C., Herman, J. G., Kinzler, K. W., Vogelstein, B., Baylin, S. B., and Schuebel, K. E. CpG methylation is maintained in human cancer cells lacking DNMT1. *Nature, 404:* 1003–1007, 2000.

50. Rhee, I., Bachman, K. E., Park, B. H., Jair, K. W., Yen, R. W., Schuebel, K. E., Cui, H., Feinberg, A. P., Lengauer, C., Kinzler, K. W., Baylin, S. B., and Vogelstein, B. DNMT1 and DNMT3b cooperate to silence genes in human cancer cells. *Nature, 416:* 552–556, 2002.

51. Graff, J. R., Herman, J. G., Myohanen, S., Baylin, S. B., and Vertino, P. M. Mapping patterns of CpG island methylation in normal and neoplastic cells implicates both upstream and downstream regions in *de novo* methylation. *J Biol Chem, 272:* 22322–22329, 1997.

52. Vertino, P. M., Yen, R. W., Gao, J., and Baylin, S. B. *de novo* methylation of CpG island sequences in human fibroblasts overexpressing DNA (cytosine-5-)-methyltransferase. *Mol Cell Biol, 16:* 4555–4565, 1996.

53. Feltus, F. A., Lee, E. K., Costello, J. F., Plass, C., and Vertino, P. M. Predicting aberrant CpG island methylation. *Proc Natl Acad Sci USA, 100:* 12253–12258, 2003.

54. Bakin, A. V. and Curran, T. Role of DNA 5-methylcytosine transferase in cell transformation by fos. *Science, 283:* 387–390, 1999.

55. Laird, P. W., Jackson-Grusby, L., Fazeli, A., Dickinson, S. L., Jung, W. E., Li, E., Weinberg, R. A., and Jaenisch, R. Suppression of intestinal neoplasia by DNA hypomethylation. *Cell, 81:* 197–205, 1995.

56. Eads, C. A., Nickel, A. E., and Laird, P. W. Complete genetic suppression of polyp formation and reduction of CpG-island hypermethylation in Apc(Min/+) Dnmt1-hypomorphic mice. *Cancer Res, 62:* 1296–1299, 2002.

57. Belinsky, S. A., Klinge, D. M., Stidley, C. A., Issa, J. P., Herman, J. G., March, T. H., and Baylin, S. B. Inhibition of DNA methylation and histone deacetylation prevents murine lung cancer. *Cancer Res, 63:* 7089–7093, 2003.

58. Eden, A., Gaudet, F., Waghmare, A., and Jaenisch, R. Chromosomal instability and tumors promoted by DNA hypomethylation. *Science , 300:* 455, 2003.

59. Gaudet, F., Hodgson, J. G., Eden, A., Jackson-Grusby, L., Dausman, J., Gray, J. W., Leonhardt, H., and Jaenisch, R. Induction of tumors in mice by genomic hypomethylation. *Science, 300:* 489–492, 2003.

60. McCabe, M. T., Davis, J. N., and Day, M. L. Regulation of DNA methyltransferase 1 by the pRb/E2F1 pathway. *Cancer Res, 65:* 3624–3632, 2005.

61. Eads, C. A., Danenberg, K. D., Kawakami, K., Saltz, L. B., Danenberg, P. V., and Laird, P. W. CpG island hypermethylation in human colorectal tumors is not associated with DNA methyltransferase overexpression. *Cancer Res, 59:* 2302–2306, 1999.

62. Agoston, A. T., Argani, P., Yegnasubramanian, S., De Marzo, A. M., Ansari-Lari, M. A., Hicks, J. L., Davidson, N. E., and Nelson, W. G. Increased protein stability causes DNA methyltransferase 1 dysregulation in breast cancer. *J Biol Chem, 280:* 18302–18310, 2005.

63. Lande-Diner, L. and Cedar, H. Silence of the genes—mechanisms of long-term repression. *Nat Rev Genet, 6:* 648–654, 2005.

64. Klose, R. J. and Bird, A. P. Genomic DNA methylation: the mark and its mediators. *Trends Biochem Sci, 31:* 89–97, 2006.

65. Roloff, T. C., Ropers, H. H., and Nuber, U. A. Comparative study of methyl-CpG-binding domain proteins. *BMC Genomics, 4:* 1, 2003.

66. Prokhortchouk, A., Hendrich, B., Jorgensen, H., Ruzov, A., Wilm, M., Georgiev, G., Bird, A., and Prokhortchouk, E. The p120 catenin partner Kaiso is a DNA methylation-dependent transcriptional repressor. *Genes Dev, 15:* 1613–1618, 2001.

67. Filion, G. J., Zhenilo, S., Salozhin, S., Yamada, D., Prokhortchouk, E., and Defossez, P. A. A family of human zinc finger proteins that bind methylated DNA and repress transcription. *Mol Cell Biol, 26:* 169–181, 2006.

68. Daniel, J. M., Spring, C. M., Crawford, H. C., Reynolds, A. B., and Baig, A. The p120(ctn)-binding partner Kaiso is a bi-modal DNA-binding protein that recognizes both a sequence-specific consensus and methylated CpG dinucleotides. *Nucleic Acids Res, 30:* 2911–2919, 2002.

69. Bird, A. P. and Wolffe, A. P. Methylation-induced repression—belts, braces, and chromatin. *Cell, 99:* 451–454, 1999.

70. Shi, Y., Lan, F., Matson, C., Mulligan, P., Whetstine, J. R., Cole, P. A., and Casero, R. A. Histone demethylation mediated by the nuclear amine oxidase homolog LSD1. *Cell, 119:* 941–953, 2004.

71. Klose, R. J., Yamane, K., Bae, Y., Zhang, D., Erdjument-Bromage, H., Tempst, P., Wong, J., and Zhang, Y. The transcriptional repressor JHDM3A demethylates trimethyl histone H3 lysine 9 and lysine 36. *Nature, 442:* 312–316, 2006.

72. Yamane, K., Toumazou, C., Tsukada, Y., Erdjument-Bromage, H., Tempst, P., Wong, J., and Zhang, Y. JHDM2A, a JmjC-containing H3K9 demethylase, facilitates transcription activation by androgen receptor. *Cell , 125:* 483–495, 2006.

73. Tsukada, Y., Fang, J., Erdjument-Bromage, H., Warren, M. E., Borchers, C. H., Tempst, P., and Zhang, Y. Histone demethylation by a family of JmjC domain-containing proteins. *Nature, 439:* 811–816, 2006.

74. Zhang, K. and Dent, S. Y. Histone modifying enzymes and cancer: going beyond histones. *J Cell Biochem, 96:* 1137–1148, 2005.

75. Jones, P. L., Veenstra, G. J., Wade, P. A., Vermaak, D., Kass, S. U., Landsberger, N., Strouboulis, J., and Wolffe, A. P. Methylated DNA and MeCP2 recruit histone deacetylase to repress transcription. *Nat Genet, 19:* 187–191, 1998.

76. Feng, Q. and Zhang, Y. The MeCP1 complex represses transcription through preferential binding, remodeling, and deacetylating methylated nucleosomes. *Genes Dev, 15:* 827–832, 2001.

77. Sarraf, S. A. and Stancheva, I. Methyl-CpG binding protein MBD1 couples histone H3 methylation at lysine 9 by SETDB1 to DNA replication and chromatin assembly. *Mol Cell, 15:* 595–605, 2004.

78. Li, H., Rauch, T., Chen, Z. X., Szabo, P. E., Riggs, A. D., and Pfeifer, G. P. The histone methyltransferase SETDB1 and the DNA methyltransferase DNMT3A interact directly and localize to promoters silenced in cancer cells. *J Biol Chem, 281:* 19489–19500, 2006.

79. Yoon, H. G., Chan, D. W., Reynolds, A. B., Qin, J., and Wong, J. N-CoR mediates DNA methylation-dependent repression through a methyl CpG binding protein Kaiso. *Mol Cell, 12:* 723–734, 2003.

80. Hendrich, B., Guy, J., Ramsahoye, B., Wilson, V. A., and Bird, A. Closely related proteins MBD2 and MBD3 play distinctive but interacting roles in mouse development. *Genes Dev, 15:* 710–723, 2001.

81. Bakker, J., Lin, X., and Nelson, W. G. Methyl-CpG binding domain protein 2 represses transcription from hypermethylated pi-class glutathione S-transferase gene promoters in hepatocellular carcinoma cells. *J Biol Chem, 277:* 22573–22580, 2002.

82. Li, E., Bestor, T. H., and Jaenisch, R. Targeted mutation of the DNA methyltransferase gene results in embryonic lethality. *Cell, 69:* 915–926, 1992.

83. Guy, J., Hendrich, B., Holmes, M., Martin, J. E., and Bird, A. A mouse Mecp2-null mutation causes neurological symptoms that mimic Rett syndrome. *Nat Genet, 27:* 322–326, 2001.

84. Klose, R. J., Sarraf, S. A., Schmiedeberg, L., McDermott, S. M., Stancheva, I., and Bird, A. P. DNA binding selectivity of MeCP2 due to a requirement for A/T sequences adjacent to methyl-CpG. *Mol Cell, 19:* 667–678, 2005.

85. Gibbons, R. J. Histone modifying and chromatin remodelling enzymes in cancer and dysplastic syndromes. *Hum Mol Genet, 14 Spec No 1:* R85–92, 2005.

86. Keshet, I., Schlesinger, Y., Farkash, S., Rand, E., Hecht, M., Segal, E., Pikarski, E., Young, R. A., Niveleau, A., Cedar, H., and Simon, I. Evidence for an instructive mechanism of *de novo* methylation in cancer cells. *Nat Genet, 38:* 149–153, 2006.

87. Vire, E., Brenner, C., Deplus, R., Blanchon, L., Fraga, M., Didelot, C., Morey, L., Van Eynde, A., Bernard, D., Vanderwinden, J. M., Bollen, M., Esteller, M., Di Croce, L., de Launoit, Y., and Fuks, F. The Polycomb group protein EZH2 directly controls DNA methylation. *Nature, 439:* 871–874, 2006.

88. Nelson, W. G., De Marzo, A. M., and Isaacs, W. B. Prostate cancer. *N Engl J Med, 349:* 366–381, 2003.

89. Tomlins, S. A., Rhodes, D. R., Perner, S., Dhanasekaran, S. M., Mehra, R., Sun, X. W., Varambally, S., Cao, X., Tchinda, J., Kuefer, R., Lee, C., Montie, J. E., Shah, R. B., Pienta, K. J., Rubin, M. A., and Chinnaiyan, A. M. Recurrent fusion of TMPRSS2 and ETS transcription factor genes in prostate cancer. *Science, 310:* 644–648, 2005.

90. Lin, X., Tascilar, M., Lee, W. H., Vles, W. J., Lee, B. H., Veeraswamy, R., Asgari, K., Freije, D., van Rees, B., Gage, W. R., Bova, G. S., Isaacs, W. B., Brooks, J. D., DeWeese, T. L., De Marzo, A. M., and Nelson, W. G. GSTP1 CpG island hypermethylation

is responsible for the absence of GSTP1 expression in human prostate cancer cells. *Am J Pathol, 159:* 1815–1826, 2001.

91. Henderson, C. J., Smith, A. G., Ure, J., Brown, K., Bacon, E. J., and Wolf, C. R. Increased skin tumorigenesis in mice lacking pi class glutathione S-transferases. *Proc Natl Acad Sci USA, 95:* 5275–5280, 1998.

92. Nelson, C. P., Kidd, L. C., Sauvageot, J., Isaacs, W. B., De Marzo, A. M., Groopman, J. D., Nelson, W. G., and Kensler, T. W. Protection against 2-hydroxyamino-1-methyl-6-phenylimidazo[4,5-b]pyridine cytotoxicity and DNA adduct formation in human prostate by glutathione S-transferase P1. *Cancer Res, 61:* 103–109, 2001.

93. Shirai, T., Sano, M., Tamano, S., Takahashi, S., Hirose, M., Futakuchi, M., Hasegawa, R., Imaida, K., Matsumoto, K., Wakabayashi, K., Sugimura, T., and Ito, N. The prostate: a target for carcinogenicity of 2-amino-1-methyl-6-phenylimidazo[4,5-b]pyridine (PhIP) derived from cooked foods. *Cancer Res, 57:* 195–198, 1997.

94. Stuart, G. R., Holcroft, J., de Boer, J. G., and Glickman, B. W. Prostate mutations in rats induced by the suspected human carcinogen 2-amino-1-methyl-6-phenylimidazo[4,5-b]pyridine. *Cancer Res, 60:* 266–268, 2000.

95. De Marzo, A. M., Marchi, V. L., Epstein, J. I., and Nelson, W. G. Proliferative inflammatory atrophy of the prostate: implications for prostatic carcinogenesis. *Am J Pathol, 155:* 1985–1992, 1999.

96. Parsons, J. K., Nelson, C. P., Gage, W. R., Nelson, W. G., Kensler, T. W., and De Marzo, A. M. GSTA1 expression in normal, preneoplastic, and neoplastic human prostate tissue. *Prostate, 49:* 30–37, 2001.

97. Zha, S., Gage, W. R., Sauvageot, J., Saria, E. A., Putzi, M. J., Ewing, C. M., Faith, D. A., Nelson, W. G., De Marzo, A. M., and Isaacs, W. B. Cyclooxygenase-2 is up-regulated in proliferative inflammatory atrophy of the prostate, but not in prostate carcinoma. *Cancer Res, 61:* 8617–8623, 2001.

98. DeMarzo, A. M., Nelson, W. G., Isaacs, W. B., and Epstein, J. I. Pathological and molecular aspects of prostate cancer. *Lancet, 361:* 955–964, 2003.

99. Nakayama, M., Bennett, C. J., Hicks, J. L., Epstein, J. I., Platz, E. A., Nelson, W. G., and De Marzo, A. M. Hypermethylation of the human glutathione S-transferase-pi gene (GSTP1) CpG island is present in a subset of proliferative inflammatory atrophy lesions but not in normal or hyperplastic epithelium of the prostate: a detailed study using laser-capture microdissection. *Am J Pathol , 163:* 923–933, 2003.

100. Brooks, J. D., Weinstein, M., Lin, X., Sun, Y., Pin, S. S., Bova, G. S., Epstein, J. I., Isaacs, W. B., and Nelson, W. G. CG island methylation changes near the GSTP1 gene in prostatic intraepithelial neoplasia. *Cancer Epidemiol Biomarkers Prev, 7:* 531–536, 1998.

101. DeWeese, T. L. and Nelson, W. G. Inadequate "caretaker" gene function and human cancer development. *Methods Mol Biol, 222:* 249–268, 2003.

102. Hmadcha, A., Bedoya, F. J., Sobrino, F., and Pintado, E. Methylation-dependent gene silencing induced by interleukin 1beta via nitric oxide production. *J Exp Med, 190:* 1595–1604, 1999.

103. Bastian, P. J., Yegnasubramanian, S., Palapattu, G. S., Rogers, C. G., Lin, X., De Marzo, A. M., and Nelson, W. G. Molecular biomarker in prostate cancer: the role of CpG island hypermethylation. *Eur Urol, 46:* 698–708, 2004.

104. Yegnasubramanian, S., Kowalski, J., Gonzalgo, M. L., Zahurak, M., Piantadosi, S., Walsh, P. C., Bova, G. S., De Marzo, A. M., Isaacs, W. B., and Nelson, W. G. Hypermethylation of CpG islands in primary and metastatic human prostate cancer. *Cancer Res, 64:* 1975–1986, 2004.

105. Santourlidis, S., Florl, A., Ackermann, R., Wirtz, H. C., and Schulz, W. A. High frequency of alterations in DNA methylation in adenocarcinoma of the prostate. *Prostate, 39:* 166–174, 1999.

106. Florl, A. R., Steinhoff, C., Muller, M., Seifert, H. H., Hader, C., Engers, R., Ackermann, R., and Schulz, W. A. Coordinate hypermethylation at specific genes in prostate carcinoma precedes LINE-1 hypomethylation. *Br J Cancer, 91:* 985–994, 2004.

107. Schulz, W. A., Elo, J. P., Florl, A. R., Pennanen, S., Santourlidis, S., Engers, R., Buchardt, M., Seifert, H. H., and Visakorpi, T. Genomewide DNA hypomethylation is associated with alterations on chromosome 8 in prostate carcinoma. *Genes Chromosomes Cancer, 35:* 58–65, 2002.

108. Varambally, S., Dhanasekaran, S. M., Zhou, M., Barrette, T. R., Kumar-Sinha, C., Sanda, M. G., Ghosh, D., Pienta, K. J., Sewalt, R. G., Otte, A. P., Rubin, M. A., and Chinnaiyan, A. M. The polycomb group protein EZH2 is involved in progression of prostate cancer. *Nature, 419:* 624–629, 2002.

109. Chen, H., Toyooka, S., Gazdar, A. F., and Hsieh, J. T. Epigenetic regulation of a novel tumor suppressor gene (hDAB2IP) in prostate cancer cell lines. *J Biol Chem, 278:* 3121–3130, 2003.

110. Chen, H., Tu, S. W., and Hsieh, J. T. Down-regulation of human DAB2IP gene expression mediated by polycomb Ezh2 complex and histone deacetylase in prostate cancer. *J Biol Chem, 280:* 22437–22444, 2005.

111. Bracken, A. P., Pasini, D., Capra, M., Prosperini, E., Colli, E., and Helin, K. EZH2 is downstream of the pRB-E2F pathway, essential for proliferation and amplified in cancer. *EMBO J, 22:* 5323–5335, 2003.

112. Soh, S., Kattan, M. W., Berkman, S., Wheeler, T. M., and Scardino, P. T. Has there been a recent shift in the pathological features and prognosis of patients treated with radical prostatectomy? *J Urol, 157:* 2212–2218, 1997.

113. Thompson, I. M., Pauler, D. K., Goodman, P. J., Tangen, C. M., Lucia, M. S., Parnes, H. L., Minasian, L. M., Ford, L. G., Lippman, S. M., Crawford, E. D., Crowley, J. J., and Coltman, C. A., Jr. Prevalence of prostate cancer among men with a prostate-specific antigen level < or = 4.0 ng per milliliter. *N Engl J Med, 350:* 2239–2246, 2004.

114. Thompson, I. M., Goodman, P. J., Tangen, C. M., Lucia, M. S., Miller, G. J., Ford, L. G., Lieber, M. M., Cespedes, R. D., Atkins, J. N., Lippman, S. M., Carlin, S. M., Ryan, A., Szczepanek, C. M., Crowley, J. J., and Coltman, C. A., Jr. The influence of finasteride on the development of prostate cancer. *N Engl J Med, 349:* 215–224, 2003.

115. Makhlouf, A. A., Krupski, T. L., Kunkle, D., and Theodorescu, D. The effect of sampling more cores on the predictive accuracy of pathological grade and tumour distribution in the prostate biopsy. *BJU Int, 93:* 271–274, 2004.

116. de la Taille, A., Antiphon, P., Salomon, L., Cherfan, M., Porcher, R., Hoznek, A., Saint, F., Vordos, D., Cicco, A., Yiou, R., Zafrani, E. S., Chopin, D., and Abbou, C. C. Prospective evaluation of a 21-sample needle biopsy procedure designed to improve the prostate cancer detection rate. *Urology, 61:* 1181–1186, 2003.

117. Sakr, W. A., Grignon, D. J., Crissman, J. D., Heilbrun, L. K., Cassin, B. J., Pontes, J. J., and Haas, G. P. High grade prostatic intraepithelial neoplasia (HGPIN) and prostatic adenocarcinoma between the ages of 20–69: an autopsy study of 249 cases. *In Vivo, 8:* 439–443, 1994.

118. Albertsen, P. C. What is the value of screening for prostate cancer in the US? *Nat Clin Pract Oncol, 2:* 536–537, 2005.

119. Sidransky, D. Emerging molecular markers of cancer. *Nat Rev Cancer, 2:* 210–219, 2002.

120. Laird, P. W. The power and the promise of DNA methylation markers. *Nat Rev Cancer, 3:* 253–266, 2003.

121. Bastian, P. J., Palapattu, G. S., Lin, X., Yegnasubramanian, S., Mangold, L. A., Trock, B., Eisenberger, M. A., Partin, A. W., and Nelson, W. G. Preoperative serum DNA GSTP1 CpG island hypermethylation and the risk of early prostate-specific antigen recurrence following radical prostatectomy. *Clin Cancer Res, 11:* 4037–4043, 2005.

122. Singer, J., Roberts-Ems, J., and Riggs, A. D. Methylation of mouse liver DNA studied by means of the restriction enzymes msp I and hpa II. *Science, 203:* 1019–1021, 1979.

123. Bird, A. P. and Southern, E. M. Use of restriction enzymes to study eukaryotic DNA methylation: I. The methylation pattern in ribosomal DNA from Xenopus laevis. *J Mol Biol, 118:* 27–47, 1978.

124. Pollack, Y., Stein, R., Razin, A., and Cedar, H. Methylation of foreign DNA sequences in eukaryotic cells. *Proc Natl Acad Sci USA, 77:* 6463–6467, 1980.

125. Singer-Sam, J., Grant, M., LeBon, J. M., Okuyama, K., Chapman, V., Monk, M., and Riggs, A. D. Use of a HpaII-polymerase chain reaction assay to study DNA methylation in the Pgk-1 CpG island of mouse embryos at the time of X-chromosome inactivation. *Mol Cell Biol, 10:* 4987–4989, 1990.

126. Singer-Sam, J., LeBon, J. M., Tanguay, R. L., and Riggs, A. D. A quantitative HpaII-PCR assay to measure methylation of DNA from a small number of cells. *Nucleic Acids Res, 18:* 687, 1990.

127. Hatada, I., Kato, A., Morita, S., Obata, Y., Nagaoka, K., Sakurada, A., Sato, M., Horii, A., Tsujimoto, A., and Matsubara, K. A microarray-based method for detecting methylated loci. *J Hum Genet , 47:* 448–451, 2002.

128. Nygren, A. O., Ameziane, N., Duarte, H. M., Vijzelaar, R. N., Waisfisz, Q., Hess, C. J., Schouten, J. P., and Errami, A. Methylation-specific MLPA (MS-MLPA): simultaneous detection of CpG methylation and copy number changes of up to 40 sequences. *Nucleic Acids Res, 33:* e128, 2005.

129. Hu, M., Yao, J., Cai, L., Bachman, K. E., van den Brule, F., Velculescu, V., and Polyak, K. Distinct epigenetic changes in the stromal cells of breast cancers. *Nat Genet, 37:* 899–905, 2005.

130. Schumacher, A., Kapranov, P., Kaminsky, Z., Flanagan, J., Assadzadeh, A., Yau, P., Virtanen, C., Winegarden, N., Cheng, J., Gingeras, T., and Petronis, A. Microarray-based DNA methylation profiling: technology and applications. *Nucleic Acids Res, 34:* 528–542, 2006.

131. Lippman, Z., Gendrel, A. V., Colot, V., and Martienssen, R. Profiling DNA methylation patterns using genomic tiling microarrays. *Nat Methods, 2:* 219–224, 2005.

132. Wang, R. Y., Gehrke, C. W., and Ehrlich, M. Comparison of bisulfite modification of 5-methyldeoxycytidine and deoxycytidine residues. *Nucleic Acids Res, 8:* 4777–4790, 1980.

133. Clark, S. J., Harrison, J., Paul, C. L., and Frommer, M. High sensitivity mapping of methylated cytosines. *Nucleic Acids Res, 22:* 2990–2997, 1994.

134. Frommer, M., McDonald, L. E., Millar, D. S., Collis, C. M., Watt, F., Grigg, G. W., Molloy, P. L., and Paul, C. L. A genomic sequencing protocol that yields a positive display of 5-methylcytosine residues in individual DNA strands. *Proc Natl Acad Sci USA, 89:* 1827–1831, 1992.

135. Herman, J. G., Graff, J. R., Myohanen, S., Nelkin, B. D., and Baylin, S. B. Methylation-specific PCR: a novel PCR assay for methylation status of CpG islands. *Proc Natl Acad Sci USA, 93:* 9821–9826, 1996.

136. Eads, C. A., Danenberg, K. D., Kawakami, K., Saltz, L. B., Blake, C., Shibata, D., Danenberg, P. V., and Laird, P. W. MethyLight: a high-throughput assay to measure DNA methylation. *Nucleic Acids Res, 28:* E32, 2000.

137. Cottrell, S. E., Distler, J., Goodman, N. S., Mooney, S. H., Kluth, A., Olek, A., Schwope, I., Tetzner, R., Ziebarth, H., and Berlin, K. A real-time PCR assay for DNA-methylation using methylation-specific blockers. *Nucleic Acids Res, 32:* e10, 2004.

138. Thomassin, H., Kress, C., and Grange, T. MethylQuant: a sensitive method for quantifying methylation of specific cytosines within the genome. *Nucleic Acids Res, 32:* e168, 2004.

139. Xiong, Z. and Laird, P. W. COBRA: a sensitive and quantitative DNA methylation assay. *Nucleic Acids Res, 25:* 2532–2534, 1997.

140. Gonzalgo, M. L. and Jones, P. A. Rapid quantitation of methylation differences at specific sites using methylation-sensitive single nucleotide primer extension (Ms-SNuPE). *Nucleic Acids Res, 25:* 2529–2531, 1997.
141. Uhlmann, K., Brinckmann, A., Toliat, M. R., Ritter, H., and Nurnberg, P. Evaluation of a potential epigenetic biomarker by quantitative methyl-single nucleotide polymorphism analysis. *Electrophoresis, 23:* 4072–4079, 2002.
142. Huang, T. H., Perry, M. R., and Laux, D. E. Methylation profiling of CpG islands in human breast cancer cells. *Hum Mol Genet, 8:* 459–470, 1999.
143. Adorjan, P., Distler, J., Lipscher, E., Model, F., Muller, J., Pelet, C., Braun, A., Florl, A. R., Gutig, D., Grabs, G., Howe, A., Kursar, M., Lesche, R., Leu, E., Lewin, A., Maier, S., Muller, V., Otto, T., Scholz, C., Schulz, W. A., Seifert, H. H., Schwope, I., Ziebarth, H., Berlin, K., Piepenbrock, C., and Olek, A. Tumour class prediction and discovery by microarray-based DNA methylation analysis. *Nucleic Acids Res, 30:* e21, 2002.
144. Gitan, R. S., Shi, H., Chen, C. M., Yan, P. S., and Huang, T. H. Methylation-specific oligonucleotide microarray: a new potential for high-throughput methylation analysis. *Genome Res, 12:* 158–164, 2002.
145. Bibikova, M., Lin, Z., Zhou, L., Chudin, E., Garcia, E. W., Wu, B., Doucet, D., Thomas, N. J., Wang, Y., Vollmer, E., Goldmann, T., Seifart, C., Jiang, W., Barker, D. L., Chee, M. S., Floros, J., and Fan, J. B. High-throughput DNA methylation profiling using universal bead arrays. *Genome Res, 16:* 383–393, 2006.
146. Ehrich, M., Nelson, M. R., Stanssens, P., Zabeau, M., Liloglou, T., Xinarianos, G., Cantor, C. R., Field, J. K., and van den Boom, D. Quantitative high-throughput analysis of DNA methylation patterns by base-specific cleavage and mass spectrometry. *Proc Natl Acad Sci USA, 102:* 15785–15790, 2005.
147. Brock, G. J., Huang, T. H., Chen, C. M., and Johnson, K. J. A novel technique for the identification of CpG islands exhibiting altered methylation patterns (ICEAMP). *Nucleic Acids Res, 29:* E123, 2001.
148. Shiraishi, M., Chuu, Y. H., and Sekiya, T. Isolation of DNA fragments associated with methylated CpG islands in human adenocarcinomas of the lung using a methylated DNA binding column and denaturing gradient gel electrophoresis. *Proc Natl Acad Sci USA, 96:* 2913–2918, 1999.
149. Rauch, T. and Pfeifer, G. P. Methylated-CpG island recovery assay: a new technique for the rapid detection of methylated-CpG islands in cancer. *Lab Invest, 85:* 1172–1180, 2005.
150. Weber, M., Davies, J. J., Wittig, D., Oakeley, E. J., Haase, M., Lam, W. L., and Schubeler, D. Chromosome-wide and promoter-specific analyses identify sites of differential DNA methylation in normal and transformed human cells. *Nat Genet, 37:* 853–862, 2005.
151. Gebhard, C., Schwarzfischer, L., Pham, T. H., Andreesen, R., Mackensen, A., and Rehli, M. Rapid and sensitive detection of CpG-methylation using methyl-binding (MB)-PCR. *Nucleic Acids Res, 34:* e82, 2006.
152. Yegnasubramanian, S., Lin, X., Haffner, M. C., DeMarzo, A. M., and Nelson, W. G. Combination of methylated-DNA precipitation and methylation-sensitive restriction enzymes (COMPARE-MS) for the rapid, sensitive and quantitative detection of DNA methylation. *Nucleic Acids Res, 34:* e19, 2006.
153. Rauch, T., Li, H., Wu, X., and Pfeifer, G. P. MIRA-assisted microarray analysis, a new technology for the determination of DNA methylation patterns, identifies frequent methylation of homeodomain-containing genes in lung cancer cells. *Cancer Res, 66:* 7939–7947, 2006.
154. Gebhard, C., Schwarzfischer, L., Pham, T. H., Schilling, E., Klug, M., Andreesen, R., and Rehli, M. Genome-wide profiling of CpG methylation identifies novel targets of aberrant hypermethylation in myeloid leukemia. *Cancer Res, 66:* 6118–6128, 2006.

155. Zhang, X., Yazaki, J., Sundaresan, A., Cokus, S., Chan, S. W., Chen, H., Henderson, I. R., Shinn, P., Pellegrini, M., Jacobsen, S. E., and Ecker, J. R. Genome-wide high-resolution mapping and functional analysis of DNA methylation in Arabidopsis. *Cell*, 126(6): 1189–201, 2006.

156. Millar, D. S., Ow, K. K., Paul, C. L., Russell, P. J., Molloy, P. L., and Clark, S. J. Detailed methylation analysis of the glutathione S-transferase pi (GSTP1) gene in prostate cancer. *Oncogene, 18:* 1313–1324, 1999.

157. Millar, D. S., Paul, C. L., Molloy, P. L., and Clark, S. J. A distinct sequence (ATAAA)n separates methylated and unmethylated domains at the 5′-end of the GSTP1 CpG island. *J Biol Chem, 275:* 24893–24899, 2000.

158. Harden, S. V., Guo, Z., Epstein, J. I., and Sidransky, D. Quantitative GSTP1 methylation clearly distinguishes benign prostatic tissue and limited prostate adenocarcinoma. *J Urol, 169:* 1138–1142, 2003.

159. Gonzalgo, M. L., Pavlovich, C. P., Lee, S. M., and Nelson, W. G. Prostate cancer detection by GSTP1 methylation analysis of postbiopsy urine specimens. *Clin Cancer Res, 9:* 2673–2677, 2003.

160. Gonzalgo, M. L., Nakayama, M., Lee, S. M., De Marzo, A. M., and Nelson, W. G. Detection of GSTP1 methylation in prostatic secretions using combinatorial MSP analysis. *Urology,* 63(2): 414–8, 2004.

161. Jeronimo, C., Henrique, R., Hoque, M. O., Ribeiro, F. R., Oliveira, J., Fonseca, D., Teixeira, M. R., Lopes, C., and Sidransky, D. Quantitative RARbeta2 hypermethylation: a promising prostate cancer marker. *Clin Cancer Res, 10:* 4010–4014, 2004.

162. Liu, L., Yoon, J. H., Dammann, R., and Pfeifer, G. P. Frequent hypermethylation of the RASSF1A gene in prostate cancer. *Oncogene, 21:* 6835–6840, 2002.

163. Zhang, J., Liu, L., and Pfeifer, G. P. Methylation of the retinoid response gene TIG1 in prostate cancer correlates with methylation of the retinoic acid receptor beta gene. *Oncogene, 23:* 2241–2249, 2004.

164. Zhu, X., Leav, I., Leung, Y. K., Wu, M., Liu, Q., Gao, Y., McNeal, J. E., and Ho, S. M. Dynamic regulation of estrogen receptor-beta expression by DNA methylation during prostate cancer development and metastasis. *Am J Pathol, 164:* 2003–2012, 2004.

165. Nelson, J. B., Chan-Tack, K., Hedican, S. P., Magnuson, S. R., Opgenorth, T. J., Bova, G. S., and Simons, J. W. Endothelin-1 production and decreased endothelin B receptor expression in advanced prostate cancer. *Cancer Res, 56:* 663–668, 1996.

166. Nelson, J. B., Hedican, S. P., George, D. J., Reddi, A. H., Piantadosi, S., Eisenberger, M. A., and Simons, J. W. Identification of endothelin-1 in the pathophysiology of metastatic adenocarcinoma of the prostate. *Nat Med, 1:* 944–949, 1995.

167. Nelson, J. B., Lee, W. H., Nguyen, S. H., Jarrard, D. F., Brooks, J. D., Magnuson, S. R., Opgenorth, T. J., Nelson, W. G., and Bova, G. S. Methylation of the 5′ CpG island of the endothelin B receptor gene is common in human prostate cancer. *Cancer Res, 57:* 35–37, 1997.

168. Nelson, J., Bagnato, A., Battistini, B., and Nisen, P. The endothelin axis: emerging role in cancer. *Nat Rev Cancer, 3:* 110–116, 2003.

169. Guise, T. A., Yin, J. J., and Mohammad, K. S. Role of endothelin-1 in osteoblastic bone metastases. *Cancer, 97:* 779–784, 2003.

170. Yin, J. J., Mohammad, K. S., Kakonen, S. M., Harris, S., Wu-Wong, J. R., Wessale, J. L., Padley, R. J., Garrett, I. R., Chirgwin, J. M., and Guise, T. A. A causal role for endothelin-1 in the pathogenesis of osteoblastic bone metastases. *Proc Natl Acad Sci USA, 100:* 10954–10959, 2003.

171. Carducci, M. A., Nelson, J. B., Bowling, M. K., Rogers, T., Eisenberger, M. A., Sinibaldi, V., Donehower, R., Leahy, T. L., Carr, R. A., Isaacson, J. D., Janus, T. J., Andre, A., Hosmane, B. S., and Padley, R. J. Atrasentan, an endothelin-receptor antagonist for refractory adenocarcinomas: safety and pharmacokinetics. *J Clin Oncol, 20:* 2171–2180, 2002.

172. Carducci, M. A., Padley, R. J., Breul, J., Vogelzang, N. J., Zonnenberg, B. A., Daliani, D. D., Schulman, C. C., Nabulsi, A. A., Humerickhouse, R. A., Weinberg, M. A., Schmitt, J. L., and Nelson, J. B. Effect of endothelin-A receptor blockade with atrasentan on tumor progression in men with hormone-refractory prostate cancer: a randomized, phase II, placebo-controlled trial. *J Clin Oncol, 21:* 679–689, 2003.

173. Kaminskas, E., Farrell, A., Abraham, S., Baird, A., Hsieh, L. S., Lee, S. L., Leighton, J. K., Patel, H., Rahman, A., Sridhara, R., Wang, Y. C., and Pazdur, R. Approval summary: azacitidine for treatment of myelodysplastic syndrome subtypes. *Clin Cancer Res, 11:* 3604–3608, 2005.

174. Jones, P. A. and Taylor, S. M. Cellular differentiation, cytidine analogs and DNA methylation. *Cell, 20:* 85–93, 1980.

175. Cheng, J. C., Matsen, C. B., Gonzales, F. A., Ye, W., Greer, S., Marquez, V. E., Jones, P. A., and Selker, E. U. Inhibition of DNA methylation and reactivation of silenced genes by zebularine. *J Natl Cancer Inst, 95:* 399–409, 2003.

176. Lin, X., Asgari, K., Putzi, M. J., Gage, W. R., Yu, X., Cornblatt, B. S., Kumar, A., Piantadosi, S., DeWeese, T. L., De Marzo, A. M., and Nelson, W. G. Reversal of GSTP1 CpG island hypermethylation and reactivation of pi-class glutathione S-transferase (GSTP1) expression in human prostate cancer cells by treatment with procainamide. *Cancer Res, 61:* 8611–8616, 2001.

177. Segura-Pacheco, B., Trejo-Becerril, C., Perez-Cardenas, E., Taja-Chayeb, L., Mariscal, I., Chavez, A., Acuna, C., Salazar, A. M., Lizano, M., and Duenas-Gonzalez, A. Reactivation of tumor suppressor genes by the cardiovascular drugs hydralazine and procainamide and their potential use in cancer therapy. *Clin Cancer Res, 9:* 1596–1603, 2003.

178. Santini, V., Kantarjian, H. M., and Issa, J. P. Changes in DNA methylation in neoplasia: pathophysiology and therapeutic implications. *Ann Intern Med, 134:* 573–586, 2001.

179. Juttermann, R., Li, E., and Jaenisch, R. Toxicity of 5-aza-2'-deoxycytidine to mammalian cells is mediated primarily by covalent trapping of DNA methyltransferase rather than DNA demethylation. *Proc Natl Acad Sci USA, 91:* 11797–11801, 1994.

180. Yang, A. S., Doshi, K. D., Choi, S. W., Mason, J. B., Mannari, R. K., Gharybian, V., Luna, R., Rashid, A., Shen, L., Estecio, M. R., Kantarjian, H. M., Garcia-Manero, G., and Issa, J. P. DNA methylation changes after 5-aza-2'-deoxycytidine therapy in patients with leukemia. *Cancer Res, 66:* 5495–5503, 2006.

181. Thibault, A., Figg, W. D., Bergan, R. C., Lush, R. M., Myers, C. E., Tompkins, A., Reed, E., and Samid, D. A phase II study of 5-aza-2'deoxycytidine (decitabine) in hormone independent metastatic (D2) prostate cancer. *Tumori, 84:* 87–89, 1998.

182. McCabe, M. T., Low, J. A., Daignault, S., Imperiale, M. J., Wojno, K. J., and Day, M. L. Inhibition of DNA methyltransferase activity prevents tumorigenesis in a mouse model of prostate cancer. *Cancer Res, 66:* 385–392, 2006.

183. Jackson-Grusby, L., Laird, P. W., Magge, S. N., Moeller, B. J., and Jaenisch, R. Mutagenicity of 5-aza-2'-deoxycytidine is mediated by the mammalian DNA methyltransferase. *Proc Natl Acad Sci USA, 94:* 4681–4685, 1997.

184. Scheinbart, L. S., Johnson, M. A., Gross, L. A., Edelstein, S. R., and Richardson, B. C. Procainamide inhibits DNA methyltransferase in a human T cell line. *J Rheumatol, 18:* 530–534, 1991.

185. Cornacchia, E., Golbus, J., Maybaum, J., Strahler, J., Hanash, S., and Richardson, B. Hydralazine and procainamide inhibit T cell DNA methylation and induce autoreactivity. *J Immunol, 140:* 2197–2200, 1988.

186. Lee, B. H., Yegnasubramanian, S., Lin, X., and Nelson, W. G. Procainamide is a specific inhibitor of DNA methyltransferase 1. *J Biol Chem, 280:* 40749–40756, 2005.

187. Quddus, J., Johnson, K. J., Gavalchin, J., Amento, E. P., Chrisp, C. E., Yung, R. L., and Richardson, B. C. Treating activated CD4+ T cells with either of two distinct DNA methyltransferase inhibitors, 5-azacytidine or procainamide, is sufficient to cause a lupus-like disease in syngeneic mice. *J Clin Invest*, 92: 38–53, 1993.

188. Kelly, W. K., Richon, V. M., O'Connor, O., Curley, T., MacGregor-Curtelli, B., Tong, W., Klang, M., Schwartz, L., Richardson, S., Rosa, E., Drobnjak, M., Cordon-Cordo, C., Chiao, J. H., Rifkind, R., Marks, P. A., and Scher, H. Phase I clinical trial of histone deacetylase inhibitor: suberoylanilide hydroxamic acid administered intravenously. *Clin Cancer Res*, 9: 3578–3588, 2003.

189. Carducci, M. A., Gilbert, J., Bowling, M. K., Noe, D., Eisenberger, M. A., Sinibaldi, V., Zabelina, Y., Chen, T. L., Grochow, L. B., and Donehower, R. C. A Phase I clinical and pharmacological evaluation of sodium phenylbutyrate on an 120-h infusion schedule. *Clin Cancer Res*, 7: 3047–3055, 2001.

190. Carducci, M. A., Nelson, J. B., Chan-Tack, K. M., Ayyagari, S. R., Sweatt, W. H., Campbell, P. A., Nelson, W. G., and Simons, J. W. Phenylbutyrate induces apoptosis in human prostate cancer and is more potent than phenylacetate. *Clin Cancer Res*, 2: 379–387, 1996.

191. Gilbert, J., Baker, S. D., Bowling, M. K., Grochow, L., Figg, W. D., Zabelina, Y., Donehower, R. C., and Carducci, M. A. A phase I dose escalation and bioavailability study of oral sodium phenylbutyrate in patients with refractory solid tumor malignancies. *Clin Cancer Res*, 7: 2292–2300, 2001.

192. Rokhlin, O. W., Glover, R. B., Guseva, N. V., Taghiyev, A. F., Kohlgraf, K. G., and Cohen, M. B. Mechanisms of cell death induced by histone deacetylase inhibitors in androgen receptor-positive prostate cancer cells. *Mol Cancer Res*, 4: 113–123, 2006.

193. Qian, D. Z., Kato, Y., Shabbeer, S., Wei, Y., Verheul, H. M., Salumbides, B., Sanni, T., Atadja, P., and Pili, R. Targeting tumor angiogenesis with histone deacetylase inhibitors: the hydroxamic acid derivative LBH589. *Clin Cancer Res*, 12: 634–642, 2006.

194. Camphausen, K., Scott, T., Sproull, M., and Tofilon, P. J. Enhancement of xenograft tumor radiosensitivity by the histone deacetylase inhibitor MS-275 and correlation with histone hyperacetylation. *Clin Cancer Res*, 10: 6066–6071, 2004.

195. Fronsdal, K. and Saatcioglu, F. Histone deacetylase inhibitors differentially mediate apoptosis in prostate cancer cells. *Prostate*, 62: 299–306, 2005.

196. Qian, D. Z., Wang, X., Kachhap, S. K., Kato, Y., Wei, Y., Zhang, L., Atadja, P., and Pili, R. The histone deacetylase inhibitor NVP-LAQ824 inhibits angiogenesis and has a greater antitumor effect in combination with the vascular endothelial growth factor receptor tyrosine kinase inhibitor PTK787/ZK222584. *Cancer Res*, 64: 6626–6634, 2004.

197. Camphausen, K., Burgan, W., Cerra, M., Oswald, K. A., Trepel, J. B., Lee, M. J., and Tofilon, P. J. Enhanced radiation-induced cell killing and prolongation of gammaH2AX foci expression by the histone deacetylase inhibitor MS-275. *Cancer Res*, 64: 316–321, 2004.

198. Rashid, S. F., Moore, J. S., Walker, E., Driver, P. M., Engel, J., Edwards, C. E., Brown, G., Uskokovic, M. R., and Campbell, M. J. Synergistic growth inhibition of prostate cancer cells by 1 alpha,25 Dihydroxyvitamin D(3) and its 19-nor-hexafluoride analogs in combination with either sodium butyrate or trichostatin A. *Oncogene*, 20: 1860–1872, 2001.

199. Butler, L. M., Webb, Y., Agus, D. B., Higgins, B., Tolentino, T. R., Kutko, M. C., LaQuaglia, M. P., Drobnjak, M., Cordon-Cardo, C., Scher, H. I., Breslow, R., Richon, V. M., Rifkind, R. A., and Marks, P. A. Inhibition of transformed cell growth and induction of cellular differentiation by pyroxamide, an inhibitor of histone deacetylase. *Clin Cancer Res*, 7: 962–970, 2001.

200. Butler, L. M., Agus, D. B., Scher, H. I., Higgins, B., Rose, A., Cordon-Cardo, C., Thaler, H. T., Rifkind, R. A., Marks, P. A., and Richon, V. M. Suberoylanilide hydroxamic

acid, an inhibitor of histone deacetylase, suppresses the growth of prostate cancer cells in vitro and in vivo. *Cancer Res, 60:* 5165–5170, 2000.

201. Cameron, E. E., Bachman, K. E., Myohanen, S., Herman, J. G., and Baylin, S. B. Synergy of demethylation and histone deacetylase inhibition in the re-expression of genes silenced in cancer. *Nat Genet, 21:* 103–107, 1999.

3 Inherited Genetic Changes in Prostate Cancer

John D. Carpten, PhD,
and Jeffrey M. Trent, PhD

CONTENTS

1. INTRODUCTION

Although the vast majority of prostate cancer cases are sporadic in nature, there is mounting evidence strongly supporting the existence of prostate cancer genetic risk factors. These genetic risk factors come in two flavors, rare highly penetrant mutations and common low-penetrant variants. Major research studies in the area of hereditary prostate cancer (HPC) are underway to help elucidate the rare highly penetrant alleles, which segregate in families with multiple affected family members. Likewise, several large cohort studies have been recently initiated to search for common low-penetrant variants in the

From: *Current Clinical Oncology: Prostate Cancer:*
Signaling Networks, Genetics, and New Treatment Strategies
Edited by: R. G. Pestell and M. T. Nevalainen © Humana Press, Totowa, NJ

general population, which are associated with increased risk of prostate cancer. Unlike other common cancers, such as breast cancer and colorectal cancer, no clear-cut high-penetrant gene has been discovered, which when mutated causes prostate cancer. It is more commonly believed that prostate cancer is a heterogenous disease, with both high- and low-penetrant genes cooperating toward the prostate cancer phenotype. It is our hope that with continued investigation, the discovery of key prostate cancer susceptibility genes will lead to earlier diagnosis of prostate cancer and will help us better comprehend the underlying etiology of this disease.

2. WHAT EVIDENCE EXISTS FOR FAMILIAL CLUSTERING OF PROSTATE CANCER?

The heritability of prostate cancer is most strikingly supported by the study of twins from Sweden, Denmark, and Finland *(1)*. In this study, concordance of cancer between monozygotic versus dizygotic twins was determined for various cancer types. As shown in Fig. 1, heritability was highest for prostate cancer. A positive correlation between familial clustering and risk of prostate cancer is evident based on a trend of increasing risk of disease with increasing

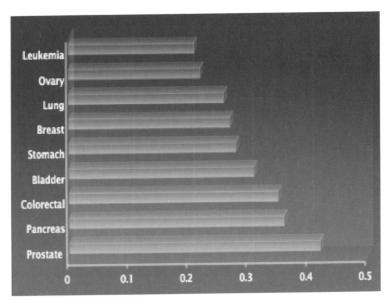

Fig. 1. Estimates of heritability for cancer types based on twin data reported by Lichtenstein et al. *(1)*.

number of affected first- or second-degree relatives. Multiple complex segregation analyses support Mendelian inheritance of prostate cancer. Carter et al. *(2)* suggested that familial clustering was best explained by autosomal dominant inheritance of a rare ($q = 0.003$) high-risk allele leading to an early onset form of prostate cancer. Penetrance estimations were 88% by age 85 for carriers versus 5% for non-carriers. This rare, highly penetrant allele was estimated to account for ~45% of early onset disease and ~9% of all prostate cancer cases. Complex segregation analysis reported by Grönberg et al. *(3)*, using a population-based sample of 2857 nuclear families ascertained from an affected father diagnosed with prostate cancer in Sweden, revealed that the observed clustering was best explained by a high-risk allele inherited in a dominant mode, with a relatively high population frequency of 1.67% and a moderate lifetime penetrance of 63%. Schaid et al. *(4)* performed complex segregation analysis on 4288 men who underwent radical prostatectomy for clinically localized prostate cancer. Although no single-gene model of inheritance clearly explained familial clustering of disease, the best fitting model for familial clustering was explained by inheritance of a rare autosomal dominant allele ($q = 0.006$), with an age-adjusted penetrance of 89% by age 85 for carriers versus 3% for non-carriers. Results from a segregation analysis of 1476 Australian prostate cancer families supported evidence for an X-linked or recessive model for prostate cancer inheritance *(5)*. More recently, complex segregation analysis showed that familial clustering of prostate cancer was equally well explained by (i) a dominant Mendelian model with a susceptibility allele frequency of 2.4%, and risk of those affected by age 80 of 75.3% and 8.2% in African-American carriers and non-carriers, respectively, or (ii) a multifactorial model with multiple genes, each having low to moderate penetrance being responsible for most inherited prostate cancer susceptibility *(6)*. These reports suggest that the vast majority of familiar prostate cancer cases are associated with rare, highly penetrant, susceptibility alleles, but data also exist supporting a multifactorial model.

3. IS PROSTATE CANCER A HERITABLE DISEASE?

Linkage analysis and positional cloning have been applied successfully to the discovery of susceptibility genes for breast cancer (*BRCA1*, *BRCA2*) *(7,8)*, colon cancer (*HNPCC* and *mut* genes) *(9,10)*, and renal cell carcinoma (*VHL*, *MET*) *(11,12)*. However, of the most common cancers, prostate cancer is the only malignancy for which a reproducible rare, high-penetrant allele has not yet been identified. Segregation analyses have prompted the collection of highly aggregated prostate cancer families for genome-wide genetic analysis in order to facilitate the identification of rare, highly penetrant HPC genes. The first reported genome-wide scan (GWS) to search for HPC genes suggested a major susceptibility locus on human chromosome 1 (designated *HPC1*) using 91 families from the USA and Sweden *(13)*. In a GWS, a set of evenly spaced genetic markers is genotyped in DNA from both prostate cancer cases and their family members. Statistical analysis is performed to determine a "logarithm (to the base 10) of the odds score," or LOD score. The LOD score is a statistical estimate determined for each genetic marker to determine the probability of the marker being linked to the disease versus it being totally independent or non-linked to the disease locus. A maximum multipoint LOD score of 5.43 was achieved for a group of markers mapping to 1q24–q31, meaning that there was nearly 1,000,000 to 1 odds that this group of markers was linked to the disease. As prostate cancer and other cancers are heterogeneous, it was assumed that only a subset of the families studied was linked to the region on chromosome 1. Therefore, the linkage analysis was performed assuming heterogeneity, with approximately 34% of the 91 families contributing to the linkage to chromosome 1. This region or locus on human chromosome 1 was named *hereditary prostate cancer 1* or *HPC1*. Several *HPC1* confirmatory studies followed, the results of which were mixed *(14–19)*. Other independent GWS studies of HPC families quickly followed with reports of linkage at several regions of the genome including Xq, 1p, 1q(q42), 8p, 16q, 17p, and 20q *(20–26)*. Confirmation of linkage to most if not all of these regions in independent data sets has been limited or controversial. More recently, a series of GWSs from multiple research groups was reported in a single issue of the journal *Prostate* (*Prostate*. Issue 57, 2003). Markers at 11 different genomic regions (with only one overlapping with previous studies) had non-parametric LOD scores (NPL) greater than 2.0 [reviewed in *(27)*]. The results of these studies strongly support significant heterogeneity in familial prostate cancer.

To deal with the issue of heterogeneity, the International Consortium for Prostate Cancer Genetics (ICPCG) was formed in 1999. This consortium is currently comprised of 11 independently collected HPC data sets. The details of the ICPCG data set are shown in Table 1, as recently reported by Xu et al. *(28)*.

Table 1

Mean age at diagnosis (years)		Number of prostate cancer cases within a family				Race		Total number of families
<65	>65	2	3	4	> 5	White	Black	
606	625	285	424	255	269	1166	48	1233

The first study reported by the ICPCG was a combined linkage analysis of 772 families, which were genotyped with the same set of markers mapping within the *HPC1* region *(19)*. A positive but marginal LOD score of 1.4 was achieved at the *HPC1* locus under the assumption of heterogeneity. The second study reported by the ICPCG was a combined analysis of the *HPC20* (20q13) linkage region, initially reported by Berry et al. *(23)*. No evidence of linkage at the *HPC20* locus was obtained by the ICPCG combined analysis *(29)*. More recently, a meta-analysis from 12 independent GWS scans was reported by the ICPCG *(28)*. This study of more than 1200 HPC families was the largest of its kind, with these data holding the most promise for identifying key prostate cancer susceptibility loci. Table 2 summarizes the results for all linkage analyses reported by the ICPCG *(28)*. Five regions of the genome (5q12, 8p21, 15q11, 17q21, and 22q12) showed LOD scores >1.86, which were suggestive of linkage (Fig. 2). A significant linkage result was seen at 22q12, where a LOD score of 3.57 was seen in 269 families with more than five affected family members. These data tell us two things: (i) prostate cancer does aggregate in families and linkage can identify putative HPC loci and (ii) there is significant heterogeneity in HPC.

Table 2
Results of ICPCG Genome-Wide Meta Analysis *(28)*

Chromosomal region (nearest marker)	LOD score	Families contributing to the LOD score
5q12 (D5S2858)	2.28	All ($N = 1233$)
8p21 (D8S1048)	1.97	All ($N = 1233$)
15q11 (D15S817)	2.10	All ($N = 1233$)
17q21 (D17S1820)	1.99	All ($N = 1233$)
22q12 (D22S283)	1.95	All ($N = 1233$)
1q25 (D1S2818)	2.62	≥5 cases ($n = 269$)
8q13 (D8S543)	2.41	≥5 cases ($n = 269$)
13q14 (D13S1807)	2.27	≥5 cases ($n = 269$)
16p13 (D16S764)	1.88	≥5 cases ($n = 269$)
17q21 (D17S1820)	2.04	≥5 cases ($n = 269$)
22q12 (D22S283)	3.57	≥5 cases ($n = 269$)
3p24 (D3S2432)	2.37	Age at dx ≤65 ($n = 606$)
5q35 (D5S1456)	2.05	Age at dx ≤65 ($n = 606$)
11q22 (D11S898	2.20	Age at dx ≤65 ($n = 606$)
Xq12 (DXS7132)	2.30	Age at dx ≤65 ($n = 606$)

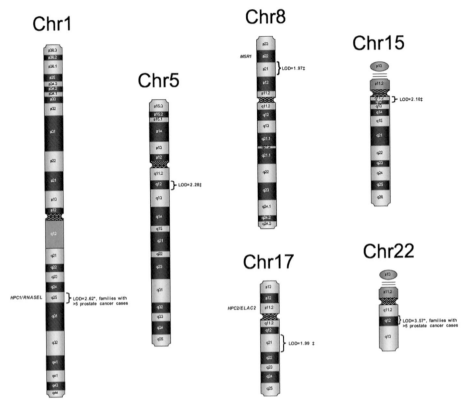

Fig. 2. Chromosomal ideograms showing regions of the human genome linked to hereditary prostate cancer.

4. HAS AN HPC GENE BEEN IDENTIFIED?

Through positional candidate cloning efforts within putative HPC loci, three genes have been reported that harbor inactivating mutations, which segregate with prostate cancer in high-risk families; including the *HPC2/ELAC2* gene, the *ribonuclease L (RNASEL)* gene, and *macrophage scavenger receptor 1 (MSR1)* gene *(25,30,31).* Map positions of these genes can be found in Fig. 2.

Positional candidate cloning of an HPC locus at 17p11 (*HPC2*) revealed germline mutations (including a frameshift and non-conservative missense changes) in the *HPC2/ELAC2* gene in multiple families, which segregated with prostate cancer in large Mormon kindreds *(25). HPC2/ELAC2* is a member of an uncharacterized gene family predicted to encode a metal-dependent hydrolase domain that is conserved among eukaryotes, archaebacteria, and eubacteria *(25).*

Positional candidate cloning within the *HPC1* locus (1q24–q31) resulted in the identification of two inactivating germline mutations in the *RNASEL* gene, which were found to segregate with prostate cancer in two *HPC1*-linked families *(30)*. This gene encodes an interferon inducible 2–5A oligoadenylate-dependent ribonuclease, which degrades double-stranded RNA during viral infection and general cellular apoptosis *(32)*. Biochemical analysis of *RNASEL* mutations shows functional consequences on enzymatic activity *(30,33)*.

More recently, germline mutations (including one nonsense mutation) were identified in the *MSR1* gene, which segregate with prostate cancer in 8p22–p23 linked families *(31)*. The *MSR1* gene encodes a member of a family of macrophage-specific trimeric integral membrane glycoproteins implicated in many macrophage-associated physiological and pathological processes including atherosclerosis, Alzheimer's disease, and host defense. Importantly, a replication study in Germany has produced two new deleterious germline mutations that track with prostate cancer in families *(34)*.

To date, these are the only genes known to harbor deleterious mutations, which seem to track specifically with prostate cancer in families. However, because of the small number of families with mutations, these genes are probably only responsible for a very small percentage of HPC. More research in this area in larger data sets is needed to discover other important genes containing deleterious mutations predisposing to prostate cancer.

5. ARE COMMON VARIANTS ASSOCIATED WITH PROSTATE CANCER RISK?

As mentioned in the previous section of this chapter, common variants in HPC genes can increase the risk for men with familial disease, as well as men with sporadic disease. Common variants, the most common of which are known as single-nucleotide polymorphisms (SNPs), in a number of genes have also been shown to increase risk of developing prostate cancer. These common variants have relatively high allele frequencies in the normal population (minor allele frequencies >5%), as opposed to rare highly penetrant alleles, which are generally not found at high frequency in the normal population (minor allele frequency <1%). The approach for discovery of these common variants relies on the analysis of cases and health controls. Generally, allele frequencies are calculated for genetic variants in both the case and control set, and minor allele frequencies are compared between the two groups. The discovery of a variant with different minor allele frequencies signifies a potential genetic risk factor. Several other factors weighing heavily on the results of these variants include the strength of association, the frequencies of the risk alleles, the sample size, and the selection of SNPs to be used in the study.

The discovery of these variants has for many years relied on *a priori* knowledge of gene function. Those genes found to harbor highly penetrant mutations have also been analyzed for their role as common risk factors for prostate cancer. Two common missense variants (Ser217Leu and Ala541Thr) in the *HPC2/ELAC2* gene have been reported, which are associated with increased risk of developing prostate cancer in unselected cases *(25,35–37)*. In the largest independent study reported using more than 400 prostate cancer cases and more than 400 controls from Europe, the authors concluded that the association with prostate cancer risk for these two variants was weak at best. A more significant association was seen in a set of more than 200 cases and more than 200 controls of Japanese decent *(38)*.

Several studies have reported the association of mutations in the *RNASEL* gene with risk of prostate cancer in patients with a positive family history of prostate cancer *(33,39–42)*. Replication has been inconsistent. Several of the RNASEL associations have been to the R462Q missense variant. Importantly, biological studies of this particular RNASEL variant show that it causes significant reduction in RNASEL protein function. Researchers have now identified the presence of gamma-retroviral sequences in prostate tumors in homozygous carriers of the R462Q variant. The viral sequences were 25 times more likely to occur in tumors from patients carrying two copies of the R462Q variant *(43)*. Although there is currently no link between this virus and tumor initiation, these findings raise the possibility for a direct relationship between a viral infection and the development of prostate cancer in genetically susceptible individuals.

Replication studies using independent data sets to assess risk associated with common *MSR1* variants have produced conflicting results to date *(44–46)*. Interestingly, one report demonstrated significant differences in allele frequencies between African-American prostate cancer cases and controls for common variants in *MSR1 (44)*. A recent meta-analysis of eight published studies produced positive association for *MSR1* common variants and increased prostate cancer risk *(47)*. Although there is some evidence for these genes being involved in prostate cancer susceptibility, the limited validation studies and controversial reports would suggest that a true, highly penetrant gene for prostate cancer has not as of yet been identified.

A number of other genes have also been studied for their role in prostate cancer susceptibility. Many of these candidate genes are selected because of their association with a pathway important in normal prostate growth and development. One obvious pathway is the steroidogenesis pathway, as it relates to the metabolism of androgens. A partial diagram of the androgen biosynthesis pathway is illustrated in Fig. 3. One family of genes intimately involved in this pathway is the cytochrome p450 family (CYPs). Members of the CYP family of genes have many roles, including metabolism of drugs and metabolism

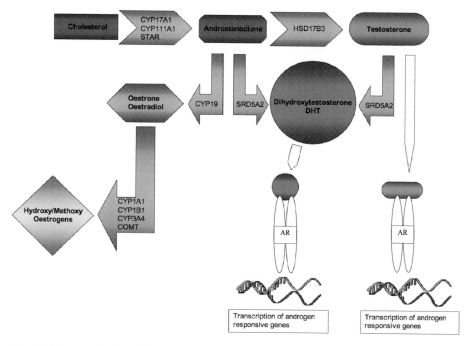

Fig. 3. Diagram of the androgen metabolism pathway.

of steroid hormones. Because of their obvious role in metabolizing steroid hormones, they have been targeted as a potentially important group of genes, which might harbor common genetic variants that modulate risk of prostate cancer.

Common variants in the genes encoding CYP1B1, CYP17, and CYP3A4 have been studied extensively. Moderate risks have been reported for SNPs within all three of these genes. The most ominous and extensively studied is the *CYP3A4* gene. An original study of variants within this gene also showed significantly different allele frequencies between African Americans and Caucasians *(48,49)*. This has also been followed up by reports showing a similar frequency trend for a specific promoter mutation following trends in prostate their cancer incidence rates in the USA *(50–52)*. This gene has been extensively studied in African Americans and some have associated variation within this gene with prostate cancer risk and aggressive prostate cancer *(53)*. However, it should be noted that one study did not associate *CYP3A4* variation with prostate cancer in African Americans and reported that positive associations may be confounded by overall differences in population allele frequencies (population stratification) *(54)*. Further studies in extremely large cohorts are needed to clarify the true risk associated with *CYP3A4* variants and risk of prostate cancer.

Another gene involved in steroid hormone metabolism that has been studied for its association with prostate cancer risk is the steroid 5-alpha reductase gene (*SRD5A2*). An original study of *SRD5A2* showed significant differences in allele frequencies across populations *(55)*. Two interesting missense variants in the *SRD5A2* gene have been shown to modulate enzymatic activity for metabolism of testosterone and are associated with increased risk of prostate cancer in several populations including African Americans *(56–58)*. The results of a recent meta-analysis of nine independent case–control studies looking at *SRD5A2* variants and their association with prostate cancer risk suggest that the true effects of these variants on prostate cancer risk are modest and likely to account for only a small percentage of prostate cancer cases *(59)*.

One of the most important genes involved in prostate growth and development is the androgen receptor (*AR*) gene. The *AR* gene has been extensively studied for its potential role in prostate cancer predisposition. Two trinucleotide repeat polymorphisms map within exon 1 of the *AR* gene, and length differences of these repeats have been shown to modulate androgen receptor transactivation *(60,61)*. Shorter repeats are strongly associated with higher AR transactivation. Furthermore, several groups have reported associations between *AR* trinucleotide repeat length and prostate cancer risk and phenotypic characteristics of disease in African Americans using case–control studies *(62–65)*. Interestingly, there are significantly different allele frequencies in trinucleotide repeat lengths, with shorter repeats being in admixture disequilibrium in African Americans *(66,67)*. These data raise an interesting question of whether the association between these *AR* variants and prostate cancer represents a true effect, or whether these associations are due to population stratification of these variants across populations.

Among the other genes commonly studied for their association with prostate cancer risk are those that encode the vitamin D receptor and the glutathione S-transferases. Results of a meta-analysis of multiple case–control studies of vitamin D receptor variants and prostate cancer risk suggest that the variants analyzed in this study are unlikely to be major risk factors for prostate cancer susceptibility *(68)*. Likewise, the results of a meta-analysis of 11 independent studies concluded that common variants in glutathione S-transferase are not likely to play a major role in prostate cancer susceptibility *(69)*.

Finally, in 2006 researchers at deCode Genetics reported significant evidence for a major prostate cancer risk gene based upon a whole-genome analysis in the homogeneous Icelandic population *(70)*. This large study relied upon several thousand prostate cancer cases and age-matched control individuals, and resulted in the discovery of an important prostate cancer risk locus on human chromosome 8 near the c-myc proto-oncogene. The results were replicated in both European and African-American data sets. The authors concluded that the population attributable risk (PAR) of the 8q24 locus was 16% in African Americans, considerably higher than the PAR for the European populations studied (5–11%) *(70)*. The authors speculate that this allele may partially account for the disparate prostate cancer incidence rates seen among African-American men. In an independent study, an admixture approach was used to search for prostate cancer susceptibility loci in a multiethnic cohort of prostate cancer cases and controls *(71)*. In this study, a set of ~1500 genome-wide admixture informative markers (AIMs) were used. These markers show marked differences in minor allele frequencies between populations. Analysis of the admixture data from ~1600 African Americans showed a statistically significant association between prostate cancer and markers mapping to 8q24, with the signal primarily associated with a late age at diagnosis (<72 years of age) *(71)*. These data provided the first independent validation for a prostate cancer susceptibility locus at 8q24. A second confirmation report from this same group, using a densely spaced set of markers across the broad 8q region, suggested at least three independent susceptibility loci within 8q24 *(72)*. Moreover, results of a large independent GWS using nearly 500,000 SNPs genotyped in ~1200 prostate cancer cases and ~1200 matched controls revealed the strongest evidence for prostate susceptibility to SNPs within the 8q24 region of the genome *(73)*. Recently, there have been four independent confirmation reports supporting the association of prostate cancer with variants mapping within 8q24, making this the first universally replicated prostate cancer susceptibility locus *(74–77)*. However, the true disease-causing variant or gene is yet to be identified.

The analysis of candidate genes and pathways for the discovery of common variants associated with risk will continue to be of extreme importance to the field. It is believed that these variants are associated with the larger set of sporadic prostate cancer cases as opposed to true hereditary cases. However, because of underpowered sample sets, the road to discovery will remain littered with false-positive and false-negative findings.

6. CAN TUMOR SUPPRESSOR GENES DOUBLE AS PROSTATE CANCER RISK FACTORS?

Candidate gene selection is based primarily on interesting functional and physiological associations between genes and disease. Another important class of genes that hold promise as potential modifiers of cancer risk is the tumor suppressor genes. Several known tumor suppressor genes have been studied for their potential role in prostate cancer susceptibility through the use of case–control studies. These include *BRCA2*, *ATM*, *KLF6*, and *EPHB2*.

BRCA2 is one of two genes unambiguously involved in inherited predisposition to breast cancer. *BRCA2* is known to play a critical role in DNA damage repair. For many years, it has been known that male carriers of deleterious *BRCA1* or *BRCA2* mutations are at higher risk of developing prostate cancer. Usually, these men belong to families with multiple breast and ovarian cancer cases as well. So these genes have not been touted as prostate cancer genes, per se. However, recent case–control studies show increased population risk of prostate cancer in men carrying *BRCA2* mutations. Edwards et al. *(78)* reported that *BRCA2* germline mutations conferred a twofold increased risk of developing early onset prostate cancer. Using a set of 250 unselected prostate cancer cases of Ashkenazi decent and more than 1400 Ashkenazi male controls, Kirchhoff and colleagues reported a significant association between prostate cancer and several Ashkenazi *BRCA2* founder mutations *(79)*. However, in a smaller study using Canadian Ashkenazi prostate cancer cases and controls, founder mutations were not found to be associated with elevated prostate cancer risk *(80)*. Another key factor in the process of DNA damage repair is *ATM*. In a study of more than 600 prostate cancer cases and more than 400 controls, common genetic variants in *ATM* were shown to be associated with increased prostate cancer risk *(81)*. These findings implicate the DNA damage pathway as possibly important in prostate cancer predisposition.

Another tumor suppressor gene, which has been studied for a potential role in prostate cancer susceptibility, encodes the KLF6 tumor suppressor. KLF6 is a krupple-like zinc finger transcription factor of yet unknown function. Somatic mutations were originally discovered in prostate tumors, and wild-type KLF6 was shown to suppress tumor growth *(82)*. There have now been reports of KLF6 mutations in multiple somatic tumor types including astrocytic gliomas, hepatocellular carcinoma, and colorectal cancer *(83–85)*; however, conflicting reports have also been published. In a large study of more than 1200 sporadic prostate cancer cases, more than 800 HPC cases, and more than 1200 controls, moderate associations were seen between prostate cancer risk and a common intronic polymorphism *(86)*. Data were also presented, which showed that the intronic polymorphism affects splicing and causes mislocalization of KLF6. The large sample size and functional data support an important role for KLF6 in prostate cancer susceptibility.

In a study published in 2004, Huusko and colleagues presented results implicating the EPHB2 tyrosine kinase as a prostate cancer tumor suppressor *(87)*. A deleterious nonsense mutation was discovered in the DU145 prostate cancer cell line, and somatic mutations were found in clinical prostate tumors at a rate of ~9% *(87)*. Furthermore, it was reported that wild-type EPHB2 significantly decreased tumorigenic growth in DU145 prostate cancer cells, which lack endogenous EPHB2. To assess a possible role in prostate cancer risk, Kittles and colleagues screened the *EPHB2* gene for germline mutations in a set of 72 African-American HPC cases *(88)*. They discovered the presence of a common nonsense mutation, which was three times more prevalent in the African-American population as compared with the Caucasian population. Further analysis of this nonsense variant in 285 African sporadic prostate cancer cases and 329 healthy controls showed a significant association between this variant and HPC in African Americans *(88)*.

Tumor suppressors are known to play a vital role in the initiation and progression of cancer. However, their role in prostate cancer predisposition remains in question. As is the case for other candidate prostate cancer risk genes, analyses need to be carried out in large, sufficiently powered sample sets using robust statistical methodologies. Until then, we may only be able to speculate about the roles of these tumor suppressor genes in prostate cancer susceptibility.

7. SUMMARY

Dating back to the mid to late 1980s, we have had evidence for a genetic role for prostate cancer. Early segregation analyses have supported the existence of highly-penetrant prostate cancer genes. Genetic linkage analyses in multiplex prostate cancer families have led to the discovery of candidate regions; however, many of these regions have not been validated across multiple independent sample sets, so their validity is in question. To help deal with this problem, the ICPCG, with its 2000 multiplex families, is sure to make a major impact in the field of HPC research. This consortium holds the most promise for discovering the more significant rare highly-penetrant mutations, which lead to early onset HPC.

The search for common low-penetrant germline variants is also critical to our overall understanding of prostate cancer susceptibility. A number of candidate gene studies have implicated various genes and pathways in prostate cancer predisposition. Aside from candidate gene approaches, which require us to know something about the genes/pathways role in prostate development, another approach is to search the entire genome for variants, which show allele frequency differences between prostate cancer cases and controls. The discovery en masse of millions of SNPs within the human genome and recent technological advances now make this type of study possible. One can now scan more than 80% of known human variation using "whole genome" SNP microarrays containing hundreds of thousands of SNPs. These studies do not rely on prior knowledge of gene function and are sure to shed light on new important pathways associated with prostate cancer development. There will still remain a great need for sufficiently powered sample sets to perform these types of studies. One major effort to deal with the issue of sample size is the National Cancer Institute Breast and Prostate Cancer Consortium. Currently, this pooled set of large cohorts contains more than 8000 prostate cancer cases and more than 9500 male controls *(89)*. With the aid of more elegantly designed studies and analytical tools to deal with issues such as gene–gene interaction, this group is poised to make a major impact in the field of prostate cancer genetic risk research.

We must remain patient and steadfast in our quest to discover important genetic risk factors for prostate cancer. Once reliable studies are completed and validated, we will be on our way toward a far better understanding of the etiology of prostate cancer. More importantly, we might be able to exploit these findings to develop more sensitive tools for early diagnosis of prostate cancer and for better treating this dreadful disease.

REFERENCES

1. Lichtenstein, P., et al. Environmental and heritable factors in the causation of cancer— analyses of cohorts of twins from Sweden, Denmark, and Finland. *N Engl J Med*, 2000. **343**(2): p. 78–85.
2. Carter, B.S., et al. Hereditary prostate cancer: epidemiologic and clinical features. *J Urol*, 1993. **150**(3): p. 797–802.
3. Gronberg, H., et al. Segregation analysis of prostate cancer in Sweden: support for dominant inheritance. *Am J Epidemiol*, 1997. **146**(7): 552–7.
4. Schaid, D.J., et al. Evidence for autosomal dominant inheritance of prostate cancer. *Am J Hum Genet*, 1998. **62**(6): p. 1425–38.
5. Cui, J., et al. Segregation analyses of 1,476 population-based Australian families affected by prostate cancer. *Am J Hum Genet*, 2001. **68**(5): p. 1207–18.
6. Gong, G., et al. Segregation analysis of prostate cancer in 1,719 white, African-American and Asian-American families in the United States and Canada. *Cancer Causes Control*, 2002. **13**(5): p. 471–82.
7. Miki, Y., et al. A strong candidate for the breast and ovarian cancer susceptibility gene BRCA1. *Science*, 1994. **266**(5182): p. 66–71.
8. Wooster, R., et al. Identification of the breast cancer susceptibility gene BRCA2. *Nature*, 1995. **378**(6559): p. 789–92.
9. Nakamura, Y., et al. Mutations of the APC (adenomatous polyposis coli) gene in FAP (familial polyposis coli) patients and in sporadic colorectal tumors. *Tohoku J Exp Med*, 1992. **168**(2): p. 141–7.
10. Groden, J., et al. Identification and characterization of the familial adenomatous polyposis coli gene. *Cell*, 1991. **66**(3): p. 589–600.
11. Latif, F., et al. Identification of the von Hippel-Lindau disease tumor suppressor gene. *Science*, 1993. **260**(5112): p. 1317–20.
12. Schmidt, L., et al. Germline and somatic mutations in the tyrosine kinase domain of the MET proto-oncogene in papillary renal carcinomas. *Nat Genet*, 1997. **16**(1): p. 68–73.
13. Smith, J.R., et al. Major susceptibility locus for prostate cancer on chromosome 1 suggested by a genome-wide search. *Science*, 1996. **274**(5291): p. 1371–4.
14. Cooney, K.A., et al. Prostate cancer susceptibility locus on chromosome 1q: a confirmatory study. *J Natl Cancer Inst*, 1997. **89**(13): p. 955–9.
15. Hsieh, C.L., et al. Re: prostate cancer susceptibility locus on chromosome 1q: a confirmatory study. *J Natl Cancer Inst*, 1997. **89**(24): p. 1893–4.

16. Eeles, R.A., et al. Linkage analysis of chromosome 1q markers in 136 prostate cancer families. The Cancer Research Campaign/British Prostate Group U.K. Familial Prostate Cancer Study Collaborators. *Am J Hum Genet*, 1998. **62**(3): p. 653–8.
17. Goode, E.L., et al. Linkage analysis of 150 high-risk prostate cancer families at 1q24–25. *Genet Epidemiol*, 2000. **18**(3): p. 251–75.
18. Cancel-Tassin, G., et al. PCAP is the major known prostate cancer predisposing locus in families from south and west Europe. *Eur J Hum Genet*, 2001. **9**(2): p. 135–42.
19. Xu, J. Combined analysis of hereditary prostate cancer linkage to 1q24–25: results from 772 hereditary prostate cancer families from the International Consortium for Prostate Cancer Genetics. *Am J Hum Genet*, 2000. **66**(3): p. 945–57.
20. Gronberg, H., et al. Early age at diagnosis in families providing evidence of linkage to the hereditary prostate cancer locus (HPC1) on chromosome 1. *Cancer Res*, 1997. **57**(21): p. 4707–9.
21. Berthon, P., et al. Predisposing gene for early-onset prostate cancer, localized on chromosome 1q42.2–43. *Am J Hum Genet*, 1998. **62**(6): p. 1416–24.
22. Gibbs, M., et al. Evidence for a rare prostate cancer-susceptibility locus at chromosome 1p36. *Am J Hum Genet*, 1999. **64**(3): p. 776–87.
23. Berry, R., et al. Evidence for a prostate cancer-susceptibility locus on chromosome 20. *Am J Hum Genet*, 2000. **67**(1): p. 82–91.
24. Suarez, B.K., et al. A genome screen of multiplex sibships with prostate cancer. *Am J Hum Genet*, 2000. **66**(3): p. 933–44.
25. Tavtigian, S.V., et al. A candidate prostate cancer susceptibility gene at chromosome 17p. *Nat Genet*, 2001. **27**(2): p. 172–80.
26. Xu, J., et al. Linkage and association studies of prostate cancer susceptibility: evidence for linkage at 8p22–23. *Am J Hum Genet*, 2001. **69**(2): p. 341–50.
27. Easton, D.F., et al. Where are the prostate cancer genes?—A summary of eight genome wide searches. *Prostate*, 2003. **57**(4): p. 261–9.
28. Xu, J., et al. A combined genomewide linkage scan of 1,233 families for prostate cancer-susceptibility genes conducted by the international consortium for prostate cancer genetics. *Am J Hum Genet*, 2005. **77**(2): p. 219–29.
29. Schaid, D.J. and B.L. Chang. Description of the International Consortium For Prostate Cancer Genetics, and failure to replicate linkage of hereditary prostate cancer to 20q13. *Prostate*, 2005. **63**(3): p. 276–90.
30. Carpten, J., et al. Germline mutations in the ribonuclease L gene in families showing linkage with HPC1. *Nat Genet*, 2002. **30**(2): p. 181–4.
31. Xu, J., et al. Germline mutations and sequence variants of the macrophage scavenger receptor 1 gene are associated with prostate cancer risk. *Nat Genet*, 2002. **32**(2): p. 321–5.
32. Silverman, R.H. Implications for RNase L in prostate cancer biology. *Biochemistry*, 2003. **42**(7): p. 1805–12.
33. Casey, G., et al. RNASEL Arg462Gln variant is implicated in up to 13% of prostate cancer cases. *Nat Genet*, 2002. **32**(4): p. 581–3.
34. Maier, C., et al. Germline mutations of the MSR1 gene in prostate cancer families from Germany. *Hum Mutat*, 2006. **27**(1): p. 98–102.
35. Rebbeck, T.R., et al. Association of HPC2/ELAC2 genotypes and prostate cancer. *Am J Hum Genet*, 2000. **67**(4): p. 1014–9.
36. Fujiwara, H., et al. Association of common missense changes in ELAC2 (HPC2) with prostate cancer in a Japanese case-control series. *J Hum Genet*, 2002. **47**(12): p. 641–8.
37. Stanford, J.L., et al. Association of HPC2/ELAC2 polymorphisms with risk of prostate cancer in a population-based study. *Cancer Epidemiol Biomarkers Prev*, 2003. **12**(9): p. 876–81.

38. Yokomizo, A., et al. HPC2/ELAC2 polymorphism associated with Japanese sporadic prostate cancer. *Prostate*, 2004. **61**(3): p. 248–52.

39. Nakazato, H., et al. Role of genetic polymorphisms of the RNASEL gene on familial prostate cancer risk in a Japanese population. *Br J Cancer*, 2003. **89**(4): p. 691–6.

40. Rennert, H., et al. A novel founder mutation in the RNASEL gene, 471delAAAG, is associated with prostate cancer in Ashkenazi Jews. *Am J Hum Genet*, 2002. **71**(4): p. 981–4.

41. Rokman, A., et al. Germline alterations of the RNASEL gene, a candidate HPC1 gene at 1q25, in patients and families with prostate cancer. *Am J Hum Genet*, 2002. **70**(5): p. 1299–304.

42. Wang, L., et al. Analysis of the RNASEL gene in familial and sporadic prostate cancer. *Am J Hum Genet*, 2002. **71**(1): p. 116–23.

43. Urisman, A., et al. Identification of a novel gammaretrovirus in prostate tumors of patients homozygous for R462Q RNASEL variant. *PLoS Pathog*, 2006. **2**(3): p. e25.

44. Miller, D.C., et al. Germ-line mutations of the macrophage scavenger receptor 1 gene: association with prostate cancer risk in African-American men. *Cancer Res*, 2003. **63**(13): p. 3486–9.

45. Wang, L., et al. No association of germline alteration of MSR1 with prostate cancer risk. *Nat Genet*, 2003. **35**(2): p. 128–9.

46. Xu, J., et al. Common sequence variants of the macrophage scavenger receptor 1 gene are associated with prostate cancer risk. *Am J Hum Genet*, 2003. **72**(1): p. 208–12.

47. Sun, J., et al. Meta-analysis of association of rare mutations and common sequence variants in the MSR1 gene and prostate cancer risk. *Prostate*, 2006. **66**(7): p. 728–37.

48. Rebbeck, T.R., et al. Modification of clinical presentation of prostate tumors by a novel genetic variant in CYP3A4. *J Natl Cancer Inst*, 1998. **90**(16): p. 1225–9.

49. Zeigler-Johnson, C.M., et al. Ethnic differences in the frequency of prostate cancer susceptibility alleles at SRD5A2 and CYP3A4. *Hum Hered*, 2002. **54**(1): p. 13–21.

50. Walker, A.H., et al. Characterization of an allelic variant in the nifedipine-specific element of CYP3A4: ethnic distribution and implications for prostate cancer risk. Mutations in brief no. 191. Online. *Hum Mutat*, 1998. **12**(4): p. 289.

51. Paris, P.L., et al. Association between a CYP3A4 genetic variant and clinical presentation in African-American prostate cancer patients. *Cancer Epidemiol Biomarkers Prev*, 1999. **8**(10): p. 901–5.

52. Loukola, A., et al. Comprehensive evaluation of the association between prostate cancer and genotypes/haplotypes in CYP17A1, CYP3A4, and SRD5A2. *Eur J Hum Genet*, 2004. **12**(4): p. 321–32.

53. Powell, I.J., et al. CYP3A4 genetic variant and disease-free survival among white and black men after radical prostatectomy. *J Urol*, 2004. **172**(5 Pt 1): p. 1848–52.

54. Kittles, R.A., et al. CYP3A4-V and prostate cancer in African Americans: causal or confounding association because of population stratification? *Hum Genet*, 2002. **110**(6): p. 553–60.

55. Reichardt, J.K., et al. Genetic variability of the human SRD5A2 gene: implications for prostate cancer risk. *Cancer Res*, 1995. **55**(18): p. 3973–5.

56. Makridakis, N., et al. A prevalent missense substitution that modulates activity of prostatic steroid 5alpha-reductase. *Cancer Res*, 1997. **57**(6): p. 1020–2.

57. Lunn, R.M., et al. Prostate cancer risk and polymorphism in 17 hydroxylase (CYP17) and steroid reductase (SRD5A2). *Carcinogenesis*, 1999. **20**(9): p. 1727–31.

58. Makridakis, N.M., et al. Association of mis-sense substitution in SRD5A2 gene with prostate cancer in African-American and Hispanic men in Los Angeles, USA. *Lancet*, 1999. **354**(9183): p. 975–8.

59. Ntais, C., A. Polycarpou, and J.P. Ioannidis. SRD5A2 gene polymorphisms and the risk of prostate cancer: a meta-analysis. *Cancer Epidemiol Biomarkers Prev*, 2003. **12**(7): p. 618–24.

60. Ding, D., et al. Effect of a short CAG (glutamine) repeat on human androgen receptor function. *Prostate*, 2004. **58**(1): p. 23–32.
61. Ding, D., et al. Effect of GGC (glycine) repeat length polymorphism in the human androgen receptor on androgen action. *Prostate*, 2005. **62**(2): p. 133–9.
62. Irvine, R.A., et al. The CAG and GGC microsatellites of the androgen receptor gene are in linkage disequilibrium in men with prostate cancer. *Cancer Res*, 1995. **55**(9): p. 1937–40.
63. Hardy, D.O., et al. Androgen receptor CAG repeat lengths in prostate cancer: correlation with age of onset. *J Clin Endocrinol Metab*, 1996. **81**(12): p. 4400–5.
64. Gilligan, T., et al. Absence of a correlation of androgen receptor gene CAG repeat length and prostate cancer risk in an African-American population. *Clin Prostate Cancer*, 2004. **3**(2): p. 98–103.
65. Powell, I.J., et al. The impact of CAG repeats in exon 1 of the androgen receptor on disease progression after prostatectomy. *Cancer*, 2005. **103**(3): p. 528–37.
66. Kittles, R.A., et al. Extent of linkage disequilibrium between the androgen receptor gene CAG and GGC repeats in human populations: implications for prostate cancer risk. *Hum Genet*, 2001. **109**(3): p. 253–61.
67. Bennett, C.L., et al. Racial variation in CAG repeat lengths within the androgen receptor gene among prostate cancer patients of lower socioeconomic status. *J Clin Oncol*, 2002. **20**(17): p. 3599–604.
68. Ntais, C., A. Polycarpou, and J.P. Ioannidis. Vitamin D receptor gene polymorphisms and risk of prostate cancer: a meta-analysis. *Cancer Epidemiol Biomarkers Prev*, 2003. **12**(12): p. 1395–402.
69. Ntais, C., A. Polycarpou, and J.P. Ioannidis. Association of GSTM1, GSTT1, and GSTP1 gene polymorphisms with the risk of prostate cancer: a meta-analysis. *Cancer Epidemiol Biomarkers Prev*, 2005. **14**(1): p. 176–81.
70. Amundadottir, L.T., et al. A common variant associated with prostate cancer in European and African populations. *Nat Genet*, 2006. **38**(6): p. 652–8.
71. Freedman, M.L., et al. Admixture mapping identifies 8q24 as a prostate cancer risk locus in African-American men. *Proc Natl Acad Sci USA*, 2006. **103**(38): p. 14068–73.
72. Haiman, C.A., et al. Multiple regions within 8q24 independently affect risk for prostate cancer. *Nat Genet*, 2007. **39**(5): p. 638–44.
73. Yeager, M., et al. Genome-wide association study of prostate cancer identifies a second risk locus at 8q24. *Nat Genet*, 2007. **39**(5): p. 645–49.
74. Severi, G., et al. The common variant rs1447295 on chromosome 8q24 and prostate cancer risk: results from an Australian population-based case-control study. *Cancer Epidemiol Biomarkers Prev*, 2007. **16**(3): p. 610–2.
75. Wang, L., et al. Two common chromosome 8q24 variants are associated with increased *risk for* prostate cancer. *Cancer Res* , 2007. **67**(7): p. 2944–50.
76. Schumacher, F.R., et al. A common 8q24 variant in prostate and breast cancer from a large nested case-control study.*Cancer Res*, 2007. **67**(7): p. 625–9.
77. Suuriniemi, M., et al. Confirmation of a positive association between prostate cancer risk and a locus at chromosome 8q24. *Cancer Epidemiol Biomarkers Prev*, 2007. **16**(4): p. 809–14.
78. Edwards, S.M., et al. Two percent of men with early-onset prostate cancer harbor germline mutations in the BRCA2 gene. *Am J Hum Genet*, 2003. **72**(1): p. 1–12.
79. Kirchhoff, T., et al. BRCA mutations and risk of prostate cancer in Ashkenazi Jews. *Clin Cancer Res*, 2004. **10**(9): p. 2918–21.
80. Hamel, N., K. Kotar, and W.D. Foulkes. Founder mutations in BRCA1/2 are not frequent in Canadian Ashkenazi Jewish men with prostate cancer. *BMC Med Genet*, 2003. **4**: p. 7.
81. Angele, S., et al. ATM polymorphisms as risk factors for prostate cancer development. *Br J Cancer*, 2004. **91**(4): p. 783–7.

82. Narla, G., et al. A germline DNA polymorphism enhances alternative splicing of the KLF6 tumor suppressor gene and is associated with increased prostate cancer risk. *Cancer Res*, 2005. **65**(4): p. 1213–22.

83. Jeng, Y.M. and H.C. Hsu. KLF6, a putative tumor suppressor gene, is mutated in astrocytic gliomas. *Int J Cancer*, 2003. **105**(5): p. 625–9.

84. Wang, S.P., X.P. Chen, and F.Z. Qiu. [A candidate tumor suppressor gene mutated in primary hepatocellular carcinoma: kruppel-like factor 6]. *Zhonghua Wai Ke Za Zhi*, 2004. **42**(20): p. 1258–61.

85. Reeves, H.L., et al. Kruppel-like factor 6 (KLF6) is a tumor-suppressor gene frequently inactivated in colorectal cancer. *Gastroenterology*, 2004. **126**(4): p. 1090–103.

86. Narla, G., et al. KLF6, a candidate tumor suppressor gene mutated in prostate cancer. *Science*, 2001. **294**(5551): p. 2563–6.

87. Huusko, P., et al. Nonsense-mediated decay microarray analysis identifies mutations of EPHB2 in human prostate cancer. *Nat Genet*, 2004. **36**(9): p. 979–83.

88. Kittles, R.A., et al. A common nonsense mutation in EphB2 is associated with prostate cancer risk in African American men with a positive family history. *J Med Genet*, 2006. **43**(6): p. 507–11.

89. Hunter, D.J., et al. A candidate gene approach to searching for low-penetrance breast and prostate cancer genes. *Nat Rev Cancer*, 2005. **5**(12): p. 977–85.

4

Prostate Molecular Oncogenesis
Gene Deletions and Somatic Mutations

Edward P. Gelmann, MD

1. INTRODUCTION

Cancers arise from individual cells that have acquired one or more mutations resulting in malignant transformation. Most of these mutations affect genes involved in signaling for cell proliferation, cell cycle control, cell death, and DNA repair. Mutations that activate the function of proteins in signal transduction, advance cell cycle progression, or inhibit apoptosis are found in dominant oncogenes that affect cell phenotype despite the presence of the contralateral normal allele. Suppressor proteins are affected by loss or disruption of genes involved in cell cycle control, apoptosis, or DNA repair. Genes coding for cell surface molecules involved in adhesion or growth inhibition may also have a tumor suppressor role. In a few instances, genes may be haploinsufficient and loss or inactivation of a single allele is sufficient to influence cancer pathogenesis.

From: *Current Clinical Oncology: Prostate Cancer:*
Signaling Networks, Genetics, and New Treatment Strategies
Edited by: R. G. Pestell and M. T. Nevalainen © Humana Press, Totowa, NJ

Carcinogenesis, particularly of epithelial cells that transform to adenocarcinomas, requires the stepwise accumulation of mutations that are characteristic of individual cancer types *(1)*. On the other hand, some hematopoietic malignancies originate after the activation of a single dominant oncogene. In these latter cases, malignant progression and drug resistance are accompanied by the accumulation of additional mutant cancer genes. Different molecular programs can cause a single malignant histologic disorder. For example, acute nonlymphocytic leukemias can have very different clinical behaviors that correlate with different sets of cytogenetic and molecular abnormalities *(2–4)*. In solid tumors such as breast cancer *(5)* and gastrointestinal stromal tumors *(6)*, the presence of specific dominant oncogenes can provide prognostic information and guide the choice of cytotoxic or targeted therapies.

Prostate cancer at presentation has one of the most diverse potential natural histories of any solid tumor. Even in a patient with a 1-cm-diameter tumor, the single best predictor of long-term clinical behavior is histologic grade *(7)*. It is likely that the broad range of clinical outcomes reflects the accumulation of different sets of oncogenic mutations in different prostate cancers. The clinical diversity of prostate cancer predicts for molecular complexity of its pathogenesis. Elucidation of the molecular programs underlying prostate cancers has been difficult because in a single prostatectomy specimen multiple transformed clones may be present, of which only one gave rise to the clinically important cancer *(8)*. Compared with other cancers, the relative inaccessibility of the prostate to image-directed biopsy and the paucity of tissue that can be obtained without organ removal have presented challenges in obtaining adequate preoperative tissue samples and in identifying the clinically important lesions for molecular analysis prior to removal of the entire gland. In about 30% of cases, histologic grade of the cancer in the prostatectomy specimen is worse than grade found in the needle biopsy cores, suggesting that sampling error of biopsy procedures may be too great for reliable molecular studies *(9)*.

Another approach to identifying genes important for prostate cancer has taken advantage of the marked influence of family history on prostate cancer risk. Extensive genetic studies have sought to map alleles that segregate with prostate cancer incidence, particularly among men under 60 years of age among whom sporadic cases are less likely to confound the analysis. A number of genes have been identified as conferring risk of hereditary prostate cancer, but their role as suppressor genes has been difficult to elucidate and each has been found disrupted in only a few pedigrees.

Other work has identified suppressor genes that are mutated or lost in a broader group of prostate cancer cases. The genes will be discussed in this chapter in detail. Also, one family of transcription factor genes is activated by chromosomal translocation in up to 70% of prostate cancers and may represent the first instance of a dominant oncogene associated with this malignancy. The malignant transformation of prostate epithelial cells resulting from loss of suppressor function and activation of a dominant transcription factor requires continuous activation of the androgen receptor (AR). AR activity is essential for the development of prostate cancer and is a target for gene activation late in the disease after treatment with androgen ablative therapy. AR is activated as the final genetic step in the development of androgen-independent prostate cancer by various mechanisms. The *AR* gene may be amplified to overexpress the protein, or it may be mutated to recognize a diverse range of steroids and antiandrogens as agonists *(10)*. AR may also be activated by interaction with cellular kinases activated during malignant progression. The identification of genes targeted for mutations in prostate cancer pathogenesis will provide insight and guidance for therapeutic strategies. This chapter will summarize the somatic mutations that have been found in human prostate cancer.

2. CYTOGENETICS OF PROSTATE CANCER

Specific chromosomal lesions that are found in a substantial number of tumor samples provide important clues to the identity of suppressor genes. Studies of chromosomal gains and losses in primary prostate cancer predominantly found losses with few consistent regions demonstrating increases in copy number *(11)*. By contrast, more regions of DNA gain were found in samples from advanced prostate cancer *(11,12)*. The most common losses in prostate cancer occur at 8p and 13q, the two chromosomal loci that are also most frequently lost in prostatic intraepithelial neoplasia (PIN) implicating these loci at the earliest phases of prostate epithelial transformation *(13)*. Two distinct regions are lost from chromosome 8 at 8p12–p21 and 8p22 *(14–17)*. Several genes in this region are candidate prostate tumor suppressors including *MSR1*, *NKX3.1*, and *TUSC3/N33 (18,19)*. The two former genes will be discussed in sections 3 and 4.2.1., respectively. The latter is a putative suppressor protein by virtue of hypermethylation or loss of heterozygosity (LOH) in a number of tumor types *(20–23)*. *N33* is related to the *OST3* gene family in yeast and may function in oligosaccharide transport. Retinoblastoma (*RB*) is a candidate tumor suppressor gene at 13q as it is frequently deleted in early prostatic tumorigenesis *(24)*. Other chromosomes are commonly involved also. Deletions in up to 45% of prostate tumors were shown to occur on 10q *(12,25)*. Putative suppressors on 10q are *MXI1 (26)* and *PTEN (27–29)*. With a comprehensive array of 1454 single-nucleotide polymorphisms distributed across the genome, 7 regions of significant loss of heterozygosity (LOH) were found by comparing tumor and normal paired DNA samples from 50 prostate cancer specimens. The chromosomal regions implicated in prostate cancer were 1p33–34, 3q27, 8p21, 10q23, 15q12 16q23–24, and 17p13 *(30)*.

Several regions of chromosomal gain have been found, most often in advanced prostate cancer. The region most commonly amplified is the entire

8q, including the locus of the *MYC* gene, that is associated with aggressive disease *(12,25)*. Interestingly, *MYC* is amplified in the LNCaP prostate cancer cell line *(31)*. Another region commonly amplified in advanced disease is *Xq11–13, a locus for AR*. Fluorescent in situ hybridization (FISH) analysis indicated an amplification of *AR* in 30% of hormone refractory prostate cancer *(32,33)* (see Section 3).

3. EPIGENETIC CHANGES THAT PREDISPOSE TO PROSTATE CANCER

Prostate cancer increases in incidence more between the ages of 55 and 80 than any other malignance *(34)*. In the course of aging, there is a cumulative effect of oxidative damage in DNA that can contribute to carcinogenesis. Simultaneously, there may be a reduced expression of genes that code for proteins protective against oxidative damage. Oxidative damage has been proposed as a frequent mechanism to initiate prostate carcinogenesis. Oxidative damage may be linked to dietary factors or to atrophic changes that accompany the aging process as discussed elsewhere in this volume *(35)*. Susceptibility to oxidative damage may be enhanced by methylation of genes that confer protection against oxidation. For example glutathione-S-transferase P1 (*GSTP1*), a gene that codes for GSTπ, an enzyme involved in catalyzing the transfer of protons from reduced glutathione, is a target for promoter hypermethylation in the prostate. Somatic hypermethylation of the *GSTP1* promoter is related to decreased expression of GSTπ in almost all prostate cancers *(36–41)*. *GSTP1* methylation is emerging as a marker for the transition of normal prostate epithelium to PIN as it was found in 6% of prostatic intraepithelial atrophy and 69% of PIN *(42)*.

Oxidative damage may also be enhanced through diminished macrophage activity due to inactivation of the macrophage scavenger receptor 1 (*MSR1*) gene. *MSR1* is located at 8p22 and codes for a receptor that is active as a cell surface trimeric protein to bind a broad range of polyanionic ligands including oxidized low-density lipoprotein *(43,44)*. Thus, compromise of *MSR1* function could expose the prostate and other organs to increased oxidative stress resulting from attenuated macrophage function. Studies have linked germline mutations and polymorphisms of *MSR1* to familial early onset prostate cancer *(14,45,46)*. Moreover, a truncating mutation in *MSR1* that codes for a protein with dominant negative effects on receptor assembly has been found in African Americans with prostate cancer *(47)*. Because *MSR1* does not affect prostate epithelial cells directly, but is presumed to predispose to oxidative damage, the LOD scores even in the positive studies are quite low *(48)* and there are studies of selected populations that failed to identify an association between *MSR1* and prostate cancer risk *(49,50)*.

4. GENETIC CHANGES OF PROSTATE EPITHELIAL CELL TRANSFORMATION

4.1. Dominant Oncogenes

Identification of dominant oncogenes, particularly those activated early in prostate cancer pathogenesis, has been slow to evolve. A report in 2005 identified the first chromosomal translocation activating two members of a transcription factor family. *AR* has long been a dominant gene for prostate cancer but is activated by mutation or amplification predominantly in androgen-independent prostate cancer. The dominant oncogenes are listed in Table 1.

4.1.1. ETS FAMILY TRANSCRIPTION FACTORS

The application of a novel analytical approach to expression array data that compared nonmalignant and malignant prostate tissues showed that two ETS family transcription factors, ERG and ETV1, were disproportionately overexpressed in prostate cancer *(51)*. The overexpression of these genes was mutually exclusive and caused by two separate translocation events that fused a 3′-segment of either gene, including the DNA-binding domain of each, with the 5′-end of *TMPRSS2*. *TMPRSS2* codes for a membrane-bound serine protease that is expressed in a wide range of tissues and is androgen-regulated in the prostate *(52–54)*. Moreover, *TMPRSS2* is overexpressed in prostate cancer cells compared with adjacent normal cells *(54)*. The 5′-end of *TMPRSS2* fused to the *ETS* family genes confers androgen-responsive expression of the chimeric oncogene compared with *ERG* or *ETV1*. *TMPRSS2* itself is overexpressed in prostate cancer and expression is androgen dependent *(54)*. In a single instance, a deletion and premature termination of *TMPRSS2* has been described in a prostate cancer specimen *(54)*. The relationship of this mutation to gene translocation in the same patient is unknown.

Table 1
Dominant Oncogenes in Prostate Cancer

Gene or locus	Chromosomal locus	Function	Citations
ERG-TMPRSS2	21q22.3–22q22.2	ETS family transcription factor	(51)
ERTV1-TMPRSS2	7p21.2–22q22.2	ETS family transcription factor	(51)
PRC17	17q12	GTPase-activating protein	(55)
AR	Xq11–12	Nuclear receptor	(33,74–87,92)

4.1.2. *PRC17*

A single report identified *PRC17* as a gene encoding a GTPase-activating protein (GAP) that was amplified in prostate cancer cell lines and tissues and in some breast cancer cell lines *(55)*. *PRC17* was amplified and overexpressed in both primary and metastatic prostate cancers. The protein was able to inactivate Rab5 GTPase activity *in vitro*. *PRC17* had *in vitro* transforming activity in that expression conferred a growth advantage to NIH/3T3 fibroblasts in medium with low serum concentrations, increased saturation density *in vitro*, and supported the growth of NIH/3T3 xenografts in nude mice. The data are highly suggestive that PRC17 is an oncoprotein in a fraction of prostate cancers, but the initial report has not been substantiated by additional published data.

4.1.3. *AR*

Androgens are essential for growth and development of the prostate, and they serve as the essential survival factor for prostate epithelial cells *(56)*. The AR is a member of the steroid and thyroid hormone receptor gene super-family and is coded by a gene located at chromosome Xq11–12. This nuclear receptor mediates hormone action by binding to hormone in the cytoplasm and translocating to the nucleus where it dimerizes in the course of DNA binding to initiate formation of a transcriptional complex at an androgen-responsive gene promoter *(57)*.

The first exon of the *AR* gene that codes for the N-terminal transcriptional activating domain of the protein contains a trinucleotide CAG repeat that varies in length as is typical for these sequences due to DNA polymerase slippage during replication. There is an inverse relationship between the length of the CAG repeat, that codes for a polyglutamine stretch, and *AR* transcriptional activity *(58–60)*. The average CAG repeat length in *AR* is 21, but there is some difference in modal number between the different races which is thought to affect prostate cancer incidence rates *(61,62)*. Shorter CAG repeat length is a risk factor for prostate cancer and advanced prostate cancer, particularly in men diagnosed before widespread use of prostate-specific antigen (PSA) screening that has increased the likely rate of overdiagnosis *(63–68)*. Reduced CAG repeat length also predisposes to prostate cancer recurrence and early onset of the disease *(69–71)*. As prostate cancer screening became more widely used, the effect of CAG repeat length on risk and prognosis diminished *(70,72,73)*.

In androgen-independent advanced prostate cancer, *AR* plays the role of a dominant oncogene. In the presence of castrate levels of androgen, *AR* activity is enhanced by gene amplification in approximately one-third of tumors *(33,74–77)*. AR expression is sustained throughout the clinical course of prostate cancer, including the most advanced phases of androgen-independent disease *(63)*. AR is also activated by missense mutation *(78–87)* that broadens the scope of hormone specificity and/or enhances hormonal response *(88–90)*. The use of antiandrogens increases the likelihood of selecting mutant receptors with resistance to the effects of these drugs *(91,92)*. In other cases, mutations can be found in the N-terminal domain that may affect the transactivating potential of the receptor *(93)*. In a few cases, the CAG repeats have been affected by mutations in androgen-independent prostate cancer that have either truncated or fragmented the repeat sequence, thereby causing hyperactivation of the receptor *(94–96)*.

Activation of AR in androgen-independent disease may also be accomplished by activation of co-activators such as p160 family members *(97)* and β-catenin, which is rarely mutated in prostate cancer *(98,99)*, but is truncated by proteolytic cleavage *(100)* and can activate AR, cause its colocalization to the nucleus *(101–103)*, and enhance hormone sensitivity *(104)*. It is likely that there will be other mechanisms of AR activation in androgen-independent prostate cancer such as phosphorylation by MAP kinase *(105)* and activation by HER family kinases *(106)*.

4.2. Tumor Suppressor Genes

Guided in part by cytogenetic data, a number of suppressor genes have been identified in substantial fractions of prostate cancers (Table 2).

Table 2
Suppressor Genes in Prostate Cancer

Gene	Locus	Function	References
NKX3.1	8p21	Encodes a prostate-specific homeobox gene that is active in regulation of prostate development	*(14,19,107–110, 112–114,117,118, 208,209)*
PTEN	10q23	Encodes a lipid phosphatase that negatively regulates the phosphatidylinositol 3-kinase pathway	*(27–29,116,130–139,141,142)*
CDKN1B	12p11–13	Encodes p27, a cell cycle inhibitor	*(144–153,155,156)*
ATBF1	16q22	Encodes a cell cycle regulator	*(157,158,162,210)*
KLF6	10p15	Encodes a Kruppel-like zinc finger transcription factor	*(171–175)*
RB	13q14.2	Cell cycle regulator	*(176–179)*
P53	17p13.1	Cell cycle checkpoint	*(173,186,196–203)*

4.2.1. *NKX3.1*

NKX3.1 is a homeobox gene with prostate-specific expression that maps to 8p21, a region that undergoes loss of heterozygosity in up to 85% of prostate cancers *(19,107,108)*. Expression of *NKX3.1* is highly organ specific and is restricted to prostatic lobes and bulbourethral gland in the mouse *(109)*. Expression of *NKX3.1* in human is found in prostate luminal epithelial cells, testis, and isolated cells in the ureter and peribronchial mucous glands *(110)*. Loss of heterozygosity in prostate cancer occurs frequently at 8p12–22 as an early event *(14,111,112)*. *NKX3.1* is located centrally within the minimally deleted region of the chromosome *(113)*. Inconsistent with the suppressor gene paradigm, *NKX3.1* does not undergo somatic mutation in human prostate cancer *(107,114,115)*.

The notion that *NKX3.1* may be haploinsufficient like some other suppressor genes came from observations of gene-targeted mice. *NKX3.1* heterozygosity predisposes to prostate epithelial dysplasia with somewhat longer latency than seen in *NKX3.1$^{-/-}$* mice and can co-operate with other oncogenic mutations to augment prostate carcinogenesis *(109,116)*. *NKX3.1* heterozygosity is accompanied by decreased expression of genes under the regulation of the *NKX3.1* homeoprotein *(117)*. In human prostate cancer, *NKX3.1* was observed to have decreased expression and, in some cases, cytoplasmic localization *(116)*. *NKX3.1* protein expression is reduced to a median level of 67% of normal in primary prostate cancer, but expression varies over a wide range with respect to adjacent normal epithelium *(118)*. Moreover, expression levels correlate with loss of heterozygosity and nonconventional gene methylation and expression was below median levels when both LOH and methylation were found. *NKX3.1* protein expression levels were also reduced in PIN comparable with reductions in adjacent invasive prostate cancer. Complete loss of *NKX3.1* expression is seen with tumor progression in metastatic prostate cancer *(110)*.

The DNA-binding domain of the homeoprotein is the 60-amino acid homeodomain whose three-dimensional structure includes two parallel and one perpendicular helix by which these proteins bind to DNA and to other proteins *(119)*. N- and C-terminal regions that flank the homeodomain are important modulators of protein function. *NKX3.1* belongs to the NK family of homeodomain proteins that have organ-specific expression in the adult. For example, *NKX2.5*, the human cardiac-specific homolog of the *Drosophila tinman* gene, is subject to a variety of autosomal dominant point mutations, many outside the homeodomain, that determine hereditary cardiac abnormalities *(120)*. *NKX2.1* is thyroid transcription factor-1 that determines lung and thyroid development. Inherited mutations resulting in *NKX2.1* haploinsufficiency determine congenital hypothyroidism, choreoathetosis, muscular hypotonia, and pulmonary problems *(121,122)*. It appears that sporadic loss of *NKX3.1* in prostate epithelial cells contributes to cell transformation due to haploinsufficiency, a characteristic that may be common to *NK* homeodomain gene family members.

Further evidence that *NKX3.1* is a haploinsufficient prostate suppressor gene comes from the finding that a family was identified with a germline *NKX3.1* (T164A) mutation that segregated with early onset prostate cancer *(123)*. The *NKX3.1* T164A mutation affects homeodomain position 41, which is occupied by a threonine and is well conserved among other NK-family homeodomain sequences *(124)*. Threonine 41 is located at the N-terminal cap position of the third helix in the *NKX3.1* homeodomain and is important for maintenance of the helical structure *(125,126)*. Even though threonine 41 is not involved in direct hydrogen bonding with the *NKX3.1* DNA recognition sequence, it is likely to affect helix 3 structure and both DNA and protein interactions. The homologous threonine 41 in the human *tinman* homolog *NKX2.5* is mutated to methionine in a family with hereditary atrial septal defect and atrioventricular block *(127)*. The *NKX2.5* (T178M), homeodomain T41M, mutation causes marked reduction in DNA-binding affinity of *NKX2.5* but does not affect complex formation with either wild-type *NKX2.5* or with *GATA4*. Similarly, *NKX3.1* (T164A) has altered physical properties and 5% of the normal affinity for the cognate *NKX3.1* DNA recognition sequence *(123,128)*.

A polymorphic *NKX3.1* allele has been described (C154T) that codes for a variant *NKX3.1*(R52C) protein with a nonconservative amino acid change at residue 52. The variant protein has altered *in vitro* and *in vivo* phosphorylation at the major phosphorylation site, serine 48. Phosphorylation of *NKX3.1* at serine 48 has been shown to affect DNA binding *in vitro* *(129)*. This polymorphism is present in 11% of the population without regard to race and is a risk factor for prostate cancer, aggressive prostate cancer, and prostatic enlargement *(129)*.

4.2.2. PTEN

The phosphatase and tensin homolog (*PTEN*) is a tumor suppresser gene that maps to 10q23. *PTEN* is a common target for deletion and downregulation in prostate cancers *(28,29,130–134)*. The PTEN protein is a lipid phosphatase that suppresses the effects of phosphoinositol-3-kinase by dephosphorylating phosphatidylinositol-3,4,5 tris phosphate, a lipid anchor for the Akt kinase at the inner plasma membrane *(135)*. Thus the action of PTEN downregulates Akt and decreases phosphorylation of its targets, inhibiting antiapoptotic and survival signals. In addition to deletion and mutation affecting *PTEN* expression, the gene is also hypermethylated in prostate cancer *(136)*. Loss of PTEN expression is more easily detected in advanced stage and high-grade prostate tumors *(137)*. Only a small fraction of primary prostate cancers have *PTEN* deletions and mutations, but more than half of all metastatic lesions have *PTEN* gene alterations *(27,138)*.

The role of *PTEN* loss in prostate transformation is underscored by targeted deletion of *PTEN* in mouse models. *PTEN* heterozygosity potentiated prostate cancer formation under influence of the SV40 T antigen *(139)*. *PTEN* heterozygosity by itself was seen to cause prostatic neoplasia in a subset of mice with longer latency than other malignancies *(140)*. Conditional deletion of *PTEN* driven by the prostate-specific probasin promoter caused a high rate of prostatic neoplasia *(141)*. Simultaneous loss of *NKX3.1* and *PTEN* results in more extensive and more aggressive prostate cancer than caused by *PTEN* loss alone *(116,142)*. Lastly, targeted expression of a constitutively active *Akt* gene in the mouse resulted in the development of prostate cancer, suggesting that the PI3 kinase pathway is a common mediator of prostatic neoplasia *(143)*.

4.2.3. CDKN1B

CDKN1B encodes the cyclin-dependent kinase inhibitor, p27^{KIP1}, a member of the Cip/Kip family of cell cycle inhibitors. p27^{KIP1} has been shown to have tumor suppressor properties in mouse model systems *(144)*. Gene targeting experiments suggest that *CDKN1B* is haploinsufficient for tumor suppression in mice *(145)*. In human prostate cancer, reduced or absent p27^{KIP1} expression correlates with high tumor grade and reduced disease-free survival *(146–148)*. Importantly, PI3K pathway activation decreases p27^{KIP1} expression, an effect that is blocked by PTEN. Conversely, PTEN suppression of Akt activation increases levels of p27^{KIP1} *(149)*. It was also shown that *PTEN* and *CDKN1B* alleles interacted by performing linkage analysis of each allele in a cohort of prostate cancer families, suggesting that these two genes may co-operate in affecting prostate cancer risk *(150)*. Further support for the association of p27^{KIP1} loss with prostate cancer arises from the finding that loss of heterozygosity of the 12p11–13 region that contains the *CDKN1B* locus in 50% of prostate cancer cases *(151,152)*. In some tissues homozygous deletion of *CDKN1B* was found. Lesions in the *CDKN1B* gene are found only infrequently, underscoring that reduced p27^{KIP1} expression is probably the main mechanism for the involvement of this tumor suppressor in prostate cancer pathogenesis, and suggesting that other mechanisms may also effect p27^{KIP1} protein downregulation *(153)*.

In mouse models, p27 is haploinsufficient such that heterozygosity is associated with prostatic epithelial hyperplasia that is more severe in *CDKN1B*$^{-/-}$ animals *(154,155)*. Loss of murine *CDKN1B* shortens the latency of prostatic hyperplasia and dysplasia caused by loss of *NKX3.1*, but combined loss of *CDKN1B* and *NKX3.1* causes only a small enhancement of prostatic neoplasia seen with loss of *NKX3.1* alone *(155)*. On the other hand, reduced p27^{kip1} is markedly potentiating of *PTEN* loss to result in aggressive prostate cancer in the mouse *(156)*.

4.2.4. *ATFB1*

Fine structure analysis of the 16q deletion in prostate cancer cell lines and xenografts led to the identification of somatic mutations in the *ATBF1* gene. The gene codes for an AT binding factor that is a multiple homeodomain-zinc finger protein that suppresses α-fetoprotein expression and interacts with the MYB transcription factor *in vitro (157,158)*. ATBF1 binds to AT-rich regions that are recognized by hepatocyte nuclear factor 1 (HNF1) and downregulates HNF1-induced expression *(157)*. Its action may also be related to activation of p21^{CIP1} *(159)*. Expression of *ATBF1* mRNA is associated with better prognosis in breast cancer *(160)* and induces growth arrest and differentiation of neuronal cells *(161)*. *In vitro* the effects of *ATBF1* are consistent with the activities of a tumor suppressor.

A wide variety of missense mutations, deletions, and genetic variations of *ATBF1* were found in prostate cancer cell lines and tissues *(162)*. Moreover, 36% of tumors tested had missense mutations likely to inactivate the protein function. Several in-frame deletions were found in glutamine-rich regions. Some mutations found in cell lines affected the generation of alternate splice variants. Knock down of *ATBF1* expression in PC-3 prostate cancer cells that have intact *ATBF1* genes resulted in growth acceleration. On the other hand, restoration of expression in *ATBF1* mutant cells decreased growth and colony formation.

4.2.5. KLF6

The Kruppel-like zinc finger transcription factor 6 (KLF6) is yet a third DNA-binding transcription factor that has been implicated in prostate cancer pathogenesis by loss of heterozygosity. KLF6 contains three carboxy-terminal zinc-binding cysteine-rich domains that interact with DNA at a GC box promoter element *(163)*. Some transcriptional targets of KLF6 include placental glycoprotein *(164)*, collagen I *(163)*, urokinase-type plasminogen activator (*uPA*) *(165)*, and transforming growth factor β1 and the types I and II TGFβ receptors *(166)*. KLF6 may also interfere with the functions of the JUN oncoprotein *(167)*. KLF6 also interacts with cyclinD1 to interfere with cell cycle signals and reduce phosphorylation of RB *(168)*.

The *KLF6* gene maps to human chromosome 10p15 and is included in a region deleted in about 55% of prostate adenocarcinomas *(169,170)*. *KLF6* undergoes loss of heterozygosity in about one fifth of prostate cancer specimens *(171)*. *KLF6* point mutations and reduced KLF6 protein expression were found in a small collection of high-grade prostate cancers *(171,172)*, but not in a second series *(173)*. Moreover, there has been substantial variation in the frequency with which *KLF6* missense coding mutations have been found in prostate cancer tissues *(171,172)*.

Further evidence for the association of *KLF6* with prostate cancer comes from the report that a *KLF6* germline polymorphism that affects alternate splicing preferences was associated with an increased risk of prostate cancer *(174)*. *In vitro* studies suggested that *KLF6* and one of its splice variants had opposite effects on cell proliferation *(175)*. The question whether *KLF6* is a tumor suppressor for prostate cancer is open pending confirmatory experiments in different laboratories and examination of the effects of *KLF6* deletion in the murine prostate.

4.2.6. RETINOBLASTOMA

The RB protein functions as a central regulator of the cell cycle and is the target of cyclin-dependent kinases 4 and 6 that phosphorylate RB to release the E2F transcription factor. *RB* probably does not have a role as a suppressor of early prostatic epithelial transformation because *RB* mutant mice are not predisposed to prostate cancer and families with inherited *RB* deficiency are not predisposed to early prostate cancer. However, loss of heterozygosity of the *RB* locus is seen in a subset of human prostate cancers and may occur sporadically as a genetic event during tumor progression *(176–179)*. The potential for *RB* loss to contribute to prostate epithelial cell transformation was underscored by the observation that conditional deletion of *RB* resulted in epithelial transformation and early invasiveness in the murine prostate *(180)*.

4.2.7. P53

P53, the suppressor most frequently lost in human cancer, has been studied extensively in prostate cancer. Like *RB*, *P53* loss does not predispose to prostate cancer either in mouse models or in Li–Fraumeni Syndrome families with one inactive *P53* allele *(181,182)*. As with many solid tumors, P53 immunohistochemical detection has been used as a surrogate for the presence of mutant P53 protein. However, mutational analysis of DNA has shown that only 50% of the P53-positive tissues could be shown to harbor *P53* mutations *(183)*. The yield can be increased by adjusting the staining conditions and analyzing mRNA, but that requires access to fresh tissue *(184)*. P53 staining in primary prostate cancer is focal and only very rarely present uniformly throughout the malignant portion of the gland *(185,186)*. Immunohistochemical staining for P53 may have biological significance even if mutations cannot be detected because a large number of reports have identified even focal P53 staining as a poor prognostic marker in prostate cancer *(187–194)*. Moreover, P53 staining in prostatic needle biopsies predicts early recurrence after radiation therapy *(195)*.

DNA analysis has suggested that the frequency of detectable *P53* mutations in primary prostate cancer is in the range of 10.5–32% *(173,186,196–203)*. The fraction of samples with detectable *P53* mutations is higher in recurrent prostate cancer *(188)*. The presence of focal P53 staining may reflect the presence of clonal variants within a prostate cancer. These variants may, in fact, be precursors of metastases or recurrent disease *(204–206)*. Because of the heterogeneity of prostate cancer and the focal nature of *P53*-positive cells, single-strand conformational polymorphism (SSCP) studies may have grossly underestimated the fraction of specimens that harbor *P53*-mutant cells. Moreover, *P53* mutant cells may have a higher likelihood of being resistant to radiation and being the precursors of metastatic foci *(207)*.

5. SUMMARY

Prostate cancer is a diverse disease with a wide range of clinical behaviors. Prostate cancer increases in incidence with age more than any other cancer type. Epigenetic events including gene methylation presage the development of prostate cancer by decreasing the cell's defenses against oxidative damage. One of the earliest oncogenic events is a decrease in expression of the tumor suppressor NKX3.1. The decline in levels of this homeoprotein may augment the effects of one of two ETS family transcription factors that in a large fraction of cases are activated by translocation and fusion with the 5′-end of the androgen responsive protease gene *TMPRSS2*. Decreased expression of the cyclin-dependent kinase inhibitor p27^{KIP1} also plays a role in prostate cancer progression. Additional suppressor genes that are inactivated are the transcription factors *ATBF1* and *KLF6*. Loss of RB expression and mutation of P53 are both late events in prostate cancer and correlate with relapsed or aggressive disease. In patients with androgen-independent prostate cancer, the AR itself acts as a dominant oncogene and through a variety of mechanisms mediates signal transduction despite castrate circulating levels of testosterone.

ACKNOWLEDGMENTS

This work was supported by US Public Health Service grants ES09888 and CA96854 to E.P.G.

REFERENCES

1. Kinzler, K. W. and Vogelstein, B. (1996) Lessons from hereditary colorectal cancer. *Cell* **87**, 159–170.
2. Mrozek, K., Heerema, N. A. and Bloomfield, C. D. (2004) Cytogenetics in acute leukemia. *Blood Rev.* **18**, 115–136.
3. Bullinger, L. and Valk, P. J. (2005) Gene expression profiling in acute myeloid leukemia. *J. Clin. Oncol.* **23**, 6296–6305.
4. Avivi, I. and Rowe, J. M. (2005) Prognostic factors in acute myeloid leukemia. *Curr. Opin. Hematol.* **12**, 62–67.

5. Muss, H. B., Thor, A. D., Berry, D. A., Kute, T., Liu, E. T., Koerner, F., Cirrincione, C. T., Budman, D. R., Wood, W. C. and Barcos, M. (1994) c-erbB-2 expression and response to adjuvant therapy in women with node-positive early breast cancer. *N. Engl. J. Med.* **330**, 1260–1266.

6. Demetri, G. D., von Mehren, M., Blanke, C. D., Van den Abbeele, A. D., Eisenberg, B., Roberts, P. J., Heinrich, M. C., Tuveson, D. A., Singer, S., Janicek, M., Fletcher, J. A., Silverman, S. G., Silberman, S. L., Capdeville, R., Kiese, B., Peng, B., Dimitrijevic, S., Druker, B. J., Corless, C., Fletcher, C. D. and Joensuu, H. (2002) Efficacy and safety of imatinib mesylate in advanced gastrointestinal stromal tumors. *N. Engl. J. Med.* **347**, 472–480.

7. Albertsen, P. C., Hanley, J. A. and Fine, J. (2005) 20-year outcomes following conservative management of clinically localized prostate cancer. *JAMA* **293**, 2095–2101.

8. Bostwick, D. G., Shan, A., Qian, J., Darson, M., Maihle, N. J., Jenkins, R. B. and Cheng, L. (1998) Independent origin of multiple foci of prostatic intraepithelial neoplasia: comparison with matched foci of prostate carcinoma. *Cancer* **83**, 1995–2002.

9. Steinberg, D. M., Sauvageot, J., Piantadosi, S. and Epstein, J. I. (1997) Correlation of prostate needle biopsy and radical prostatectomy Gleason grade in academic and community settings. *Am. J. Surg. Pathol.* **21**, 566–576.

10. Gelmann, E. P. (2002) Molecular biology of the androgen receptor. *J. Clin. Oncol.* **20**, 3001–3015.

11. Visakorpi, T., Kallioniemi, A. H., Syvanen, A. C., Hyytinen, E. R., Karhu, R., Tammela, T., Isola, J. J. and Kallioniemi, O. P. (1995) Genetic changes in primary and recurrent prostate cancer by comparative genomic hybridization. *Cancer Res.* **55**, 342–347.

12. Nupponen, N. N., Kakkola, L., Koivisto, P. and Visakorpi, T. (1998) Genetic alterations in hormone-refractory recurrent prostate carcinomas. *Am. J. Pathol.* **153**, 141–148.

13. Nupponen, N. N. and Visakorpi, T. (2000) Molecular cytogenetics of prostate cancer. *Microsc. Res. Tech.* **51**, 456–463.

14. Bova, G. S., Carter, B. S., Bussemakers, J. G., Emi, M., Fujiwara, Y., Kyprianon, N., Jacobs, S. C., Robinson, J. C., Epstein, J. I., Walsh, P. C. and Issacs, W. B. (1993) Homozygous deletion and frequent allelic loss of chromosome 8p22 loci in human prostate cancer. *Cancer Res.* **53**, 3869–3873.

15. Trapman, J., Sleddens, H. F., vanderWeiden, M. M., Dinjens, W. N., Konig, J. J., Schroder, F. H., Faber, P. W. and Bosman, F. T. (1994) Loss of heterozygosityof chromosome 8 microsatellite loci implicates a candidate tumor suppressor gene between the loci D8S87 and D8S133 in human prostate cancer. *Cancer Res.* **54**, 6061–6064.

16. Emmert-Buck, M. R., Vocke, C. D., Pozzatti, R. O., Duray, P. H., Jennings, S. B., Florence, C. D., Bostwick, D. G., Liotta, L. A. and Linehan, W. M. (1995) Allelic loss on chromosome 8p12–21 in microdissected prostate intraepithelial neoplasia. *Cancer Res.* **55**, 2959–2962.

17. Micale, M. A., Mohamed, A., Sakr, W., Powell, I. J. and Wolman, S. R. (1992) Cytogenetics of primary prostate adenocarcinoma Clonality and chromosome instability. *Cancer Genet. Cytogenet.* **61**, 165–173.

18. Bookstein, R., Bova, G. S., MacGrogan, D., Levy, A. and Isaacs, W. B. (1997) Tumour-suppressor genes in prostatic oncogenesis: a positional approach. *Br. J. Urol.* **79 Suppl 1**, 28–36.

19. He, W. W., Sciavolino, P. J., Wing, J., Augustus, M., Hudson, P., Meissner, P. S., Curtis, R. T., Shell, B. K., Bostwick, D. G., Tindall, D. J., Gelmann, E. P., Abate-Shen, C. and Carter, K. C. (1997) A novel human prostate-specific, androgen-regulated homeobox gene (NKX3.1) that maps to 8p21, a region frequently deleted in prostate cancer. *Genomics* **43**, 69–77.

20. Li, Q., Jedlicka, A., Ahuja, N., Gibbons, M. C., Baylin, S. B., Burger, P. C. and Issa, J. P. (1998) Concordant methylation of the ER and N33 genes in glioblastoma multiforme. *Oncogene* **16**, 3197–3202.

21. Pils, D., Horak, P., Gleiss, A., Sax, C., Fabjani, G., Moebus, V. J., Zielinski, C., Reinthaller, A., Zeillinger, R. and Krainer, M. (2005) Five genes from chromosomal band 8p22 are significantly down-regulated in ovarian carcinoma. *Cancer* **104**, 2417–2429.

22. Ahmed, M. N., Kim, K., Haddad, B., Berchuck, A. and Qumsiyeh, M. B. (2000) Comparative genomic hybridization studies in hydatidiform moles and choriocarcinoma: amplification of 7q21–q31 and loss of 8p12–p21 in choriocarcinoma. *Cancer Genet. Cytogenet.* **116**, 10–15.

23. Levy, A., Dang, U. C. and Bookstein, R. (1999) High-density screen of human tumor cell lines for homozygous deletions of loci on chromosome arm 8p. *Genes Chromosomes Cancer* **24**, 42–47.

24. Phillips, S. M., Barton, C. M., Lee, S. J., Morton, D. G., Wallace, D. M., Lemoine, N. R. and Neoptolemos, J. P. (1994) Loss of the retinoblastoma susceptibility gene (RB1) is a frequent and early event in prostatic tumorigenesis. *Br. J. Cancer* **70**, 1252–1257.

25. Cher, M. L., Bova, G. S., Moore, D. H., Small, E. J., Carroll, P. R., Pin, S. S., Epstein, J. I., Isaacs, W. B. and Jensen, R. H. (1996) Genetic alterations in untreated metastases and androgen-independent prostate cancer detected by comparative genomic hybridization and allelotyping. *Cancer Res.* **56**, 3091–3102.

26. Eagle, L. R., Yin, X., Brothman, A. R., Williams, B. J., Atkin, N. B. and Prochownik, E. V. (1995) Mutation of the MXI1 gene in prostate cancer. *Nat. Genet.* **9**, 249–255.

27. Cairns, P., Okami, K., Halachmi, S., Halachmi, N., Esteller, M., Herman, J. G., Jen, J., Isaacs, W. B., Bova, G. S. and Sidransky, D. (1997) Frequent inactivation of PTEN/MMAC1 in primary prostate cancer. *Cancer Res.* **57**, 4997–5000.

28. Feilotter, H. E., Nagai, M. A., Boag, A. H., Eng, C. and Mulligan, L. M. (1998) Analysis of PTEN and the 10q23 region in primary prostate carcinomas. *Oncogene* **16**, 1743–1748.

29. Pesche, S., Latil, A., Muzeau, F., Cussenot, O., Fournier, G., Longy, M., Eng, C. and Lidereau, R. (1998) PTEN/MMAC1/TEP1 involvement in primary prostate cancers. *Oncogene* **16**, 2879–2883.

30. Lieberfarb, M. E., Lin, M., Lechpammer, M., Li, C., Tanenbaum, D. M., Febbo, P. G., Wright, R. L., Shim, J., Kantoff, P. W., Loda, M., Meyerson, M. and Sellers, W. R. (2003) Genome-wide loss of heterozygosity analysis from laser capture microdissected prostate cancer using single nucleotide polymorphic allele (SNP) arrays and a novel bioinformatics platform dChipSNP. *Cancer Res.* **63**, 4781–4785.

31. Nag, A. and Smith, R. G. (1989) Amplification, rearrangement, and elevated expression of c-myc in the human prostatic carcinoma cell line LNCaP. *Prostate* **15**, 115–122.

32. Koivisto, P., Kononen, J., Palmberg, C., Tammela, T., Hyytinen, E., Isola, J., Trapman, J., Cleutjens, K., Noordzij, A., Visakorpi, T. and Kallioniemi, O. P. (1997) Androgen receptor gene amplification: a possible molecular mechanism for androgen deprivation therapy failure in prostate cancer. *Cancer Res.* **57**, 314–319.

33. Visakorpi, T., Hyytinen, E., Koivisto, P., Tanner, M., Keinanen, R., Palmberg, C., Palotie, A., Tammela, T., Isola, J. and Kallioniemi, O. P. (1995) In vivo amplification of the androgen receptor gene and progression of human prostate cancer. *Nat. Genet.* **9**, 401–406.

34. Edwards, B. K., Brown, M. L., Wingo, P. A., Howe, H. L., Ward, E., Ries, L. A., Schrag, D., Jamison, P. M., Jemal, A., Wu, X. C., Friedman, C., Harlan, L., Warren, J., Anderson, R. N. and Pickle, L. W. (2005) Annual report to the nation on the status of cancer, 1975–2002, featuring population-based trends in cancer treatment. *J. Natl. Cancer Inst.* **97**, 1407–1427.

35. Nelson, W. G., De Marzo, A. M. and Isaacs, W. B. (2003) Prostate cancer. *N. Engl. J. Med.* **349**, 366–381.

36. Miller, J. R., Hocking, A. M., Brown, J. D. and Moon, R. T. (1999) Mechanism and function of signal transduction by the Wnt/beta-catenin and Wnt/Ca2+ pathways. *Oncogene* **18,** 7860–7872.

37. Lin, X., Tascilar, M., Lee, W. H., Vles, W. J., Lee, B. H., Veeraswamy, R., Asgari, K., Freije, D., van Rees, B., Gage, W. R., Bova, G. S., Isaacs, W. B., Brooks, J. D., DeWeese, T. L., De Marzo, A. M. and Nelson, W. G. (2001) GSTP1 CpG island hypermethylation is responsible for the absence of GSTP1 expression in human prostate cancer cells. *Am. J. Pathol.* **159,** 1815–1826.

38. De Marzo, A. M., Meeker, A. K., Zha, S., Luo, J., Nakayama, M., Platz, E. A., Isaacs, W. B. and Nelson, W. G. (2003) Human prostate cancer precursors and pathobiology. *Urology* **62,** 55–62.

39. Lee, W. H., Morton, R. A., Epstein, J. I., Brooks, J. D., Campbell, P. A., Bova, G. S., Hsieh, W. S., Isaacs, W. B. and Nelson, W. G. (1994) Cytidine methylation of regulatory sequences near the pi-class glutathione S-transferase gene accompanies human prostatic carcinogenesis. *Proc. Natl. Acad. Sci. U.S.A.* **91,** 11733–11737.

40. Santourlidis, S., Florl, A., Ackermann, R., Wirtz, H. C. and Schulz, W. A. (1999) High frequency of alterations in DNA methylation in adenocarcinoma of the prostate. *Prostate* **39,** 166–174.

41. Woodson, K., Hayes, R., Wideroff, L., Villaruz, L. and Tangrea, J. (2003) Hypermethylation of GSTP1, CD44, and E-cadherin genes in prostate cancer among US Blacks and Whites. *Prostate* **55,** 199–205.

42. Bastian, P. J., Ellinger, J., Schmidt, D., Wernert, N., Wellmann, A., Muller, S. C. and von, R. A. (2004) GSTP1 hypermethylation as a molecular marker in the diagnosis of prostatic cancer: is there a correlation with clinical stage, Gleason grade, PSA value or age? *Eur. J. Med. Res.* **9,** 523–527.

43. Kodama, T., Freeman, M., Rohrer, L., Zabrecky, J., Matsudaira, P. and Krieger, M. (1990) Type I macrophage scavenger receptor contains alpha-helical and collagen-like coiled coils. *Nature* **343,** 531–535.

44. Platt, N. and Gordon, S. (2001) Is the class A macrophage scavenger receptor (SR-A) multifunctional?–the mouse's tale. *J. Clin. Invest.* **108,** 649–654.

45. Xu, J., Zheng, S. L., Komiya, A., Mychaleckyj, J. C., Isaacs, S. D., Chang, B., Turner, A. R., Ewing, C. M., Wiley, K. E., Hawkins, G. A., Bleecker, E. R., Walsh, P. C., Meyers, D. A. and Isaacs, W. B. (2003) Common sequence variants of the macrophage scavenger receptor 1 gene are associated with prostate cancer risk. *Am. J. Hum. Genet.* **72,** 208–212.

46. Xu, J., Zheng, S. L., Komiya, A., Mychaleckyj, J. C., Isaacs, S. D., Hu, J. J., Sterling, D., Lange, E. M., Hawkins, G. A., Turner, A., Ewing, C. M., Faith, D. A., Johnson, J. R., Suzuki, H., Bujnovszky, P., Wiley, K. E., DeMarzo, A. M., Bova, G. S., Chang, B., Hall, M. C., McCullough, D. L., Partin, A. W., Kassabian, V. S., Carpten, J. D., Bailey-Wilson, J. E., Trent, J. M., Ohar, J., Bleecker, E. R., Walsh, P. C., Isaacs, W. B. and Meyers, D. A. (2002) Germline mutations and sequence variants of the macrophage scavenger receptor 1 gene are associated with prostate cancer risk. *Nat. Genet.* **32,** 321–325.

47. Miller, D. C., Zheng, S. L., Dunn, R. L., Sarma, A. V., Montie, J. E., Lange, E. M., Meyers, D. A., Xu, J. and Cooney, K. A. (2003) Germ-line mutations of the macrophage scavenger receptor 1 gene: association with prostate cancer risk in African-American men. *Cancer Res.* **63,** 3486–3489.

48. Wiklund, F., Jonsson, B. A., Goransson, I., Bergh, A. and Gronberg, H. (2003) Linkage analysis of prostate cancer susceptibility: confirmation of linkage at 8p22–23. *Hum. Genet.* **112,** 414–418.

49. Seppala, E. H., Ikonen, T., Autio, V., Rokman, A., Mononen, N., Matikainen, M. P., Tammela, T. L. and Schleutker, J. (2003) Germ-line alterations in MSR1 gene and prostate cancer risk. *Clin. Cancer Res.* **9,** 5252–5256.

50. Wang, L., McDonnell, S. K., Cunningham, J. M., Hebbring, S., Jacobsen, S. J., Cerhan, J. R., Slager, S. L., Blute, M. L., Schaid, D. J. and Thibodeau, S. N. (2003) No association of germline alteration of MSR1 with prostate cancer risk. *Nat. Genet.* **35**, 128–129.

51. Tomlins, S. A., Rhodes, D. R., Perner, S., Dhanasekaran, S. M., Mehra, R., Sun, X. W., Varambally, S., Cao, X., Tchinda, J., Kuefer, R., Lee, C., Montie, J. E., Shah, R. B., Pienta, K. J., Rubin, M. A. and Chinnaiyan, A. M. (2005) Recurrent fusion of TMPRSS2 and ETS transcription factor genes in prostate cancer. *Science* **310**, 644–648.

52. Paoloni-Giacobino, A., Chen, H., Peitsch, M. C., Rossier, C. and Antonarakis, S. E. (1997) Cloning of the TMPRSS2 gene, which encodes a novel serine protease with transmembrane, LDLRA, and SRCR domains and maps to 21q22.3. *Genomics* **44**, 309–320.

53. Lin, B., Ferguson, C., White, J. T., Wang, S., Vessella, R., True, L. D., Hood, L. and Nelson, P. S. (1999) Prostate-localized and androgen-regulated expression of the membrane-bound serine protease TMPRSS2. *Cancer Res.* **59**, 4180–4184.

54. Vaarala, M. H., Porvari, K., Kyllonen, A., Lukkarinen, O. and Vihko, P. (2001) The TMPRSS2 gene encoding transmembrane serine protease is overexpressed in a majority of prostate cancer patients: detection of mutated TMPRSS2 form in a case of aggressive disease. *Int. J. Cancer* **94**, 705–710.

55. Pei, L., Peng, Y., Yang, Y., Ling, X. B., Van Eyndhoven, W. G., Nguyen, K. C., Rubin, M., Hoey, T., Powers, S. and Li, J. (2002) PRC17, a Novel Oncogene Encoding a Rab GTPase-activating Protein, Is Amplified in Prostate Cancer. *Cancer Res.* **62**, 5420–5424.

56. Kokontis, J. M. and Liao, S. (1999) Molecular action of androgen in the normal and neoplastic prostate. *Vitam. Horm.* **55**, 219–307.

57. Shang, Y., Myers, M. and Brown, M. (2002) Formation of the androgen receptor transcription complex. *Mol. Cell* **9**, 601–610.

58. Beilin, J., Ball, E. M., Favaloro, J. M. and Zajac, J. D. (2000) Effect of the androgen receptor CAG repeat polymorphism on transcriptional activity: specificity in prostate and non-prostate cell lines. *J. Mol. Endocrinol.* **25**, 85–96.

59. Kazemi-Esfarjani, P., Trifiro, M. A. and Pinsky, L. (1995) Evidence for a repressive function of the long polyglutamine tract in the human androgen receptor: possible pathogenetic relevance for the (CAG)n-expanded neuronopathies. *Hum. Mol. Genet.* **4**, 523–527.

60. Chamberlain, N. L., Driver, E. D. and Miesfeld, R. L. (1994) The length and location of CAG trinucleotide repeats in the androgen receptor N-terminal domain affect transactivation function. *Nucleic Acids Res.* **22**, 3181–3186.

61. Sartor, O., Zheng, Q. and Eastham, J. A. (1999) Androgen receptor gene CAG repeat length varies in a race-specific fashion in men without prostate cancer. *Urology* **53**, 378–380.

62. Bennett, C. L., Price, D. K., Kim, S., Liu, D., Jovanovic, B. D., Nathan, D., Johnson, M. E., Montgomery, J. S., Cude, K., Brockbank, J. C., Sartor, O. and Figg, W. D. (2002) Racial Variation in CAG Repeat Lengths Within the Androgen Receptor Gene Among Prostate Cancer Patients of Lower Socioeconomic Status. *J. Clin. Oncol.* **20**, 3599–3604.

63. Heinlein, C. A. and Chang, C. (2004) Androgen receptor in prostate cancer. *Endocr. Rev.* **25**, 276–308.

64. Tayeb, M. T., Clark, C., Murray, G. I., Sharp, L., Haites, N. E. and McLeod, H. L. (2004) Length and somatic mosaicism of CAG and GGN repeats in the androgen receptor gene and the risk of prostate cancer in men with benign prostatic hyperplasia. *Ann. Saudi Med.* **24**, 21–26.

65. Tsujimoto, Y., Takakuwa, T., Takayama, H., Nishimura, K., Okuyama, A., Aozasa, K. and Nonomura, N. (2004) In situ shortening of CAG repeat length within the androgen receptor gene in prostatic cancer and its possible precursors. *Prostate* **58**, 283–290.

66. Giovannucci, E., Stampfer, M. J., Krithivas, K., Brown, M., Dahl, D., Brufsky, A., Talcott, J., Hennekens, C. H. and Kantoff, P. W. (1997) The CAG repeat within the androgen receptor gene and its relationship to prostate cancer. *Proc. Natl. Acad. Sci. U.S.A.* **94**, 3320–3323.

67. Kantoff, P., Giovannucci, E. and Brown, M. (1998) The androgen receptor CAG repeat polymorphism and its relationship to prostate cancer. *Biochim. Biophys. Acta* **1378,** C1–C5.
68. Hsing, A. W., Gao, Y. T., Wu, G., Wang, X., Deng, J., Chen, Y. L., Sesterhenn, I. A., Mostofi, F. K., Benichou, J. and Chang, C. (2000) Polymorphic CAG and GGN repeat lengths in the androgen receptor gene and prostate cancer risk: a population-based case-control study in China. *Cancer Res.* **60,** 5111–5116.
69. Nam, R. K., Elhaji, Y., Krahn, M. D., Hakimi, J., Ho, M., Chu, W., Sweet, J., Trachtenberg, J., Jewett, M. A. and Narod, S. A. (2000) Significance of the CAG repeat polymorphism of the androgen receptor gene in prostate cancer progression. *J. Urol.* **164,** 567–572.
70. Bratt, O., Borg, A., Kristoffersson, U., Lundgren, R., Zhang, Q. X. and Olsson, H. (1999) CAG repeat length in the androgen receptor gene is related to age at diagnosis of prostate cancer and response to endocrine therapy, but not to prostate cancer risk. *Br. J. Cancer* **81,** 672–676.
71. Hardy, D. O., Scher, H. I., Bogenreider, T., Sabbatini, P., Zhang, Z. F., Nanus, D. M. and Catterall, J. F. (1996) Androgen receptor CAG repeat lengths in prostate cancer: correlation with age of onset. *J. Clin. Endocrinol. Metab.* **81,** 4400–4405.
72. Correa-Cerro, L., Wohr, G., Haussler, J., Berthon, P., Drelon, E., Mangin, P., Fournier, G., Cussenot, O., Kraus, P., Just, W., Paiss, T., Cantu, J. M. and Vogel, W. (1999) (CAG)nCAA and GGN repeats in the human androgen receptor gene are not associated with prostate cancer in a French-German population. *Eur. J. Hum. Genet.* **7,** 357–362.
73. Giovannucci, E. (2002) Is the androgen receptor CAG repeat length significant for prostate cancer? *Cancer Epidemiol. Biomarkers Prev.* **11,** 985–986.
74. Ford, O. H., III, Gregory, C. W., Kim, D., Smitherman, A. B. and Mohler, J. L. (2003) Androgen Receptor Gene Amplification and Protein Expression in Recurrent Prostate Cancer. *J. Urol.* **170,** 1817–1821.
75. Koivisto, P., Kononen, J., Palmberg, C., Tammela, T., Hyytinen, E., Isola, J., Trapman, J., Cleutjens, K., Noordzij, A., Visakorpi, T. and Kallioniemi, O. P. (1997) Androgen receptor gene amplification: a possible molecular mechanism for androgen deprivation therapy failure in prostate cancer. *Cancer Res.* **57,** 314–319.
76. Koivisto, P., Visakorpi, T. and Kallioniemi, O. P. (1996) Androgen receptor gene amplification: a novel molecular mechanism for endocrine therapy resistance in human prostate cancer. *Scand. J. Clin. Lab. Invest. Suppl.* **226,** 57–63.
77. Linja, M. J., Savinainen, K. J., Saramaki, O. R., Tammela, T. L., Vessella, R. L. and Visakorpi, T. (2001) Amplification and overexpression of androgen receptor gene in hormone- refractory prostate cancer. *Cancer Res.* **61,** 3550–3555.
78. Bentel, J. M. and Tilley, W. D. (1996) Androgen receptors in prostate cancer. *J. Endocrinol.* **151,** 1–11.
79. Buchanan, G., Greenberg, N. M., Scher, H. I., Harris, J. M., Marshall, V. R. and Tilley, W. D. (2001) Collocation of androgen receptor gene mutations in prostate cancer. *Clin. Cancer Res.* **7,** 1273–1281.
80. Culig, Z., Hobisch, A., Hittmair, A., Cronauer, M. V., Radmayr, C., Bartsch, G. and Klocker, H. (1997) Androgen receptor gene mutations in prostate cancer. Implications for disease progression and therapy. *Drugs Aging* **10,** 50–58.
81. Marcelli, M., Ittmann, M., Mariani, S., Sutherland, R., Nigam, R., Murthy, L., Zhao, Y., DiConcini, D., Puxeddu, E., Esen, A., Eastham, J., Weigel, N. L. and Lamb, D. J. (2000) Androgen receptor mutations in prostate cancer. *Cancer Res.* **60,** 944–949.
82. Newmark, J. R., Hardy, D. O., Tonb, D. C., Carter, B. S., Epstein, J. I., Isaacs, W. B., Brown, T. R. and Barrack, E. R. (1992) Androgen receptor gene mutations in human prostate cancer. *Proc. Natl. Acad. Sci. U.S.A.* **89,** 6319–6323.

83. Suzuki, H., Sato, N., Watabe, Y., Masai, M., Seino, S. and Shimazaki, J. (1993) Androgen receptor gene mutations in human prostate cancer. *J. Steroid Biochem. Mol. Biol.* **46,** 759–765.
84. Taplin, M. E., Rajeshkumar, B., Halabi, S., Werner, C. P., Woda, B. A., Picus, J., Stadler, W., Hayes, D. F., Kantoff, P. W., Vogelzang, N. J. and Small, E. J. (2003) Androgen receptor mutations in androgen-independent prostate cancer: Cancer and Leukemia Group B Study 9663. *J. Clin. Oncol.* **21,** 2673–2678.
85. Taplin, M. E., Bubley, G. J., Ko, Y. J., Small, E. J., Upton, M., Rajeshkumar, B. and Balk, S. P. (1999) Selection for androgen receptor mutations in prostate cancers treated with androgen antagonist. *Cancer Res.* **59,** 2511–2515.
86. Wallen, M. J., Linja, M., Kaartinen, K., Schleutker, J. and Visakorpi, T. (1999) Androgen receptor gene mutations in hormone-refractory prostate cancer. *J. Pathol.* **189,** 559–563.
87. Taplin, M. E., Bubley, G. J., Shuster, T. D., Frantz, M. E., Spooner, A. E., Ogata, G. K., Keer, H. N. and Balk, S. P. (1995) Mutation of the androgen-receptor gene in metastatic androgen- independent prostate cancer. *N. Engl. J. Med.* **332,** 1393–1398.
88. Tan, J., Sharief, Y., Hamil, K. G., Gregory, C. W., Zang, D. Y., Sar, M., Gumerlock, P. H., deVere White, R. W., Pretlow, T. G., Harris, S. E., Wilson, E. M., Mohler, J. L. and French, F. S. (1997) Dehydroepiandrosterone activates mutant androgen receptors expressed in the androgen-dependent human prostate cancer xenograft CWR22 and LNCaP cells. *Mol. Endocrinol.* **11,** 450–459.
89. Veldscholte, J., Ris-Stalpers, C., Kuiper, G. G. J., Jenster, G., van Rooij, H. C. J., Trapman, J., Brinkmann, A. O. and Mulder, E. (1990) A mutation in the ligand binding domain of the androgen receptor of human LNCaP cells affects steroid binding characteristics and response to anti-androgens. *Biochem. Biophys. Res. Commun.* **173,** 534–540.
90. Veldscholte, J., Voorhorst-Ogink, M. M., Bolt-de Vries, J., van Rooij, H. C., Trapman, J. and Mulder, E. (1990) Unusual specificity of the androgen receptor in the human prostate tumor cell line LNCaP: high affinity for progestagenic and estrogenic steroids. *Biochim. Biophys. Acta* **1052,** 187–194.
91. Taplin, M. E., Bubley, G. J., Ko, Y. J., Small, E. J., Upton, M., Rajeshkumar, B. and Balk, S. P. (1999) Selection for androgen receptor mutations in prostate cancers treated with androgen antagonist. *Cancer Res.* **59,** 2511–2515.
92. Yoshida, T., Kinoshita, H., Segawa, T., Nakamura, E., Inoue, T., Shimizu, Y., Kamoto, T. and Ogawa, O. (2005) Antiandrogen bicalutamide promotes tumor growth in a novel androgen-dependent prostate cancer xenograft model derived from a bicalutamide-treated patient. *Cancer Res.* **65,** 9611–9616.
93. Tilley, W. D., Buchanan, G., Hickey, T. E. and Bentel, J. M. (1996) Mutations in the androgen receptor gene are associated with progression of human prostate cancer to androgen independence. *Clin. Cancer Res.* **2,** 277–285.
94. Alvarado, C., Beitel, L. K., Sircar, K., Aprikian, A., Trifiro, M. and Gottlieb, B. (2005) Somatic mosaicism and cancer: a micro-genetic examination into the role of the androgen receptor gene in prostate cancer. *Cancer Res.* **65,** 8514–8518.
95. Buchanan, G., Yang, M., Cheong, A., Harris, J. M., Irvine, R. A., Lambert, P. F., Moore, N. L., Raynor, M., Neufing, P. J., Coetzee, G. A. and Tilley, W. D. (2004) Structural and functional consequences of glutamine tract variation in the androgen receptor. *Hum. Mol. Genet.* **13,** 1677–1692.
96. Schoenberg, M. P., Hakimi, J. M., Wang, S., Bova, G. S., Epstein, J. I., Fischbeck, K. H., Isaacs, W. B., Walsh, P. C. and Barrack, E. R. (1994) Microsatellite mutation (CAG24->18) in the androgen receptor gene in human prostate cancer. *Biochem. Biophys. Res. Commun.* **198,** 74–80.

97. Slagsvold, T., Kraus, I., Fronsdal, K. and Saatcioglu, F. (2001) DNA binding-independent transcriptional activation by the androgen receptor through triggering of coactivators. *J. Biol. Chem.* **276**, 31030–31036.

98. Voeller, H. J., Truica, C. I. and Gelmann, E. P. (1998) Beta-catenin mutations in human prostate cancer. *Cancer Res.* **58**, 2520–2523.

99. Chesire, D. R., Ewing, C. M., Sauvageot, J., Bova, G. S. and Isaacs, W. B. (2000) Detection and analysis of beta-catenin mutations in prostate cancer. *Prostate* **45**, 323–334.

100. Rios-Doria, J., Kuefer, R., Ethier, S. P. and Day, M. L. (2004) Cleavage of beta-catenin by calpain in prostate and mammary tumor cells. *Cancer Res.* **64**, 7237–7240.

101. Pawlowski, J. E., Ertel, J. R., Allen, M. P., Xu, M., Butler, C., Wilson, E. M. and Wierman, M. E. (2002) Liganded androgen receptor interaction with beta-catenin: nuclear co-localization and modulation of transcriptional activity in neuronal cells. *J. Biol. Chem.* **277**, 20702–20710.

102. Mulholland, D. J., Cheng, H., Reid, K., Rennie, P. S. and Nelson, C. C. (2002) The androgen receptor can promote beta-catenin nuclear translocation independently of adenomatous polyposis coli. *J. Biol. Chem.* **277**, 17933–17943.

103. Yang, F., Li, X., Sharma, M., Sasaki, C. Y., Longo, D. L., Lim, B. and Sun, Z. (2002) Linking beta-catenin to androgen-signaling pathway. *J. Biol. Chem.* **277**, 11336–11344.

104. Song, L. N., Herrell, R., Byers, S., Shah, S., Wilson, E. M. and Gelmann, E. P. (2003) Beta-catenin binds to the activation function 2 region of the androgen receptor and modulates the effects of the N-terminal domain and TIF2 on ligand-dependent transcription. *Mol. Cell Biol.* **23**, 1674–1687.

105. Bakin, R. E., Gioeli, D., Sikes, R. A., Bissonette, E. A. and Weber, M. J. (2003) Constitutive Activation of the Ras/Mitogen-activated Protein Kinase Signaling Pathway Promotes Androgen Hypersensitivity in LNCaP Prostate Cancer Cells. *Cancer Res.* **63**, 1981–1989.

106. Mellinghoff, I. K., Vivanco, I., Kwon, A., Tran, C., Wongvipat, J. and Sawyers, C. L. (2004) HER2/neu kinase-dependent modulation of androgen receptor function through effects on DNA binding and stability. *Cancer Cell* **6**, 517–527.

107. Voeller, H. J., Augustus, M., Madlike, V., Bova, G. S., Carter, K. C. and Gelmann, E. P. (1997) Coding region of NKX3.1, prostate-specific homeobox gene on 8p21, is not mutated in human prostate cancers. *Cancer Res.* **57**, 4455–4459.

108. Vocke, C. D., Pozzatti, R. O., Bostwick, D. G., Florence, C. D., Jennings, S. B., Strup, S. E., Duray, P. H., Liotta, L. A., Emmert-Buck, M. R. and Linehan, W. M. (1996) Analysis of 99 microdissected prostate carcinomas reveals a high frequency of allelic loss on chromosome 8p21–22. *Cancer Res.* **56**, 2411–2416.

109. Bhatia-Gaur, R., Donjacour, A. A., Sciavolino, P. J., Kim, M., Desai, N., Norton, C. R., Gridley, T., Cardiff, R. D., Cunha, G. R., Abate-Shen, C. and Shen, M. M. (1999) Roles for Nkx3.1 in prostate development and cancer. *Genes Dev.* **13**, 966–977.

110. Bowen, C., Bubendorf, L., Voeller, H. J., Slack, R., Willi, N., Sauter, G., Gasser, T. C., Koivisto, P., Lack, E. E., Kononen, J., Kallioniemi, O. P. and Gelmann, E. P. (2000) Loss of NKX3.1 expression in human prostate cancers correlates with tumor progression. *Cancer Res.* **60**, 6111–6115.

111. MacGrogan, D., Levy, A., Bostwick, D., Wagner, M., Wells, D. and Bookstein, R. (1994) Loss of chromosome arm 8p loci in prostate cancer: mapping by quantitative allelic imbalance. *Genes Chromosomes Cancer* **10**, 151–159.

112. Matsuyama, H., Pan, Y., Skoog, L. and Tribukait, B. (1994) Deletion mappting of chromosome 8p in prostate cancer by fluorescence in situ hybridization. *Oncogene* **9**, 3071–3076.

113. Swalwell, J. I., Vocke, C. D., Yang, Y., Walker, J. R., Grouse, L., Myers, S. H., Gillespie, J. W., Bostwick, D. G., Duray, P. H., Linehan, W. M. and Emmert-Buck, M. R.

(2002) Determination of a minimal deletion interval on chromosome band 8p21 in sporadic prostate cancer. *Genes Chromosomes Cancer* **33**, 201–205.

114. Ornstein, D. K., Cinquanta, M., Weiler, S., Duray, P. H., Emmert-Buck, M. R., Vocke, C. D., Linehan, W. M. and Ferretti, J. A. (2001) Expression studies and mutational analysis of the androgen regulated homeobox gene nkx3.1 in benign and malignant prostate epithelium. *J. Urol.* **165**, 1329–1334.

115. Xu, L. L., Srikantan, V., Sesterhenn, I. A., Augustus, M., Dean, R., Moul, J. W., Carter, K. C. and Srivastava, S. (2000) Expression profile of an androgen regulated prostate specific homeobox gene NKX3.1 in primary prostate cancer. *J. Urol.* **163**, 972–979.

116. Kim, M. J., Cardiff, R. D., Desai, N., Banach-Petrosky, W. A., Parsons, R., Shen, M. M. and Abate-Shen, C. (2002) Cooperativity of Nkx3.1 and Pten loss of function in a mouse model of prostate carcinogenesis. *Proc. Natl. Acad. Sci. U.S.A.* **99**, 2884–2889.

117. Magee, J. A., Abdulkadir, S. A. and Milbrandt, J. (2003) Haploinsufficiency at the Nkx3.1 locus. A paradigm for stochastic, dosage-sensitive gene regulation during tumor initiation. *Cancer Cell* **3**, 273–283.

118. Asatiani, E., Huang, W. X., Wang, A., Rodriguez, O. E., Cavalli, L. R., Haddad, B. R. and Gelmann, E. P. (2005) Deletion, methylation, and expression of the NKX3.1 suppressor gene in primary human prostate cancer. *Cancer Res.* **65**, 1164–1173.

119. Tsao, D. H., Gruschus, J. M., Wang, L. H., Nirenberg, M. and Ferretti, J. A. (1995) The three-dimensional solution structure of the NK-2 homeodomain from Drosophila. *J. Mol. Biol.* **251**, 297–307.

120. Benson, D. W., Silberbach, G. M., Kavanaugh-McHugh, A., Cottrill, C., Zhang, Y., Riggs, S., Smalls, O., Johnson, M. C., Watson, M. S., Seidman, J. G., Seidman, C. E., Plowden, J. and Kugler, J. D. (1999) Mutations in the cardiac transcription factor NKX2.5 affect diverse cardiac developmental pathways. *J. Clin. Invest.* **104**, 1567–1573.

121. Krude, H., Schutz, B., Biebermann, H., von, M. A., Schnabel, D., Neitzel, H., Tonnies, H., Weise, D., Lafferty, A., Schwarz, S., DeFelice, M., von, D. A., van, L. F., DiLauro, R. and Gruters, A. (2002) Choreoathetosis, hypothyroidism, and pulmonary alterations due to human NKX2–1 haploinsufficiency. *J. Clin. Invest.* **109**, 475–480.

122. Iwatani, N., Mabe, H., Devriendt, K., Kodama, M. and Miike, T. (2000) Deletion of NKX2.1 gene encoding thyroid transcription factor-1 in two siblings with hypothyroidism and respiratory failure. *J. Pediatr.* **137**, 272–276.

123. Zheng, S. L., Ju, J. H., Chang, B. L., Ortner, E., Sun, J., Isaacs, S. D., Sun, J., Wiley, K. E., Liu, W., Zemedkun, M., Walsh, P. C., Ferretti, J., Gruschus, J., Isaacs, W. B., Gelmann, E. P. and Xu, J. (2006) Germ-line mutation of NKX3.1 cosegregates with hereditary prostate cancer and alters the homeodomain structure and function. *Cancer Res.* **66**, 69–77.

124. Sciavolino, P. J., Abrams, E. W., Yang, L., Austenberg, L. P., Shen, M. M. and Abate-Shen, C. (1997) Tissue-specific expression of murine Nkx3.1 in the male urogenital sinus. *Dev. Dyn.* **209**, 127–138.

125. Kapp, G. T., Richardson, J. S. and Oas, T. G. (2004) Kinetic role of helix caps in protein folding is context-dependent. *Biochemistry* **43**, 3814–3823.

126. Koscielska-Kasprzak, K., Cierpicki, T. and Otlewski, J. (2003) Importance of alpha-helix N-capping motif in stabilization of betabetaalpha fold. *Protein Sci.* **12**, 1283–1289.

127. Kasahara, H. and Benson, D. W. (2004) Biochemical analyses of eight NKX2.5 homeodomain missense mutations causing atrioventricular block and cardiac anomalies. *Cardiovasc. Res.* **64**, 40–51.

128. Steadman, D. J., Giuffrida, D. and Gelmann, E. P. (2000) DNA-binding sequence of the human prostate-specific homeodomain protein NKX3.1. *Nucleic Acids Res.* **28**, 2389–2395.

129. Gelmann, E. P., Steadman, D. J., Ma, J., Ahronovitz, N., Voeller, H. J., Swope, S., Abbaszadegan, M., Brown, K. M., Strand, K., Hayes, R. B. and Stampfer, M. J. (2002)

Occurrence of NKX3.1 C154T Polymorphism in Men with and without Prostate Cancer and Studies of Its Effect on Protein Function. *Cancer Res.* **62,** 2654–2659.

130. Li, J., Yen, C., Liaw, D., Podsypanina, K., Bose, S., Wang, S. I., Puc, J., Miliaresis, C., Rodgers, L., McCombie, R., Bigner, S. H., Giovanella, B. C., Ittman, M., Tycko, B., Hibshoosh, H., Wigler, M. H. and Parsons, R. (1997) PTEN, a putative protein tyrosine phosphatase gene mutated in human brain, breast, and prostate cancer. *Science* **275,** 1943–1947.

131. Dong, J. T., Sipe, T. W., Hyytinen, E. R., Li, C. L., Heise, C., McClintock, D. E., Grant, C. D., Chung, L. W. and Frierson, H. F., Jr. (1998) PTEN/MMAC1 is infrequently mutated in pT2 and pT3 carcinomas of the prostate. *Oncogene* **17,** 1979–1982.

132. Dong, J. T., Li, C. L., Sipe, T. W. and Frierson, H. F., Jr. (2001) Mutations of PTEN/MMAC1 in primary prostate cancers from Chinese patients. *Clin. Cancer Res.* **7,** 304–308.

133. Orikasa, K., Fukushige, S., Hoshi, S., Orikasa, S., Kondo, K., Miyoshi, Y., Kubota, Y. and Horii, A. (1998) Infrequent genetic alterations of the PTEN gene in Japanese patients with sporadic prostate cancer. *J. Hum. Genet.* **43,** 228–230.

134. Wang, S. I., Parsons, R. and Ittmann, M. (1998) Homozygous deletion of the PTEN tumor suppressor gene in a subset of prostate adenocarcinomas. *Clin. Cancer Res.* **4,** 811–815.

135. Myers, M. P., Pass, I., Batty, I. H., Van der Kaay, J., Stolarov, J. P., Hemmings, B. A., Wigler, M. H., Downes, C. P. and Tonks, N. K. (1998) The lipid phosphatase activity of PTEN is critical for its tumor suppressor function. *Proc. Natl. Acad. Sci. U.S.A.* **95,** 13513–13518.

136. Whang, Y. E., Wu, X., Suzuki, H., Reiter, R. E., Tran, C., Vessella, R. L., Said, J. W., Isaacs, W. B. and Sawyers, C. L. (1998) Inactivation of the tumor suppressor PTEN/MMAC1 in advanced human prostate cancer through loss of expression. *Proc. Natl. Acad. Sci. U.S.A.* **95,** 5246–5250.

137. McMenamin, M. E., Soung, P., Perera, S., Kaplan, I., Loda, M. and Sellers, W. R. (1999) Loss of PTEN expression in paraffin-embedded primary prostate cancer correlates with high Gleason score and advanced stage. *Cancer Res.* **59,** 4291–4296.

138. Suzuki, H., Freije, D., Nusskern, D. R., Okami, K., Cairns, P., Sidransky, D., Isaacs, W. B. and Bova, G. S. (1998) Interfocal heterogeneity of PTEN/MMAC1 gene alterations in multiple metastatic prostate cancer tissues. *Cancer Res.* **58,** 204–209.

139. Kwabi-Addo, B., Giri, D., Schmidt, K., Podsypanina, K., Parsons, R., Greenberg, N. and Ittmann, M. (2001) Haploinsufficiency of the Pten tumor suppressor gene promotes prostate cancer progression. *Proc. Natl. Acad. Sci. U.S.A.* **98,** 11563–11568.

140. Podsypanina, K., Ellenson, L. H., Nemes, A., Gu, J., Tamura, M., Yamada, K. M., Cordon-Cardo, C., Catoretti, G., Fisher, P. E. and Parsons, R. (1999) Mutation of Pten/Mmac1 in mice causes neoplasia in multiple organ systems. *Proc. Natl. Acad. Sci. U.S.A.* **96,** 1563–1568.

141. Wang, S., Gao, J., Lei, Q., Rozengurt, N., Pritchard, C., Jiao, J., Thomas, G. V., Li, G., Roy-Burman, P., Nelson, P. S., Liu, X. and Wu, H. (2003) Prostate-specific deletion of the murine Pten tumor suppressor gene leads to metastatic prostate cancer. *Cancer Cell* **4,** 209–221.

142. Abate-Shen, C., Banach-Petrosky, W. A., Sun, X., Economides, K. D., Desai, N., Gregg, J. P., Borowsky, A. D., Cardiff, R. D. and Shen, M. M. (2003) Nkx3.1; Pten mutant mice develop invasive prostate adenocarcinoma and lymph node metastases. *Cancer Res.* **63,** 3886–3890.

143. Majumder, P. K., Yeh, J. J., George, D. J., Febbo, P. G., Kum, J., Xue, Q., Bikoff, R., Ma, H., Kantoff, P. W., Golub, T. R., Loda, M. and Sellers, W. R. (2003) Prostate intraepithelial neoplasia induced by prostate restricted Akt activation: the MPAKT model. *Proc. Natl. Acad. Sci. U.S.A.* **100,** 7841–7846.

144. Philipp-Staheli, J., Payne, S. R. and Kemp, C. J. (2001) p27(Kip1): regulation and function of a haploinsufficient tumor suppressor and its misregulation in cancer. *Exp. Cell Res.* **264,** 148–168.

145. Fero, M. L., Randel, E., Gurley, K. E., Roberts, J. M. and Kemp, C. J. (1998) The murine gene p27kip1 is haplo-insufficient for tumour suppression. *Nature* **396,** 177–180.

146. Cote, R. J., Shi, Y., Groshen, S., Feng, A. C., Cordon-Cardo, C., Skinner, D. and Lieskovosky, G. (1998) Association of p27Kip1 levels with recurrence and survival in patients with stage C prostate carcinoma. *J. Natl. Cancer Inst.* **90,** 916–920.

147. Guo, Y., Sklar, G. N., Borkowski, A. and Kyprianou, N. (1997) Loss of the cyclin-dependent kinase inhibitor p27(Kip1) protein in human prostate cancer correlates with tumor grade. *Clin. Cancer Res.* **3,** 2269–2274.

148. Yang, R. M., Naitoh, J., Murphy, M., Wang, H. J., Phillipson, J., deKernion, J. B., Loda, M. and Reiter, R. E. (1998) Low p27 expression predicts poor disease-free survival in patients with prostate cancer. *J. Urol.* **159,** 941–945.

149. Nakamura, N., Ramaswamy, S., Vazquez, F., Signoretti, S., Loda, M. and Sellers, W. R. (2000) Forkhead transcription factors are critical effectors of cell death and cell cycle arrest downstream of PTEN. *Mol. Cell Biol.* **20,** 8969–8982.

150. Xu, J., Langefeld, C. D., Zheng, S. L., Gillanders, E. M., Chang, B. L., Isaacs, S. D., Williams, A. H., Wiley, K. E., Dimitrov, L., Meyers, D. A., Walsh, P. C., Trent, J. M. and Isaacs, W. B. (2004) Interaction effect of PTEN and CDKN1B chromosomal regions on prostate cancer linkage. *Hum. Genet.* **115,** 255–262.

151. Kibel, A. S., Schutte, M., Kern, S. E., Isaacs, W. B. and Bova, G. S. (1998) Identification of 12p as a region of frequent deletion in advanced prostate cancer. *Cancer Res.* **58,** 5652–5655.

152. Kibel, A. S., Freije, D., Isaacs, W. B. and Bova, G. S. (1999) Deletion mapping at 12p12–13 in metastatic prostate cancer. *Genes Chromosomes Cancer* **25,** 270–276.

153. Bloom, J. and Pagano, M. (2003) Deregulated degradation of the cdk inhibitor p27 and malignant transformation. *Semin. Cancer Biol.* **13,** 41–47.

154. Cordon-Cardo, C., Koff, A., Drobnjak, M., Capodieci, P., Osman, I., Millard, S. S., Gaudin, P. B., Fazzari, M., Zhang, Z. F., Massague, J. and Scher, H. I. (1998) Distinct altered patterns of p27KIP1 gene expression in benign prostatic hyperplasia and prostatic carcinoma. *J. Natl. Cancer Inst.* **90,** 1284–1291.

155. Gary, B., Azuero, R., Mohanty, G. S., Bell, W. C., Eltoum, I. E. and Abdulkadir, S. A. (2004) Interaction of Nkx3.1 and p27kip1 in prostate tumor initiation. *Am. J. Pathol.* **164,** 1607–1614.

156. Di Cristofano, A., De Acetis, M., Koff, A., Cordon-Cardo, C. and Pandolfi, P. P. (2001) Pten and p27KIP1 cooperate in prostate cancer tumor suppression in the mouse. *Nat. Genet.* **27,** 222–224.

157. Yasuda, H., Mizuno, A., Tamaoki, T. and Morinaga, T. (1994) ATBF1, a multiple-homeodomain zinc finger protein, selectively down-regulates AT-rich elements of the human alpha-fetoprotein gene. *Mol. Cell Biol.* **14,** 1395–1401.

158. Kaspar, P., Dvorakova, M., Kralova, J., Pajer, P., Kozmik, Z. and Dvorak, M. (1999) Myb-interacting protein, ATBF1, represses transcriptional activity of Myb oncoprotein. *J. Biol. Chem.* **274,** 14422–14428.

159. Kataoka, H., Bonnefin, P., Vieyra, D., Feng, X., Hara, Y., Miura, Y., Joh, T., Nakabayashi, H., Vaziri, H., Harris, C. C. and Riabowol, K. (2003) ING1 represses transcription by direct DNA binding and through effects on p53. *Cancer Res.* **63,** 5785–5792.

160. Zhang, Z., Yamashita, H., Toyama, T., Sugiura, H., Ando, Y., Mita, K., Hamaguchi, M., Kawaguchi, M., Miura, Y. and Iwase, H. (2005) ATBF1-a messenger RNA expression is correlated with better prognosis in breast cancer. *Clin. Cancer Res.* **11,** 193–198.

161. Jung, C. G., Kim, H. J., Kawaguchi, M., Khanna, K. K., Hida, H., Asai, K., Nishino, H. and Miura, Y. (2005) Homeotic factor ATBF1 induces the cell cycle arrest associated with neuronal differentiation. *Development* **132**, 5137–5145.

162. Sun, X., Frierson, H. F., Chen, C., Li, C., Ran, Q., Otto, K. B., Cantarel, B. L., Vessella, R. L., Gao, A. C., Petros, J., Miura, Y., Simons, J. W. and Dong, J. T. (2005) Frequent somatic mutations of the transcription factor ATBF1 in human prostate cancer. *Nat. Genet.* **37**, 407–412.

163. Ratziu, V., Lalazar, A., Wong, L., Dang, Q., Collins, C., Shaulian, E., Jensen, S. and Friedman, S. L. (1998) Zf9, a Kruppel-like transcription factor up-regulated in vivo during early hepatic fibrosis. *Proc. Natl. Acad. Sci. U.S.A.* **95**, 9500–9505.

164. 164. Koritschoner, N. P., Bocco, J. L., Panzetta-Dutari, G. M., Dumur, C. I., Flury, A. and Patrito, L. C. (1997) A novel human zinc finger protein that interacts with the core promoter element of a TATA box-less gene. *J. Biol. Chem.* **272**, 9573–9580.

165. Kojima, S., Hayashi, S., Shimokado, K., Suzuki, Y., Shimada, J., Crippa, M. P. and Friedman, S. L. (2000) Transcriptional activation of urokinase by the Kruppel-like factor Zf9/COPEB activates latent TGF-beta1 in vascular endothelial cells. *Blood* **95**, 1309–1316.

166. Kim, Y., Ratziu, V., Choi, S. G., Lalazar, A., Theiss, G., Dang, Q., Kim, S. J. and Friedman, S. L. (1998) Transcriptional activation of transforming growth factor beta1 and its receptors by the Kruppel-like factor Zf9/core promoter-binding protein and Sp1. Potential mechanisms for autocrine fibrogenesis in response to injury. *J. Biol. Chem.* **273**, 33750–33758.

167. Slavin, D. A., Koritschoner, N. P., Prieto, C. C., Lopez-Diaz, F. J., Chatton, B. and Bocco, J. L. (2004) A new role for the Kruppel-like transcription factor KLF6 as an inhibitor of c-Jun proto-oncoprotein function. *Oncogene* **23**, 8196–8205.

168. Benzeno, S., Narla, G., Allina, J., Cheng, G. Z., Reeves, H. L., Banck, M. S., Odin, J. A., Diehl, J. A., Germain, D. and Friedman, S. L. (2004) Cyclin-dependent kinase inhibition by the KLF6 tumor suppressor protein through interaction with cyclin D1. *Cancer Res.* **64**, 3885–3891.

169. Ittmann, M. (1996) Allelic loss on chromosome 10 in prostate adenocarcinoma. *Cancer Res.* **56**, 2143–2147.

170. Trybus, T. M., Burgess, A. C., Wojno, K. J., Glover, T. W. and Macoska, J. A. (1996) Distinct areas of allelic loss on chromosomal regions 10p and 10q in human prostate cancer. *Cancer Res.* **56**, 2263–2267.

171. Narla, G., Heath, K. E., Reeves, H. L., Li, D., Giono, L. E., Kimmelman, A. C., Glucksman, M. J., Narla, J., Eng, F. J., Chan, A. M., Ferrari, A. C., Martignetti, J. A. and Friedman, S. L. (2001) KLF6, a candidate tumor suppressor gene mutated in prostate cancer. *Science* **294**, 2563–2566.

172. Chen, C., Hyytinen, E. R., Sun, X., Helin, H. J., Koivisto, P. A., Frierson, H. F., Jr., Vessella, R. L. and Dong, J. T. (2003) Deletion, mutation, and loss of expression of KLF6 in human prostate cancer. *Am. J. Pathol.* **162**, 1349–1354.

173. Muhlbauer, K. R., Grone, H. J., Ernst, T., Grone, E., Tschada, R., Hergenhahn, M. and Hollstein, M. (2003) Analysis of human prostate cancers and cell lines for mutations in the TP53 and KLF6 tumour suppressor genes. *Br. J. Cancer* **89**, 687–690.

174. Narla, G., DiFeo, A., Reeves, H. L., Schaid, D. J., Hirshfeld, J., Hod, E., Katz, A., Isaacs, W. B., Hebbring, S., Komiya, A., McDonnell, S. K., Wiley, K. E., Jacobsen, S. J., Isaacs, S. D., Walsh, P. C., Zheng, S. L., Chang, B. L., Friedrichsen, D. M., Stanford, J. L., Ostrander, E. A., Chinnaiyan, A. M., Rubin, M. A., Xu, J., Thibodeau, S. N., Friedman, S. L. and Martignetti, J. A. (2005) A germline DNA polymorphism enhances alternative splicing of the KLF6 tumor suppressor gene and is associated with increased prostate cancer risk. *Cancer Res.* **65**, 1213–1222.

175. Narla, G., DiFeo, A., Yao, S., Banno, A., Hod, E., Reeves, H. L., Qiao, R. F., Camacho-Vanegas, O., Levine, A., Kirschenbaum, A., Chan, A. M., Friedman, S. L. and Martignetti, J. A. (2005) Targeted inhibition of the KLF6 splice variant, KLF6 SV1, suppresses prostate cancer cell growth and spread. *Cancer Res.* **65**, 5761–5768.

176. Bookstein, R., Rio, P., Madreperla, S. A., Hong, F., Allred, C., Grizzle, W. E. and Lee, W. H. (1990) Promoter deletion and loss of retinoblastoma gene expression in human prostate carcinoma. *Proc. Natl. Acad. Sci. U.S.A.* **87**, 7762–7766.

177. Phillips, S. M., Morton, D. G., Lee, S. J., Wallace, D. M. and Neoptolemos, J. P. (1994) Loss of heterozygosity of the retinoblastoma and adenomatous polyposis susceptibility gene loci and in chromosomes 10p, 10q and 16q in human prostate cancer. *Br. J. Urol.* **73**, 390–395.

178. Kubota, Y., Fujinami, K., Uemura, H., Dobashi, Y., Miyamoto, H., Iwasaki, Y., Kitamura, H. and Shuin, T. (1995) Retinoblastoma gene mutations in primary human prostate cancer. *Prostate* **27**, 314–320.

179. Tricoli, J. V., Gumerlock, P. H., Yao, J. L., Chi, S. G., D'Souza, S. A., Nestok, B. R. and Vere White, R. W. (1996) Alterations of the retinoblastoma gene in human prostate adenocarcinoma. *Genes Chromosomes Cancer* **15**, 108–114.

180. Maddison, L. A., Sutherland, B. W., Barrios, R. J. and Greenberg, N. M. (2004) Conditional deletion of rb causes early stage prostate cancer. *Cancer Res.* **64**, 6018–6025.

181. Li, F. P. and Fraumeni, J. F. (1969) Soft-tissue sarcomas, breast cancer, and other neoplasms: a familial syndrome? *Ann. Intern. Med.* **71**, 747–752.

182. Malkin, D., Li, F. P., Strong, L. C., Fraumeni, J. F., Nelson, C. E., Kim, D. H., Kassel, J., Fryka, M. A., Bischoff, F. Z., Tainsky, M. A. and Friend, S. H. (1990) Germ line p53 mutations in a familial syndrome of breast cancer, sarcomas and other neoplasms. *Science* **250**, 1233–1238.

183. Berner, A., Geitvik, G., Karlsen, F., Foss˜a, S. D., Nesland, J. M. and Borresen, A. L. (1995) TP53 mutations in prostatic cancer. Analysis of pre- and post- treatment archival formalin-fixed tumour tissue. *J. Pathol.* **176**, 299–308.

184. Wertz, I. E., Deitch, A. D., Gumerlock, P. H., Gandour-Edwards, R., Chi, S. G. and deVereWhite, R. W. (1996) Correlation of genetic and immunodetection of TP53 mutations in malignant and benign prostate tissues. *Hum. Pathol.* **27**, 573–580.

185. Yang, G., Stapleton, A. M. F., Wheeler, T. M., Truong, L. D., Timme, T. L., Scardino, P.'T. and Thompson, T. C. (1996) Clustered p53 immunostaining: a novel pattern associated with prostate cancer progression. *Clin. Cancer Res.* **2**, 399–401.

186. Griewe, G. L., Dean, R. C., Zhang, W., Young, D., Sesterhenn, I. A., Shanmugam, N., McLeod, D. G., Moul, J. W. and Srivastava, S. (2003) p53 Immunostaining guided laser capture microdissection (p53-LCM) defines the presence of p53 gene mutations in focal regions of primary prostate cancer positive for p53 protein. *Prostate Cancer Prostatic Dis.* **6**, 281–285.

187. Visakorpi, T., Kallioniemi, O. P., Heikkinen, A., Koivula, T. and Isola, J. (1992) Small subgroup of aggressive, highly proliferative prostatic carcinomas defined by p53 accumulation. *J. Natl. Cancer Inst.* **84**, 883–887.

188. Navone, N. M., Troncoso, P., Pisters, L. L., Goodrow, T. L., Palmer, J. L., Nichols, W. W., von Eschenbach, A. C. and Conti, C. J. (1993) p53 Protein accumulation and gene mutation in the progression of human prostate carcinoma. *J. Natl. Cancer Inst.* **85**, 1657–1669.

189. Bauer, J. J., Sesterhenn, I. A., Mostofi, K. F., McLeod, D. G., Srivastava, S. and Moul, J. W. (1995) p53 nuclear protein expression is an independent prognostic marker in clinically localized prostate cancer patients undergoing radical prostatectomy. *Clin. Cancer Res.* **1**, 1295–1300.

190. Grignon, D. J., Caplan, R., Sarkar, F. H., Lawton, C. A., Hammond, E. H., Pilepich, M. V., Forman, J. D., Mesic, J., Fu, K. K., Abrams, R. A., Pajak, T. F., Shipley, W. U. and

Cox, J. D. (1997) p53 status and prognosis of locally advanced prostatic adenocarcinoma: a study based on RTOG 8610. *J. Natl. Cancer Inst.* **89,** 158–165.

191. Stackhouse, G. B., Sesterhenn, I. A., Bauer, J. J., Mostofi, F. K., Connelly, R. R., Srivastava, S. K. and Moul, J. W. (1999) p53 and bcl-2 immunohistochemistry in pretreatment prostate needle biopsies to predict recurrence of prostate cancer after radical prostatectomy. *J. Urol.* **162,** 2040–2045.

192. Quinn, D. I., Henshall, S. M., Head, D. R., Golovsky, D., Wilson, J. D., Brenner, P. C., Turner, J. J., Delprado, W., Finlayson, J. F., Stricker, P. D., Grygiel, J. J. and Sutherland, R. L. (2000) Prognostic significance of p53 nuclear accumulation in localized prostate cancer treated with radical prostatectomy. *Cancer Res.* **60,** 1585–1594.

193. Borre, M., Stausbol-Gron, B. and Overgaard, J. (2000) p53 accumulation associated with bcl-2, the proliferation marker MIB-1 and survival in patients with prostate cancer subjected to watchful waiting. *J. Urol.* **164,** 716–721.

194. Leibovich, B. C., Cheng, L., Weaver, A. L., Myers, R. P. and Bostwick, D. G. (2000) Outcome prediction with p53 immunostaining after radical prostatectomy in patients with locally advanced prostate cancer. *J. Urol.* **163,** 1756–1760.

195. Scherr, D. S., Vaughan, E. D., Jr., Wei, J., Chung, M., Felsen, D., Allbright, R. and Knudsen, B. S. (1999) BCL-2 and p53 expression in clinically localized prostate cancer predicts response to external beam radiotherapy. *J. Urol.* **162,** 12–16.

196. Brooks, J. D., Bova, G. S., Ewing, C. M., Piantadosi, S., Carter, B. S., Robinson, J. C., Epstein, J. I. and Isaacs, W. B. (1996) An uncertain role for p53 gene alteration in human prostate cancers. *Cancer Res.* **56,** 3814–3822.

197. Dahiya, R., Deng, G., Chen, K. M., Chui, R. M., Haughney, P. C. and Narayan, P. (1996) P53 tumour-suppressor gene mutations are mainly localised on exon 7 in human primary and metastatic prostate cancer. *Br. J. Cancer* **74,** 264–268.

198. Chi, S. G., deVereWhite, R. W., Meyers, F. J., Siders, D. B., Lee, F. and Gumerlock, P. H. (1994) p53 in prostate cancer: frequent expressed transition mutations. *J. Natl. Cancer Inst.* **86,** 926–933.

199. Massenkeil, G., Oberhuber, H., Hailemariam, S., Sulser, T., Diener, P. A., Bannwart, F., Schafer, R. and Schwarte-Waldhoff, I. (1994) P53 mutations and loss of heterozygosity on chromosomes 8p, 16q, 17p, and 18q are confined to advanced prostate cancer. *Anticancer Res.* **14,** 2785–2790.

200. Voeller, H. J., Sugars, L. Y., Pretlow, T. and Gelmann, E. P. (1994) p53 oncogene mutations in human prostate cancer specimens. *J. Urol.* **151,** 492–495.

201. Kallakury, B. V., Jennings, T. A., Ross, J. S., Breese, K., Figge, H. L., Fisher, H. A. and Figge, J. (1994) Alteration of the p53 locus in benign hyperplastic prostatic epithelium associated with high-grade prostatic adenocarcinoma. *Diagn. Mol. Pathol.* **3,** 227–232.

202. Watanabe, M., Ushijima, T., Kakiuchi, H., Shiraishi, T., Yatani, R., Shimazaki, J., Kotake, T., Sugimura, T. and Nagao, M. (1994) p53 gene mutations in human prostate cancers in Japan: different mutation spectra between Japan and western countries. *Jpn. J. Cancer Res.* **85,** 904–910.

203. Bookstein, R., MacGrogan, D., Hilsenbeck, S. G., Sharkey, F. and Allred, D. C. (1993) p53 is mutated in a subset of advanced-stage prostate cancers. *Cancer Res.* **53,** 3369–3373.

204. Effert, P., Neubauer, A., Walther, P. J. and Liu, E. T. (1992) Alterations of the p53 gene are associated with the progression of a human prostate carcinoma. *J. Urol.* **147,** 789–793.

205. Effert, P. J., McCoy, R. H., Walther, P. J. and Liu, E. T. (1993) p53 gene alterations in human prostate carcinoma. *J. Urol.* **150,** 257–261.

206. Eastham, J. A., Stapleton, A. M. F., Gousse, A. E., Timme, T. L., Yang, G., Slawin, K. M., Wheeler, T. M., Scardino, P. T. and Thompson, T. C. (1995) Association of p53 mutations with metastatic prostate cancer. *Clin. Cancer Res.* **1,** 1111–1118.

207. Grossfeld, G. D., Olumi, A. F., Connolly, J. A., Chew, K., Gibney, J., Bhargava, V., Waldman, F. M. and Carroll, P. R. (1998) Locally recurrent prostate tumors following either radiation therapy or radical prostatectomy have changes in Ki-67 labeling index, p53 and bcl- 2 immunoreactivity. *J. Urol.* **159,** 1437–1443.
208. MacGrogen, D., Levy, A., Bostwick, D., Wagner, M., Wells, D. and Bookstein, R. (1994) Loss of chromosome arm 8p loci in prostate cancer: mapping by quantitative allelic imbalance. *Genes Chromosomes Cancer* **10,** 151–159.
209. Abdulkadir, S. A., Magee, J. A., Peters, T. J., Kaleem, Z., Naughton, C. K., Humphrey, P. A. and Milbrandt, J. (2002) Conditional loss of Nkx3.1 in adult mice induces prostatic intraepithelial neoplasia. *Mol. Cell Biol.* **22,** 1495–1503.
210. Morinaga, T., Yasuda, H., Hashimoto, T., Higashio, K. and Tamaoki, T. (1991) A human alpha-fetoprotein enhancer-binding protein, ATBF1, contains four homeodomains and seventeen zinc fingers. *Mol. Cell Biol.* **11,** 6041–6049.

5 Somatic Genetic Changes in Prostate Cancer

Androgen Receptor Alterations

Merja A. Helenius, PhD,
Kati K. Waltering, MSc,
and Tapio Visakorpi, MD, PhD

CONTENTS

1. INTRODUCTION

It has been known for decades that the growth of prostate cancer (PC) is dependent on androgens. Men castrated early in their life do not develop PC *(1)*. The significance of androgens in the development of PC was recently demonstrated once again in a prevention trial with 5α-reductase inhibitor, finasteride, in which a 25% reduction in the number of PC cases was found in men treated with finasteride *(2)*. 5α-Reductase is an enzyme that converts testosterone into dihydrotestosterone (DHT), the most potent form of androgens in prostate.

Already, in the early 1940s, Huggins and Hodges *(3)* showed that castration is an effective treatment in PC *(3)*. Today, endocrine therapy is the standard

From: *Current Clinical Oncology: Prostate Cancer:*
Signaling Networks, Genetics, and New Treatment Strategies
Edited by: R. G. Pestell and M. T. Nevalainen © Humana Press, Totowa, NJ

treatment for advanced PC. More than 90% of patients initially show a biochemical response to the therapy *(4)*, and clinical response rates of 80% have been reported *(5)*. However, androgen ablation ultimately fails, and so-called androgen-independent or hormone-refractory PC emerges. There are at present no effective therapies available for hormone-refractory PC, although recent clinical trials on docetaxel have indicated that PC is sensitive to at least some chemotherapeutics *(6,7)*.

Androgen action is mediated by a specific receptor molecule, androgen receptor (AR), which is a nuclear transcription factor and a member of the steroid hormone receptor superfamily *(8,9)*. Without the ligand, AR is inactive and bound to heat-shock chaperone proteins *(10)*. AR is activated when lipid-soluble androgens diffuse into cells and bind to ARs. This binding induces a conformation change in the AR, dissociation from chaperone proteins, receptor dimerization, and nuclear translocation of the complex *(9,11)*. In the nucleus, the dimerized receptor complex binds to a palindromic AR response element (ARE) in target genes, influencing their expression. Androgens regulate the expression of hundreds of target genes in the prostate gland *(12,13)* including prostate-specific antigen *(PSA) (14)*, prostate acid phosphatase *(PAP) (15)*, and many growth factors [e.g., vascular endothelial growth factor *(VEGF)* and keratinocyte growth factor *(KGF)*] *(16,17)*, but also genes involved in cell cycle control such as *cdk2* and *cdk4*, and cdk inhibitors, such as *p16, p21, (18,19)*, and *p27 (20)*. Many genes involved in apoptosis and invasion such as *integrins (21–23)*, *prostin-1 (24)*, and *matrix metalloprotease-2 (25)* are also AR-regulated. However, which of the androgen-regulated genes are crucial for the development of the prostate gland or PC are not known. Interestingly, it has now been shown that some of the androgen-regulated genes are upregulated in the progression of PC during androgen withdrawal, indicating re-activation of AR signaling *(26–28)*.

2. ANDROGEN RECEPTOR

2.1. Structure of AR

The human *AR* gene is located on chromosome Xq11–12 *(29)*. It is over 90 kb long and contains eight exons *(30–32)*. The *AR* gene has two transcription initiation sites, the regulatory role of which is still unclear *(33)*. Because of different splicing of the 3′-untranslated region (3′-UTR), two mRNAs of 10.6 kb (the major form) and 7 kb in length *(34)* are produced. The first exon is about 1580 bp long and contains trinucleotide repeat regions, whose lengths are highly polymorphic. The N-terminal CAG repeat coding for the amino acid glutamine is located within the region that is required for full ligand-inducible transcription *(30,35,36)*. The length of this triplet varies from 14 to 35 repeats, an average being 21 ± 2 repeats *(37)*. Decrease in the length of this repeat increases the transcriptional activity of AR *(38)*. The length of the repeat may also vary with ethnicity and race *(37,39,40)*. In some studies, the short germline CAG repeat has been shown to be associated with increased risk of PC *(41)*. However, a number of studies have failed to confirm this association *(42–46)*.

The other trinucleotide, the polyglycine (GGN) repeat, has an average of 16 repeats *(37)*. It lies at the C-terminal of the CAG repeat *(38,47,48)* and shows a lesser degree of polymorphism than the CAG repeat. Short length of the GGN repeat has also been reported to be associated with increased PC risk in two studies *(42,49)*, whereas one study reported an increased risk of PC recurrence and increased risk of death with long GGN repeat lengths (GGN >16) *(44)*. It has been suggested that the most common 16-repeat-long GGC-triplet encodes an AR with an optimal length poly-G tract for normal receptor function *(50)*. This 16-repeat GGC allele has been shown to be least prevalent among African Americans (20%) having a high risk of PC and most prevalent in low-risk Asians (70%). However, because of conflicting data, it is not possible to conclude whether either or both of the polymorphisms are truly associated with the risk of PC.

The AR protein has three functional domains. The amino-terminal ligand-independent transcription activation function domain (AF1) is a primary effector region of AR and is coded by exon 1. Deletion of this AF1 domain diminishes the transactivation potential of the receptor *(35,51)*. The second domain contains a typical DNA-binding domain (DBD) of two zinc fingers and is coded by exons 2–3. It is followed by a hinge region (the end of exon 3 and beginning of exon 4) containing the major nuclear localization signal (NLS) sequence. The carboxy-terminal domain is coded by exons 4–8 and forms a ligand-binding domain (LBD). It also contains a ligand-dependent transcription activation function domain (AF2). The LBD folds into 12 helices that form a typical ligand-binding pocket as in other known members of the steroid hormone receptor superfamily. In AR, the primary contact regions for ligands are helices 4, 5, and 10 *(52–54)*. The LBD interaction with N-terminal trans-activation domain is also needed for the full ligand-dependent transactivation function of AR *(54–56)*.

2.2. Regulation of AR Activity

The AR activates the expression of target genes by facilitating the initiation of transcription. This also requires several auxiliary protein complexes. The transcriptional activity of ligand-bound AR is modulated by interaction of AR with coregulators and by phosphorylation of AR and these coregulators. N- and C-terminal interaction of AR has been reported to be important for the full AR function by influencing receptor dimerization, stabilization of ligand in the ligand-binding pocket, retardation of AR degradation, and binding of AR to chromatin *(57–60)*. Two transcription activation units (TAUs) have been identified in N-terminal domain. The first (TAU-1) is responsible for wild-type AR transactivation capacity and the other (TAU-5) for the transactivation capacity of the constitutively active AR *(51,61)*. Wang et al. *(62)* reported recently that the occupancy of AR coactivation complex on the PSA promoter occurs slowly, peaking 16 h after DHT administration. Post-translational modifications, such as acetylation *(63,64)*, sumoylation *(65)*, and phosphorylation *(66)*, can regulate the action of steroid hormone receptors. AR is phosphorylated in at least seven major sites in living cells *(54,57,66,67)*. One site, Ser94, is constitutively phosphorylated and six sites are phosphorylated in response to androgen. One of them, Ser650, is also phosphorylated in response to a number of non-steroidal agonists [epidermal growth factor (EGF), phorbol myristate acetate and forskolin] *(66,67)*. Phosphorylation of S650 in the hinge region is known to be required for the full activity of AR *(57)*. AR can be sumoylated in the N-terminal domain in an androgen-enhanced fashion, and this sumoylation represses AR-mediated transcription *(65,68)*. In addition, the activity of AR can be prevented by influencing its cellular localization, e.g., DAX-1 represses AR by affecting its localization in the cell *(69)*.

3. ANDROGEN RECEPTOR AND PROSTATE CANCER

3.1. Alterations of AR Expression in Prostate Cancer

Almost all PCs express AR, an exception being rare small cell carcinoma of the prostate *(70,71)*. It has been suggested that high AR expression is associated with short, the recurrent-free survival *(72)* although the finding has not been confirmed. Expression of AR is maintained during prostate carcinogenesis from primary PC to hormone-refractory PC. However, several alterations occur in the AR-signaling pathway during the development and progression of PC. In the normal prostate gland, androgen-stimulated proliferation of the epithelium requires paracrine involvement of stromal cells. The direct effect of androgens on the epithelium seems to be differentiation. In malignant cells, the androgen-mediated signaling has been converted into autocrine mode and no interaction with stroma is needed for the mitogenic effect of androgens *(73,74)*.

It has been shown that androgens inhibit expression of AR mRNA, but this effect is compensated by the stabilization of the AR protein *(75)*. The level of both AR mRNA and protein is increased during long-term androgen ablation *(76–78)*. Expression of AR is also influenced by post-translational and epigenetic modifications. Jarrard et al. *(79)* showed that methylation of the AR promoter CpG island is associated with loss of AR expression *in vitro*. Nakayama et al. *(80)* later reported that AR gene is hypermethylated in about 30% of hormone-refractory PCs. However, the vast majority of prostate tumors expressed AR, at least heterogeneously. In fact, it has now been shown that AR is overexpressed in the vast majority of hormone-refractory PCs as compared with untreated tumors both in mRNA and in protein levels *(81–84)*. Interestingly, the localization of AR in clinical tumors is always nuclear, despite the androgen levels, whether the patient is hormone-naïve, recently castrated, or hormone-refractory, indicating that AR is activated throughout PC progression *(85* Laitinen and Visakorpi, unpublished data).

The significance of the overexpression of AR in hormone-refractory PC was recently demonstrated by Chen and co-authors *(86)*, who first showed that the common denominator in gene expression profiles of androgen-independent PC xenograft models as compared with androgen-dependent counterparts was the increased expression of AR in the independent xenografts. They subsequently showed that overexpression of AR, by transfection, was both necessary and sufficient to convert the growth of the xenografts from androgen-dependent to independent, and to convert androgen antagonist, bicalutamide, to agonist.

3.2. Somatic Mutations of AR in Prostate Cancer

Han and co-authors *(59)* have recently shown that mutated AR causes oncogenic transformation of prostate. They generated transgenic mice expressing a mutated (E231G) form of AR. While the wild-type littermates did not develop cancer, the expression of AR-E231G caused development of prostatic intraepithelial neoplasia that progressed to invasive and metastatic disease in 100% of mice examined. The findings directly confirm the oncogenic properties of AR.

Somatic mutations in *AR* have been extensively screened in PC [*(87)*, see Tables 1 and 2]. Mutations seem to be rare in early-stage, untreated tumors and become more common in late-stage hormone-independent prostate tumors (Table 1). In untreated PC, *AR* mutations have been found in only a few percentages of tumors. There are two reports *(88,89)* in which point mutations have been found in a substantial part of untreated tumors. Gaddipati and co-authors *(88)* found the LNCaP mutation (T877A) in 25% of the transurethral resections of prostate (TURP) specimens of the patients with untreated metastatic PC. Tilley et al. *(89)* reported that about 50% of PCs have a mutated *AR*. However, these findings have not been confirmed by others. Instead, the highest frequency of mutations seems to be in PCs treated with antiandrogen, flutamide or bicalutamide. Mutation frequencies of 10–30% have been reported in such cases *(90–92)*. AR mutations are generally missense point mutations and seem to cluster in discrete hot spots located mostly at C-terminal LBD. The predominant effect of the AR substitutions has been the relaxation of AR ligand specificity, thereby increasing the number of ligands capable of inducing AR transcriptional activity. In addition to LBD of *AR*, some point mutations have been reported to occur in the N-terminal domain where coregulators bind *(89,92–94)*, some in the DBD *(89,95,96)*, and at the boundary of hinge and LBD *(96)*. Somatic contraction of the CAG repeat has also been

Table 1
Somatic Mutations of Androgen Recepter (AR) in Untreated Prostate Cancer

NTD* region (exon1)			DBD** and Hinge region			LBD*** region			All regions		Stage	Reference
Mutation	%	N	Mutation	%	N	Mutation	%	N	%	N		
x	x	x	x	x	x	V730M	4%	1/26	4%	1/26	Stage B	(110)
x	x	x		0%	0/7		0%	0/7	0%	0/7	Stage B	(101)
x	x	x	x	x	x	T877A	25%	6/24	25%	6/24	Stage C or D	(88)
CAG 24->18	3%	1/40	x	x	x	x	x	x	3%	1/40	Most stages A-B	(97)
x	x	x		0%	0/23		0%	0/23	0%	0/23		(173)
	0%	0/31		0%	0/31		0%	0/31	0%	0/31	All stages	(174)
x	x	x		0%	0/30		0%	0/30	0%	0/30	Stage B or C	(175)
M265T:P268S, L54S, L57Q, Q111H, Q64A, D526G, K179R	28%	7/25	L572P	4%	1/25	I670T, S780N, Q668R, S789P, W794X, L828P, F889L	24%	6/25	44%	11/25	Stage C or D	(89)

104

CAG repeat changes									
9%	2/23					9%	2/23		Stage AC (108)
0%	0/84	0%	0/99	0%	0/99	0%	0/99		Stage B (176)
x	x	13	5/38	8%	3/38	21%	8/38		Stage D (176)
		T575A, K580R, A586V, A587S, C619Y		A748T, S865P, Q867X, Q919R, R846G, V757A					
						0%	0/42		(177)
0%	0/10	0%	0/10	0%	0/10	0%	0/10		Stage D (178)

(Continued)

105

Table 1
(Continued)

NTD* region (exon1)			DBD** and Hinge region			LBD*** region			All regions		Stage	Reference
Mutation	%	N	Mutation	%	N	Mutation	%	N	%	N		
P12A	14%	1/7		0%	0/7		0%	0/7	14%	1/7	Metastasis	(179)
G142E	7%	1/14		0%	0/14	E653L, S646F, E654D	21%	3/14	29%	4/14	High grade tumors	(179)
CAG mosaicism	100%	72/72	x	x	x	x	x	x	100%	72/72		(99)

*NTD N-terminal domain; **DBD, DNA binding domain, ***LBD, ligand binding domain a; x, region not analyzed in the study.

Table 2
Somatic Mutations of Androgen Receptor (AR) in Hormone Refractory Prostate Cancer

NTD region (exon1)			DBD and hinge region			LBD region			All regions		Treatment	Reference
Mutation	%	N	Mutation	%	N	Mutation	%	N	%	N		
	0	0/7		0	0/7	V715M	14	1/7	14	1/7	Orchiectomy and flutamide	(109)
x	x	x	x	0	0/8	L701H[a] & T877A	13	1/8	13	1/8	Orchiectomy	(101)
only CAG screened	0	0/18		0	0/18	T877A	0	0/18	0	0/18	Orchiectomy or LHRH analog	(70)
x	x	x	x	x	x	only T877A screened	0	0/23	0	0/23	Orchiectomy or LHRH analog	(85)
x	x	x	x	0	0/10	T877S, H874Y, Q902R, A721T, S647N, G724D, L880Q, A896T	50	5/10	50	5/10	Orchiectomy and flutamide	(102)

(Continued)

Table 2
(Continued)

NTD region (exon1)			DBD and hinge region			LBD region			All regions		Treatment	Reference
Mutation	%	N	Mutation	%	N	Mutation	%	N	%	N		
x	x	x	x	0	0/6	All	0	0/6	0	0/6	castration and antiandrogens	(173)
x	x	x	x	0	0/22	All T877A	14	3/22	14	3/22		(175)
	0	0/13		0	0/13		0	0/13	0	0/13	Orchiectomy or LHRH analog	(174)
x	x	x		0	0/13	G674A	8	1/13	8	1/13	Orchiectomy hormonal or hormonal and antiandrogens	(81)
CAG21->22	1	1/13		0	0/13	L701H[a], K910R	15	2/13	23	3/13		(108)
CAG20->18	3	1/32		0	0/32	G674A	3	1/32	6	2/32	Orchiectomy (most) and/or oestrogens	(98)

NTD			DBD			LBD					Treatment	Ref.
x	x	x	x	x	x	T877A, T877S	31	5/16	31	5/16	orch/LHRH and flutamide	(90)
x	x	x	x	x	x	D890N	6	1/17	6	1/17	Androgen ablation only	(90)
G166S	9	1/11	0	0	0/11	M749I, Q867Q, W741C, D732D	27	3/11	27	4/11	orchiectomy and bicalutamide	(92)
CAG24->20, G524D,G524S, W526X, P514S	19	4/21	P533S	3	1/21	R726L, V757I, V866M	14	3/21	43	7/21	orchiectomy (most) and estramustine or estrogens	(94)
CAG22->21	2	1/48	0	0	0/48	N756D, T877A, W741C, D879G	10	5/48	13	5/48	Orch/LHRH and anti-androgens	(91)
CAG20-> 26	10	1/10	0	0	0	0/10	10	1/10	10	1/10	orchiectomy or LHRH analog	(178)

NTD, N-terminal domain; DBD, DNA binding domain; LBD = ligand binding domain; x, region not analyzed in the study. [a]also found in partial AIS,

109

reported in some cases of both androgen-dependent and hormone-refractory, castration-treated, tumors *(97,98)*. In a recent study, heterogeneity of CAG repeats was found in 100% of PC tissue samples, which was not observed in normal prostate tissue *(99)*. N-terminal domain mutations and CAG repeat alterations could affect the N-terminal and C-terminal interaction needed for ligand binding, and therefore the binding capacity for other ligands, and/or increase the affinity of special AR cofactors binding to N-terminal domain.

The most frequently found point mutation of *AR* is a T877A mutation (threonine at position 877 is substituted to alanine), which was originally characterized in the LNCaP cell line *(100)*. It is thus often called a LNCaP mutation. The same mutation has been detected later in a number of PC patients treated with the antiandrogen flutamide *(101–103)*. This amino acid is located on a helix 11 at the ligand-binding pocket and contacts ligand directly. It alters the stereochemistry of the binding pocket and broadens the ligand binding of AR *(104,105)*. In these mutant ARs, other nuclear hormones (estrogen and progestins), antiandrogens (cyproterone and hydroxyflutamide, but not bicalutamide) *(100,105,106)*, and pregnenolone, a common precursor of steroids *(107)*, can all bind to and activate AR.

Several other mutations at the *AR* LBD, [e.g., L701H, *(101,108)*, V715M *(109)*, V730M *(110)*, and H874Y *(102)*,] that confer enhanced transcriptional sensitivity to antiandrogens and adrenal androgens have been identified. The mutation at position 874 (H874Y) was identified in the PC xenograft CWR22 *(111)*. This site is located distant from the ligand-binding pocket and affects the binding of coregulator proteins (enhanced p160 transactivation) and thus, indirectly, ligand specificity. In this mutant AR, the adrenal androgen dehydroepiandrosterone (DHEA), several other steroid hormones, and flutamide can activate AR *(111)*. The W741C mutation, found in bicalutamide-treated patients *(91,92)*, has been shown to lead to bicalutamide-resistant growth in the LNCaP-subline containing the mutation *(112)*. Bohl et al. *(113)* recently solved the X-ray crystal structure of the mutant W741L AR LBD bound to *R*-bicalutamide and postulated a structural explanation for bicalutamide withdrawal syndrome. Also a double mutation (T877A together with L701H), allowing these cells to respond to a broader spectrum of ligands than either of these point mutations alone, has been detected *(114,115)*. Many of the circulating corticosteroids at the concentration found *in vivo* are able to effectively activate this mutant AR *(115)*.

A smaller number of missense mutations has been detected in other domains of *AR*. A N-terminal missense mutation (K179R) found in untreated primary PCs *(89)* has been reported to lead to a more potent AR *(61)*. A C619Y mutation located near the cysteines forming the zinc-finger motif for DNA binding has been found to cause inactivation and mislocation of the receptor with concomitant sequestration of coactivator SRC-1 (steroid receptor coacti-

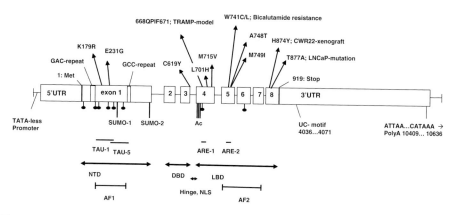

Fig. 1. Structure of androgen receptor *(AR)* gene and the functional domains. Somatic mutations found in clinical prostate cancer (PC) samples with functional significance are shown above the structure. *Phosphorylation sites *(8)* are marked with pinheads. NTD, N-terminal domain; DBD, DNA-binding domain; NLS, nuclear localization signal; LBD, ligand-binding domain; AF1 and 2, activation function 1 and 2; TAU-1 and 5, transactivation units 1 and 5; ARE1 and 2, androgen responsive elements 1 and 2; SUMO-1 and 2, sumoylation sites 1 and 2; Ac, acetylation sites.

vator 1) *(95)*. Buchanan et al. *(96)* demonstrated that the naturally occurring mutations in PC, and also in the transgenic adenocarcinoma mouse prostate (TRAMP) mouse model, at the boundary of the hinge and LBD regions (668QPIF671) result in AR variants with two to fourfold greater transactivation capacity and a wider spectrum of ligands capable of transactivating AR. A list of AR mutations found in PC is presented in Tables 1 and 2 as well as in Fig. 1, and a catalog of all human AR mutations and single-nucleotide polymorphisms (SNPs) identified can be found on the Internet (http://www.mcgill.cs/androgendb.)

The above-mentioned mutations are all located in the coding regions of the *AR* gene. Mutations in non-coding regions of *AR*, which could affect, for example, the expression of AR, have been much less studied. In a recent investigation of non-coding regulatory regions of *AR*, no recurrent alterations were identified to explain overexpression of AR in the hormone-refractory PCs *(116)*. It seems therefore that mutations in these regions are not very common and not likely to be the mechanisms for AR overexpression.

3.3. AR Gene Amplification in Prostate Cancer

Gain and amplification of the *AR* gene is one of the most frequent chromosomal gains in late-stage PC *(117)*. Nearly 80% of hormone-refractory PCs have been reported to carry an elevated *AR* gene copy number *(84)*. About 30% (13–50%) of hormone-refractory PCs contain *AR* gene amplification,

Table 3
Androgen Receptor (AR) Gene Amplifications in Prostate Cancer

Frequency%	N	Reference
Untreated tumors		
0	0/16	(85)
0	0/10	(180)
0	0/26	(81)
1	2/205	(118)
0	0/37	(181)
0	0/22	(182)
5	1/20	(183)
0	0/33	(82)
0	0/9	(184)
2	1/48	(84)
Hormone-refractory tumors		
30	7/23	(85)
30	3/10	(180)
28	15/54	(81)
23	11/47	(118)
22	12/59	(118)
20	1/5	(181)
56	10/18	(119)
13	10/77	(121)
36	23/63	(182)
15	3/20	(183)
0	0/11	(92)
31	4/13	(82)
20	2/10	(82)
25	4/16	(94)
50	9/18	(184)
18	4/22	(185)
20	10/49	(84)
33	8/24	(120)

whereas untreated primary tumors rarely if ever contain amplification of the *AR* gene *(84,85,118–120,* lists in Table 3). Even in cases where the *AR* gene was amplified in a hormone-refractory tumor and the tumor specimens prior to treatment have been available, only one case of amplification has been detected in the untreated samples. This indicates that the amplification is selected for during the emergence of hormone-refractory disease *(81,84,85,119,120).* The *AR* gene amplification is associated with increased expression of *AR* as expected for a target gene of the amplification *(81,82,84,119)*. The low level gain of *AR* copy number probably has only a marginal effect on *AR* expression levels *(82,84,120)*. There is some evidence that *AR* amplification is more often found in hormone-refractory tumors that originally responded well to androgen ablation than in poorly responding tumors *(81)*. The finding suggests that tumors with *AR* gene amplifications are hypersensitive to androgens remaining in the body during castration. Indeed, patients with *AR* gene amplification respond better to second-line maximal androgen blockade, also preventing the effects of adrenal androgens, than patients without the amplification *(121)*.

3.4. Alteration of AR Transcriptional Activity and Prostate Cancer

The binding of AR to DNA is the first step in the assembly of an active protein complex on DNA needed for the transcription initiation. AR action is mediated by a growing group of coregulatory proteins (either coactivators or corepressors) that bind to AR and influence the initiation of transcription. Both types of coregulators are needed for efficient modulation of AR-regulated gene expression *(122–125)*.

Most of the AR coregulators identified so far are coactivators. They modify chromatin structure and interact with the basal transcription machinery. They enhance the AR transactivation potential in a ligand-dependent manner. They bind to AF1 in N-terminal end of AR and some in the N-terminal signature sequence in AF-2. AR coactivators either function as mediator proteins in transcription initiation bridging the transcriptional machinery to nuclear receptor (p300/CBP), or control the susceptibility of chromatin to transcription through, for example, histone acetyltransferase (p160 family) or ATPase (ARIP4) functions *(122,126)*. Overexpression of several coactivators such as SRC-1, transcriptional intermediary factor 2/glucocorticoid-receptor-interacting protein 1 (TIF-2/GRIP1), ARA55, or ARA70 has been shown to increase the transcriptional activity of AR in response to low concentration of DHT or to low-affinity ligands like DHEA and androstenedione *(122,127,128)*. Several coregulators such as SRC-1, CREB-binding protein (CBP), and p300 have been identified as targets of signaling pathways like mitogen-activated protein kinase (MAPK). For example, EGF increases the transcriptional activity of AR through MAPK pathway by enhancing the phosphorylation of coactivator GRIP1 *(129)*. Recently, Metzger et al. *(130)* reported a ligand-dependent association of lysine-specific demethylase 1 (LSD1) and AR at the PSA promoter and a specific demethylation of repressive histone marks resulting in a strong activation of the target gene. In addition, LSD1 was reported to affect androgen-induced cell proliferation. β-Catenin, a multifunctional adaptor protein and oncoprotein, is reported to function as a transcriptional coactivator of AR *(131–133)*. β-Catenin can also bind to coactivator GRIP1 and synergistically enhance the activity of AR *(134)*. In addition, β-catenin is able to increase AR responsiveness to unspecific ligands such as androstenedione and estradiol, and diminish the action of bicalutamide *(86,131,134–136)*.

Corepressors mediate their action using several different strategies *(137)*. They can *(1)* inhibit the DNA binding or nuclear translocation of AR (e.g., calreticulin, DAX-1, PAK6), *(2)* recruit histone deacetylases (HDACs) (e.g., HDAC1, SMRT/NCoR, TGIF), *(3)* interrupt the interaction between AR and its coactivators (e.g., SHP, cyclin D1), *(4)* interrupt the interaction between the N- and C-terminus of AR (e.g., FLNa, GSK3β), *(5)* function as scaffolds for other AR coregulators (e.g., PATZ), and *(6)* target the basal transcriptional machinery (e.g., AES). In addition, there are several other AR corepressors identified whose mechanisms of inhibition are not yet solved (e.g., HBOI, SRY, Ebpl). Corepressors can bind to both liganded and unliganded AR depending on the particular corepressor. Antagonist-bound AR is shown to recruit corepressor–HDAC complexes at the target gene promoter, inducing target gene repression *(138)*. Corepressor proteins SMRT (silencing mediator of retinoid and thyroid hormone receptor) and N-CoR (nuclear receptor corepressor) have been demonstrated to be involved in transcriptional suppression by both agonist-bound and antagonist-bound AR. They regulated the magnitude of hormone response, at least partly, by competing with coactivators *(138,139)*. In addition, there are evidences that SMRT and NCoR can repress AR transcription through other mechanisms *(137,140)*.

Because of their function, it has been suggested that AR coregulators could play a role in PC tumorigenesis. Still, few investigations on the alterations of the coregulators in PC have been reported. However, expression of a large number of coregulators has been studied *(127,141,142)*. Overexpression of SRC-1 protein in hormone-refractory tumors has been demonstrated *(126,142)*, whereas paradoxically the *SRC-1* mRNA seems to be downregulated in hormone-refractory PC *(141)*. Alterations in the expression of TIF-2/GRIP1 have also been suggested *(127)*, but not confirmed by all *(141)*. Most importantly, no common genetic alterations in the AR coregulator genes have, so far, been reported in PC. For example, unlike in breast cancer, where amplification of *AIB1* (alias nuclear receptor coactivator 3, NCOA3) gene has been detected *(143)*, no amplifications of this gene have been found in PC. Recently, *SRC-1* was screened for mutations and gene amplification in PC *(141,142)*. Only one case of *SRC-1* gene amplification was found and no recurrent sequence variations were detected. Obviously, more analyses of coregulators are needed to ascertain whether they are targeted by genetic alterations. However, there are some evidences that *AR* mutations occurring in regions that influence interactions with AR coregulators can cause oncogenic transformation of the prostate *(59,111)*.

3.5. *Ligand-Independent Activation of the AR in Prostate Cancer*

One potential AR activation mechanism is a ligand-independent activation. In the absence of a ligand, AR can be activated by many growth factors [EGF, KGF, insulin-like growth factor 1 (IGF-1) *(144)*, a growth factor receptor (Her2/neu) *(145)*, cytokines (interleukin-6) *(146)*, and protein kinase-A *(147,148)*]. In addition, components of the MAPK pathway *(145,149,150)* and PI3K/Akt signaling pathway *(135,151)* are able to activate AR in a ligand-independent manner. These factors and their signaling pathways can increase intracellular kinase activity or decrease phosphatase activity affecting the phosphorylation of AR and its coregulators *(152,153)*. Thus, stimulation with increased level of growth factors may decrease the requirement for androgens in AR activation in PC cells.

Overexpression of peptide growth factors and their receptors is believed to activate AR without the ligand. Thus, they could also contribute to the failure of androgen deprivation therapy. The level of EGF, transforming growth factor-α (TGF-α), KGF, basic fibroblast growth factor (bFGF), and IGF-I and their receptors have been reported to be overexpressed in advanced hormone-independent PCs *(144,154,155)*. In model systems, overexpression, for example, of HER2/neu (ERBB2), a member of the EGF receptor family receptor, enhanced the AR-activated gene expression (such as PSA) in a ligand-dependent manner and increased cell survival *(145,156)*. HER2/neu signaling through MAPK and Akt pathways can also promote ligand-independent activation of AR *(157)* and modulate the DNA binding and stability of AR *(158)*. However, the expression of HER2/neu is not altered in PC in humans *(159)*.

It has been demonstrated that decreased PTEN and GSK3β activity leads to activation of both PI3K/Akt and AR signaling, providing necessary survival signals to prostate epithelial cells to escape the apoptotic response associated with androgen ablation therapy *(160)*. Inactivating mutations in *PTEN* and amplification of *AR* belong to the most commonly known genetic alterations in hormone-refractory PCs and, interestingly, there seems to be crosstalk between these two pathways.

4. FUTURE DIRECTIONS

It is now clear that AR signaling plays a major role in PC from early development to the lethal progression of the disease. AR signaling is affected by several forms of genetic alterations (Fig. 2). The crucial question is why the currently available endocrine therapies (androgen withdrawal and antiandrogens) do not inhibit AR signaling better. Obviously, new, more potent pharmaceuticals to inhibit AR-signaling pathway are needed. It has already been shown that hammerhead ribozymes *(161)* and antisense oligonucleotides

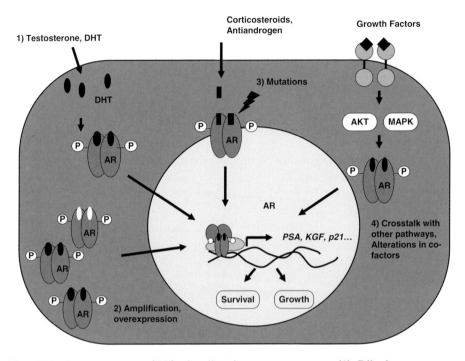

Fig. 2. Androgen receptor (AR) signaling in prostate cancer. (1) Dihydrotestosterone (DHT), which is the active form of testosterone (T) in the prostate, binds to AR, which is then released from chaperones to form a compact homodimer, and is translocated into the nucleus to regulate transcription of several genes involved in growth and differentiation. Changes in AR signaling leads to androgen-independent growth of prostate cancer cells. (2) Overexpression and amplification of AR gene results in high protein levels of AR leading to hypersensitivity to androgens. (3) Point mutations of the AR gene widen the spectrum of ligands capable to bind and transactivate AR. (4) Alterations of AR cofactors and/or altered growth factor signaling crosstalking with AR could mediate the growth signals for a cell.

(162) can be used to inhibit AR. In addition, small molecules such as resveratrol, vitamin E, selenium, some non-steroidal anti-inflammatory drugs, and geldanamycin decrease the expression of AR target genes and downregulate cell proliferation *(163–168)*. However, more potent molecules still need to be designed. As the recent findings from the novel targeted drugs such as trastuzumab *(169)*, gefitinib *(170,171)*, and imatinib *(172)* indicate, genetic alterations pinpoint the weak-spot of cancer. Thus, one would expect that at least hormone-refractory tumors with *AR* gene amplification or mutation should be highly susceptible to targeted drugs against AR.

5. SUMMARY

AR is a critical mediator of PC, transmitting growth signals to tumor cells throughout the onset and progression of PC. Binding of androgens to AR induces its transcriptional activity, influencing the expression of androgen responsive genes. The importance of AR in PC is exemplified by the facts that *AR* gene is expressed in almost all prostate carcinomas and that the gene is targeted both by gene amplification and mutations. Although androgen withdrawal and antiandrogens, such as flutamide and bicalutamide, inhibit AR-mediated signaling and initially prevent tumor growth, it is now clear that the AR signaling is later re-activated during the treatment and the development of hormone-refractory PC. Thus, new, more potent drugs to inhibit AR are still needed.

REFERENCES

1. Isaacs JT. (1994) Role of androgens in prostatic cancer. *Vitam Horm* **49**, 433–502.
2. Thompson IM, Goodman PJ, Tangen CM, Lucia MS, Miller GJ, Ford LG, Lieber MM, Cespedes RD, Atkins JN, Lippman SM, Carlin SM, Ryan A, Szczepanek CM, Crowley JJ, Coltman CA Jr. (2003) The influence of finasteride on the development of prostate cancer. *N Engl J Med* **349**(3), 215–24.
3. Huggins C, Hodges CV. (1941) The effect of castration, of estrogen and of androgen injection on serum phosphatases in metastatic carcinoma of the prostate. *Cancer Res* **1**, 293–7.
4. Palmberg C, Koivisto P, Visakorpi T, Tammela TL. (1999) PSA decline is an independent prognostic marker in hormonally treated prostate cancer. *Eur Urol* **36**(3), 191–6.
5. Gittes RF. (1991) Carcinoma of the prostate. *N Engl J Med* **324**(4), 236–45.
6. Tannock IF, de Wit R, Berry WR, Horti J, Pluzanska A, Chi KN, Oudard S, Theodore C, James ND, Turesson I, Rosenthal MA, Eisenberger MA; TAX 327 Investigators. (2004) Docetaxel plus prednisone or mitoxantrone plus prednisone for advanced prostate cancer. *N Engl J Med* **351**(15), 1502–12.
7. Petrylak DP, Tangen CM, Hussain MH, Lara PN Jr, Jones JA, Taplin ME, Burch PA, Berry D, Moinpour C, Kohli M, Benson MC, Small EJ, Raghavan D, Crawford ED. (2004) Docetaxel and estramustine compared with mitoxantrone and prednisone for advanced refractory prostate cancer. *N Engl J Med* **351**(15), 1513–20.
8. Huang H, Tindall DJ. (2002) The role of the androgen receptor in prostate cancer. *Crit Rev Eukaryot Gene Expr* **12**(3), 193–207.
9. Black BE, Paschal BM. (2004) Intranuclear organization and function of the androgen receptor. *Trends Endocrinol Metab* **15**(9), 411–7.
10. Prescott J, Coetzee GA. (2006) Molecular chaperones throughout the life cycle of the androgen receptor. *Cancer Lett* **231**(1), 12–9.
11. Moras D, Gronemeyer H. (1998) The nuclear receptor ligand-binding domain: structure and function. *Curr Opin Cell Biol* **10**(3), 384–91.
12. Nelson PS, Clegg N, Arnold H, Ferguson C, Bonham M, White J, Hood L, Lin B. (2002) The program of androgen-responsive genes in neoplastic prostate epithelium. *Proc Natl Acad Sci USA* **99**(18), 11890–5.
13. Dehm SM, Tindall DJ. (2006) Molecular regulation of androgen action in prostate cancer. *J Cell Biochem* **99**(2), 333–44.

14. Young CY, Andrews PE, Montgomery BT, Tindall DJ. (1992) Tissue-specific and hormonal regulation of human prostate-specific glandular kallikrein. *Biochemistry* **31**(3), 818–24.

15. Perry JE, Grossmann ME, Tindall DJ. (1996) Androgen regulation of gene expression. *Prostate Suppl* **6**, 79–81.

16. Shibata Y, Kashiwagi B, Arai S, Fukabori Y, Suzuki K, Honma S, Yamanaka H. (2004) Direct regulation of prostate blood flow by vascular endothelial growth factor and its participation in the androgenic regulation of prostate blood flow in vivo. *Endocrinology* **145**(10), 4507–12.

17. Fasciana C, van der Made AC, Faber PW, Trapman J. (1996) Androgen regulation of the rat keratinocyte growth factor (KGF/FGF7) promoter. *Biochem Biophys Res Commun* **220**(3), 858–63.

18. Lu S, Tsai SY, Tsai MJ. (1997) Regulation of androgen-dependent prostatic cancer cell growth: androgen regulation of CDK2, CDK4, and CKI p16 genes. *Cancer Res* **57**(20), 4511–6.

19. Lu S, Tsai SY, Tsai MJ. (1999) Molecular mechanisms of androgen-independent growth of human prostate cancer LNCaP-AI cells. *Endocrinology* **140**(11), 5054–9.

20. Tsihlias J, Zhang W, Bhattacharya N, Flanagan M, Klotz L, Slingerland J. (2000) Involvement of p27Kip1 in G1 arrest by high dose 5 alpha-dihydrotestosterone in LNCaP human prostate cancer cells. *Oncogene* **19**(5), 670–9.

21. Bonaccorsi L, Carloni V, Muratori M, Salvadori A, Giannini A, Carini M, Serio M, Forti G, Baldi E. (2000) Androgen receptor expression in prostate carcinoma cells suppresses alpha6beta4 integrin-mediated invasive phenotype. *Endocrinology* **141**(9), 3172–82.

22. Bonaccorsi L, Muratori M, Marchiani S, Forti G, Baldi E. (2006) The androgen receptor and prostate cancer invasion. *Mol Cell Endocrinol* **246**(1–2), 157–62.

23. Evangelou A, Letarte M, Marks A, Brown TJ. (2002) Androgen modulation of adhesion and antiadhesion molecules in PC-3 prostate cancer cells expressing androgen receptor. *Endocrinology* **143**(10), 3897–904.

24. Manos EJ, Kim ML, Kassis J, Chang PY, Wells A, Jones DA. (2001) Dolichol-phosphate-mannose-3 (DPM3)/prostin-1 is a novel phospholipase C-gamma regulated gene negatively associated with prostate tumor invasion. *Oncogene* **20**(22), 2781–90.

25. Liao X, Thrasher JB, Pelling J, Holzbeierlein J, Sang QX, Li B. (2003) Androgen stimulates matrix metalloproteinase-2 expression in human prostate cancer. *Endocrinology* **144**(5), 1656–63.

26. Amler LC, Agus DB, LeDuc C, Sapinoso ML, Fox WD, Kern S, Lee D, Wang V, Leysens M, Higgins B, Martin J, Gerald W, Dracopoli N, Cordon-Cardo C, Scher HI, Hampton GM. (2000) Dysregulated expression of androgen-responsive and nonresponsive genes in the androgen-independent prostate cancer xenograft model CWR22-R1. *Cancer Res* **60**(21), 6134–41.

27. Mousses S, Wagner U, Chen Y, Kim JW, Bubendorf L, Bittner M, Pretlow T, Elkahloun AG, Trepel JB, Kallioniemi OP. (2001) Failure of hormone therapy in prostate cancer involves systematic restoration of androgen responsive genes and activation of rapamycin sensitive signaling. *Oncogene* **20**(46), 6718–23.

28. Holzbeierlein J, Lal P, LaTulippe E, Smith A, Satagopan J, Zhang L, Ryan C, Smith S, Scher H, Scardino P, Reuter V, Gerald WL. (2004) Gene expression analysis of human prostate carcinoma during hormonal therapy identifies androgen-responsive genes and mechanisms of therapy resistance. *Am J Pathol* **164**(1), 217–27.

29. Brown CJ, Goss SJ, Lubahn DB, Joseph DR, Wilson EM, French FS, Willard HF. (1989) Androgen receptor locus on the human X chromosome: regional localization to Xq11–12 and description of a DNA polymorphism. *Am J Hum Genet* **44**, 264–9.

30. Chang C, Kokontis J, Liao S. (1988) Molecular cloning of human and rat complementary DNA encoding androgen receptors. *Science* **240**, 324–6.

31. Lubahn DB, Joseph DR, Sullivan PM, Willard HF, French FS, Wilson EM. (1988) Cloning of human androgen receptor complementary DNA and localization to the X chromosome. *Science* **240**, 327–30.

32. Lubahn DB, Joseph DR, Sar M, Tan J, Higgs HN, Larson RE, French FS, Wilson EM. (1988) The human androgen receptor: complementary deoxyribonucleic acid cloning, sequence analysis and gene expression in prostate. *Mol Endocrinol* **2**, 1265–75.

33. Faber PW, van Rooij HC, Schipper HJ, Brinkmann AO, Trapman J. (1993) Two different, overlapping pathways of transcription initiation are active on the TATA-less human androgen receptor promoter. The role of Sp1. *J Biol Chem* **268**(13), 9296–301.

34. Faber PW, van Rooij HC, van der Korput HA, Baarends WM, Brinkmann AO, Grootegoed JA, Trapman J. (1991) Characterization of the human androgen receptor transcription unit. *J Biol Chem* **266**(17), 10743–9.

35. Jenster G, van der Korput HA, van Vroonhoven C, van der Kwast TH, Trapman J, Brinkmann AO. (1991) Domains of the human androgen receptor involved in steroid binding, transcriptional activation, and subcellular localization. *Mol Endocrinol* **5**(10), 1396–404.

36. Simental JA, Sar M, Lane MV, French FS, Wilson EM. (1991) Transcriptional activation and nuclear targeting signals of the human androgen receptor. *J Biol Chem* **266**(1), 510–8.

37. Irvine RA, Yu MC, Ross RK, Coetzee GA. (1995) The CAG and GGC microsatellites of the androgen receptor gene are in linkage disequilibrium in men with prostate cancer. *Cancer Res* **55**(9), 1937–40.

38. Chamberlain NL, Driver ED, Miesfeld RL. (1994) The length and location of CAG trinucleotide repeats in the androgen receptor N-terminal domain affect transactivation function. *Nucleic Acids Res* **22**(15), 3181–6.

39. Sartor O, Zheng Q, Eastham JA. (1999) Androgen receptor gene CAG repeat length varies in a race-specific fashion in men without prostate cancer. *Urology* **53**(2), 378–80.

40. Buchanan G, Yang M, Cheong A, Harris JM, Irvine RA, Lambert PF, Moore NL, Raynor M, Neufing PJ, Coetzee GA, Tilley WD. (2004) Structural and functional consequences of glutamine tract variation in the androgen receptor. *Hum Mol Genet* **13**(16), 1677–92.

41. Giovannucci E, Stampfer MJ, Krithivas K, Brown M, Dahl D, Brufsky A, Talcott J, Hennekens CH, Kantoff PW. (1997) The CAG repeat within the androgen receptor gene and its relationship to prostate cancer. *Proc Natl Acad Sci USA* **94**(7), 3320–3.

42. Stanford JL, Just JJ, Gibbs M, Wicklund KG, Neal CL, Blumenstein BA, Ostrander EA. (1997) Polymorphic repeats in the androgen receptor gene: molecular markers of prostate cancer risk. *Cancer Res* **57**(6), 1194–8.

43. Correa-Cerro L, Wohr G, Haussler J, Berthon P, Drelon E, Mangin P, Fournier G, Cussenot O, Kraus P, Just W, Paiss T, Cantu JM, Vogel W. (1999) (CAG)nCAA and GGN repeats in the human androgen receptor gene are not associated with prostate cancer in a French-German population. *Eur J Hum Genet* **7**(3), 357–62.

44. Edwards SM, Badzioch MD, Minter R, Hamoudi R, Collins N, Ardern-Jones A, Dowe A, Osborne S, Kelly J, Shearer R, Easton DF, Saunders GF, Dearnaley DP, Eeles RA. (1999) Androgen receptor polymorphisms: association with prostate cancer risk, relapse and overall survival. *Int J Cancer* **84**(5), 458–65.

45. Salinas CA, Austin MA, Ostrander EO, Stanford JL. (2005) Polymorphisms in the androgen receptor and the prostate-specific antigen genes and prostate cancer risk. *Prostate* **65**(1), 58–65.

46. Mononen N, Ikonen T, Autio V, Rokman A, Matikainen MP, Tammela TL, Kallioniemi OP, Koivisto PA, Schleutker J. (2002) Androgen receptor CAG polymorphism and prostate cancer risk. *Hum Genet* **111**(2), 166–71.

47. Gao T, Marcelli M, McPhaul MJ. (1996) Transcriptional activation and transient expression of the human androgen receptor. *J Steroid Biochem Mol Biol* **59**(1), 9–20.

48. Kazemi-Esfarjani P, Trifiro MA, Pinsky L. (1995) Evidence for a repressive function of the long polyglutamine tract in the human androgen receptor: possible pathogenetic relevance for the (CAG)n-expanded neuronopathies. *Hum Mol Genet* **4**(4), 523–7.

49. Hakimi JM, Schoenberg MP, Rondinelli RH, Piantadosi S, Barrack ER. (1997) Androgen receptor variants with short glutamine or glycine repeats may identify unique subpopulations of men with prostate cancer. *Clin Cancer Res* **3**(9), 1599–608.

50. Buchanan G, Irvine RA, Coetzee GA, Tilley WD. (2001) Contribution of the androgen receptor to prostate cancer predisposition and progression. *Cancer Metastasis Rev* **20**(3–4), 207–23.

51. Jenster G, van der Korput HA, Trapman J, Brinkmann AO. (1995) Identification of two transcription activation units in the N-terminal domain of the human androgen receptor. *J Biol Chem* **270**(13), 7341–6.

52. Matias PM, Donner P, Coelho R, Thomaz M, Peixoto C, Macedo S, Otto N, Joschko S, Scholz P, Wegg A, Basler S, Schafer M, Egner U, Carrondo MA. (2000) Structural evidence for ligand specificity in the binding domain of the human androgen receptor. Implications for pathogenic gene mutations. *J Biol Chem* **275**(34), 26164–71.

53. Gobinet J, Poujol N, Sultan C. (2002) Molecular action of androgens. *Mol Cell Endocrinol* **198**(1–2), 15–24.

54. Gelmann EP. (2002) Molecular biology of the androgen receptor. *J Clin Oncol* **20**(13), 3001–15.

55. Ikonen T, Palvimo JJ, Janne OA. (1997) Interaction between the amino- and carboxyl-terminal regions of the rat androgen receptor modulates transcriptional activity and is influenced by nuclear receptor coactivators. *J Biol Chem* **272**(47), 29821–8.

56. Schaufele F, Carbonell X, Guerbadot M, Borngraeber S, Chapman MS, Ma AA, Miner JN, Diamond MI. (2005) The structural basis of androgen receptor activation: intramolecular and intermolecular amino-carboxy interactions. *Proc Natl Acad Sci USA* **102**(28), 9802–7.

57. Zhou ZX, Kemppainen JA, Wilson EM. (1995) Identification of three proline-directed phosphorylation sites in the human androgen receptor. *Mol Endocrinol* **9**(5), 605–15.

58. He B, Kemppainen JA, Voegel JJ, Gronemeyer H, Wilson EM. (1999) Activation function 2 in the human androgen receptor ligand binding domain mediates interdomain communication with the NH(2)-terminal domain. *J Biol Chem* **274**(52), 37219–25.

59. Han G, Buchanan G, Ittmann M, Harris JM, Yu X, Demayo FJ, Tilley W, Greenberg NM. (2005) Mutation of the androgen receptor causes oncogenic transformation of the prostate. *Proc Natl Acad Sci USA* **102**(4), 1151–6.

60. Li J, Fu J, Toumazou C, Yoon HG, Wong J. (2006) A role of the amino-terminal (N) and carboxyl-terminal (C) interaction in binding of androgen receptor to chromatin. *Mol Endocrinol* **20**(4), 776–85.

61. Callewaert L, Van Tilborgh N, Claessens F. (2006) Interplay between two hormone-independent activation domains in the androgen receptor. *Cancer Res* **66**(1), 543–53.

62. Wang Q, Carroll JS, Brown M. (2005) Spatial and temporal recruitment of androgen receptor and its coactivators involves chromosomal looping and polymerase tracking. *Mol Cell* **19**(5), 631–42.

63. Fu M, Rao M, Wang C, Sakamaki T, Wang J, Di Vizio D, Zhang X, Albanese C, Balk S, Chang C, Fan S, Rosen E, Palvimo JJ, Janne OA, Muratoglu S, Avantaggiati ML, Pestell RG. (2003) Acetylation of androgen receptor enhances coactivator binding and promotes prostate cancer cell growth. *Mol Cell Biol* **23**(23), 8563–75.

64. Fu M, Wang C, Zhang X, Pestell RG. (2004) Acetylation of nuclear receptors in cellular growth and apoptosis. *Biochem Pharmacol* **68**(6), 1199–208.

65. Poukka H, Karvonen U, Janne OA, Palvimo JJ. (2000) Covalent modification of the androgen receptor by small ubiquitin-like modifier 1 (SUMO-1). *Proc Natl Acad Sci USA* **97**(26), 14145–50.

66. Gioeli D, Ficarro SB, Kwiek JJ, Aaronson D, Hancock M, Catling AD, White FM, Christian RE, Settlage RE, Shabanowitz J, Hunt DF, Weber MJ. (2002) Androgen receptor phosphorylation. Regulation and identification of the phosphorylation sites. *J Biol Chem* **277**(32), 29304–14.

67. Gioeli D. (2005) Signal transduction in prostate cancer progression. *Clin Sci (Lond)* **108**(4), 293–308.

68. Nishida T, Yasuda H. (2002) PIAS1 and PIASxalpha function as SUMO-E3 ligases toward androgen receptor and repress androgen receptor-dependent transcription. *J Biol Chem* **277**(44), 41311–7.

69. Holter E, Kotaja N, Makela S, Strauss L, Kietz S, Janne OA, Gustafsson JA, Palvimo JJ, Treuter E. (2002) Inhibition of androgen receptor (AR) function by the reproductive orphan nuclear receptor DAX-1. *Mol Endocrinol* **16**(3), 515–28.

70. Ruizeveld de Winter JA, Janssen PJ, Sleddens HM, Verleun-Mooijman MC, Trapman J, Brinkmann AO, Santerse AB, Schroder FH, van der Kwast TH. (1994) Androgen receptor status in localized and locally progressive hormone refractory human prostate cancer. *Am J Pathol* **144**(4), 735–46.

71. Helpap B, Kollermann J. (1999) Undifferentiated carcinoma of the prostate with small cell features: immunohistochemical subtyping and reflections on histogenesis. *Virchows Arch* **434**(5), 385–91.

72. Lee DK, Chang C. (2003) Endocrine mechanisms of disease: Expression and degradation of androgen receptor: mechanism and clinical implication. *J Clin Endocrinol Metab* **88**(9), 4043–54.

73. Gao J, Arnold JT, Isaacs JT. (2001) Conversion from a paracrine to an autocrine mechanism of androgen-stimulated growth during malignant transformation of prostatic epithelial cells. *Cancer Res* **61**(13), 5038–44.

74. Henshall SM, Quinn DI, Lee CS, Head DR, Golovsky D, Brenner PC, Delprado W, Stricker PD, Grygiel JJ, Sutherland RL. (2001) Altered expression of androgen receptor in the malignant epithelium and adjacent stroma is associated with early relapse in prostate cancer. *Cancer Res* **61**(2), 423–7.

75. Krongrad A, Wilson CM, Wilson JD, Allman DR, McPhaul MJ. (1991) Androgen increases androgen receptor protein while decreasing receptor mRNA in LNCaP cells. *Mol Cell Endocrinol* **76**(1–3), 79–88.

76. Kokontis J, Takakura K, Hay N, Liao S. (1994) Increased androgen receptor activity and altered c-myc expression in prostate cancer cells after long-term androgen deprivation. *Cancer Res* **54**(6), 1566–73.

77. Culig Z, Hoffmann J, Erdel M, Eder IE, Hobisch A, Hittmair A, Bartsch G, Utermann G, Schneider MR, Parczyk K, Klocker H. (1999) Switch from antagonist to agonist of the androgen receptor bicalutamide is associated with prostate tumour progression in a new model system. *Br J Cancer* **81**(2), 242–51.

78. Hara T, Nakamura K, Araki H, Kusaka M, Yamaoka M. (2003) Enhanced androgen receptor signaling correlates with the androgen-refractory growth in a newly established MDA PCa 2b-hr human prostate cancer cell subline. *Cancer Res* **63**(17), 5622–8.

79. Jarrard DF, Kinoshita H, Shi Y, Sandefur C, Hoff D, Meisner LF, Chang C, Herman JG, Isaacs WB, Nassif N. (1998) Methylation of the androgen receptor promoter CpG island is associated with loss of androgen receptor expression in prostate cancer cells. *Cancer Res* **58**(23), 5310–4.

80. Nakayama T, Watanabe M, Suzuki H, Toyota M, Sekita N, Hirokawa Y, Mizokami A, Ito H, Yatani R, Shiraishi T.(2000) Epigenetic regulation of androgen receptor gene expression in human prostate cancers. *Lab Invest* **80**(12), 1789–96.

81. Koivisto P, Kononen J, Palmberg C, Tammela T, Hyytinen E, Isola J, Trapman J, Cleutjens K, Noordzij A, Visakorpi T, Kallioniemi OP. (1997) Androgen receptor gene amplification:

a possible molecular mechanism for androgen deprivation therapy failure in prostate cancer. *Cancer Res* **57**(2), 314–9.

82. Linja MJ, Savinainen KJ, Saramaki OR, Tammela TL, Vessella RL, Visakorpi T. (2001) Amplification and overexpression of androgen receptor gene in hormone-refractory prostate cancer. *Cancer Res* **61**(9), 3550–5.

83. Latil A, Bieche I, Vidaud D, Lidereau R, Berthon P, Cussenot O, Vidaud M. (2001) Evaluation of androgen, estrogen (ER alpha and ER beta), and progesterone receptor expression in human prostate cancer by real-time quantitative reverse transcription-polymerase chain reaction assays. *Cancer Res* **61**(5), 1919–26.

84. Edwards J, Krishna NS, Grigor KM, Bartlett JM. (2003) Androgen receptor gene amplification and protein expression in hormone refractory prostate cancer. *Br J Cancer* **89**(3), 552–6.

85. Visakorpi T, Hyytinen E, Koivisto P, Tanner M, Keinanen R, Palmberg C, Palotie A, Tammela T, Isola J, Kallioniemi OP. (1995) In vivo amplification of the androgen receptor gene and progression of human prostate cancer. *Nat Genet* **9**(4), 401–6.

86. Chen CD, Welsbie DS, Tran C, Baek SH, Chen R, Vessella R, Rosenfeld MG, Sawyers CL. (2004) Molecular determinants of resistance to antiandrogen therapy. *Nat Med* **10**(1), 33–9.

87. Linja MJ, Visakorpi T. (2004) Alterations of androgen receptor in prostate cancer. *J Steroid Biochem Mol Biol* **92**(4), 255–64.

88. Gaddipati JP, McLeod DG, Heidenberg HB, Sesterhenn IA, Finger MJ, Moul JW, Srivastava S. (1994) Frequent detection of codon 877 mutation in the androgen receptor gene in advanced prostate cancers. *Cancer Res* **54**(11), 2861–4.

89. Tilley WD, Buchanan G, Hickey TE, Bentel JM. (1996) Mutations in the androgen receptor gene are associated with progression of human prostate cancer to androgen independence. *Clin Cancer Res* **2**(2), 277–85.

90. Taplin ME, Bubley GJ, Ko YJ, Small EJ, Upton M, Rajeshkumar B, Balk SP. (1999) Selection for androgen receptor mutations in prostate cancers treated with androgen antagonist. *Cancer Res* **59**(11), 2511–5.

91. Taplin ME, Rajeshkumar B, Halabi S, Werner CP, Woda BA, Picus J, Stadler W, Hayes DF, Kantoff PW, Vogelzang NJ, Small EJ; Cancer and Leukemia Group B Study 9663. (2003) Androgen receptor mutations in androgen-independent prostate cancer: Cancer and Leukemia Group B Study 9663. *J Clin Oncol* **21**(14), 2673–8.

92. Haapala K, Hyytinen ER, Roiha M, Laurila M, Rantala I, Helin HJ, Koivisto PA. (2001) Androgen receptor alterations in prostate cancer relapsed during a combined androgen blockade by orchiectomy and bicalutamide. *Lab Invest* **81**(12), 1647–51.

93. Buchanan G, Greenberg NM, Scher HI, Harris JM, Marshall VR, Tilley WD. (2001) Collocation of androgen receptor gene mutations in prostate cancer. *Clin Cancer Res* **7**(5), 1273–81.

94. Hyytinen ER, Haapala K, Thompson J, Lappalainen I, Roiha M, Rantala I, Helin HJ, Janne OA, Vihinen M, Palvimo JJ, Koivisto PA. (2002) Pattern of somatic androgen receptor gene mutations in patients with hormone-refractory prostate cancer. *Lab Invest* **82**(11), 1591–8.

95. Nazareth LV, Stenoien DL, Bingman WE 3rd, James AJ, Wu C, Zhang Y, Edwards DP, Mancini M, Marcelli M, Lamb DJ, Weigel NL. (1999) A C619Y mutation in the human androgen receptor causes inactivation and mislocalization of the receptor with concomitant sequestration of SRC-1 (steroid receptor coactivator 1). *Mol Endocrinol* **13**(12), 2065–75.

96. Buchanan G, Yang M, Harris JM, Nahm HS, Han G, Moore N, Bentel JM, Matusik RJ, Horsfall DJ, Marshall VR, Greenberg NM, Tilley WD. (2001) Mutations at the boundary of the hinge and ligand binding domain of the androgen receptor confer increased transactivation function. *Mol Endocrinol* **15**(1), 46–56.

97. Schoenberg MP, Hakimi JM, Wang S, Bova GS, Epstein JI, Fischbeck KH, Isaacs WB, Walsh PC, Barrack ER. (1994) Microsatellite mutation (CAG24?18) in the androgen receptor gene in human prostate cancer. *Biochem Biophys Res Commun* **198**(1), 74–80.

98. Wallen MJ, Linja M, Kaartinen K, Schleutker J, Visakorpi T. (1999) Androgen receptor gene mutations in hormone-refractory prostate cancer. *J Pathol* **189**(4), 559–63.

99. Alvarado C, Beitel LK, Sircar K, Aprikian A, Trifiro M, Gottlieb B. (2005) Somatic mosaicism and cancer: a micro-genetic examination into the role of the androgen receptor gene in prostate cancer. *Cancer Res* **65**(18), 8514–8.

100. Veldscholte J, Ris-Stalpers C, Kuiper GG, Jenster G, Berrevoets C, Claassen E, van Rooij HC, Trapman J, Brinkmann AO, Mulder E. (1990) A mutation in the ligand binding domain of the androgen receptor of human LNCaP cells affects steroid binding characteristics and response to anti-androgens. *Biochem Biophys Res Commun* **173**(2), 534–40.

101. Suzuki H, Sato N, Watabe Y, Masai M, Seino S, Shimazaki J. (1993) Androgen receptor gene mutations in human prostate cancer. *J Steroid Biochem Mol Biol* **46**(6), 759–65.

102. Taplin ME, Bubley GJ, Shuster TD, Frantz ME, Spooner AE, Ogata GK, Keer HN, Balk SP. (1995) Mutation of the androgen-receptor gene in metastatic androgen-independent prostate cancer. *N Engl J Med* **332**(21), 1393–8.

103. Gottlieb B, Beitel LK, Lumbroso R, Pinsky L, Trifiro M. (1999) Update of the androgen receptor gene mutations database. *Hum Mutat* **14**(2), 103–14.

104. Sack JS, Kish KF, Wang C, Attar RM, Kiefer SE, An Y, Wu GY, Scheffler JE, Salvati ME, Krystek SR Jr, Weinmann R, Einspahr HM. (2001) Crystallographic structures of the ligand-binding domains of the androgen receptor and its T877A mutant complexed with the natural agonist dihydrotestosterone. *Proc Natl Acad Sci USA* **98**(9), 4904–9.

105. Veldscholte J, Berrevoets CA, Zegers ND, van der Kwast TH, Grootegoed JA, Mulder E. (1992) Hormone-induced dissociation of the androgen receptor-heat-shock protein complex: use of a new monoclonal antibody to distinguish transformed from nontransformed receptors. *Biochemistry* **31**(32), 7422–30

106. McDonald S, Brive L, Agus DB, Scher HI, Ely KR. (2000) Ligand responsiveness in human prostate cancer: structural analysis of mutant androgen receptors from LNCaP and CWR22 tumors. *Cancer Res* **60**(9), 2317–22.

107. Grigoryev DN, Long BJ, Njar VC, Brodie AH. (2000) Pregnenolone stimulates LNCaP prostate cancer cell growth via the mutated androgen receptor. *J Steroid Biochem Mol Biol* **75**(1), 1–10.

108. Watanabe M, Ushijima T, Shiraishi T, Yatani R, Shimazaki J, Kotake T, Sugimura T, Nagao M. (1997) Genetic alterations of androgen receptor gene in Japanese human prostate cancer. *Jpn J Clin Oncol* **27**(6), 389–93.

109. Culig Z, Hobisch A, Cronauer MV, Cato AC, Hittmair A, Radmayr C, Eberle J, Bartsch G, Klocker H. (1993) Mutant androgen receptor detected in an advanced-stage prostatic carcinoma is activated by adrenal androgens and progesterone. *Mol Endocrinol* **7**(12), 1541–50.

110. Newmark JR, Hardy DO, Tonb DC, Carter BS, Epstein JI, Isaacs WB, Brown TR, Barrack ER. (1992) Androgen receptor gene mutations in human prostate cancer. *Proc Natl Acad Sci USA* **89**(14), 6319–23.

111. Tan J, Sharief Y, Hamil KG, Gregory CW, Zang DY, Sar M, Gumerlock PH, deVere White RW, Pretlow TG, Harris SE, Wilson EM, Mohler JL, French FS. (1997) Dehydroepiandrosterone activates mutant androgen receptors expressed in the androgen-dependent human prostate cancer xenograft CWR22 and LNCaP cells. *Mol Endocrinol* **11**(4), 450–9.

112. Hara T, Miyazaki J, Araki H, Yamaoka M, Kanzaki N, Kusaka M, Miyamoto M. (2003) Novel mutations of androgen receptor: a possible mechanism of bicalutamide withdrawal syndrome. *Cancer Res* **63**(1), 149–53.

113. Bohl CE, Gao W, Miller DD, Bell CE, Dalton JT. (2005) Structural basis for antagonism and resistance of bicalutamide in prostate cancer. *Proc Natl Acad Sci USA* **102**(17), 6201–6.

114. Zhao XY, Boyle B, Krishnan AV, Navone NM, Peehl DM, Feldman D. (1999) Two mutations identified in the androgen receptor of the new human prostate cancer cell line MDA PCa 2a. *J Urol* **162**(6), 2192–9.

115. Zhao XY, Malloy PJ, Krishnan AV, Swami S, Navone NM, Peehl DM, Feldman D. (2000) Glucocorticoids can promote androgen-independent growth of prostate cancer cells through a mutated androgen receptor. *Nat Med* **6**(6), 703–6.

116. Kati K. Waltering, Mika Wallén, Teuvo Tammela, Robert L. Vessella, and Tapio Visakorpi. (2006) Mutation screening of the androgen receptor promoter and untranslated regions in prostate cancer. *Prostate* **66**(15), 1585–91.

117. Nupponen N, Visakorpi T. (1999) Molecular biology of progression of prostate cancer. *Eur Urol* **35**(5–6), 351–4.

118. Bubendorf L, Kononen J, Koivisto P, Schraml P, Moch H, Gasser TC, Willi N, Mihatsch MJ, Sauter G, Kallioniemi OP. (1999) Survey of gene amplifications during prostate cancer progression by high-throughout fluorescence in situ hybridization on tissue microarrays. *Cancer Res* **59**(4), 803–6.

119. Hernes EH, Linja M, Fossa SD, Visakorpi T, Berner A, Winderen M, Koivisto PA. (2000) Hormone-resistant prostate cancer with symptomatic pelvic tumours: patient survival and prognostic factors. *BJU Int* **86**(3), 240–7.

120. Ford OH 3rd, Gregory CW, Kim D, Smitherman AB, Mohler JL. (2003) Androgen receptor gene amplification and protein expression in recurrent prostate cancer. *J Urol* **170**(5), 1817–21.

121. Palmberg C, Koivisto P, Kakkola L, Tammela TL, Kallioniemi OP, Visakorpi T. (2000) Androgen receptor gene amplification at primary progression predicts response to combined androgen blockade as second line therapy for advanced prostate cancer. *J Urol* **164**(6), 1992–5.

122. Heinlein CA, Chang C. (2002) Androgen receptor (AR) coregulators: an overview. *Endocr Rev* **23**(2), 175–200.

123. Gao X, Loggie BW, Nawaz Z. (2002) The roles of sex steroid receptor coregulators in cancer. *Mol Cancer* **1**, 7.

124. Wang L, Hsu CL, Chang C. (2005) Androgen receptor corepressors: an overview. *Prostate* **63**(2), 117–30.

125. He B, Gampe RT Jr, Hnat AT, Faggart JL, Minges JT, French FS, Wilson EM. (2006) Probing the functional link between androgen receptor coactivator and ligand-binding sites in prostate cancer and androgen insensitivity. *J Biol Chem* **281**(10), 6648–63.

126. Lee DK, Chang C. (2003) Molecular communication between androgen receptor and general transcription machinery. *J Steroid Biochem Mol Biol* **84**(1), 41–9.

127. Gregory CW, Johnson RT Jr, Mohler JL, French FS, Wilson EM. (2001) Androgen receptor stabilization in recurrent prostate cancer is associated with hypersensitivity to low androgen. *Cancer Res* **61**(7), 2892–8.

128. Yeh S, Kang HY, Miyamoto H, Nishimura K, Chang HC, Ting HJ, Rahman M, Lin HK, Fujimoto N, Hu YC, Mizokami A, Huang KE, Chang C. (1999) Differential induction of androgen receptor transactivation by different androgen receptor coactivators in human prostate cancer DU145 cells. *Endocrine* **11**(2), 195–202.

129. Gregory CW, Fei X, Ponguta LA, He B, Bill HM, French FS, Wilson EM. (2004) Epidermal growth factor increases coactivation of the androgen receptor in recurrent prostate cancer. *J Biol Chem* **279**(8), 7119–30.

130. Metzger E, Wissmann M, Yin N, Muller JM, Schneider R, Peters AH, Gunther T, Buettner R, Schule R. (2005) LSD1 demethylates repressive histone marks to promote androgen-receptor-dependent transcription. *Nature* **437**(7057), 436–9.

131. Truica CI, Byers S, Gelmann EP. (2000) Beta-catenin affects androgen receptor transcriptional activity and ligand specificity. *Cancer Res* **60**(17), 4709–13.

132. Yang F, Li X, Sharma M, Sasaki CY, Longo DL, Lim B, Sun Z. (2002) Linking beta-catenin to androgen-signaling pathway. *J Biol Chem* **277**(13), 11336–44.

133. Mulholland DJ, Cheng H, Reid K, Rennie PS, Nelson CC. (2002) The androgen receptor can promote beta-catenin nuclear translocation independently of adenomatous polyposis coli. *J Biol Chem* **277**(20), 17933–43.

134. Li H, Kim JH, Koh SS, Stallcup MR. (2004) Synergistic effects of coactivators GRIP1 and beta-catenin on gene activation: cross-talk between androgen receptor and Wnt signaling pathways. *J Biol Chem* **279**(6), 4212–20.

135. Mulholland DJ, Dedhar S, Coetzee GA, Nelson CC. (2005) Interaction of nuclear receptors with the Wnt/beta-catenin/Tcf signaling axis: Wnt you like to know? *Endocr Rev* **26**(7), 898–915.

136. Masiello D, Chen SY, Xu Y, Verhoeven MC, Choi E, Hollenberg AN, Balk SP. (2004) Recruitment of beta-catenin by wild-type or mutant androgen receptors correlates with ligand-stimulated growth of prostate cancer cells. *Mol Endocrinol* **18**(10), 2388–401.

137. Wang L, Hsu CL, Chang C. (2005) Androgen receptor corepressors: an overview. *Prostate* **63**(2), 117–30.

138. Yoon HG, Wong J. (2006) The corepressors SMRT and N-CoR are involved in agonist- and antagonist-regulated transcription by androgen receptor. *Mol Endocrinol* **20**(5): 1048–60.

139. Hodgson MC, Astapova I, Cheng S, Lee LJ, Verhoeven MC, Choi E, Balk SP, Hollenberg AN. (2005) The androgen receptor recruits nuclear receptor CoRepressor (N-CoR) in the presence of mifepristone via its N and C termini revealing a novel molecular mechanism for androgen receptor antagonists. *J Biol Chem* **280**(8), 6511–9.

140. Liao G, Chen LY, Zhang A, Godavarthy A, Xia F, Ghosh JC, Li H, Chen JD. (2003) Regulation of androgen receptor activity by the nuclear receptor corepressor SMRT. *J Biol Chem* **278**(7), 5052–61.

141. Linja MJ, Porkka KP, Kang Z, Savinainen KJ, Janne OA, Tammela TL, Vessella RL, Palvimo JJ, Visakorpi T. (2004) Expression of androgen receptor coregulators in prostate cancer. *Clin Cancer Res* **10**(3), 1032–40.

142. Mäki HE, Waltering KK, Wallén MJ, Martikainen PM, Tammela TLJ, van Weerden WM, Vessella RL, Visakorpi T. (2006) Screening of genetic and expression alterations of SRC1 gene in prostate cancer. *Prostate* **66**(13):1391–8.

143. Anzick SL, Kononen J, Walker RL, Azorsa DO, Tanner MM, Guan XY, Sauter G, Kallioniemi OP, Trent JM, Meltzer PS. (1997) AIB1, a steroid receptor coactivator amplified in breast and ovarian cancer. *Science* **277**(5328), 965–8.

144. Culig Z, Hobisch A, Cronauer MV, Radmayr C, Trapman J, Hittmair A, Bartsch G, Klocker H. (1994) Androgen receptor activation in prostatic tumor cell lines by insulin-like growth factor-I, keratinocyte growth factor, and epidermal growth factor. *Cancer Res* **54**(20), 5474–8.

145. Craft N, Shostak Y, Carey M, Sawyers CL. (1999) A mechanism for hormone-independent prostate cancer through modulation of androgen receptor signaling by the HER-2/neu tyrosine kinase. *Nat Med* **5**(3), 280–5.

146. Hobisch A, Eder IE, Putz T, Horninger W, Bartsch G, Klocker H, Culig Z. (1998) Interleukin-6 regulates prostate-specific protein expression in prostate carcinoma cells by activation of the androgen receptor. *Cancer Res* **58**(20), 4640–5.

147. Nazareth LV, Weigel NL. (1996) Activation of the human androgen receptor through a protein kinase A signaling pathway. *J Biol Chem* **271**(33), 19900–7.

148. Sadar MD. (1999) Androgen-independent induction of prostate-specific antigen gene expression via cross-talk between the androgen receptor and protein kinase A signal transduction pathways. *J Biol Chem* **274**(12), 7777–83.

149. Ikonen T, Palvimo JJ, Kallio PJ, Reinikainen P, Janne OA. (1994) Stimulation of androgen-regulated transactivation by modulators of protein phosphorylation. *Endocrinology* **135**(4), 1359–66.

150. Reinikainen P, Palvimo JJ, Janne OA. (1996) Effects of mitogens on androgen receptor-mediated transactivation. *Endocrinology* **137**(10), 4351–7.

151. Sharma M, Chuang WW, Sun Z. (2002) Phosphatidylinositol 3-kinase/Akt stimulates androgen pathway through GSK3beta inhibition and nuclear beta-catenin accumulation. *J Biol Chem* **277**(34), 30935–41.

152. Jenster G. (2000) Ligand-independent activation of the androgen receptor in prostate cancer by growth factors and cytokines. *J Pathol* **191**(3), 227–8.

153. Culig Z, Steiner H, Bartsch G, Hobisch A. (2005) Mechanisms of endocrine therapy-responsive and -unresponsive prostate tumours. *Endocr Relat Cancer* **12**(2), 229–44.

154. Culig Z, Hobisch A, Bartsch G, Klocker H. (2000) Androgen receptor-an update of mechanisms of action in prostate cancer. *Urol Res* **28**(4), 211–9.

155. Scher HI, Sarkis A, Reuter V, Cohen D, Netto G, Petrylak D, Lianes P, Fuks Z, Mendelsohn J, Cordon-Cardo C. (1995) Changing pattern of expression of the epidermal growth factor receptor and transforming growth factor alpha in the progression of prostatic neoplasms. *Clin Cancer Res* **1**(5), 545–50.

156. Yeh S, Lin HK, Kang HY, Thin TH, Lin MF, Chang C. (1999) From HER2/Neu signal cascade to androgen receptor and its coactivators: a novel pathway by induction of androgen target genes through MAP kinase in prostate cancer cells. *Proc Natl Acad Sci USA* **96**(10), 5458–63.

157. Wen Y, Hu MC, Makino K, Spohn B, Bartholomeusz G, Yan DH, Hung MC. (2000) HER-2/neu promotes androgen-independent survival and growth of prostate cancer cells through the Akt pathway. *Cancer Res* **60**(24), 6841–5.

158. Mellinghoff IK, Vivanco I, Kwon A, Tran C, Wongvipat J, Sawyers CL. (2004) HER2/neu kinase-dependent modulation of androgen receptor function through effects on DNA binding and stability. *Cancer Cell* **6**(5), 517–27.

159. Savinainen KJ, Saramaki OR, Linja MJ, Bratt O, Tammela TL, Isola JJ, Visakorpi T. (2002) Expression and gene copy number analysis of ERBB2 oncogene in prostate cancer. *Am J Pathol* **160**(1), 339–45.

160. Mulholland DJ, Dedhar S, Wu H, Nelson CC. (2006) PTEN and GSK3beta: key regulators of progression to androgen-independent prostate cancer. *Oncogene* **25**(3), 329–37.

161. Chen S, Song CS, Lavrovsky Y, Bi B, Vellanoweth R, Chatterjee B, Roy AK. (1998) Catalytic cleavage of the androgen receptor messenger RNA and functional inhibition of androgen receptor activity by a hammerhead ribozyme. *Mol Endocrinol* **12**(10), 1558–66.

162. Eder IE, Culig Z, Ramoner R, Thurnher M, Putz T, Nessler-Menardi C, Tiefenthaler M, Bartsch G, Klocker H. (2000) Inhibition of LNCaP prostate cancer cells by means of androgen receptor antisense oligonucleotides. *Cancer Gene Ther* **7**(7), 997–1007.

163. Mitchell SH, Zhu W, Young CY. (1999) Resveratrol inhibits the expression and function of the androgen receptor in LNCaP prostate cancer cells. *Cancer Res* **59**(23), 5892–5.

164. Zhu W, Smith A, Young CY. (1999) A nonsteroidal anti-inflammatory drug, flufenamic acid, inhibits the expression of the androgen receptor in LNCaP cells. *Endocrinology* **140**(11), 5451–4.

165. Zhang Y, Ni J, Messing EM, Chang E, Yang CR, Yeh S. (2002) Vitamin E succinate inhibits the function of androgen receptor and the expression of prostate-specific antigen in prostate cancer cells. *Proc Natl Acad Sci USA* **99**(11), 7408–13.

166. Lim JT, Piazza GA, Pamukcu R, Thompson WJ, Weinstein IB. (2003) Exisulind and related compounds inhibit expression and function of the androgen receptor in human prostate cancer cells. *Clin Cancer Res* **9**(13), 4972–82.

167. Dong Y, Lee SO, Zhang H, Marshall J, Gao AC, Ip C. (2004) Prostate specific antigen expression is down-regulated by selenium through disruption of androgen receptor signaling. *Cancer Res* **64**(1), 19–22.

168. Solit DB, Scher HI, Rosen N. (2003) Hsp90 as a therapeutic target in prostate cancer. *Semin Oncol* **30**(5), 709–16.

169. Pegram M, Slamon D. (2000) Biological rationale for HER2/neu (c-erbB2) as a target for monoclonal antibody therapy. *Semin Oncol* **27**(5 Suppl 9), 13–9.

170. Lynch TJ, Bell DW, Sordella R, Gurubhagavatula S, Okimoto RA, Brannigan BW, Harris PL, Haserlat SM, Supko JG, Haluska FG, Louis DN, Christiani DC, Settleman J, Haber DA. (2004) Activating mutations in the epidermal growth factor receptor underlying responsiveness of non-small-cell lung cancer to gefitinib. *N Engl J Med* **350**(21), 2129–39.

171. Paez JG, Janne PA, Lee JC, Tracy S, Greulich H, Gabriel S, Herman P, Kaye FJ, Lindeman N, Boggon TJ, Naoki K, Sasaki H, Fujii Y, Eck MJ, Sellers WR, Johnson BE, Meyerson M. (2004) EGFR mutations in lung cancer: correlation with clinical response to gefitinib therapy. *Science* **304**(5676), 1497–500.

172. Sawyers C. (2004) Targeted cancer therapy. *Nature* **432**, 294–7.

173. Elo JP, Kvist L, Leinonen K, Isomaa V, Henttu P, Lukkarinen O, Vihko P. (1995) Mutated human androgen receptor gene detected in a prostatic cancer patient is also activated by estradiol. *J Clin Endocrinol Metab.* **80**,(12), 3494–500.

174. Evans BA, Harper ME, Daniells CE, Watts CE, Matenhelia S, Green J, Griffiths K. (1996) Low incidence of androgen receptor gene mutations in human prostatic tumors using single strand conformation polymorphism analysis. *Prostate.* **28**(3), 162–71.

175. Suzuki H, Akakura K, Komiya A, Aida S, Akimoto S, Shimazaki J. (1996) Codon 877 mutation in the androgen receptor gene in advanced prostate cancer: relation to antiandrogen withdrawal syndrome. *Prostate.* **29**(3), 153–8.

176. Marcelli M, Ittmann M, Mariani S, Sutherland R, Nigam R, Murthy L, Zhao Y, DiConcini D, Puxeddu E, Esen A, Eastham J, Weigel NL, Lamb DJ. (2000) Androgen receptor mutations in prostate cancer. *Cancer Res.* **60**(4), 944–9.

177. Segawa N, Nakamura M, Shan L, Utsunomiya H, Nakamura Y, Mori I, Katsuoka Y, Kakudo K. (2002) Expression and somatic mutation on androgen receptor gene in prostate cancer. *Int J Urol.* **9**(10), 545–53.

178. Lamb DJ, Puxeddu E, Malik N, Stenoien DL, Nigam R, Saleh GY, Mancini M, Weigel NL, Marcelli M. (2003) Molecular analysis of the androgen receptor in ten prostate cancer specimens obtained before and after androgen ablation. *J Androl.* **24**(2), 215–25.

179. Thompson J, Hyytinen ER, Haapala K, Rantala I, Helin HJ, Janne OA, Palvimo JJ, Koivisto PA. (2003) Androgen receptor mutations in high-grade prostate cancer before hormonal therapy. *Lab Invest.* **83**(12), 1709–13.

180. Koivisto P, Hyytinen E, Palmberg C, Tammela T, Visakorpi T, Isola J, Kallioniemi OP. (1995) Analysis of genetic changes underlying local recurrence of prostate carcinoma during androgen deprivation therapy. *Am J Pathol.* **147**(6), 1608–14.

181. Miyoshi Y, Uemura H, Fujinami K, Mikata K, Harada M, Kitamura H, Koizumi Y, Kubota Y. (2000) Fluorescence in situ hybridization evaluation of c-myc and androgen receptor gene amplification and chromosomal anomalies in prostate cancer in Japanese patients. *Prostate.* **43**(3), 225–32.

182. Kaltz-Wittmer C, Klenk U, Glaessgen A, Aust DE, Diebold J, Lohrs U, Baretton GB. (2000) FISH analysis of gene aberrations (MYC, CCND1, ERBB2, RB, and AR) in advanced prostatic carcinomas before and after androgen deprivation therapy. *Lab Invest.* **80**(9): 1455–64.

183. Edwards J, Krishna NS, Mukherjee R, Watters AD, Underwood MA, Bartlett JM. (2001) Amplification of the androgen receptor may not explain the development of androgen-independent prostate cancer. *BJU Int.* **88**(6), 633–7.

184. Brown RS, Edwards J, Dogan A, Payne H, Harland SJ, Bartlett JM, Masters JR. (2002) Amplification of the androgen receptor gene in bone metastases from hormone-refractory prostate cancer. *J Pathol.* **198**(2), 237–44.
185. Edwards J, Krishna NS, Grigor KM, Bartlett JM. (2003) Androgen receptor gene amplification and protein expression in hormone refractory prostate cancer. *Br J Cancer.* **89**(3), 552–6.

6 Androgen Receptor in Prostate Cancer Progression

Hiroshi Miyamoto, MD, PhD,
Saleh Altuwaijri, DVM, PhD,
and Chawnshang Chang, PhD

1. INTRODUCTION

The growth of prostate cancer, as well as embryonic development and adult function of the prostate, is generally dependent on androgens mediated through the androgen receptor (AR). Therefore, hormonal manipulation, such as castration and/or the use of antiandrogens, remains the critical therapeutic option for advanced forms of prostate cancer [reviewed in *(1)*]. Although this treatment produces a significant clinical response in most patients, the majority of responders eventually develop recurrences, which have been termed

From: *Current Clinical Oncology: Prostate Cancer:*
Signaling Networks, Genetics, and New Treatment Strategies
Edited by: R. G. Pestell and M. T. Nevalainen © Humana Press, Totowa, NJ

androgen-independent or hormone-refractory prostate cancer. Recent research on the AR in androgen-dependent and androgen-independent prostate cancers has led to discoveries that significantly changed our understanding of tumor biology. This article reviews clinical and molecular evidence indicating distinct roles of the AR in prostate cancer, particularly in cancer cell proliferation, apoptosis, and invasion/metastasis. A key molecule, cyclooxygenase (COX)-2, which is involved in cell proliferation, apoptosis, differentiation, cell adhesion, invasion, and angiogenesis, is also discussed in relation to its role in prostate cancer.

2. THE AR AND ITS ROLE IN PROSTATE FUNCTION

The AR, a member of the nuclear receptor superfamily, is a transcription factor that regulates gene expression in response to ligands in target cells (2,3). The AR gene is composed of eight exons that encode four functional domains: the NH_2-terminal transactivation domain, the DNA-binding domain, a hinge region, and the COOH-terminal ligand-binding domain (LBD). The AR, in its inactive state, associates with heat-shock proteins, and unliganded AR is mainly located in the cytoplasm. Upon binding of androgens, the AR undergoes a conformational change within the LBD and dissociates from the heat-shock proteins. Activated ARs then form homodimers and translocate to the nucleus. In the nucleus, the complex initiates gene transcription of androgen-regulated genes, such as prostate-specific antigen (PSA), by binding to specific DNA sequences termed androgen-responsive elements (AREs). The dimerized AR-ligand complex also interacts with coregulatory proteins, and these transcriptional coregulators can recruit general transcription factors and acetylate core histones, leading to transcription activation or repression (4). In this process, the AR NH_2-/COOH-terminal interaction that stabilizes bound ligand has also been identified (5), and several AR coactivators have been shown to facilitate this interaction (6,7).

Androgens and the AR have been shown to regulate the early embryological differentiation and later growth cycles of the prostate. In the adult, the prostate gland is about the size of a walnut and secretes a milky, alkaline fluid that helps to nourish and protect the sperm during intercourse. The prostate consists of luminal and basal epithelial components. AR expression is detected in differentiated luminal glandular epithelial cells (8). The basal compartment, composed of various distinct subsets of epithelial cells, is a small component of the progenitor/stem cell. The AR has been reported to be absent, weak, or strongly expressed in epithelial basal cells (9–11). Androgen supplementation following castration in male rats or mice results in up-regulation of the AR in basal cells and increase in numbers of these stem cells (8,10,11).

3. CELL CYCLE REGULATORS AND PROSTATE CANCER

Increased proliferation indexes (e.g., Ki-67, proliferating cell nuclear antigen, bromodeoxyuridine) in prostate cancer cells have been shown to correlate with increased tumor grade or advanced stage of disease *(12)*. Similarly, apoptotic index within a prostate cancer has been shown to be a prognostic parameter. It is known that androgen-sensitive LNCaP prostate cancer LNCaP cells display a striking biphasic growth response to androgens *(13,14)*. At low concentrations of androgens [e.g., 0.1 nM dihydrotestosterone (DHT) or 0.1 nM synthetic androgen R1881] proliferation is stimulated, whereas high concentrations (e.g., 10 nM DHT or 10 nM R1881) inhibit proliferation and further promote differentiation. Recent studies have improved the understanding of the mechanisms by which activation of the AR in prostate cancer leads to complex proliferation and apoptotic events.

Genetic aberrations in the control of the cell cycle are present in virtually all human cancers. Key molecules that involve cell cycle regulation and apoptosis in prostate cancer cells include cyclins, cyclin-dependent kinases (CDKs), retinoblastoma protein (pRb), $p27^{Kip1}$, $p21^{Waf1/Cip1}$, p16, c-myc, p53, and bcl-2.

The kinase family that is critical in the transitions through each phase of the cell cycle is the cyclin or CDK family, each of which is dependent on a discrete protein partner for activity. For instance, CDK1–cyclin B1 complex governs the G2/M transition, and CDK4/6–cyclin D complexes or CDK2–cyclins A/E complexes are required for the G1/S transition and progression through S phase *(15)*. There are several reports that suggest the relationship between expression levels of cyclins and outcome in prostate cancer *(12)*.

The pRb plays a central role in cell cycle regulation. Phosphorylation of pRb by CDK–cyclin D complexes in G1 phase inactivates the pRb. The pRb also serves as a transcription regulator that activates certain differentiation transcription factors to promote cellular differentiation *(16)*. *Retinoblastoma* gene mutations that may closely correlate with loss of pRb expression have been reported in human prostate cancer *(12,17)*.

The p27 and p21, belonging to the Kip/Cip family of CDK inhibitors, inhibit all CDK–cyclin complexes and may not be specific for a particular phase of cell cycle *(18)*. It is noteworthy that a number of human prostate cancers, particularly high-grade tumors, lack p27 expression *(19)*. The p27 is a target of Akt and has been proposed as a downstream mediator through which *PTEN*, a tumor suppressor gene whose inactivation with deletions is frequently found in prostate cancer, negatively regulates cell cycle progression *(20)*. Interestingly, *PTEN* can interact with the AR and modulate AR transcriptional activity in LNCaP cells *(21)*. A loss of p21 was also reported to correlate with a high risk for progression of prostate cancer *(22)*.

The p16 (MTS-1/CDKN2/INK4A) is a member of a family of inhibitors specific for CDKs (e.g., CDK4, CDK6) and is deleted and/or inactivated in a variety of human malignancies (23). The mechanisms for its tumor suppression may not only be prevention of Rb phosphorylation but also be required for p53-independent G1 arrest in response to DNA-damaging agents. Although several studies fail to detect p16 gene mutations in prostate cancer, p16 overexpression correlates with earlier biochemical (PSA) relapse after prostatectomy (24). Hypermethylation of p16 was detected in AR-negative prostate cancer PC-3 cells (25).

The c-myc protein activates a variety of known genes, including cyclin D2 and CDK2, and is thus known to involve the G1/S transition. Amplification and overexpression of c-myc are reported to correlate with prognosis of prostate cancer patients (26,27).

The p53 functions by regulating the transcription of genes involved in G1-phase growth arrest of cells and also plays a role in the G2/M transition and chromosome partition (the spindle checkpoint) (28). The effects of p53 related to cancer include cell cycle regulation and apoptosis as well as angiogenesis. The *p53* gene is the most frequently mutated gene found in multiple malignancies, including prostate cancer (29). Nuclear accumulation of p53 could be a prognostic indicator in several human cancers, including advanced forms of prostate cancer, although its value in localized prostate cancer has been debated (12).

The Bcl-2 is part of an expanding family of apoptosis-regulatory molecules, which may act as either death antagonists (e.g., Bcl-2, Bcl-xL, Mcl-1) or death agonists (e.g., Bax, Bak, Bcl-xS, Bad, Bid) (30). The *Bcl-2* gene is not expressed in normal prostatic secretary epithelial cells but is expressed in prostate cancer cells. Overexpression of Bcl-2 protects cancer cells from apoptosis in vitro and confers resistance to androgen ablation therapy in vivo (31), suggesting that androgen independence may result from Bcl-2 expression.

4. THE AR AND PROSTATE CANCER CELL PROLIFERATION/APOPTOSIS

Effects of androgens on the expression of each molecule in AR-positive/negative, androgen-sensitive/insensitive prostate cancer cells are summarized in Table 1. Androgens enhance CDK2 and CDK4 at both mRNA and protein levels and inhibit p16 expression in cells expressing the AR, but not in androgen-independent cells (32). In LNCaP cells, androgen (0.3, 1, and 10 nM DHT) significantly increases cyclin D1 expression (33). In the CWR22 human prostate cancer xenograft model with a mutant AR, mRNA expression of cyclins A and B1, CDK1, and CDK2 is down-regulated after castration and is up-regulated after androgen supplementation (34). CDK inhibitors more

Table 1
Summary of Reported Androgen Effects on the Expression of Key Molecules Related to Cell Proliferation/Apoptosis in Androgen-Sensitive and Androgen-Insensitive Prostate Cancer Cells

Molecules	+ Androgen in AR(+) AS cells[1]	– Androgen in AR(+) AI cells[1]	+ Androgen in AR(+) AI cells[2]	– Androgen in AR(–) AI cells[1]
Cyclins	Up	Down	NC	Up
CDKs	Up (Low[a]-Up, High[b]-Down)	NC/Down	Down	Up
pRb	Up (Low[a]-Up, High[b]-Down)	Down (Low[a]-NC, High[b]-Down)		
p27	Down (Low[a]-Up, High[b]-Down)	Up (Low[a]-Up, High[b]-Down)	Up	Down
p21	Up	Up/Down (Low[a]-NC, High[b]-Down)	Up	Down
p16	Down	Down		Down
c-myc	Low[a]-Up, High[b]-Down	Low[a]-NC, High[b]-Down	Low[a]-Up, High[b]-Down	Up
p53	Down (Flu[c]-NC)	Up/Down (Low[a]-NC, High[b]-Down)	NC	Down
bcl-2	NC (Low[a]-NC, High[b]-Down)	Up/Down		Up

AS, androgen-sensitive; AI, androgen-insensitive; Up, up-regulated; Down, down-regulated; NC, no significant change (compared with the expression in AR(+) AS cells without androgen treatment[1] or AR(+) AI cells without androgen treatment[2]).
[a]Low: low doses of androgen (e.g., 0.1 nM DHT/R1881).
[b]High: high doses of androgen (e.g., 10 nM DHT/R1881).
[c]Flu: Flutamide treatment.

effectively inhibit the growth of androgen-sensitive LNCaP cells than that of AR-negative DU 145 cells *(35)*. Synergistic inhibitory effects of a CDK inhibitor olomoucine and an antiandrogen bicalutamide on LNCaP cell proliferation were also observed *(36)*. Low androgenic concentrations, known to promote LNCaP cell proliferation, induce an increase of pRb phosphorylation, accompanied by increased expression of the transcription factor E2F-1 and its target gene product cyclin A *(14)*. Similarly, Ye et al. *(37)* showed that androgen, as well as epidermal growth factor (EGF), stimulates proliferation of androgen-responsive MDA PCa prostate cancer cells, associated with increased

CDK2 activity and decreased levels of p27. By contrast, high androgenic concentrations inhibit cell proliferation, induce pRb hypophosphorylation and E2F-1 down-regulation, and increase expression of a cell cycle inhibitor p27. Tsihlias et al. *(38)* suggested that the inhibition of cyclin E/CDK2 by p27 most significantly contributes to the G1 arrest of LNCaP following high doses of androgen. In the CWR22 xenograft model, androgen ablation results in tumor regression with sustained increases in p27 and p16 protein expression and a decrease in pRb expression *(39)*. In the same study, the authors suggest that following androgen ablation, increased p53 protein induces a cell cycle arrest, without activation of p53-mediated apoptosis *(39)*. Induction of p27 by androgen was also observed in long-term androgen-ablated LNCaP cells *(40)*. In androgen-sensitive LNCaP cells, androgens enhance the expression of p21 that is revealed to possess a functional ARE *(41)*, suggesting anti-apoptotic activity induced by androgen in prostate cancer cells. An androgen-insensitive LNCaP subline maintained in long-term androgen-depleted conditions expresses a much higher level of p21 as well as bcl-2 compared with parental LNCaP. Androgen (1 nM R1881) further enhances p21 expression, but not bcl-2, in both androgen-sensitive and androgen-insensitive LNCaP cells *(42)*. By contrast, Wang et al. *(43)* showed that an androgen-independent LNCaP subline with an overexpressed AR down-regulates p21 and is resistant to apoptosis. In this subline, a reduction of AR level by antisense treatment was associated with increased p21 expression and partial reversion of androgen-sensitive phenotype. Similarly, down-regulation of the AR by a small interference RNA (siRNA) molecule led to a marked cell growth inhibition, associated with significant up-regulation of p21 and cyclin D1, in both parental LNCaP and its androgen-independent subline *(44)*. R1881 at low concentrations stimulates LNCaP cell proliferation with no or slight increase in c-myc expression, whereas high doses of R1881 inhibit proliferation with significant decrease in c-myc expression *(45,46)*. In androgen-insensitive sublines of LNCaP, low concentrations of androgen enhance and high concentrations of androgen reduce c-myc expression. Repression of cell proliferation was blocked by retroviral overexpression of c-myc in these cells *(45)*. Zhou et al. *(47)* showed that androgen-insensitive LNCaP xenograft tumors developed in castrated male mice have increased proliferation index, decreased apoptotic index, and increased expression of p53, bcl-2, and p21, compared with parental LNCaP xenograft tumors. However, a recent cDNA microarray analysis in LNCaP cells revealed that treatment with the antiandrogen flutamide, which is known to act as a strong agonist in this cell line *(48)*, up-regulates p21 expression, but not p53 expression *(49)*.

In addition, we showed that the pRb can physically interact with the AR and stimulate AR-mediated transactivation *(50)*. Interestingly, overexpression of pRb in DU 145 cells result in an apoptotic activity only when the AR is co-expressed, suggesting that pRb-induced apoptosis in prostate cancer cells is dependent on AR transcriptional activity *(51)*.

5. THE AR AND PROSTATE CANCER INVASION/METASTASIS

Prostate cancer preferentially metastasizes to the bone, as well as lymph nodes. Although the whole picture of mechanisms responsible for tumor progression and metastasis is not fully understood, a number of invasion- or metastasis-related molecules that may exhibit unique functions associated with modulation of AR activity in prostate cancer have been identified. Effects of androgens on such molecules in AR-positive/negative prostate cancer cells are summarized in Table 2.

Interaction between cells and the extracellular matrix (ECM) is mediated by a family of transmembrane glycoproteins termed integrins, heterodimers composed of non-covalently associated α and β subunits. The integrin $\alpha6\beta4$ is a receptor for laminins and plays a pivotal role in migration and invasion of cancer cells *(52)*. Integrin $\alpha6\beta4$ expression is much higher in AR-negative PC-3 cells than in LNCaP cells *(53)*. Stable expression of the AR in PC-3 cells

Table 2
Summary of Reported Androgen Effects on the Expression of Key Molecules Related to Cancer Invasion/Metastasis in Prostate Cancer Cells

Molecules	+ Androgen in AR(+) AS cells	– Androgen in AR(–) AI cells
Integrin $\alpha6\beta4$	Down	Up
MUC-1	Up	Down
EGFR	Down	Up
MMPs	Up/Down	Up
ET-1	Down	Up
NEP	Up	Down
VEGF	Up/Down	Up/NC
bFGF	Up	Down
TGF-β	Up	Up
IGF-1	Up	Up
IL-6	Up/Down	Up
Caveolin-1	Up	Up
PSCA	Up	

AS, androgen-sensitive; AI, androgen-insensitive; Up, up-regulated; Down, down-regulated; NC, no significant change (compared with the expression in AR(+) AS cells without androgen treatment).

results in decrease in integrin α6β4 expression associated with lower invasion ability, and androgens further down-regulate integrin α6β4 in both LNCaP and PC-3 expressing the AR *(53)*. A recent study confirmed these findings in AR-negative prostate cancer DU 145 cells with stable AR transfection *(54)*. DHT has also been shown to modulate the expression of another integrin, α2β1, and an anti-adhesive mucin MUC-1, both of which are involved in metastatic process of prostate cancer cells *(55)*. Only a few studies were performed on integrin expression in prostate cancer tissues showed an apparent loss of expression of integrin α6β4 and its ligand, laminin-5, in invasive prostate cancer as compared with normal prostate or prostatic intraepithelial neoplasia (PIN) *(56)*.

It has been shown that integrin α6β4 interacts with erbB-2 and the EGF receptor (EGFR) to promote cell migration and invasion in response to EGF *(57)*. In PC-3/AR cells, auto-transphosphorylation of EGFR, which then activates a downstream phosphatidylinositol 3-kinase (PI3K)/Akt pathway, is reduced, compared with parental PC-3 cells, and is further reduced following androgen treatment, despite similar EGFR expression levels in the two cell lines *(58)*. The AR and EGFR co-localize in prostate cancer cells, and treatment with EGF in the presence of androgen enhances this co-localization *(58)*. Co-immunoprecipitation studies confirmed the interaction between the AR and EGFR in LNCaP and PC-3/AR cells. Expression of EGFR has been shown to correlate with disease relapse and/or progression to androgen-independent disease in patients with prostate cancer *(59)*.

Matrix metalloproteinases (MMPs) constitute a broad family of zinc-binding endopeptidases that play a key role in the degradation of the ECM and basement membrane. A substantial amount of evidence supports the hypothesis that MMPs play a key role in multiple steps of tumor progression, including tumor promotion, angiogenesis, invasion, and metastasis. Previous studies have demonstrated a positive correlation between increased expression/activity of MMPs, especially MMP-2 and MMP-9, and malignant potential of prostate cancer in prostate cancer tissues *(60,61)*. Several studies analyzing primary cultures of prostatic epithelial and stromal cells also support the contention that MMP-2 and MMP-9 are differentially expressed/secreted in normal prostate epithelial and prostate cancer cells. Co-cultures of prostate cancer and stromal cells also showed significantly enhanced expression of MMP-9 *(62)*, confirming the importance of tumor–stromal interactions in tumor progression. Moreover, several growth factors, such as transforming growth factor (TGF)-β *(63)* and fibroblast growth factors (FGFs) *(64)*, have been shown to up-regulate expression/secretion of MMPs in prostate cancer cells. Other *in vivo* studies have also shown enhanced expression/activity of MMP-2 or MMP-9 in metastatic prostate tumors *(62,65)*, indicating the significance of MMPs in the process of metastasis. In prostate cancer cells, androgens have been shown to mediate gene expression of MMPs. Schneikert et al. (66) showed that the AR, through the interaction with the Ets protein, down-regulates expression of MMPs, including MMP-1, MMP-3, and MMP-7. By contrast, other groups recently demonstrated that treatment of LNCaP cells with androgens resulted in increased expression and activity of MMP-2 or MMP-9 (67,68). However, we have more recently found that androgen supplementation of the androgen-depleted media significantly reduces secretion and activity of MMP-9 in AR-positive prostate cancer cells and flutamide blocks this effect (unpublished data). It is noted that two potential ARE-like motifs located in the promoter region of the MMP-2 gene, with one nucleotide mismatch, have been found (67). Animal studies have reported a similar phenomenon in which androgen deprivation leads to increased expression of MMPs (69). For example, Powell et al. (70) showed that expression of MMP-2 and MMP-7 significantly increased during involution of rat ventral prostate induced by castration. MMP-9 has been shown to be expressed at high levels in PC-3 and DU 145 cells and at low levels in LNCaP cells. PC-3 cells are capable of forming osteolytic lesions and MMP-9 may contribute to this activity. We have shown that androgens can inhibit cell invasion of a stable PC-3/AR cell line and modulate the activity of MMP-9 via the nuclear factor (NF)-κB-signaling pathway, with little effect on cell proliferation. These data suggest that androgens/AR may inhibit the invasiveness of prostate cancer cells by modulating MMP-9. In addition, radiograph analyses reveal that PC-3 cells induce a significant osteolytic response in xenograft models, whereas PC-3/AR cells produce only a slight osteolytic response. MMP-9 activity decreases in sera from severe combined immunodeficiency (SCID) mice bearing xenograft tumors from PC-3/AR cells. Importantly, androgenic stimulation of prostate cancer cells expressing the AR diminishes the secretion and activity of MMP-9 and NF-κB, predicting that androgen deprivation stimulates MMP-9 activity in prostate cancer.

Endothelin (ET)-1 is produced by endothelial cells as well as many types of cancer cells and plays an important role in vasoconstriction as well as bone formation. Specifically, ET-1 has been shown to stimulate osteoblast activity and inhibit osteoclast activity, suggesting a role in metastatic progression of prostate cancer to the bone (71). In LNCaP cells, the ET-1 pathways appear to be turned-off, compared with PC-3 and DU 145 cells (72). In AR-negative prostate cancer cells, ET-1 gene expression and secretion are up-regulated by factors involved in tumor progression, such as TGF-β, EGF, and interleukin (IL)-1. By contrast, in PC-3/AR cells, androgens reduce mRNA expression and secretion of ET-1 (72). In this study, antiandrogens are shown to antagonize the androgen effects and even increase ET-1 in the absence of androgens. In addition, Padley et al. showed that surgical castration in the male dog results in a significant increase in ET receptors in the prostate (73).

Neutral endopeptidase 24.11 (NEP) is a cell surface metallopeptidase expressed by prostatic epithelial cells and inactivates various bioactive peptides including ET-1 (74). NEP expression is decreased in AR-negative prostate cancer cell lines and in most of metastatic androgen-independent prostate cancer specimens (75). Expression of *NEP* gene is transcriptionally activated by androgen and reduced with androgen depletion in LNCaP cells (75). NEP contains an ARE that binds AR, as well as progesterone and glucocorticoid receptors, and a unique androgen responsive region, which only binds AR, with a homology to the one identified in the promoter of PSA (76).

Vascular endothelial growth factor (VEGF) is a key factor in the regulation of tumor angiogenesis, which is produced by tumor and stromal cells. Androgen depletion in LNCaP culture and castration in male mice bearing LNCaP tumors were initially shown to result in down-regulation of VEGF expression (77). By contrast, Li et al. (78) recently reported VEGF-C up-regulation by androgen ablation in LNCaP cells. Interestingly, in CWR22Rv1 androgen-responsive but androgen-independent prostate cancer model, androgen ablation inhibits tumor growth associated with decrease in angiogenesis and VEGF expression (79). In radical prostatectomy specimens which are presumably androgen-dependent tumors, a significant correlation between expressions of AR and VEGF was observed (80).

Among the FGF family, several members including basic FGF (bFGF or FGF2) and FGF7 (also known as keratinocyte growth factor) have been implicated in prostate cancer (81). In addition to the known effects of FGFs on cancer cell proliferation, motility, and angiogenesis, FGFs may function as an essential survival factor in prostate cancer. In LNCaP or PC-3/AR cells, FGF2 production appears to be regulated by androgen, as androgens increase FGF2 expression and its bioavailability, as well as expression of FGF-binding protein, which binds FGF1 and FGF2 (82).

Androgen regulation of other growth factors/their receptors in prostate cancer cells and cross-talk between AR signaling and some of such growth factors' pathways have been reported. These growth factors include TGF-β (83), insulin-like growth factor-I (IGF-I) (84), and hepatocyte growth factor (85). One of the potential mechanisms for androgen-independent progression of prostate cancer is ligand-independent activation of the AR (1). In vitro studies have demonstrated that several growth factors, such as IGF-I, FGF7, and EGF, as well as IL-6, increase AR transcriptional activity in the absence of androgens in prostate cancer cells. These growth factors serve as ligands for receptor tyrosine kinases, such as EGF receptor and erbB-2, mediated through signal transduction pathways, such as mitogen-activated protein kinase (MAPK) and Akt, which can specifically bind to and phosphorylate the AR. In addition, a number of prostate cancer metastasis-related genes have been identified. Among them, caveolin-1 *(86)* and prostate stem cell antigen (PSCA) *(87)* have been shown to be likely under androgen control in prostate cancer cells.

6. COX-2 AND THE AR IN PROSTATE CANCER

COX is a key enzyme in the production of prostaglandins (PGs) and other eicosanoids from arachidonic acid. Two COX isozymes have been identified: COX-1, which is constitutively expressed in many tissues and is considered a housekeeping enzyme; and COX-2, which is not normally detected in most tissues, but is expressed by inflammatory cells such as macrophages and monocytes. It has been suggested that potential mechanistic roles of COX-2 in tumorigenesis and tumor progression include (i) decreased apoptosis; (ii) increased angiogenesis; (iii) increased tumor invasiveness; and (iv) decreased immune surveillance [reviewed in *(88)*]. Indeed, overexpression of COX-2 has been observed in many cancers, including prostate cancer. Consequently, suppression of COX-2 by specific inhibitors or other types of non-steroidal anti-inflammatory drugs (NSAIDs), which inhibit the activity of COX-2 (and COX-1), may lead to inhibition of tumor growth.

A number of studies showed increased expression of COX-2 in prostate cancer and PIN compared with normal or hyperplastic prostate *(89–92)*. In most of these studies, increased expression of COX-2 correlated with the higher grade or Gleason score of the tumors. By contrast, a recent report showed no significant differences in COX-2 expression between benign and malignant (cancer or PIN) prostate tissues *(93)*. Instead, consistently increased expression of COX-2 was observed only in proliferative inflammatory atrophy of the prostate, which has been postulated as a putative precursor lesion for prostate cancer *(94)*. In vitro and animal studies have suggested the involvement of COX-2 in prostate tumorigenesis and cancer progression. The enzyme activity and protein expression of COX-2 were found to be significantly higher in the dorsolateral prostate of the transgenic adenocarcinoma of the mouse prostate (TRAMP) model at 8, 16, and 24 weeks of age, compared with their non-transgenic littermates *(95)*. LNCaP cells with stably overexpressed COX-2 increased cell proliferation in vitro and tumor growth in vivo, as compared with parental LNCaP cells, which was associated with increased expression of VEGF *(96)*.

COX-2 expression in the LNCaP cell line is lower than that in the PC-3 cell line *(89)*. However, our recent studies show that in LNCaP cells cultured in charcoal-stripped serum medium, androgen treatment slightly reduces COX-2 expression, whereas an antiandrogen flutamide treatment induces it. Moreover, reporter gene assays demonstrate that androgens down-regulate the promoter activity of the COX-2 gene but flutamide, showing marginal effects by itself, antagonizes the androgen effect (unpublished data). These results indicate that androgen ablation (together with antiandrogen treatment) induces COX-2 overexpression/activation in prostate cancer cells.

7. SUMMARY

In the last decade, there has been a considerable change in understanding the roles of the AR in prostate cancer. Particularly, the evidence showing that most of androgen-independent prostate cancers are still AR-dependent for growth has greatly stimulated research on AR functions in prostate cancer. A better understanding of the molecular mechanisms of AR functions might also lead to new ideas for therapeutic targets for prostate cancer. Here we have described the effects of androgens or modulation of AR activity on various key molecules that are involved in cell proliferation, apoptosis, invasion, and metastasis of prostate cancer. In patients with androgen-dependent tumors, modulation of expression/activity of these molecules by androgen deprivation therapy might be a mechanism for tumor regression. On the contrary, in patients with androgen-independent tumors, these molecules could be novel therapeutic targets. Indeed, some strategies that target these molecules are being assessed in clinical settings. These potential therapeutic approaches in combination with androgen deprivation therapy, which might exhibit better inhibitory effects as well as might prolong the androgen-dependent state (97), also need to be considered.

REFERENCES

1. Miyamoto, H., Messing, E. M., and Chang, C. (2004) Androgen deprivation therapy for prostate cancer: current status and future prospects. *Prostate* **61,** 332–353.
2. Chang, C., Kokontis, J., and Liao, S. (1988) Molecular cloning of human and rat complementary DNA encoding androgen receptors. *Science* **240,** 324–326.
3. Heinlein, C. A. and Chang, C. (2004) Androgen receptor in prostate cancer. *Endocr. Rev.* **25,** 276–308.
4. Heinlein, C. A. and Chang, C. (2002) Androgen receptor (AR) coregulators: an overview. *Endocr. Rev.* **23,** 175–200.
5. He, B., Kemppainen, J. A., Voegel, J. J., Gronemeyer, H., and Wilson, E. M. (1999) Activation function 2 in the human androgen receptor ligand binding domain mediates interdomain communication with NH$_2$-terminal domain. *J. Biol. Chem.* **274,** 37219–37225.
6. Ikonen, T., Palvimo, J. J., and Jänne, O. A. (1997) Interaction between the amino- and carboxyl-terminal regions of the rat androgen receptor modulates transcriptional activity and is influenced by nuclear receptor coactivators. *J. Biol. Chem.* **272,** 29821–29828.
7. Rahman, M., Miyamoto, H., and Chang, C. (2004) Androgen receptor coregulators in prostate cancer: mechanisms and clinical implications. *Clin. Cancer Res.* **10,** 2208–2219.
8. Wang, Y., Hayward, S., Cao, M., Thayer, K., and Cunha, G. (2001) Cell differentiation lineage in the prostate. *Differentiation* **68,** 270–279.

Chapter 6 / Androgen Receptor and Prostate Cancer

141

9. Bonkhoff, H. and Remberger, K. (1996) Differentiation pathways and histogenetic aspects of normal and abnormal prostatic growth: a stem cell model. *Prostate* **28**, 98–106.
10. Mirosevich, J., Bentel, J. M., Zeps, N., Redmond, S. L., D'Antuono, M. F., and Dawkins, H. J. S. (1999) Androgen receptor expression of proliferating basal and luminal cells in adult murine ventral prostate. *J. Endocrinol.* **162**, 341–350.
11. Tokar, E. J., Ancrill, B. B., Cunha, G. R., and Webber, M. M. (2005) Stem/progenitor and intermediate cell types and the origin of human prostate cancer. *Differentiation* **73**, 463–473.
12. Quinn, D. I., Henshall, S. M., and Sutherland, R. L. (2005) Molecular markers of prostate cancer outcome. *Eur. J. Cancer* **41**, 858–887.
13. de Launoit, Y., Veilleux, R., Dufour, M., Simard, J., and Labrie, F. (1991) Characteristics of the biphasic action of androgens and of the potent antiproliferative effects of the new pure antiestrogen EM-139 on cell cycle kinetic parameters in LNCaP human prostatic cancer cells. *Cancer Res.* **51**, 5165–5170.
14. Hofman, K., Swinnen, J. V., Verhoeven, G., and Heyns, W. (2001) E2F activity is biphasically regulated by androgens in LNCaP cells. *Biochem. Biophys. Res. Commun.* **283**, 97–101.
15. Sánchez, I. and Dynlacht, B. D. (2005) New insights into cyclins, CDKs, and cell cycle control. *Semin. Cell Dev. Biol.* **16**, 311–321.
16. Zhu, L. (2005) Tumor suppressor retinoblastoma protein Rb: a transcriptional regulator. *Eur. J. Cancer* **41**, 2415–2427.
17. Kubota, Y., Fujinami, K., Uemura, H., Dobashi, Y., Miyamoto, H., Iwasaki, Y., Kitamura, H., and Shuin, T. (1995) Retinoblastoma gene mutations in primary prostate cancer. *Prostate* **27**, 314–320.
18. Coqueret, O. (2003) New roles for p21 and p27 cell-cycle inhibitors: a function for each cell compartment? *Trends Cell Biol.* **13**, 65–70.
19. Guo, Y., Sklar, N., Borkowski, A., and Kyprianou, N. (1997) Loss of the cyclin-dependent kinase inhibitor p27 (kip1) protein in human prostate cancer correlates with tumor grade. *Clin. Cancer Res.* **3**, 2269–2274.
20. Li, D. M. and Sun, H. (1998) PTEN/MMAC1/TEP1 suppresses the tumorigenicity and induces G1 cell cycle arrest in human glioblastoma cells. *Proc. Natl. Acad. Sci. U.S.A.* **95**, 15406–15411.
21. Lin, H.-K., Hu, Y.-C., Lee, D. K., and Chang, C. (2004) Regulation of androgen receptor signaling by PTEN (phosphatase and tensin homolog deleted on chromosome 10) tumor suppressor through distinct mechanisms in prostate cancer cells. *Mol. Endocrinol.* **18**, 2409–2423.
22. Gao, X., Chen, Y. Q., Wu, N., Grignon, D. J., Sakr, W., Porter, A. T., and Honn, K. V. (1995) Somatic mutations of the WAF1/CIP1 gene in primary prostate cancer. *Oncogene* **11**, 1395–1398.
23. Rocco, J. W. and Sidransky, D. (2001) p16 (MTS-1/CDKN2/INK4a) in cancer progression. *Exp. Cell Res.* **264**, 42–55.
24. Lee, C. T., Capodieci, P., Osman, I., Fazari, M., Ferrara, J., Scher, H. I., and Cordon-Cardo, C. (1999) Overexpression of the cyclin-dependent kinase inhibitor p16 is associated with tumor recurrence in human prostate cancer. *Clin. Cancer Res.* **5**, 977–983.
25. Herman, J. G., Merlo, A., Mao, L., Lapidus, R. G., Issa, J. P., Davidson, N. E., Sidransky, D., and Baylin, S. B. (1995) Inactivation of the CDKN2/p16/MTS1 gene is frequently associated with aberrant DNA methylation in all common human cancers. *Cancer Res.* **55**, 4525–4530.
26. Jenkins, R. B., Qian, J., Lieber, M. M., and Bostwick, D. G. (1997) Detection of c-myc oncogene amplification and chromosomal anomalies in metastatic prostatic carcinoma by fluorescence in situ hybridization. *Cancer Res.* **57**, 524–531.

27. Fleming, W. H., Hamel, A., MacDonald, R., Ramsey, E., Pettigrew, N. M., Johnston, B., Dodd, J. G., and Matusik, R. J. (1986) Expression of c-myc protooncogene in human prostatic carcinoma and benign prostatic hyperplasia. *Cancer Res.* **46,** 1535–1538.

28. Stewart, Z. A. and Pietenpol, J. A. (2001) p53 signaling and cell cycle checkpoints. *Chem. Res. Toxicol.* **14,** 243–263.

29. Kubota, Y., Shuin, T., Uemura, H., Fujinami, K., Miyamoto, H., Torigoe, S., Dobashi, Y., Kitamura, H., Iwasaki, Y., Danenberg, K., and Danenberg, P. V. (1995) Tumor suppressor gene p53 mutations in human prostate cancer. *Prostate* **27,** 18–24.

30. Coultas, L. and Strasser, A. (2003) The role of the Bcl-2 protein family in cancer. *Semin. Cancer Biol.* **13,** 115–123.

31. Raffo, A. J., Perlman, H., Chen, M. W., Day, M. L., Stretman, J. S., and Buttyan, R. (1995) Overexpression of bcl-2 protects prostate cancer cells from apoptosis *in vitro* and confers resistance to androgen depletion *in vivo. Cancer Res.* **55,** 4438–4445.

32. Lu, S., Tsai, S. Y., and Tsai, M.-J. (1997) Regulation of androgen-dependent prostate cancer cell growth: androgen regulation of CDK2, CDK4, and CKI p16 genes. *Cancer Res.* **57,** 4111–4116.

33. Chen, Y., Martinez, L. A., LaCava, M., Coghlan, L., and Conti, C. J. (1998) Increased cell growth and tumorigenicity in human prostate LNCaP cells by overexpression to cyclin D1. *Oncogene* **16,** 1913–1920.

34. Gregory, C. W., Johnson, R. T. J., Presnell, S. C., Mohler, J. L., and French, F. C. (2001) Androgen receptor regulation of G1 cyclin and cyclin-dependent kinase function in the CWR human prostate cancer xenograft. *J. Androl.* **22,** 537–548.

35. Mad'arová, J., Lukešová, M., Hlobilková, A., Strnad, M., Vojtešek, B., Lenobel, R., Hajdúch, M., Murray, P. G., Perera, S., and Kolár, Z. (2002) Synthetic inhibitors of CDKs induce different responses in androgen sensitive and androgen insensitive prostatic cancer cell lines. *Mol. Pathol.* **55,** 227–234.

36. Knillová, J., Bouchal, J., Hlobilková, A., Strnad, M., and Kolár, Z. (2004) Synergistic effects of the cyclin-dependent kinase (CDK) inhibitor olomoucine and androgen-antagonist bicalutamide on prostatic cancer cell lines. *Neoplasia* **51,** 358–367.

37. Ye, D., Mendelson, J., and Fan, Z. (1999) Androgen and epidermal growth factor down-regulate cyclin-dependent kinase inhibitor p27[Kip1] and costimulate proliferation of MDA PCa 2a and MDA PCa 2b prostate cancer cells. *Clin. Cancer Res.* **5,** 2171–2177.

38. Tsihlias, J., Zhang, W., Bhattacharya, N., Flanagan, M., Klotz, L., and Slingerland, J. (2000) Involvement of p27[Kip1] in G1 arrest by high dose 5α-dihydrotestosterone in LNCaP human prostate cancer cells. *Oncogene* **19,** 670–679.

39. Agus, D. B., Cordon-Cardo, C., Fox, W., Drobnjak, M., Koff, A., Golde, D. W., and Scher, H. I. (1999) Prostate cancer cell cycle regulators: response to androgen withdrawal and development of androgen independence. *J. Natl. Cancer Inst.* **91,** 1869–1876.

40. Kokontis, J. M., Hay, N., and Liao, S. (1998) Progression of LNCaP prostate tumor cells during androgen deprivation: hormone-independent growth, repression of proliferation by androgen, and role for p27[Kip1] in androgen-induced cell cycle arrest. *Mol. Endocrinol.* **12,** 941–953.

41. Lu, S., Liu, M., Epner, D. E., Tsai, S. Y., and Tsai, M.-J. (1999) Androgen regulation of the cyclin-dependent kinase inhibitor p21 gene through an androgen response element in the proximal promoter. *Mol. Endocrinol.* **13,** 376–384.

42. Lu, S., Tsai, S. Y., and Tsai, M.-J. (1999) Molecular mechanisms of androgen-independent growth of human prostate cancer LNCaP-AI cells. *Endocrinology* **140,** 5054–5059.

43. Wang, L. G., Ossowski, L., and Ferrari, A. C. (2001) Overexpressed androgen receptor linked to p21[WAF1] silencing may be responsible for androgen independence and resistance to apoptosis of a prostate cancer cell line. *Cancer Res.* **61,** 7544–7551.

44. Hååg, P., Bektic, J., Bartsch, G., Klocker, H., and Eder, I. E. (2005) Androgen receptor down regulation by small interference RNA induces cell growth inhibition in androgen sensitive as well as in androgen independent prostate cancer cells. *J. Steroid Biochem. Mol. Biol.* **96,** 251–258.

45. Kokontis, J., Takakura, K., Hay, N., and Liao, S. (1994) Increased androgen receptor activity and altered c-myc expression in prostate cancer cells after long-term androgen deprivation. *Cancer Res.* **54,** 1566–1573.

46. Foury, O., Nicolas, B., Joly-Pharaboz, M. O., and André, J. (1998) Control of the proliferation of prostate cancer cells by an androgen and two antiandrogens. Cell specific sets of responses. *J. Steroid Biochem. Mol. Biol.* **66,** 235–240.

47. Zhou, J.-R., Yu, L., Zerbini, L. F., Libermann, T. A., and Blackburn, G. L. (2004) Progression to androgen-independent LNCaP human prostate tumors: cellular and molecular alterations. *Int. J. Cancer* **110,** 800–806.

48. Miyamoto, H., Rahman, M. M., and Chang, C. (2004) Molecular basis for the antiandrogen withdrawal syndrome. *J. Cell. Biochem.* **91,** 3–12.

49. Wang, Y., Shao, C., Shi, C.-H., Zhang, L., Yue, H.-H., Wang, P.-F., Yang, B., Zhang, Y.-T., Liu, F., Qin, W.-J., Wang, H., and Shao, G.-X. (2005) Change of the cell cycle after flutamide treatment in prostate cancer cells and its molecular mechanism. *Asian J. Androl.* **7,** 375–380.

50. Yeh, S., Miyamoto, H., Nishimura, K., Kang, H., Ludlow, J., Hsiao, P., Wang, C., Su, C., and Chang, C. (1998) Retinoblastoma, a tumor suppressor, is a coactivator for the androgen receptor in human prostate cancer DU145 cells. *Biochem. Biophys. Res. Commun.* **248,** 361–367.

51. Wang, X., Deng, H., Basu, I., and Zhu, L. (2004) Induction of androgen receptor-dependent apoptosis in prostate cancer cells by the retinoblastoma protein. *Cancer Res.* **64,** 1377–1385.

52. Meredith, J. E., Winitz, S., McArthur Lewis, J., Hess, S., Ren, X.-D., Renshaw, M. W., and Schwartz, M. A. (1996) The regulation of growth and intracellular signaling by integrins. *Endocr. Rev.* **17,** 207–220.

53. Bonaccorsi, L., Carloni, V., Muratori, M., Salvadori, A., Giannini, A., Carini, M., Serio, M., Forti, G., and Baldi, E. (2000) Androgen receptor expression in prostate carcinoma cells suppresses α6β4 integrin-mediated invasive phenotype. *Endocrinology* **141,** 3172–3182.

54. Nakagawa, O., Akashi, T., Hayakawa, Y., Junicho, A., Koizumi, K., Fujiuchi, Y., Furuya, Y., Matsuda, T., Fuse, H., and Saiki, I. (2004) Differential expression of integrin subunits in DU-145/AR prostate cancer cells. *Oncol. Rep.* **12,** 837–841.

55. Evangelou, A., Letarte, M., Marks, A., and Brown, T. J. (2002) Androgen modulation of adhesion and antiadhesion molecules in PC-3 prostate cancer cells expressing androgen receptor. *Endocrinology* **143,** 3897–3904.

56. Davis, T. L., Cress, A. E., Dalkin, B. L., and Nagle, R. B. (2001) Unique expression pattern of the α6β4 integrin α6β4 and laminin-5 in human prostate carcinoma. *Prostate* **46,** 240–248.

57. Gambaletta, D., Marchetti, A., Benedetti, L., Mercurio, A. M., Sacchi, A., and Falcioni, R. (2000) Cooperative signaling between α6β4 integrin and ErbB-2 receptor is required to promote phosphatidylinositol 3-kinase-dependent invasion. *J. Biol. Chem.* **275,** 10604–10610.

58. Bonaccorsi, L., Carloni, V., Muratori, M., Formigli, L., Zecchi, S., Forti, G., and Baldi, E. (2004) EGF receptor (EGFR) signaling promoting invasion id disrupted in androgen-sensitive prostate cancer cells by an interaction between EGFR and androgen receptor (AR). *Int. J. Cancer* **112,** 78–86.

59. Di Lorenzo, G., Tortora, G., D'Armiento, F. P., De Rosa, G., Staibano, S., Autorino, R., D'Armiento, M, De Laurentiis, M., De Placido, S., Catalano, G., Bianco, A. R., and Ciardiello, F. (2002) Expression of epidermal growth factor receptor correlates with disease

relapse and progression to androgen-independence in human prostate cancer. *Clin. Cancer Res.* **8**, 3438–3444.

60. Wood, M., Fudge, K., Mohler, J. L., Frost, A. R., Garcia, F., Wang, M., and Stearns, M. E. (1997) In situ hybridization studies of metalloproteinases 2 and 9 and TIMP-1 and TIMP-2 expression in human prostate cancer. *Clin. Exp. Metastasis* **15**, 246–258.

61. Sauer, C. G., Kappeler, A., Späth, M., Kaden, J. J., Michel, M. S., Mayer, D., Bleyl, U., and Grobholz, R. (2004) Expression and activity of matrix metalloproteinases-2 and -9 in serum, core needle biopsies and tissue specimens of prostate cancer patients. *Virchows Arch.* **444**, 518–526.

62. Dong, Z., Nemeth, J. A., Cher, M. L., Palmer, K. C., Bright, R. C., and Fridman, R. (2001) Differential regulation of matrix metalloproteinase-9, tissue inhibitor of metalloproteinase-1 (TIMP-1) and TIMP-2 expression in co-cultures of prostate cancer and stromal cells. *Int. J. Cancer* **93**, 507–515.

63. Wilson, M. J., Sellers, R. G., Wiehr, C., Melamud, O., Pei, D., and Peehl, D. M. (2002) Expression of matrix metalloproteinase-2 and -9 and their inhibitors, tissue inhibitor of metalloproteinase-1 and -2, in primary cultures of human prostatic stromal and epithelial cells. *J. Cell. Physiol.* **191**, 208–216.

64. Klein, R. D., Maliner-Jongewaard, M. S., Udayakumar, T. S., Boyd, J. L., Nagle, R. B., and Bowden, G. T. (1999) Promatrilysin expression is induced by fibroblast growth factors in the prostatic carcinoma cell line LNCaP but not in normal primary prostate epithelial cells. *Prostate* **41**, 215–223.

65. Quax, P. H. A., de Bart, A. C. W., Schalken, J. A., and Verheijen, J. H. (1997) Plasminogen activator and matrix metalloproteinase production and extracellular matrix degradation by rat prostate cancer cells in vitro: correlation with metastatic behavior in vivo. *Prostate* **32**, 196–204.

66. Schneikert, J., Peterziel, H., Defossez, P. A., Klocker, H., Launoit, Y., and Cato, A. C. (1996) Androgen receptor-Ets protein interaction is a novel mechanism for steroid hormone-mediated down-modulation of matrix metalloproteinase expression. *J. Biol. Chem.* **271**, 23907–23913.

67. Liao, X., Thrasher, J. B., Pelling, J., Holzbeierlein, J., Sang, Q.-X. A., and Li, B. (2003) Androgen stimulates matrix metalloproteinase-2 expression in human prostate cancer. *Endocrinology* **144**, 1656–1663.

68. Vayalil, P. K., Mittal, A., and Katiyar, S. K. (2004) Proanthocyanidins from grape seeds inhibit expression of matrix metalloproteinases in human prostate carcinoma cells, which is associated with the inhibition of activation of MAPK and NFκB. *Carcinogenesis* **25**, 987–995.

69. Lokeshwar, B. L. (1999) MMP inhibition in prostate cancer. *Ann. N. Y. Acad. Sci.* **878**, 271–289.

70. Powell, W. C., Domann, F. E., Jr., Mitchen, J. M., Matrisian, L. M., Nagle, R. B., and Bowden, G. T. (1996) Matrilysin expression in the involuting rat ventral prostate. *Prostate* **29**, 159–168.

71. Nelson, J. B., Hedican, S. P., George, D. J., Reddi, A. H., Piantadosi, S., Eisenberger, M. A., and Simons, J. W. (1995) Identification of endothelin-1 in the pathophysiology of metastatic adenocarcinoma of the prostate. *Nat. Med.* **1**, 944–949.

72. Granchi, S., Brocchi, S., Bonaccorsi, L., Baldi, E., Vinci, M. C., Forti, G., Serio, M., and Maggi, M. (2001) Endothelin-1 production by prostate cancer cell lines is up-regulated by factors involved in cancer progression and down-regulated by androgens. *Prostate* **49**, 267–277.

73. Padley, R. J., Dixon, D. B., and Wu-Wong, J. R. (2002) Effect of castration on endothelin receptors. *Clin. Sci.* **103**, 442S–445S.

74. Aprikian, A. G., Han, K., Guy, L., Landry, F., Begin, L. R., and Chevalier, S. (1998) Neuroendocrine differentiation and the bombesin/gastrin releasing peptide family of neuropeptides in the progression of human prostate cancer. *Prostate* **Suppl 8,** 52–61.

75. Papandreou, C. N., Usami, B., Geng, Y. P., Bogenrieder, T., Freeman, R. H., Wilk, S., Finstad, C. L., Reuter, V. E., Powell, C. T., Scheinberg, D., Magill, C., Scher, H. I., Albino, A. P., and Nanus, D. M. (1998) Neutral endopeptidase 24.11 loss in metastatic human prostate cancer contributes to androgen-independent progression. *Nat. Med.* **4,** 50–57.

76. Shen, R., Sumitomo, M., Dai, J., Hardy, D. O., Navarro, D., Usami, B., Papandreou, C. N., Hersh, L. B., Shipp, M. A., Freedman, L. P., and Nanus, D. M. (2000) Identification and characterization of two androgen responsive regions in the human neutral endopeptidase gene. *Mol. Cell. Endocrinol.* **25,** 131–142.

77. Stewart, R. J., Panigrahy, D., Flynn, E., and Folkman, J. (2001) Vascular endothelial growth factor expression and tumor angiogenesis are regulated by androgens in hormone responsive prostate carcinoma: evidence for androgen dependent destabilization of vascular endothelial growth factor transcripts. *J. Urol.* **165,** 688–693.

78. Li, J., Wang, E., Rinaldo, F., and Datta, K. (2005) Upregulation of VEGF-C by androgen depletion: the involvement of IGF-IR-FOXO pathway. *Oncogene* **24,** 5510–5520.

79. Cheng, L., Zhang, S., Sweeney, C. J., Kao, C., Gardner, T. A., and Eble, J. N. (2004) Androgen withdrawal inhibits tumor growth and is associated with decrease in angiogenesis and VEGF expression in androgen-independent CWR22Rv1 human prostate cancer model. *Anticancer Res.* **24,** 2135–2140.

80. Boddy, J. L., Fox, S. B., Han, C., Campo, L., Turley, H., Kanga, S., Malone, P. R., and Harris, A. L. (2005) Endothelial growth factor and hypoxia sensing via hypoxia-inducible factros HIF-1a, HIF-2a, and the prolyl hydroxylases in human prostate cancer. *Clin. Cancer Res.* **11,** 7658–7663.

81. Kwabi-Addo, B., Ozen, M., and Ittmann, M. (2004) The role of fibroblast growth factors and their receptors in prostate cancer. *Endocr. Relat. Cancer* **11,** 709–724.

82. Rosini, P., Bonaccorsi, L., Baldi, E., Chiasserini, C., Forti, G., De Chiara, G., Lucibello, M., Mongiat, M., Iozzo, R. V., Garaci, E., Cozzolino, F., and Torcia, M. G. (2002) Androgen receptor expression induces FGF2, FGF-binding protein production, and FGF2 release in prostate carcinoma cells: role of FGF2 in growth, survival, and androgen receptor down-modulation. *Prostate* **53,** 310–321.

83. Blanchère, M., Saunier, E., Mestayer, C., Broshuis, M., and Mowszowicz, I. (2002) Alterations of expression and regulation of transforming growth factor β in human cancer prostate cell lines. *J. Steroid Biochem. Mol. Biol.* **82,** 297–304.

84. Pandini, G., Mineo, R., Frasca, F., Roberts, C. T., Jr., Marcelli, M., Vigneri, R., and Belfiore, A. (2005) Androgens up-regulate the insulin-like growth factor-I receptor in prostate cancer cells. *Cancer Res.* **65,** 1849–1857.

85. Knudsen, B. S., Lucas, J. M., Fazli, L., Hawley, S., Falcon, S., Coleman, I. M., Martin, D. B., Xu, C., True, L. D., Gleave, M. E., Nelson, P. S., and Ayala, G. E. (2005) Regulation of hepatocyte activator inhibitor-1 expression by androgen and oncogenic transformation in the prostate. *Am. J. Pathol.* **167,** 255–266.

86. Li, L., Yang, G., Ebara, S., Satoh, T., Nasu, Y., Timme, T. L., Ren, C., Wang, J., Tahir, S. A., and Thompson, T. C. (2001) Caveolin-1 mediates testosterone-stimulated survival/clonal growth and promotes metastatic activities in prostate cancer cells. *Cancer Res.* **61,** 4386–4392.

87. Jain, A., Lam, A., Vivanco, I., Carey, M. F., and Reiter, R. E. (2002) Identification of an androgen-dependent enhancer within the prostate stem cell antigen gene. *Mol. Endocrinol.* **16,** 2323–2337.

88. Pruthi, R. S., Derksen, E., and Gaston, K. (2003) Cyclooxygenase-2 as a potential target in the prevention and treatment of genitourinary tumors: a review. *J. Urol.* **169,** 2352–2359.

89. Gupta, S., Srivastava, M., Ahmad, N., Bostwick, D. G., and Mukhtar, H. (2000) Over-expression of cyclooxygenase-2 in human prostate adenocarcinoma. *Prostate* **42**, 73–78.
90. Kirschenbaum, A., Klausner, A. P., Lee, R., Unger, P., Yao, S., Liu, X. H., and Levine, A. C. (2000) Expression of cyclooxygenase-1 and cyclooxygenase-2 in the human prostate. *Urology* **56**, 671–676.
91. Madaan, S., Abel, P. D., Chaudhary, K. S., Hewitt, R., Stott, M. A., Stamp, G. W., and Lalani, E. N. (2000) Cytoplasmic induction and over-expression of cyclooxygenase-2 in human prostate cancer: implications for prevention and treatment. *BJU Int.* **86**, 736–741.
92. Yoshimura, R., Sano, H., Masuda, C., Kawamura, M., Tsubouchi, Y., Chargui, J., Yoshimura, N., Hla, T., and Wada, S. (2000) Expression of cyclooxygenase-2 in prostate carcinoma. *Cancer* **89**, 589–596.
93. Zha, S., Gage, W. R., Sauvageot, J., Saria, E. A., Putzi, M. J., Ewing, C. M., Faith, D. A., Nelson, W. G., De Marzo, A. M., and Isaacs, W. B. (2001) Cyclooxygenase-2 is up-regulated in proliferative inflammatory atrophy of the prostate, but not in prostate carcinoma. *Cancer Res.* **61**, 8617–8623.
94. Litvinov, I. V., De Marzo, A. M., and Isaacs, J. T. (2003) Is the Achilles' heel for prostate cancer therapy a gain of function in androgen receptor signaling? *J. Clin. Endocrinol. Metab.* **88**, 2972–2982.
95. Gupta, S., Adhami, V. M., Subbarayan, M., MacLennan, G. T., Lewin, J. S., Hanfeli, U. O., Fu, P., and Mukhtar, H. (2004) Suppression of prostate carcinogenesis by dietary supplementation of celecoxib in transgenic adenocarcinoma of the mouse prostate model. *Cancer Res.* **64**, 3334–3343.
96. Fujita, H., Koshida, K., Keller, E. T, Takahashi, Y., Yoshimoto, T., Namiki, M., and Mizokami, A. (2002) Cyclooxygenase-2 promotes prostate cancer progression. *Prostate* **53**, 232–240.
97. Miyamoto, H., Altuwaijri, S., Cai, Y., Messing, E. M., and Chang, C. (2005) Inhibition of the Akt, cyclooxygenase-2, and matrix metalloproteinase-9 pathways in combination with androgen deprivation therapy: potential therapeutic approaches for prostate cancer. *Mol. Carcinogen.* **44**, 1–10.

7 Epigenetic Modification and Acetylation of Androgen Receptor Regulate Prostate Cellular Growth

Michael J. Powell, BS, Shengwen Li, PhD,
Michael P. Lisanti, MD, PhD,
Marja T. Nevalainen, MD, PhD,
Chenguang Wang, PhD, and
Richard G. Pestell, MD, PhD, MBBS, FRACP

CONTENTS

1. INTRODUCTION

The androgen receptor (AR) is modified by histone acetyltransferases (HATs) within its hinge region (HR) at a conserved lysine motif. AR acetylation governs a subset of AR functions including ligand specificity, cellular DNA synthesis, and prostate cancer cell apoptosis. The ligand dihydrotestosterone (DHT) augments AR acetylation, which in turn regulates association

From: *Current Clinical Oncology: Prostate Cancer:*
Signaling Networks, Genetics, and New Treatment Strategies
Edited by: R. G. Pestell and M. T. Nevalainen © Humana Press, Totowa, NJ

between the AR and its co-repressor complexes. AR acetylation site gain-of-function mutants show increased association with p300 and decreased association with NCoR/HDAC/Smad3 co-repressor complexes. The AR acetylation site serves as a substrate for HATs and both trichostatin A (TSA)- and nicotinamide adenine dinucleotide (NAD)-dependent deacetylases, the sirtuins. The finding that Sirt1 deacetylates AR *(1)* provides evidence for a mechanism by which intracellular metabolic substrates that regulate Sirt activity *(2)* may, in turn, regulate hormone signaling and thereby prostate cancer cellular growth. Recent evidence suggests that sirtuins play a role in cellular aging and transcription factor function (Fig. 1). The finding that Sirt1 represses AR activity raises the prospect that dysregulation of AR function with increasing age may be a consequence of defective Sirt-mediated repression. An enduring model showing that the AR is associated with a family of histone deacetylases (HDACs) has been proposed and thoroughly discussed.

1.1. Androgen Receptor

The AR is a member of the structurally related nuclear receptor superfamily which includes receptors for steroid and thyroid hormones, retinoids, vitamins, and other proteins for which the native ligands have not yet been found. The nuclear receptor functional domains are conserved within the superfamily members and include the activation function (AF) region, DNA-binding domain (DBD), HR, and the ligand-binding domain (LBD) *(3)*. The binding

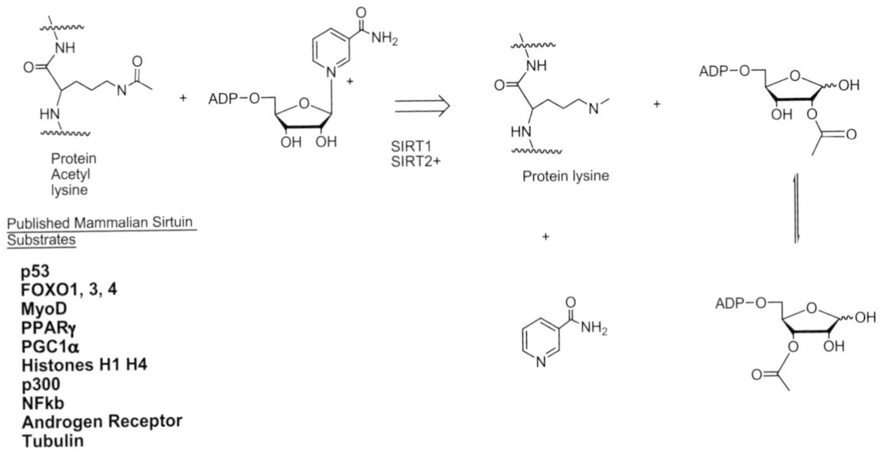

Fig. 1. Reaction of sirtuins. Nicotinamide adenine dinucleotide (NAD) and acetylated protein substrate are converted to nicotinamide, free lysine side chain, and 2′-O-acetyl-ADPR, which is in equilibrium with the 3′-O-acetyl-ADPR ester. Substrates for the human sirtuins SIRT1 and SIRT2 are listed.

of the AR to specific DNA sequences occurs through the DBD and may occur as a homodimer or as a heterodimer with other nuclear receptors or transcription factors. The LBD, upon binding the cognate hormone, adopts an active conformation that facilitates the dissociation of heat shock proteins and reassociation with co-regulatory proteins. The N-terminal and LBDs together interact with transcriptional co-regulators. The AR is unusual among nuclear receptors as most of its activity is mediated via constitutive activation in the N-terminus. Deletion of the LBD results in a molecule with constitutive activity. There are two discrete overlapping activation domains in the N-terminus and their usage is cell context-dependent.

The AR physically interacts with transcriptional co-regulators, which encode both intrinsic and associated histone acetylase activities, and have the ability to recruit additional enzymes that modulate gene expression in response to hormonal signals. The co-activator adaptors for nuclear receptor family members have been characterized by a number of laboratories around the world over the last decade and a half. The Steroid Receptor Co-activator-1 (SRC1) protein, also known as *a*mplified *in b*reast cancer 1/*ac*tivator of the *t*hyroid and retinoid acid *r*eceptor (AIB1/ACTR), augments nuclear receptor activity. The p300/CBP (*C*RE-*b*inding *p*rotein) and P/CAF (p300/*C*BP-*a*ssociated *f*actor) augment AR activity through distinct functions. The co-integrator protein p300/CBP augments AR activity, in part through their HAT activity and in part through serving as a molecular bridge between the basal transcription apparatus and the nuclear receptor itself. The physical association between AR and p300 is regulated through several distinct contact points of AR. The p160 co-activator SRC1 interacts directly with the N-terminus of the AR in a ligand-dependent manner via a conserved glutamine-rich region between residues 1053 and 1123 of SRC which is both necessary and sufficient for recruitment of SRC1 to AR *(4)*. Modification of a conserved lysine motif (residues K630–633) of AR alters the affinity for p300 both in vitro and in vivo.

Histone acetylation intrinsic to the co-integrator proteins is thought to augment transcriptional activity by facilitating access of transcriptional factors to the local hormone responsive element of the gene promoter. This activity is in part contributed by nucleosome destabilization. Co-activator molecules contain a well-conserved LXXLL motif (nuclear receptor box) that mediates their interaction with the nuclear hormone receptor. The p160 co-activator family members interact predominantly with the amino terminus of the AR. The AR co-activator interactive protein 60 kDa (TIP60) contains a single nuclear receptor box in its extreme carboxyl terminus that interacts predominantly with the amino terminus of the AR. The LXXLL motif of TIP60 is sufficient for AR interaction and, like p300, TIP60 functions to directly acetylate the AR.

In addition, nuclear receptor co-repressors (NCoRs) regulate AR activity. The NCoR and its homolog SMRT function as co-repressors through the recruitment of HDACs. NCoR and SMRT share structural homology within the N-terminal repression domain that interacts with HDAC complexes and a transducin-like molecule that interacts with histones. It has been proposed that a single co-repressor binds to the AR in a DNA-bound dimer. A variety of distinct AR co-repressor complexes have been identified and recently comprehensively reviewed (5). Initially implicated through yeast two-hybrid analysis, a variety of distinct co-repressor molecules have been identified and characterized primarily using in vitro tissue culture approaches (5). These studies have been complemented by a more recent proteomic analysis. AR co-repressors function by inhibiting DNA binding following nuclear translocation, recruiting co-repressors with HDAC activity, interrupting the N-terminus and C-terminus of the AR by functioning as scaffolds, and by targeting the basal transcriptional machinery or through new mechanisms involving methylases or other enzymatic functions.

1.2. Epigenetic Modification

DNA is packaged into nucleosomes consisting of histone proteins including H2A, H2B, H3, and H4, together with the 147 base pairs of DNA to form eukaryotic nucleosomes. The DNA wrapped around an octamer of histones containing two copies of each of the four core histone proteins together with a fifth class of histone, the linker histone-H1, together facilitates the packing of genomic DNA into chromatin. Several classes of proteins have been identified that remodel local chromatin complexes including those with either ATPase function or HDAC activity. ATP-dependent complexes directly change chromatin structure. The energy derived from ATP hydrolysis is used by these complexes to alter the position and stability of the nucleosomes in a non-covalent manner. The catalytic subunit of the ATP-dependent nucleosome remodeling complex belongs to the SWI2/SNF-2 superfamily of DNA helicases and also five major families of ATP-dependent remodeling complexes. The ATP-dependent remodeling complexes, based on their distinct ATP kinase subunits, include SWI/SNF, ISWI, Mi-2/NuRD, INO80, and SWR1 (6).

The two human gene orthologs of the SWI/SNF complex are the human BRM protein and the BRM-related gene-1 (BRG-1) protein. The SWI/SNF proteins contain a bromodomain that recognizes and binds acetylated lysine residues in histones and related proteins. The BRG/BRM1 complex co-elutes in multi-protein complexes with other histone-modifying enzymes. The Mi-3/NuRD complex, for example, contains HDAC1/2, methylated DNA-binding proteins, and members of the metastasis associated protein family (MTA1, MTA2, and MTA3) (7).

An additional class of chromatin-modeling complexes consists of histone-modifying enzymes, which include HATs, HDACs, histone methyltransferases, kinases, and ubiquitin ligases. Covalent modification of amino acid residues in histone tails including acetylation, phosphorylation, methylation, and ubiqui-tination occurs in a dynamic manner. A series of inter-dependent interactions have been documented on which phosphorylation, acetylation, and methylation cascades occur within histone tails. Acetylation, which occurs at the conserved lysine residues present in the amino terminal tails of all four core histones, is thought to neutralize the basic charge of histone tails, thereby reducing their affinity for negatively charged chromosomal DNA. This results in less densely packed "euchromatin," which is transcriptionally active. The notion that temporarily coordinated dynamic interactions may occur within histone tails to coordinate signal transduction is now referred to as the "histone code hypothesis."

The HAT enzymes are thought of as either type A, located in the nucleus, or type B, located primarily in the cytoplasm. Although exceptions to these locations are well characterized, the type B HATs conduct housekeeping roles in the cell, acetylating newly synthesized free histones in the cytoplasm. By contrast, type A HATs acetylate nucleosomal histones within the chromatin of the nucleus. The type A HATs consists of five families: the Gcn5-related acetylated transferases (GNATs), the MYST (MOZ, Ybf2/Sas3, Sas2, and Tip60)-related HATs, p300/CBP HATs; the general transcription factor HATs (TAFII 250), and the nuclear hormone-related HATs (SRC-p160 co-activators).

In addition to histones, the HATs modify an array of distinct substrates, including structural proteins such as tubulin, transport proteins such as Importin α, transcription factors such as p53, EKLF, HMG box architectural factor UBF, and nuclear hormone receptors *(8)*.

1.3. Androgen Receptor Acetylation

1.3.1. Nuclear Receptor and AR Acetylation

Nuclear receptors are post-translationally modified by several distinct enzymes. In addition to phosphorylation, sumoylation, and ubiquitination, the AR is acetylated by HATs. The initial description of nuclear receptor acety-lation identified a motif within the estrogen receptor-α which served as a direct substrate for acetylation by p300 *(9)*. The candidate acetylation motif KXKK was conserved amongst different species including vertebrates, arthropods, and nematodes. The motif was identified within related nuclear receptor members including AR, TR, RARα, PPARγ, LXR, FXR, GR, NGF4, and SF1. The first study demonstrating AR serves as a substrate for HATs was from Fu et al. *(10)*. In these studies, AR was modified in vitro by p/CAF and p300 *(10)* and in subsequent studies also by TIP60 *(11)*. More importantly, immuno-precipitation and western blotting demonstrated that the AR was acetylated *in vivo (12)*. Both CBP and p300 were shown to enhance ligand-dependent AR activity using androgen-responsive reporter genes *(10)*. These findings were consistent with observations that the HDAC inhibitor TSA enhanced androgen responsive reporter gene activity. Through point mutational analyses and *in vitro* acetylation studies, a minimal region of the AR was identified to function as a substrate of SirT1 deacetylation *(1)*. The AR acetylation motif is well conserved in different species, and mutations of the AR lysine residues 630, 632, and 633 abolished p300-dependent activation or DHT-induced AR activity. It has been hypothesized that acetylation of lysine residues in the histone tails alters their charge. Acetylation mimic mutants of the AR lysine residues, either glutamine or threonine substitutions, were shown to augment the ligand-dependent activation of the AR's activity.

A series of further studies demonstrated that the acetylation state of AR determines its affinity for co-regulator proteins. Using anti-acetyl-lysine antibodies for immunoprecipitation and western blot analysis, AR was shown to be acetylated upon treatment with DHT or addition of stimuli known to induce AR activity such as bombesin *(13)*. To determine the mechanisms by which the AR gain-of-function acetylation mutants enhance androgen-responsive reporter gene activity, analyses were conducted on the co-activator proteins associated with the AR in cultured cells. The AR was shown to bind p300 and in the NCoR/HDAC/Smad3 co-repressor complex in cultured cells *(12)* (also see Fig. 2). The AR acetylation gain-of-function mutants that showed enhanced transactivation activity also showed enhanced co-activator-binding capacity in cultured cells. Conversely, acetylation dead substitution mutations (alanine or arginine substitutions) demonstrated reduced p300 binding and enhanced co-repressor binding associated with reduced transactivation activity. Studies *in vitro* demonstrated that the physical association between AR and p300, which occurred in an acetyl-CoA-dependent manner, was regulated by the charge of lysine residues within the AR acetylation motif. These findings were consistent with prior observations showing that the transcription factor acety-lation site lysine residues are regulated relative to the affinity for co-activator proteins such as p300 *(14)*.

Functional consequences of the AR acetylation have been addressed. Although the acetylation site regulates co-activator binding, the acetylation motif did not affect post-translational modification by either sumoylation or regulation of transrepression activities. AR is known to regulate NFκB, Sp1, and AP-1 activities; however, AR acetylation site mutants showed no alter-ation in their ability to regulate gene expression through NFκB, Sp1, or AP1 activities *(15)*. AR acetylation mutants, however, did alter their affinity with androgen-responsive gene promoters in the context of local chromatin structure in chromatin-immunoprecipitation (ChIP) assays *(12–16)*. As expected, AR acetylation sites' "dead" mutants were defective in TSA-dependent recruitment of the AR to the androgen response element of target genes in ChIP assays *(16)*. It appears that, although the AR acetylation site regulates its binding activity to an endogenous ARE, AR acetylation does not regulate the AR function on NFκB, Sp1, or AP-1 *(15)*.

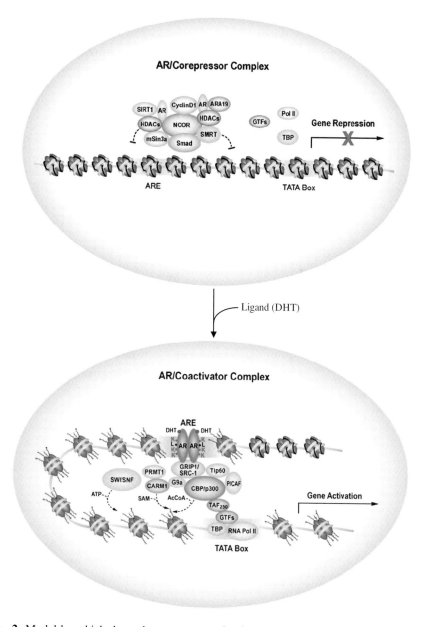

Fig. 2. Model by which the androgen receptor (AR) acetylation site particpates in ligand-induced gene expression. Schematic representation of AR bound to co-repressors in the absence of ligand in the context of local chromatin dependent in part upon the AR lysine motif. SIRT1 inhibits AR activity. SIRT1 is shown to associate with nuclear receptor co-repressor (NCoR)/histone deacetylases (HDACs) repression complex. Upon the addition of ligand (DHT), disengagement of the co-repressor complex and recruitment of the co-activator complex engage gene expression, dependent in part, upon the AR lysine motif modifications.

The downstream effect of the AR acetylation on cell cycle control gene promoters was assessed by ChIP assay. The AR acetylation site gain-of-function mutants showed enhanced recruitment to the promoter of the *cyclin D1* gene *(12)*. Consistent with this observation is that the *cyclin D1* gene is known to be induced by DHT and to promote growth of prostate cancer cells in culture and in nude mice *(17)*.

1.3.2. ACETYLATION OF THE AR PROMOTES PROSTATE CANCER CELLULAR GROWTH IN VITRO AND IN VIVO

The studies on the regulatory effects of AR acetylation site on the growth of human prostate cancer cell lines demonstrated that AR acetylation has an anti-apoptotic and pro-proliferative function *(12)*. The comparison of growth characteristics of stable cell lines showed that an increase in the size and

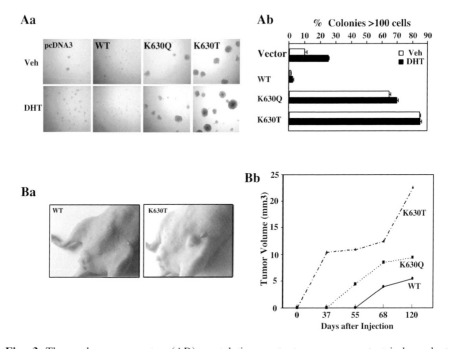

Fig. 3. The androgen receptor (AR) acetylation mutants convey contact-independent growth. (**Aa,b**) DU145 cells stably expressing the ARwt or AR acetylation site mutants were seeded in soft agar. Phase contrast images of the colonies from a representative experiment are shown (×100). Colony numbers and size (percentage of colonies with more than 100 cells) determined on day 14. (**Ba,b**) Nude mice were implanted with 1 × 10⁶ cells of stable lines expressing either the ARwt or AR acetylation site mutants. Mean volume of DU145 tumors grown in nude mice are shown at each time point. Reproduced with permission *(12)*.

number of colonies in AR acetylation mimic mutants (ARK630Q, ARK630T) compared with the wild-type AR (Fig. 3). As the relative abundance of the wild-type and mutant ARs was similar between cell lines, the data suggest that the single residue substitutions function as a key molecular switch in prostate cancer cellular growth. When implanted in nude mice, these prostate cancer cells expressing the AR acetylation site gain-of-function mutants grew more rapidly (Fig. 3). Careful analysis of apoptosis demonstrated a reduction in apoptotic nuclei in tumors from nude mice harboring the AR acetylation gain-of-function mutants (15). Molecular analysis demonstrated that JNK-dependent AR-mediated apoptosis was reduced by the AR acetylation site mutations (15). Collectively, these studies were consistent with the model in which the AR acetylation site regulates cellular growth in part through a reduction in cellular apoptosis. Analysis of tumors derived from the AR acetylation site mimic mutants showed an increased expression of a subset of cell cycle proteins. Such proteins that exhibited an expression increase in the AR acetylation site mutant cell lines included cyclin D1 and cyclin E (12). The promoters of both *cyclin D1* and *cyclin E* gene were activated by the AR acetylation mimic mutants proportionately more than by the wild-type AR. Together, these studies were consistent with a model in which the AR acetylation site mimic mutants convey both an anti-apoptotic and pro-proliferative function.

1.3.3. Expression of AR Co-Activators in Prostate Cancer

Abnormal expression of AR co-activators in prostate cancer has been observed, which induce prostate cancer growth. The expression of p300 is higher in prostate cancer tissue and correlates with a higher Gleason score, larger tumor volumes, and extra prostatic extension. These findings suggest that upregulation of p300 may participate in apoptotic and phenotypic cellular changes that promote prostate cancer progression (18). The AR co-activators SRC1 and SRC2 are expressed in androgen-independent prostate cancer. In the majority of recurrent prostate cancers overexpression of SRC1 and SRC2 is exhibited. Nuclear staining of SRC1 and SRC2 is more intense in prostate cancer compared with benign prostate hyperplasia (19). In addition, broad changes in chromatin acetylation are observed in prostate cancer specimens, which correlate with tumor progression and expression of histone demethylases (LSD1). This demethylase binds the AR and is overexpressed in human prostate cancer samples (20,21).

2. ANDROGEN RECEPTOR DEACETYLATION

2.1. Histone Deacetylation

Post-translational acetylation of ε amino lysine residues is reversible. In humans, some ATP-dependent deacetylases have been identified. These deacetylases regulate the removal of acetyl groups and maintain a dynamic equilibrium of both histone acetylation and the acetylation of non-histones, including transcription factors. HDACs are recruited into the context of local chromatin through hormone signaling *(22)*. HDAC activity is coordinated at nuclear-receptor-binding sites in the context of local chromatin structure. HDACs consist of two broad protein families: those with NAD-dependent HDAC activity and those with TSA-sensitive HDAC activity.

According to their homology to yeast transcriptional repressors, the HDAC family is divided into three distinct classes with Class I HDACs (HDAC-1, -2, -3, -8) and Class II HDACs (HDAC-4, -8, -9, -11) homologous to Rpd3P and Hda1P, respectively. The Class I HDACs are localized primarily in the nucleus and are relatively ubiquitously expressed. The Class II HDACs are located in both the nucleus and the cytoplasm. The expression patterns of Class II HDACs are relatively more tissue-specific with HDAC-4, HDAC-8, and HDAC-9 showing more expression in tumor tissues than in normal cells. HDAC-11 contains a unique catalytic domain at the N-terminus which has HDAC activity and, by homology, most closely resembles a Class I HDAC.

The Class III HDACs are homologs of the yeast transcriptional repressor Sirt2p. In mammalian cells, seven related genes have been identified and comprise the Sirt family. The mammalian sirtuins share a common NAD-dependent histone/protein deacetylase activity. The NAD-dependent deacetylase enzymes are widely distributed in nature and are conserved from bacteria to humans *(2)*. The sirtuins are NAD^+-dependent deacetylases that require NAD^+ stoichiometrically to deacetylate acetyl-lysine residues of multiple downstream substrates (Fig. 1). The deacetylation reaction forms the products 2´- and 3´-*O*-acetyl-ADP-ribose. These products are considered candidate second messengers of signaling pathways *(23)*. In yeast, these metabolites function in heterochromatin formation and maintenance. The enzyme is in turn inhibited by nicotinamide, a product of the sirtuin reaction. The reduced dinucleotide, NADH, also inhibits the production of NAD. As nicotinamide serves as a metabolic product in the cell, it has been proposed that sirtuins may function through NAD to integrate metabolic signaling to gene expression via deacetylation of key acetylated lysine residues on local histone and transcription factors (Fig. 1) *(2)*.

The sirtuins regulate insulin-1/insulin-like growth factor (IGF)-1-signaling activities in *Caenorhabditis elegans* and in mice *(24)*. In yeast, Sirt plays a key role in DNA repair. Acetylation and deacetylation of histones are important in double-stranded DNA damage repair in mice and in cultured cells. Sirt2 affects DNA double-strand repair of the non-homologous end-joining type through an indirect mechanism *(25–27)*. TSA-sensitive HDACs also play a role, and Sin3p deficiency reduces the efficiency of non-homologous end-joining repair. Sin3p-dependent hyperacetylation of lysine 6 histone H4 occurs in the vicinity of double-strand breaks suggesting a role for Sin3p in DSB in *Saccharomyces cerevisiae* cells *(28)*. In mammalian cells, sirtuins may affect DNA damage through two mechanisms: first, by modifying chromatin structure at the damaged site via altering the access of repair enzymes and, second, by deacetylating p53, thereby blocking p53-dependent apoptosis.

Sirtuins play an evolutionarily conserved role in regulating insulin signaling in *C. elegans*, mice, and humans. In *C. elegans*, Sir2 opposes insulin signaling *(29)* and Sirt1 regulates FOXO1 activity *(30)*. FOXO1 is phosphorylated in mammalian systems and inhibited by insulin via the PI-3 kinase/Akt-dependent signaling pathway *(30)*. Sirt1 is an important mediator of gluconeogenic responses in the liver, potentiating FOXO1 activity in hepatocytes to direct glucose metabolism toward gluconeogenesis. Sirt1 upregulates PGC1α through deacetylation of PGC1 *(31)*. Sirt1 modulates insulin secretion in the pancreas and regulates fat metabolism *(2)*.

2.2. Androgen Receptor Deacetylation

Recent studies examined the ability of Sirt1 to regulate AR function through AR acetylation *(1)*. Immunoprecipitation and western blotting demonstrated the co-association of the AR with Sirt1 in cultured cells. The expression and function of the AR was repressed by expression of Sirt1, and AR acetylation was enhanced by inhibitors of Sirt1. The inhibition of Sirt function by Sirtinol or nicotinamide induced AR expression and AR activity prostate in cancer cells. Inhibition of endogenous Sirt activity enhanced androgen-responsive reporter gene activity to a similar level as the addition of DHT *(1)*. The ability of Sirt1 to inhibit AR-mediated gene expression was dependent upon the catalytic function of Sirt1. A single point mutation of hSIRT1 (SIRT1H363Y) abrogated SIRT1-dependent repression of the AR (Fig. 3).

It has been shown that AR associates both *in vitro* and *in vivo* with Sirt1. AR co-localized with Sirt1 in the presence of DHT, forming nuclear, extranucleolar cord-like structures. In biochemical analysis, direct comparison was made between the ability of the Sirt enzyme to deacetylate an acetylated AR peptide. The AR served as an equivalent substrate for Sirt to either acetylated p53 or p300, both of which are known to serve as Sirt substrates *(15)*. The acetylated AR served as a substrate for yeast, bacterial, murine, and human SIRT in biochemical assays, indicating the conservation of this molecular interaction. Sirt1 transduction of AR-expressing prostate cancer cells inhibited cellular growth *(1)*. Growth repression by Sirt1 was dependent upon the presence of the AR as exhibited by Sirt1's failure to inhibit AR-deficient prostate cancer cells lines. Sirt1 inhibited androgen-dependent prostate cancer cellular growth *(1)*. The minimized structure of the hSIRT1 AR peptide/NAD complex was shown as a model (Fig. 4) generated by using the molecular display program Chimera from the University of San Francisco. In this model, the KLKK

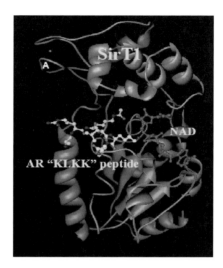

Fig. 4. Homology model of androgen receptor (AR)-"KLKK" peptide with nicotinamide adenine dinucleotide (NAD)-bound SIRT1. The structure of the hSirT1 and AR "KLKK" peptide–NAD complex is shown as the ribbon model generated using the molecular display program Chimera. The "KLKK" peptide (yellow) and the NAD molecule (cyan) are shown as the "ball-and-stick" model. Reproduced with permission *(1)*.

peptide (yellow) and the NAD molecule (cyan) is shown as "a ball and stick model." The hydrophobic part of the lysine K630 side chain packs favorably against the aromatic ring in the side chain at F309.

Endogenous levels of nicotinamide may inhibit Sirt activity *(1)* indicating that the concentration of nicotinamide in response to physiological changes could affect Sirt function. Androgens are known to maintain muscle mass and induce cellular gene expression in muscle *(32,33)*. Sirt regulation of AR activity may play a role in maintaining normal muscle mass. Sirt activity is regulated by NAD/NADH. Thus, it is known that in the resting muscle, increases in the NAD/NADH ratio inhibit muscle gene expression, whereas lactate reduces the NAD/NADH ratio and stimulates muscle gene expression. Lactate-mediated inhibition of Sirt1 activity must be anticipated to induce AR function and enhance muscle anabolism. The role of altered NAD/NADH ratios in prostate cancer onset and progression remains speculative. However, it is known that during prostate cancer progression, metabolism shifts toward cytosolic glycolysis *(34,35)*. Because of this shift, the increased production of lactate that occurs during prostate cancer progression *(36,37)* is predicted to inhibit Sirt activity and may thereby enhance AR function. Global histone gene modifications occur in prostate cancer tissue including acetylation of histone H3K18 and H4K12 and acetylation of H3K9 *(38)*, which are all targets of Sirt1 *(39)*. The role of SIRT1 as a risk factor for prostate cancer occurrence and the role of Sirt1 activity in regulating histone H3 lysine 9 during prostate cancer progression remain to be determined.

3. CONCLUSIVE REMARK AND PERSPECTIVE

An evolving model now shows that the AR is associated with a family of HDACs that are disengaged upon the addition of ligand (Fig. 2). The class III HDAC, Sirt1, deacetylates the AR. Like the AR, the estrogen receptor is acetylated at conserved residues, suggesting that acetylation is a general and well-conserved mechanism governing the function of the nuclear receptor superfamily. Future studies will be required to determine processes regulating nuclear receptor acetylation and deacetylation in normal physiology and the effects of NR acetylation in pathological conditions including breast and prostate cancer. Therapeutic intervention may become feasible with further understanding of the dynamic process governing nuclear receptor acetylation and deacetylation and their role in breast and prostate cancer progression.

ACKNOWLEDGMENTS

This work was supported in part by R01CA70896, R01CA75503, R01CA86072, and R01CA86071 (R.G.P.), a generous support grant from the Marian Falk Foundation. The Kimmel Cancer Center was supported by the NIH Cancer Center Core grant P30CA56036 (R.G.P.). This project is funded in part under a grant with the Pennsylvania Department of Health. The Department specifically disclaims responsibility for an analysis, interpretations, or conclusions. Dr. Lisanti (M.P.L.) was supported by grants from the NIH/NCI and the Pennsylvania Department of Health.

REFERENCES

1. Fu, M., Sauve, A. A., Liu, M., Jiao, X., Zhang, X., Powell, M., Yang, T., Gu, W., Avantaggiati, M. L., Pattabiraman, N., Pestell, T. G., Wang, C., Wang, F., Quong, A., and Pestell, R. G. Hormonal control of androgen receptor function through SIRT1. *Mol Biol Cell, 26:* 8122–8135, 2006.
2. Yang, T., Fu, M., Pestell, R., and Sauve, A. A. SIRT1 and endocrine signaling. *Trends Endocrinol Metab, 17:* 186–191, 2006.
3. Leader, J. E., Wang, C., Fu, M., and Pestell, R. G. Epigenetic regulation of nuclear steroid receptors. *Biochem Pharmacol,* 2006.
4. Bevan, C. L., Hoare, S., Claessens, F., Heery, D. M., and Parker, M.G. The AF1 and AF2 domains of the androgen receptor interact with distinct regions of SRC1. *Mol Cell Biol, 19:* 8383–8392, 1999.
5. Wang, L., Hsu, C. L., and Chang, C. Androgen receptor corepressors: an overview. *Prostate, 63:* 117–130, 2005.
6. Eberharter, A. and Becker, P. B. ATP-dependent nucleosome remodelling: factors and functions. *J Cell Sci, 117:* 3707–3711, 2004.

7. Chen, J., Kinyamu, H. K., and Archer, T. K. Changes in attitude, changes in latitude: nuclear receptors remodeling chromatin to regulate transcription. *Mol Endocrinol, 20:* 1–13, 2006.

8. Fu, M., Wang, C., Wang, J., Zafonte, B. T., Lisanti, M. P., and Pestell, R. G. Acetylation in hormone signaling and the cell cycle. *Cytokine Growth Factor Rev, 13:* 259–276, 2002.

9. Wang, C., Fu, M., Angeletti, R. H., Siconolfi-Baez, L., Reutens, A. T., Albanese, C., Lisanti, M. P., Katzenellenbogen, B. S., Kato, S., Hopp, T., Fuqua, S. A., Lopez, G. N., Kushner, P. J., and Pestell, R. G. Direct acetylation of the estrogen receptor alpha hinge region by p300 regulates transactivation and hormone sensitivity. *J Biol Chem, 276:* 18375–18383, 2001.

10. Fu, M., Wang, C., Reutens, A. T., Wang, J., Angeletti, R. H., Siconolfi-Baez, L., Ogryzko, V., Avantaggiati, M. L., and Pestell, R. G. p300 and p300/cAMP-response element-binding protein-associated factor acetylate the androgen receptor at sites governing hormone-dependent transactivation. *J Biol Chem, 275:* 20853–20860, 2000.

11. Gaughan, L., Logan, I. R., Cook, S., Neal, D. E., and Robson, C. N. Tip60 and histone deacetylase 1 regulate androgen receptor activity through changes to the acetylation status of the receptor. *J Biol Chem, 277:* 25904–25913, 2002.

12. Fu, M., Wang, C., Wang, J., Sakamaki, T., Di Vizio, D., Zhang, X., Albanese, C., Balk, S., Chang, C., Fan, S., Rosen, E., Palvimo, J. J., Janne, O. A., Muratoglu, S., Avantaggiati, M., and Pestell, R. G. Acetylation of the androgen receptor enhances coactivator binding and promotes prostate cancer cell growth. *Mol Cell Biol, 23:* 8563–8575, 2003.

13. Gong, J., Zhu, J., Goodman, O. B., Pestell, R. G., Schlegel, P. N., Nanus, D. M., and Shen, R. Activation of p300 histone acetyltransferase activity and acetylation of the androgen receptor by bombesin in prostate cancer cells. *Oncogene, 25:* 2011–2021, 2006.

14. Polesskaya, A., Naguibneva, I., Duquet, A., Bengal, E., Robin, P., and Harel-Bellan, A. Interaction between acetylated MyoD and the bromodomain of CBP and/or p300. *Mol Cell Biol, 21:* 5312–5320, 2001.

15. Fu, M., Wang, C., Wang, J., Sakamaki, T., Zhang, X., Yeung, Y.-G., Chang, C., Hopp, T., Fuqua, S. A. W., Jaffray, E., Hay, R. T., Palvimo, J. J., Jänne, O. A., and Pestell, R. G. The Androgen Receptor Acetylation governs transactivation and MEKK1-induced apoptosis without affecting in vitro sumoylation and transrepression function. *Mol Cell Biol, 22:* 3373–3388, 2002.

16. Fu, M., Rao, M., Wu, K., Wang, C., Zhang, X., Hessien, M., Yeung, Y. G., Gioeli, D., Weber, M. J., and Pestell, R. G. The androgen receptor acetylation site regulates cAMP and AKT but not ERK-induced activity. *J Biol Chem, 279:* 29436–29449, 2004.

17. Ananthaswamy, H. N., Ouhtit, A., Evans, R. L., Gorny, A., Khaskina, P., Sands, A. T., and Conti, C. J. Persistence of p53 mutations and resistance of keratinocytes to apoptosis are associated with the increased susceptibility of mice lacking the XPC gene to UV carcinogenesis. *Oncogene, 18:* 7395–7398, 1999.

18. Debes, J. D., Sebo, T. J., Lohse, C. M., Murphy, L. M., Haugen de, A. L., and Tindall, D. J. p300 in prostate cancer proliferation and progression. *Cancer Res, 63:* 7638–7640, 2003.

19. Gregory, C. W., He, B., Johnson, R. T., Ford, O. H., Mohler, J. L., French, F. S., and Wilson, E. M. A mechanism for androgen receptor-mediated prostate cancer recurrence after androgen deprivation therapy. *Cancer Res, 61:* 4315–4319, 2001.

20. Metzger, E., Wissmann, M., Yin, N., Muller, J. M., Schneider, R., Peters, A. H., Gunther, T., Buettner, R., and Schule, R. LSD1 demethylates repressive histone marks to promote androgen-receptor-dependent transcription. *Nature, 437:* 436–439, 2005.

21. Shi, Y., Lan, F., Matson, C., Mulligan, P., Whetstine, J. R., Cole, P. A., Casero, R. A., and Shi, Y. Histone demethylation mediated by the nuclear amine oxidase homolog LSD1. *Cell, 119:* 941–953, 2004.

22. Fu, M., Rao, M., Bouras, T., Wang, C., Wu, K., Zhang, X., Li, Z., Yao, T.-P., and Pestell, R. G. Cyclin D1 inhibits PPARgamma-mediated adipogenesis through HDAC recruitment. *J Biol Chem*, *280:* 16934–16941, 2005.

23. Sauve, A. A. and Schramm, V. L. SIR2: the biochemical mechanism of NAD(+)-dependent protein deacetylation and ADP-ribosyl enzyme intermediates. *Curr Med Chem*, *11:* 807–826, 2004.

24. Kruszewski, M. and Szumiel, I. Sirtuins (histone deacetylases III) in the cellular response to DNA damage—facts and hypotheses. *DNA Repair (Amst)*, *4:* 1306–1313, 2005.

25. Tsukamoto, Y., Kato, J., and Ikeda, H. Silencing factors participate in DNA repair and recombination in Saccharomyces cerevisiae. *Nature*, *388:* 900–903, 1997.

26. Valencia, M., Bentele, M., Vaze, M. B., Herrmann, G., Kraus, E., Lee, S. E., Schar, P., and Haber, J. E. NEJ1 controls non-homologous end joining in Saccharomyces cerevisiae. *Nature*, *414:* 666–669, 2001.

27. Frank-Vaillant, M. and Marcand, S. NHEJ regulation by mating type is exercised through a novel protein, Lif2p, essential to the ligase IV pathway. *Genes Dev*, *15:* 3005–3012, 2001.

28. Jazayeri, A., McAinsh, A. D., and Jackson, S. P. Saccharomyces cerevisiae Sin3p facilitates DNA double-strand break repair. *Proc Natl Acad Sci USA*, *101:* 1644–1649, 2004.

29. Tissenbaum, H. A. and Guarente, L. Increased dosage of a sir-2 gene extends lifespan in Caenorhabditis elegans. *Nature*, *410:* 227–230, 2001.

30. Al-Regaiey, K. A., Masternak, M. M., Bonkowski, M., Sun, L., and Bartke, A. Long-lived growth hormone receptor knockout mice: interaction of reduced insulin-like growth factor i/insulin signaling and caloric restriction. *Endocrinology*, *146:* 851–860, 2005.

31. Rodgers, J. T., Lerin, C., Haas, W., Gygi, S. P., Spiegelman, B. M., and Puigserver, P. Nutrient control of glucose homeostasis through a complex of PGC-1alpha and SIRT1. *Nature*, *434:* 113–118, 2005.

32. Grinspoon, S., Corcoran, C., Lee, K., Burrows, B., Hubbard, J., Katznelson, L., Walsh, M., Guccione, A., Cannan, J., Heller, H., Basgoz, N., and Klibanski, A. Loss of lean body and muscle mass correlates with androgen levels in hypogonadal men with acquired immunodeficiency syndrome and wasting. *J Clin Endocrinol Metab*, *81:* 4051–4058, 1996.

33. Schroeder, E. T., Terk, M., and Sattler, F. R. Androgen therapy improves muscle mass and strength but not muscle quality: results from two studies. *Am J Physiol Endocrinol Metab*, *285:* E16–24, 2003.

34. Altenberg, B. and Greulich, K. O. Genes of glycolysis are ubiquitously overexpressed in 24 cancer classes. *Genomics*, *84:* 1014–1020, 2004.

35. Chowdhury, S. K., Gemin, A., and Singh, G. High activity of mitochondrial glycerophosphate dehydrogenase and glycerophosphate-dependent ROS production in prostate cancer cell lines. *Biochem Biophys Res Commun*, 2005.

36. Baron, A., Migita, T., Tang, D., and Loda, M. Fatty acid synthase: a metabolic oncogene in prostate cancer? *J Cell Biochem*, *91:* 47–53, 2004.

37. Rossi, S., Graner, E., Febbo, P., Weinstein, L., Bhattacharya, N., Onody, T., Bubley, G., Balk, S., and Loda, M. Fatty acid synthase expression defines distinct molecular signatures in prostate cancer. *Mol Cancer Res*, *1:* 707–715, 2003.

38. Seligson, D. B., Horvath, S., Shi, T., Yu, H., Tse, S., Grunstein, M., and Kurdistani, S. K. Global histone modification patterns predict risk of prostate cancer recurrence. *Nature*, *435:* 1262–1266, 2005.

39. Vaquero, A., Scher, M., Lee, D., Erdjument-Bromage, H., Tempst, P., and Reinberg, D. Human SirT1 interacts with histone H1 and promotes formation of facultative heterochromatin. *Mol Cell*, *16:* 93–105, 2004.

8

Estrogen Receptor α and β in the Regulation of Normal and Malignant Prostate Epithelium

Otabek Imamov, MD,
Nikolai A. Lopatkin, MD, PhD,
and Jan-Åke Gustafsson, MD, PhD

From: *Current Clinical Oncology: Prostate Cancer:*
Signaling Networks, Genetics, and New Treatment Strategies
Edited by: R. G. Pestell and M. T. Nevalainen © Humana Press, Totowa, NJ

1. INTRODUCTION: HISTORICAL PERSPECTIVE

In 1941, Huggins et al. *(1–3)* established a basis for the hormonal treatment of prostate cancer in a paper describing the beneficial effect of endocrine treatment for locally advanced prostate adenocarcinoma. In 1946 and 1950, Nesbit *(4,5)* conducted a thorough retrospective analysis of 1818 prostate cancer cases investigating the outcomes of various treatment strategies. He found that a combination of the non-steroidal estrogen, diethylstilbestrol (DES), with bilateral orchiectomy was the best treatment option for the patients with locally advanced disease. The hypothetical mechanism of action of estrogens was speculated to be suppression of testosterone (T) synthesis through negative feedback on the hypothalamo–pituitary axis. Interestingly, the idea of Huggins that there could be a direct estrogenic influence on the prostatic epithelium was ignored. Of possible significance in this context is that the synthetic estrogen Chlorotrianisene, introduced into the clinic in 1951, was shown to be efficient although it did not lower T levels to castrated values *(6)*.

The first evidence of cardiovascular side effects of DES treatment came with the results of Veterans' Administration Cooperative Urological Research Group (VACURG) in 1967 *(7)*. The study was conducted in a randomized fashion, comparing 5 mg DES with placebo, and included more than 2000 cases. One of the major conclusions of the study was that DES at a dose of 5 mg/day caused an extremely high cardiovascular mortality rate. As clinicians became aware of the cardiovascular complications accompanying the use of DES, its use in the treatment of prostate cancer began to decline and two alternatives were introduced to the clinic. Canadian endocrinologist Ferdinand Labrie was, and still is a strong proponent of the use of luteinizing hormone-releasing hormone (LHRH) agonists. He introduced them into clinical treatment of prostate cancer. Then came the first non-steroidal antiandrogen, flutamide, marking the beginning of the era of a combined hormonal treatment (CHT), also known as maximal androgen blockade (MAB). In terms of median survival, the method was superior to LHRH agonist monotherapy *(8)* and soon got recognition in the urological world. The estrogens were relegated to a small niche in the urological armamentarium.

2. DISCOVERY OF ESTROGEN RECEPTORS

Whereas pharmaceutical research was more focused on androgens and anti-androgens in the treatment of prostate cancer, basic science was trying to clarify the actions of estrogen in the prostate. In 1962 Elwood Jensen and colleagues *(9)* identified the first estrogen receptor (ER)α in uterine cytosol. Identification was based on the high affinity of ERα for 17β-estradiol (E_2) and this was only possible when highly radioactive [^{3}H] E_2 was synthesized. More than 20 years after its discovery, ERα was cloned with the help of very specific antibodies raised in the Jensen laboratory *(10–12)*. ERα turned out to be a member of the nuclear receptor supergene family, the first member of which, the glucocorticoid receptor, had been cloned in 1984 *(13)*. There followed quite rapidly the cloning of the other members of this superfamily, including androgen, progesterone, mineralocorticoid, and vitamin D and vitamin A receptors *(14–20)*. With cross-hybridization techniques, using the conserved DNA-binding domain (DBD) as a hybridization probe, a surprising number of novel members of this family were discovered, and for some of these receptors, no known ligand has yet been discovered. In 1995, during a search for nuclear receptors in the prostate, the second ER, ERβ, was discovered *(21)*.

The structural architecture of ERα and ERβ is typical for all other members of the NR family, namely an N-terminal region (A/B domain), containing a constitutively active transactivation region (AF-1); a DBD (C domain), which contains the P-box, a short motif responsible for DNA-binding specificity and involved in dimerization of ER; a D domain that behaves as a flexible hinge between the C and E domains and contains the nuclear localization signal (NLS); and an E domain, whose secondary structure of 12 α-helices is responsible for ligand binding. Comparisons of the protein structure of the two ERs showed 96% identity in DBD domain and 59% homology in LBD domain *(21)*. Hypothetically that means that both ERs would bind to the same response elements on the DNA, but the ligands, activating receptors, would be different. The N-terminal region, containing AF-1, and the C-terminus with AF-2 were even less conserved. This would predict different co-regulator binding properties between two receptors.

Although all of the details of the signaling through ER have not been worked out, the following is a simplified scheme as to what happens: ERs are expressed in the cells of a target tissue in an inactive form in complexes with chaperons. Once the E_2 binds to ER, one of the chaperons, HSP-90, dissociates from the complex and ERs are released; ER monomers then associate to form dimers and the dimers bind to specific regions of DNA, attracting co-modulator proteins and influencing the transcription of different target genes.

Soon after ERα was found, virtually all tissues in the body were screened for its presence. As predicted, ERα was found in the mammary gland, uterus, placenta, liver, central nervous system, cardiovascular system, and bones. These tissues expressed high levels of ERα and responded to E_2 by increasing the transcription of certain estrogen-controlled genes. In other tissues, including prostate, testis, gall bladder, skin, lymphatic and erythropoietic systems, which do respond to estrogen, ERα expression was not detectable *(22)*. This is how the classification of target tissues into "classical target tissues" and "non-classical target tissues" began.

The prostate gland was classified as a non-classical E_2-target tissue. ERα was found in the prostate, but it was localized exclusively in the stromal part of the gland *(23–26)*. This finding led to the concept that all direct influences of estrogen on the prostate are mediated by stromal ER and possibly via growth factor signaling pathways *(27,28)*. An obligatory role for stroma in the estrogen actions in the prostate became an obsolete concept soon after ERβ was discovered. Not only was ERβ cloned from the rodent prostate, it is abundantly expressed in prostatic epithelium. Interestingly, ERβ is expressed in both the epithelium and stroma of the human prostate.

3. CREATION OF KNOCKOUT MOUSE MODELS

The next big challenge was to discover the function of ERβ. Very early it was found that ERβ does not elicit the classical estrogen actions in the uterus and pituitary. To many endocrinologists, this simply meant that ERβ was a vestigial receptor. However, to other investigators, this led to the idea that ERβ selective agonists could act on the prostate and breast without having the worst side effects of estradiol, i.e., chemical castration and uterine cancer. Study of ER knockout mice was to prove very fruitful in the question of the functions of ERβ.

Silencing of a gene of interest with unknown function by creating genetically modified mice became a classical method in modern science *(29,30)*. A mouse with inactivated ERα (ERα$^{-/-}$) was created in 1993 *(31)*. It was reported that male ERα$^{-/-}$ mice have altered spermatogenesis with reduced fertility, but that the prostate was not affected morphologically *(32)*. ERβ$^{-/-}$ mice were generated in 1998 *(33)*. Interestingly, in ERβ$^{-/-}$ mice at 2–3 months of age, the prostates were morphologically normal. However, as mice aged, the ventral prostate developed foci of epithelial hyperplasia. Our laboratory published several papers describing mouse prostatic epithelial hyperplasia in the absence of ERβ signaling *(33–35)*. The incidence of this phenotype increased with age; very seldom seen in 6-month-old mice, it could be found almost in every mouse at the age of 24 months. These lesions were reminiscent of a well-recognized morphological precursor of prostate cancer in humans, namely prostatic intraepithelial neoplasia (PIN) *(36)*. However, we have never seen cellular atypia in PIN-like lesions in mice and never detected any signs of *carcinoma in situ* (CIS). Mouse hyperplastic lesions showed only mild tissue atypia, similar to low-grade PIN (LG PIN) in humans. This phenotypical feature of ERβ$^{-/-}$ mice became one of the corner stones in the ERβ research. The two other large laboratories studying ERβ$^{-/-}$ mice saw no abnormalities in the prostate and called our findings a "cutting artifact" *(33,34,37)*.

At about the same time, a new knockout mouse model useful to study estrogen signaling became available. In 1998, Evan Simpson's *(38)* group reported the creation of aromatase knockout (ArKO) mouse. Aromatase

(P450arom) is the enzyme that catalyzes metabolic transformation of C19 steroids into estrogens. ArKO mice cannot produce estrogens but do express both ERs. The prostates of these mice were reported to be enlarged and hyperplastic (39). Thus by the end of 1990s, researchers in the estrogen field had three mouse knockout models to use for the dissection of ERα and ERβ function. This dissection is not complete and our knowledge base is growing everyday with new data coming from different laboratories. However, the accumulated data on estrogen signaling is already at the stage when new ER-modulating agents are close to being introduced into the clinics.

4. TESTOSTERONE–5α-DIHYDROTESTOSTERONE–5α-ANDROSTANE-3β, 17β-DIOL (3βADIOL) PATHWAY

The reduction of T to dihydrotestosterone (DHT) is a very well characterized biochemical reaction, catalyzed by the enzymes 5α-reductase type I and II. The type II enzyme is in fact used as a target for pharmacological therapy of benign prostatic hyperplasia (BPH). Blocking 5α-reductase in the prostate is considered beneficial, because DHT is a more potent agonist for AR than T. Examples of 5α-reductase inhibitors are medications finasteride (Proscar, MSD, Whitehouse Station, NJ, USA) and dutasteride (Avodart, GSK UK) widely used in the clinical practice. We have been trying to make clinicians aware of the fact the 5α-reductase inhibitors do have an unwanted side effect. Metabolites of DHT participate in the complex system of hormonal control, and their absence, when DHT synthesis is blocked, can result in loss of growth control and loss of differentiation of the prostatic epithelium. Figure 1 outlines the T–DHT pathway. As is clear from the figure, DHT

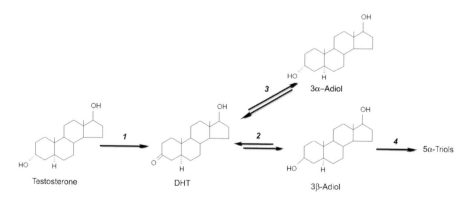

Fig. 1. Testosterone metabolic pathway in the prostate. 1: 5α-reductase type II; 2: 3β-hydroxysteroid dehydrogenase and Δ^5, Δ^4-isomerase; 3: 3α-hydroxysteroid dehydrogenase and Δ^5, Δ^4-isomerase; 4: 3β-Adiol hydroxylase.

can be metabolized to two stereoisomers: 5α-androstane-3α, 17β-diol (3α-Adiol) and 5α-androstane-3β, 17β-diol (3α-Adiol). This biochemical reaction is reversible and catalyzed by 3α- and 3β-hydroxysteroid dehydrogenases (3α-HSD and 3β-HSD). 5α-Androstane-3α, 17β-diol (3α-Adiol) is an androgenic compound, possibly serving as a depot for DHT. 5α-androstane-3β, 17β-diol (3β-Adiol) is an estrogenic compound, capable of activating both ERα and ERβ. Moreover, the concentration of 3β-Adiol in the prostate is higher than that of E_2, making it a perfect candidate to be a natural ligand for ERβ (40). 3β-Adiol undergoes further transformation to inactive triols, a reaction catalyzed by 3β-Adiol hydroxylase (CYP7B1).

5. ERS AND PROSTATIC EPITHELIUM

5.1. ERα

The tragic outcome of the use of DES in pregnant women is one of the dramas in the modern history of medicine. DES is a synthetic non-steroidal estrogen that was prescribed to about 4 million women in the USA between 1938 and 1971 to prevent miscarriages [reviewed in (41)]. The drug was ineffective in preventing miscarriages, but *in utero* DES exposure caused vaginal adenocarcinoma and cervical cancer in the daughters of the women taking medication (42). The sons of DES-treated mothers are also reported to have higher incidence of genital abnormalities, testicular cancer (43,44), and squamous metaplasia in the prostate (45). This process of hormonal programming of a developing organ predisposing it to the changes in adulthood is called imprinting.

In order to understand the toxicity of DES in human fetuses, it was quite natural to test the effects of DES in mice. Administration of DES during the neonatal period resulted in prostate enlargement and increased risk of dysplasia in adulthood (27,46,47). Similar effects have been reported for rats, undergoing *in utero* estrogenization (48). The question, which of the ERs mediates these effects, was open until the definitive study from the group of Korach was published (28). It was shown that ERα$^{-/-}$ mouse prostates are resistant to prenatal estrogenization whereas ERβ$^{-/-}$ and wild type (Wt) are equally sensitive. At this point, it became clear that negative effects of prenatal estrogenization are mediated by ERα. Because ERα is localized in the prostatic stroma, it was concluded that prostatic epithelium receives signals from stromal ERα through some growth factor signaling pathways.

In 2005 we published a study (49), showing that ERα is abundantly expressed in the epithelium of the developing prostate so it is not necessary to invoke a mechanism involving growth factors from the stroma. During the specific time frame of second to fourth weeks of postnatal life, ERα and not ERβ is predominantly expressed in the prostatic epithelium. This transient expression coincides with high proliferative activity and branching morphogenesis of the prostatic epithelium. Around the fourth week of postnatal life, marking the end of proliferation and beginning of differentiation and functional activation of epithelium, ERα is switched off and ERβ becomes dominant. Even with the naked eye, it is obvious that once removed from its site at the base of the bladder, the ERα$^{-/-}$ prostate has an overall appearance quite different from Wt prostates. The gland does not maintain its shape but tends to spread out as though the structure is weak. The reason for this apparently fragile structure became clear when the gland was examined under a microscope. Overall, the ductal system of the ERα$^{-/-}$ mouse ventral prostate is composed of two main very long primary ducts with no branching at the bifurcation of the two ducts. These observations suggest a role for ERα in branching morphogenesis of the prostate.

5.2. ERβ

ERβ$^{-/-}$ mice, despite showing signs of hyperplasia with aging, still have functionally active prostate glands. Studies from our group demonstrated that in the absence of ERβ, proliferation of prostatic epithelium is increased and apoptosis is suppressed. We have postulated that the general function of ERβ in the prostate is repression of proliferation. Our observations are supported by reports from other groups, testing this hypothesis on the cell lines *(50)*.

On the basis of our observations on altered terminal differentiation of mammary gland epithelium in the absence of ERβ signaling *(51)*, we hypothesized that ERβ plays a general role in the regulation in epithelial differentiation. Epithelial cells in the prostate form a time-line or a continuum of cells in different stages of differentiation. The continuum can be roughly divided into three groups or stages of differentiation: basal cells, intermediate cells, and luminal cells. The intermediate cell group is also known as transiently proliferating/amplifying pool of cells, because of the capability of these cells for rapid proliferation. TP/A group of cells is also divided into basal intermediate and luminal intermediate cells (Fig. 2).

These cellular pools are characterized by a specific protein expression profile that makes it possible to separate one from the other. This protein expression pattern is also known as a cytokeratin profile *(52–54)*. Basal cells are localized in the basal cellular layer, attached to the basement membrane. These cells express AR and are believed to include a pool of prostatic stem cells *(55,56)*. Basal cells are androgen sensitive, but independent of androgens for survival. Upon stimulation with androgen, they undergo a slow division process that can be symmetrical—giving rise to two similar basal cells—or asymmetrical, when one of daughter cells is entering the differentiation process. Intermediate cells (also known as transiently proliferating/amplified cells) represent an in-between stage of prostatic epithelial differentiation. These cells are androgen

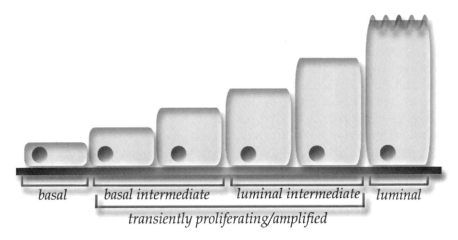

Fig. 2. Differentiation continuum of prostatic epithelium.

sensitive but also dependent upon androgens for survival. Upon androgen stimulation, they proliferate, but upon androgen withdrawal, they die. The luminal cell pool consists of highly specialized secretory cells, located at the luminal side of the duct. These cells produce components of prostatic secretion and eventually die by shedding off into the lumen. Luminal cells are androgen sensitive and dependent, but because they are highly specialized, they have lost the ability to proliferate. Androgen stimulates secretory activity of these cells.

Studies in our group showed that, in the absence of ERβ signaling, mouse prostatic epithelial differentiation is altered, resulting in the accumulation of cells in TP/A stage, most probably belonging to the basal intermediate group *(35)*. These cells are capable of rapid proliferation upon androgen stimulation. We used cytokeratin profiles to characterize the cellular composition of ERβ$^{-/-}$ prostates. We found that the differentiation pattern in the absence of ERβ signaling was altered in that the ratio between the three cell pools is shifted toward cells in the TP/A stage. That means that there are less cells that possess luminal and basal phenotype and more cells in the intermediate pool, capable of rapid proliferation process, hence the increased proliferation rate in ERβ$^{-/-}$ mouse prostates.

6. PROSTATE CANCER PREVENTION TRIAL FROM ERβ POINT OF VIEW

After publication of our study, the results of the Prostate Cancer Prevention Trial (PCPT) were published. This was a multicenter prospective double-blinded study of finasteride (Proscar, MSD) as a preventive agent for prostate cancer. The rationale for the study was that a 5α-reductase inhibitor could reduce the incidence of prostate cancer (CaP) by decreasing intraprostatic levels of DHT. More than 18,000 healthy volunteers aged 55 or above were randomized into two arms: finasteride 5 mg daily and placebo. After 7 years of treatment, as anticipated, the incidence of prostate cancer in finasteride arm was reduced (18.4% vs. 24.8% in the placebo arm). However, there was a big surprise when the histology of the cancers was examined. In the finasteride-treated group, the incidence of low-differentiated, aggressive Gleason 7–10 tumors was 67% higher (finasteride, 280/757, 37%, vs. placebo 237/1068, 22.2%; $P < 0.001$). Despite the reduction in cancer risk in the finasteride arm, any urologist would admit that increased incidence in Gleason score above 7 is totally unacceptable because it is a strong indication of very poor survival.

The authors of the original paper provided several possible explanations of the phenomenon. First, they attributed the higher incidence of aggressive tumors in the finasteride arm to likely "treatment effect"; clonal selection of tumors more sensitive to low androgen environment; or selective killing of the low-grade tumors. There were several editorials questioning the interpretation of the results. One problem in particular is the fact that treatment with finasteride changes prostate morphology (57). It is thought that finasteride treatment causes similar morphological alterations as seen in LHRH-treated prostates (58–60). Thus, according to some authors, application of Gleason grading system is not applicable, because essentially non-malignant morphological alterations can lead to over grading. Some analysts propose that because the same number of high Gleason tumors were detected in the placebo group, it is the ratio of low-Gleason tumors that is changed after finasteride treatment, suggesting that finasteride is only effective preventing low-Gleason tumors (61).

We contributed to the discussion of the PCPT with our own view of what has happened after 7 years of blocking 5α-reductase. It is our hypothesis that higher incidence of low differentiated tumors in finasteride-treated arm of the PCPT is caused by lack of the natural ERβ ligand, 3β-Adiol (62). From the clinical point of view, blocking 5α-reductase activity and altering T–DHT pathway has obvious benefits that have been discussed elsewhere. However, one unforeseen casualty lies downstream of such biochemical intervention. It is 3β-Adiol which, as a natural ligand for ERβ, is pro-differentiative. We think that finasteride should not be given as monotherapy, but should be given in combination with ERβ agonists. Interestingly, even before PCPT, a small study was carried out in 52 men with prostate-specific antigen (PSA) higher than 4 ng/ml, but no morphological evidence of prostate cancer. After randomization into active and placebo groups and 12 months' treatment, it showed a significantly higher cancer incidence in finasteride group, while the incidence of PIN was not different (63). A second report from the PCPT study retracts the evidence that the Gleason scores were indeed higher in the finasteride arm of the study and this has led to further debate about the design and interpretation of the study. The debate about the PCPT results is still ongoing.

At the moment, a study of a new, dual 5α-reductase inhibitor, dutasteride, is being performed (64). The study, named Reduction by Dutasteride of Prostate Cancer Events (REDUCE), will analyze not only Gleason score, but the cancer aggressiveness by using specific markers for aggressive growth. However, one has to keep in mind that dutasteride, because of its 45-fold higher potency compared with finasteride and the ability to inhibit both isoforms of 5α-reductase, type I and II, would result in much lower serum and intraprostatic concentrations of DHT. These observations would suggest that the amount of 3β-Adiol available for ERβ activation would be even less than under finasteride treatment, possibly resulting in more prominent changes in prostatic morphology. Phytoestrogens seem to be good candidates for combination with finasteride. New synthetic compounds capable to activate ERβ without affecting ERα are on their way to the clinics. One should expect the appearance of a new class of pro-differentiative agents, perhaps useful for cancer prevention, as neo-adjuvant therapy and in combination treatment of prostate cancer.

7. ERS AND PROSTATIC STROMA

It is known that an imbalance in the ratio between estrogen and androgen is involved in the pathogenesis of BPH. Estrogenic activity in the prostate is the sum of the actions of both E_2 and 3β-Adiol (40) as well as the responses of ERα and ERβ. At the same time, intraprostatic levels of E_2 are very low compared with those of 3β-Adiol (40). DHT and 3β-Adiol are both synthesized in the stroma where 5α-reductase type II (65) and 17β-HSD type 7 are primarily located. Stromal 3β-Adiol can activate both epithelial/stromal ERβ and stromal ERα (66). In the prostate stroma, ERα seems to be proliferative.

At present, the clinical rationale for the use of 5α-reductase inhibitors in the treatment of BPH is that stromal AR is responsible for stromal overgrowth, and indeed, administration of 5α-reductase inhibitors and castration effectively lead to prostatic involution. There is one puzzling fact, which does not fit with the explanation that 5α-reductase inhibitors are efficient in BPH by reducing androgens in the aging prostate. The fact is that the estrogen/androgen ratio in the aging prostate is shifted in favor of estrogen, because of the decline in androgen (67).

We propose an alternative explanation for the beneficial use of 5α-reductase inhibition in the treatment of BPH. It involves ERα, 3β-Adiol, and CYP7B1. It is as follows: DHT is converted into 3β-Adiol in the stroma where it activates ERα. The estrogenicity of 3β-Adiol is regulated by cellular levels of CYP7B1. If CYP7B1 levels are low, ERα is activated and stromal growth ensues. We think that 5α-reductase inhibitors work because, by reducing DHT, they also decrease the level of stromal 3β-Adiol. This role of 5α-reductase inhibition helps us to solve the problem of why the incidence of prostate cancer increases as the T levels decline with age and why androgen replacement therapy does not increase BPH or development of prostate cancer. Interestingly, a recently performed large cohort study showed an inverse relationship between serum levels of T and cancer aggressiveness *(68)*. The same phenomenon was reported for circulating androgen bioactivity *(69)*.

There is one other important player to be considered if we are to understand estrogen action in the prostate, ERβcxβ. ERβcx does not bind to E_2 or to 3β-Adiol, but if expressed in the same cell with ERα, it acts as a dominant repressor *(70,71)*. If ERβcx is expressed in stromal cells with ERα, stromal growth should be repressed. Questions about cellular localization and regulation of the expression levels of ERβcx and CYP7B1 need to be answered for an appropriate understanding of estrogen signaling in prostate disease.

8. ESTROGEN IMPRINTING OF PROSTATE

We have already touched the subject of estrogen imprinting of the prostate and the role of ERα in this phenomenon. Here we would like to discuss this amazing phenomenon trying to put ERβ in the context. The term "estrogen imprinting" was first suggested by Rajfer and Coffey in their hallmark publication in 1978 *(72)*.

The DES catastrophe and studies of estrogen administration in rodents resulted in negative effects of estrogenization *in utero*. Could there be positive effects of *in utero* estrogenization? Mice do not spontaneously develop prostate cancer and therefore it is impossible to study the anti-cancer protection that might occur during *in utero* treatment. Such protection surely exists, even though only indirect evidence on the matter is available right now. Soy phytoestrogens in the Asian-style diet are considered the key factor in low incidence of prostate cancer in certain countries. Phytoestrogen-containing products were known to be healthy long before the discovery of ERs. Population studies of the influence of Western-style diet on incidence of prostate cancer provided evidence that Asian food does protect against prostate cancer. However, it is not only the beneficial properties of an Asian-style diet but also the unhealthy properties of the Western-style diet that have to be taken into consideration. At the same time, it is known from the literature that a soybean isoflavone mixture suppresses the development of invasive carcinomas of the rat prostate/seminal vesicles *(73)*, and genistein and daidzein possess anti-cancer effects on relatively early stages of prostate cancer development *(74,75)*.

However, according to population studies, based on careful retrospective analysis of migrated Japanese and Chinese, the protective effect of the Asian-style diet lasts in the first and also in a second generation of emigrants, despite their changed life-style *(76)*. Interesting evidence came from the analysis of Multiethnic Cohort Study (MEC), performed between 1993 and 1996, which involved 215,000 cases. After careful statistical analysis of the obtained incidence data, authors postulate the existence of "residual effects of exposures during childhood," which had an important influence on natural history of hormone-dependent cancers, including prostate cancer in adulthood *(77)*.

Thus, we can hypothetically distinguish two types of imprinting: a negative imprinting, well documented in DES administration studies, and positive imprinting lying behind the Asian-style diet anti-cancer protection. Following this logic, it is easy to hypothesize about the role of various ERs in different types of imprinting. This approach is a rather straightforward one, and cannot be omitted until proved otherwise. We hypothesize that the two types of imprinting are mediated independently by the two ERs, ERα and ERβ, and that they occur at distinct windows in time during development. Although we have clear indications as to when these windows occur in rodents, nothing is known about them in human development. It is possible that activation of ERα signaling during a certain time frame in childhood would predispose for prostate cancer, whereas activation of ERβ at some other time point would offer protection.

The possible scenario behind negative DES imprinting can be hyperactivation of ERα during a specific period of tissue remodeling. It is not clear how the results of this event are being maintained for years before resulting in squamous metaplasia of prostate. Positive imprinting by soy phytoestrogens, mediated by ERβ, occurs perhaps also during specific periods of ERβ activity. Because some phytoestrogens, like genistein, show higher affinity to ERβ than to ERα *(66,78)*, one can speculate that, throughout prenatal development, the plasma concentration of such phytoestrogens is not enough to stimulate ERα, and just enough to stimulate ERβ. However, it is very difficult to speculate over the cellular substrate of positive imprinting. Our speculations on negative and positive imprinting do, of course, require future research to dissect the exact mechanisms of this amazing phenomenon. Our division of imprinting into positive and negative is also quite subjective. The DES catastrophe caused a great number of diseases, and the Asian-style diet does protect against prostate cancer. However, imprinting is a polyorganic event, and protection against prostate cancer can be associated with a negative influence on other organs and systems. It is too early to use the concept of positive imprinting as a pharmacological strategy. At the same time, ERβ signaling can be used as a target for medical treatment of prostatic diseases.

9. SELECTIVE ER MODULATORS

One of the pioneers of chemotherapy, Paul Ehrlich, was the first to suggest that a good medication should influence the parasite without affecting the host. The concept of a medication, designed to be receptor-specific, became a rule of a thumb in designing new pharmacological agents. As is clear now, there are two major ERs expressed in different tissues like breast and prostate. Moreover, these receptors are often expressed in the same tissue and oppose each other's actions while the same ligands can modulate the activity of both of them. In cell lines *(79)* and in some tissues *(34,51)*, E_2 in the presence of ERα elicits proliferation, but in the presence of ERβ, it inhibits proliferation providing the perfect example of opposite effects caused by the same hormone. Likewise, prostate has epithelial/stromal ERβ and stromal ERα. According to our studies, activation of ERβ can be beneficial in the treatment of prostate cancer and possibly cancer prevention.

Initially the main mechanism for estrogen in the treatment of prostate cancer was believed to be through the hypothalamo–pituitary–gonadal axis with subsequent inhibition of T synthesis. However, several estrogenic compounds are shown to act independently of this pathway. Many hypothetical mechanisms have been described in the literature, including disruption of apoptotic regulators *(80)*, depolymerization of microtubules *(81)*, inhibition of DNA synthesis *(82)*, induction of apoptosis *(83)*, and interruption of the cell cycle *(84)*. Interestingly, all these effects have been attributed to ERβ and not ERα signaling. Moreover, all the above-mentioned effects, seen in the experimental settings, can be a part of the same signaling mechanism downstream of ERβ. This concept might rationalize the ongoing search for selective ER modulators (SERMs) for the treatment of prostatic diseases.

The standard test for an estrogen stimulation of growth of the uterus is, of course, still a good test for an ERα agonist, but there is no single good test for an ERβ agonist. In fact, there may not be such a thing as a single good ERβ agonist. What is emerging is an array of ERβ-selective agonists, each with a specific profile of genes, which they influence *(85–90)*. Although we know what a consensus ERE is, most estrogen-responsive genes do not contain perfect consensus sequences and the transcriptional activity of ERα or ERβ on such sequences is influenced by the chemical structure of the estrogenic ligand. Hall and Korach *(91)* have evaluated the activities of ERα and ERβ on four different EREs (vitellogenin A2, human pS2, lactoferrin, and complement 3) in the presence of estradiol, phytoestrogens, and xenoestrogens. In terms of transactivation by ERα and ERβ, the vitellogenin and lactoferrin promoters were not discriminatory. The pS2 and complement 3 were most responsive to ERβ. In addition, the transcriptional activity of either receptor on any promoter varied with the ligand. Another factor influencing selectivity of ER ligands is that the influence of ERs on transcription is not confined to EREs. ERs modulate transcription at AP-1 and Sp1 sites and interact with the nuclear factor (NF)-κB pathway *(92)*. The action of the two receptors at these sites can be opposite to each other, but this depends on cellular context and it is not possible to predict how ERα and ERβ will influence transcription at these sites. Selective ERα and ERβ ligands have already been developed which have actions on selective target tissues and even selective target genes *(93)*. One ERβ agonist developed by Eli Lilly is a great inhibitor of prostate growth but has no effect on the immune system *(94)*, whereas one developed by Wyeth is a powerful immunosuppressant but has no influence on the prostate *(86)*.

10. SUMMARY

Long before the discovery of ERs, it was well known to clinicians that estrogen has profound effects on the prostate. As early as in 1941 Charles Huggins, one of the pioneers in the research of hormonal regulation of prostatic growth, reported that in prostate cancer patients, oral estrogens had the same effect as castration *(1)*. Thus began the era of hormonal treatment of prostate cancer. Huggins et al. suggested many common sites of estrogen action as possible explanations for the mechanism of action of estrogen in cancer patients. These included increased inactivation of androgens, depression of gonadotropic agents of the anterior pituitary, and depression of interstitial cells of testis, but Huggins also introduced the idea of a "direct action on prostatic epithelium." However, until recently the study of the direct effects of estrogen on prostatic epithelium received very little attention. For about three decades, until the late 1960s, estrogens were an accepted treatment for prostate cancer. After the discovery and synthesis of anti-androgenic compounds, use of estrogen fell into disfavor. Now, 65 years after Huggins' historical publication, the role of estrogens in the development and functioning of the prostate is once more the focus of attention. In this chapter, we try to explain the renewed optimism for the use of estrogens in the treatment of prostatic diseases. We review the most recent data on estrogen signaling in the normal and diseased prostate and propose the use of ER-selective therapy for treatment of prostatic diseases. We have formulated a hypothesis for the mechanism behind the phenomenon of estrogen imprinting of the prostate, in an attempt to understand the puzzling results of PCPT suggesting a pro-differentiative role for ERβ and DHT metabolite.

ACKNOWLEDGMENTS

The studies are supported by the Swedish Cancer Society and KaroBioAB.

REFERENCES

1. Huggins, C. and Hodges, C. V. (1941) *Cancer Res* **1**, 293.
2. Huggins, C., Scott, W. W. and Hodges, C. V. (1941) *J Urol* **46**, 997.
3. Huggins, C., Stevens, R. E., Jr. and Hodges, C. V. (1941) *Arch Surg* **43**, 209.
4. Nesbit, R. M. and Plumb, R. T. (1946) *Surgery* **20**.
5. Nesbit, R. M. and Baum, W. C. (1950) *JAMA* **143**, 1317.
6. Baba, S., Janetschek, G., Pollow, K., Hahn, K. and Jacobi, G. H. (1982) *Br J Urol* **54**, 393–8.
7. (1967) *Surg Gynecol Obstet* **124**, 1011–7.
8. Labrie, F., Dupont, A., Giguere, M., Borsanyi, J. P., Belanger, A., Lacourciere, Y., Emond, J. and Monfette, G. (1986) *J Steroid Biochem* **25**, 877–83.
9. Jensen, E. V. and Jacobson, H. I. (1962) *Recent Prog Horm Res* **18**, 387–414.
10. Greene, G. L., Nolan, C., Engler, J. P. and Jensen, E. V. (1980) *Proc Natl Acad Sci USA* **77**, 5115–9.
11. Green, S., Walter, P., Greene, G., Krust, A., Goffin, C., Jensen, E., Scrace, G., Waterfield, M. and Chambon, P. (1986) *J Steroid Biochem* **24**, 77–83.
12. Walter, P., Green, S., Greene, G., Krust, A., Bornert, J. M., Jeltsch, J. M., Staub, A., Jensen, E., Scrace, G., Waterfield, M. et al. (1985) *Proc Natl Acad Sci USA* **82**, 7889–93.
13. Miesfeld, R., Okret, S., Wikstrom, A. C., Wrange, O., Gustafsson, J. A. and Yamamoto, K. R. (1984) *Nature* **312**, 779–81.
14. Trapman, J., Klaassen, P., Kuiper, G. G., van der Korput, J. A., Faber, P. W., van Rooij, H. C., Geurts van Kessel, A., Voorhorst, M. M., Mulder, E. and Brinkmann, A. O. (1988) *Biochem Biophys Res Commun* **153**, 241–8.
15. Jeltsch, J. M., Krozowski, Z., Quirin-Stricker, C., Gronemeyer, H., Simpson, R. J., Garnier, J. M., Krust, A., Jacob, F. and Chambon, P. (1986) *Proc Natl Acad Sci USA* **83**, 5424–8.
16. Conneely, O. M., Sullivan, W. P., Toft, D. O., Birnbaumer, M., Cook, R. G., Maxwell, B. L., Zarucki-Schulz, T., Greene, G. L., Schrader, W. T. and O'Malley, B. W. (1986) *Science* **233**, 767–70.
17. Arriza, J. L., Weinberger, C., Cerelli, G., Glaser, T. M., Handelin, B. L., Housman, D. E. and Evans, R. M. (1987) *Science* **237**, 268–75.
18. Petkovich, M., Brand, N. J., Krust, A. and Chambon, P. (1987) *Nature* **330**, 444–50.
19. Brand, N., Petkovich, M., Krust, A., Chambon, P., de The, H., Marchio, A., Tiollais, P. and Dejean, A. (1988) *Nature* **332**, 850–3.
20. Baker, A. R., McDonnell, D. P., Hughes, M., Crisp, T. M., Mangelsdorf, D. J., Haussler, M. R., Pike, J. W., Shine, J. and O'Malley, B. W. (1988) *Proc Natl Acad Sci USA* **85**, 3294–8.
21. Kuiper, G. G., Enmark, E., Pelto-Huikko, M., Nilsson, S. and Gustafsson, J. A. (1996) *Proc Natl Acad Sci USA* **93**, 5925–30.
22. Ciocca, D. R. and Roig, L. M. (1995) *Endocr Rev* **16**, 35–62.
23. Bashirelahi, N., Kneussl, E. S., Vassil, T. C., Young, J. D., Jr., Sanefugi, H. and Trump, B. (1979) *Prog Clin Biol Res* **33**, 65–84.
24. Chaisiri, N. and Pierrepoint, C. G. (1980) *Prostate* **1**, 357–66.
25. Kozak, I., Bartsch, W., Krieg, M. and Voigt, K. D. (1982) *Prostate* **3**, 433–8.
26. Swaneck, G. E., Alvarez, J. M. and Sufrin, G. (1982) *Biochem Biophys Res Commun* **106**, 1441–7.
27. Prins, G. S., Birch, L., Habermann, H., Chang, W. Y., Tebeau, C., Putz, O. and Bieberich, C. (2001) *Reprod Fertil Dev* **13**, 241–52.
28. Prins, G. S., Birch, L., Couse, J. F., Choi, I., Katzenellenbogen, B. and Korach, K. S. (2001) *Cancer Res* **61**, 6089–97.

29. Thomas, K. R. and Capecchi, M. R. (1987) *Cell* **51,** 503–12.
30. Doetschman, T., Gregg, R. G., Maeda, N., Hooper, M. L., Melton, D. W., Thompson, S. and Smithies, O. (1987) *Nature* **330,** 576–8.
31. Lubahn, D. B., Moyer, J. S., Golding, T. S., Couse, J. F., Korach, K. S. and Smithies, O. (1993) *Proc Natl Acad Sci USA* **90,** 11162–6.
32. Eddy, E. M., Washburn, T. F., Bunch, D. O., Goulding, E. H., Gladen, B. C., Lubahn, D. B. and Korach, K. S. (1996) *Endocrinology* **137,** 4796–805.
33. Krege, J. H., Hodgin, J. B., Couse, J. F., Enmark, E., Warner, M., Mahler, J. F., Sar, M., Korach, K. S., Gustafsson, J. A. and Smithies, O. (1998) *Proc Natl Acad Sci USA* **95,** 15677–82.
34. Weihua, Z., Makela, S., Andersson, L. C., Salmi, S., Saji, S., Webster, J. I., Jensen, E. V., Nilsson, S., Warner, M. and Gustafsson, J. A. (2001) *Proc Natl Acad Sci USA* **98,** 6330–5.
35. Imamov, O., Morani, A., Shim, G. J., Omoto, Y., Thulin-Andersson, C., Warner, M. and Gustafsson, J. A. (2004) *Proc Natl Acad Sci USA* **101,** 9375–80.
36. Bostwick, D. G. and Brawer, M. K. (1987) *Cancer* **59,** 788–94.
37. Dupont, S., Krust, A., Gansmuller, A., Dierich, A., Chambon, P. and Mark, M. (2000) *Development* **127,** 4277–91.
38. Fisher, C. R., Graves, K. H., Parlow, A. F. and Simpson, E. R. (1998) *Proc Natl Acad Sci USA* **95,** 6965–70.
39. Jarred, R. A., McPherson, S. J., Jones, M. E., Simpson, E. R. and Risbridger, G. P. (2003) *Prostate* **56,** 54–64.
40. Weihua, Z., Lathe, R., Warner, M. and Gustafsson, J. A. (2002) *Proc Natl Acad Sci USA* **99,** 13589–94.
41. Schrager, S. and Potter, B. E. (2004) *Am Fam Physician* **69,** 2395–400.
42. (1976) *JAMA* **236,** 1107–9.
43. Henderson, B. E., Benton, B., Cosgrove, M., Baptista, J., Aldrich, J., Townsend, D., Hart, W. and Mack, T. M. (1976) *Pediatrics* **58,** 505–7.
44. Docimo, S. G., Silver, R. I. and Cromie, W. (2000) *Am Fam Physician* **62,** 2037–44, 2047–8.
45. Driscoll, S. G. and Taylor, S. H. (1980) *Obstet Gynecol* **56,** 537–42.
46. vom Saal, F. S., Timms, B. G., Montano, M. M., Palanza, P., Thayer, K. A., Nagel, S. C., Dhar, M. D., Ganjam, V. K., Parmigiani, S. and Welshons, W. V. (1997) *Proc Natl Acad Sci USA* **94,** 2056–61.
47. Strauss, L., Makela, S., Joshi, S., Huhtaniemi, I. and Santti, R. (1998) *Mol Cell Endocrinol* **144,** 83–93.
48. Prinsac, G. S., Birch, L., Habermann, H., Chang, W. Y., Tebeau, C., Putz, O. and Bieberich, C. (2001) *Reprod Fertil Dev* **13,** 241–52.
49. Omoto, Y., Imamov, O., Warner, M. and Gustafsson, J. A. (2005) *Proc Natl Acad Sci USA* **102,** 1484–9.
50. Cheng, J., Lee, E. J., Madison, L. D. and Lazennec, G. (2004) *FEBS Lett* **566,** 169–72.
51. Forster, C., Makela, S., Warri, A., Kietz, S., Becker, D., Hultenby, K., Warner, M. and Gustafsson, J. A. (2002) *Proc Natl Acad Sci USA* **99,** 15578–83.
52. Wang, Y., Hayward, S., Cao, M., Thayer, K. and Cunha, G. (2001) *Differentiation* **68,** 270–9.
53. Isaacs, J. T. and Coffey, D. S. (1989) *Prostate Suppl* **2,** 33–50.
54. Bonkhoff, H., Stein, U. and Remberger, K. (1994) *Hum Pathol* **25,** 42–6.
55. Hayward, S. W., Brody, J. R. and Cunha, G. R. (1996) *Differentiation* **60,** 219–27.
56. English, H. F., Drago, J. R. and Santen, R. J. (1985) *Prostate* **7,** 41–51.
57. Civantos, F., Soloway, M. S. and Pinto, J. E. (1996) *Semin Urol Oncol* **14,** 22–31.
58. Rubin, M. A. and Kantoff, P. W. (2003) *N Engl J Med* **349,** 1569–72; author reply 1569–72.
59. Rubin, M. A. and Kantoff, P. W. (2004) *J Cell Biochem* **91,** 478–82.

60. Rubin, M. A., Allory, Y., Molinie, V., Leroy, X., Faucon, H., Vacherot, F., Huang, W., Kuten, A., Salomon, L., Rebillard, X., Cussenot, O., Abbou, C. and de la Taille, A. (2005) *Urology* **66**, 930–4.

61. Andriole, G., Bostwick, D., Civantos, F., Epstein, J., Lucia, M. S., McConnell, J. and Roehrborn, C. G. (2005) *J Urol* **174**, 2098–104.

62. Imamov, O., Lopatkin, N. A. and Gustafsson, J. A. (2004) *N Engl J Med* **351**, 2773–4.

63. Cote, R. J., Skinner, E. C., Salem, C. E., Mertes, S. J., Stanczyk, F. Z., Henderson, B. E., Pike, M. C. and Ross, R. K. (1998) *Br J Cancer* **78**, 413–8.

64. Andriole, G., Bostwick, D., Brawley, O., Gomella, L., Marberger, M., Tindall, D., Breed, S., Somerville, M. and Rittmaster, R. (2004) *J Urol* **172**, 1314–7.

65. Luo, J., Dunn, T. A., Ewing, C. M., Walsh, P. C. and Isaacs, W. B. (2003) *Prostate* **57**, 134–9.

66. Kuiper, G. G., Lemmen, J. G., Carlsson, B., Corton, J. C., Safe, S. H., van der Saag, P. T., van der Burg, B. and Gustafsson, J. A. (1998) *Endocrinology* **139**, 4252–63.

67. Schatzl, G., Brossner, C., Schmid, S., Kugler, W., Roehrich, M., Treu, T., Szalay, A., Djavan, B., Schmidbauer, C. P., Soregi, S. and Madersbacher, S. (2000) *Urology* **55**, 397–402.

68. Massengill, J. C., Sun, L., Moul, J. W., Wu, H., McLeod, D. G., Amling, C., Lance, R., Foley, J., Sexton, W., Kusuda, L., Chung, A., Soderdahl, D. and Donahue, T. (2003) *J Urol* **169**, 1670–5.

69. Raivio, T., Santti, H., Schatzl, G., Gsur, A., Haidinger, G., Palvimo, J. J., Janne, O. A. and Madersbacher, S. (2003) *Prostate* **55**, 194–8.

70. Moore, J. T., McKee, D. D., Slentz-Kesler, K., Moore, L. B., Jones, S. A., Horne, E. L., Su, J. L., Kliewer, S. A., Lehmann, J. M. and Willson, T. M. (1998) *Biochem Biophys Res Commun* **247**, 75–8.

71. Ogawa, S., Inoue, S., Watanabe, T., Orimo, A., Hosoi, T., Ouchi, Y. and Muramatsu, M. (1998) *Nucleic Acids Res* **26**, 3505–12.

72. Rajfer, J. and Coffey, D. S. (1978) *Invest Urol* **16**, 186–90.

73. Onozawa, M., Kawamori, T., Baba, M., Fukuda, K., Toda, T., Sato, H., Ohtani, M., Akaza, H., Sugimura, T. and Wakabayashi, K. (1999) *Jpn J Cancer Res* **90**, 393–8.

74. Kato, K., Takahashi, S., Cui, L., Toda, T., Suzuki, S., Futakuchi, M., Sugiura, S. and Shirai, T. (2000) *Jpn J Cancer Res* **91**, 786–91.

75. Jarred, R. A., Keikha, M., Dowling, C., McPherson, S. J., Clare, A. M., Husband, A. J., Pedersen, J. S., Frydenberg, M. and Risbridger, G. P. (2002) *Cancer Epidemiol Biomarkers Prev* **11**, 1689–96.

76. Cook, L. S., Goldoft, M., Schwartz, S. M. and Weiss, N. S. (1999) *J Urol* **161**, 152–5.

77. Kolonel, L. N., Altshuler, D. and Henderson, B. E. (2004) *Nat Rev Cancer* **4**, 519–27.

78. Miller, C. P., Collini, M. D. and Harris, H. A. (2003) *Bioorg Med Chem Lett* **13**, 2399–403.

79. Strom, A., Hartman, J., Foster, J. S., Kietz, S., Wimalasena, J. and Gustafsson, J. A. (2004) *Proc Natl Acad Sci USA* **101**, 1566–71.

80. Rafi, M. M., Rosen, R. T., Vassil, A., Ho, C. T., Zhang, H., Ghai, G., Lambert, G. and DiPaola, R. S. (2000) *Anticancer Res* **20**, 2653–8.

81. Dahllof, B., Billstrom, A., Cabral, F. and Hartley-Asp, B. (1993) *Cancer Res* **53**, 4573–81.

82. Kuwajerwala, N., Cifuentes, E., Gautam, S., Menon, M., Barrack, E. R. and Reddy, G. P. (2002) *Cancer Res* **62**, 2488–92.

83. Qadan, L. R., Perez-Stable, C. M., Anderson, C., D'Ippolito, G., Herron, A., Howard, G. A. and Roos, B. A. (2001) *Biochem Biophys Res Commun* **285**, 1259–66.

84. Kumar, A. P., Garcia, G. E. and Slaga, T. J. (2001) *Mol Carcinog* **31**, 111–24.

85. Merchenthaler, I., Hoffman, G. E. and Lane, M. V. (2005) *Endocrinology* **146**, 2760–5.

86. Elloso, M. M., Phiel, K., Henderson, R. A., Harris, H. A. and Adelman, S. J. (2005) *J Endocrinol* **185**, 243–52.

87. Harris, H. A., Bruner-Tran, K. L., Zhang, X., Osteen, K. G. and Lyttle, C. R. (2005) *Hum Reprod* **20,** 936–41.
88. Benvenuti, S., Luciani, P., Vannelli, G. B., Gelmini, S., Franceschi, E., Serio, M. and Peri, A. (2005) *J Clin Endocrinol Metab* **90,** 1775–82.
89. Lund, T. D., Rovis, T., Chung, W. C. and Handa, R. J. (2005) *Endocrinology* **146,** 797–807.
90. Hillisch, A., Peters, O., Kosemund, D., Muller, G., Walter, A., Schneider, B., Reddersen, G., Elger, W. and Fritzemeier, K. H. (2004) *Mol Endocrinol* **18,** 1599–609.
91. Hall, J. M. and Korach, K. S. (2002) *J Biol Chem* **277,** 44455–61.
92. Paech, K., Webb, P., Kuiper, G. G., Nilsson, S., Gustafsson, J., Kushner, P. J. and Scanlan, T. S. (1997) *Science* **277,** 1508–10.
93. Harrington, W. R., Sheng, S., Barnett, D. H., Petz, L. N., Katzenellenbogen, J. A. and Katzenellenbogen, B. S. (2003) *Mol Cell Endocrinol* **206,** 13–22.
94. Neubauer, B. L., McNulty, A. M., Chedid, M., Chen, K., Goode, R. L., Johnson, M. A., Jones, C. D., Krishnan, V., Lynch, R., Osborne, H. E. and Graff, J. R. (2003) *Cancer Res* **63,** 6056–62.

9 Estrogen Action in Normal Prostate Epithelium and in Prostate Cancer

Gail S. Prins, PhD,
and Kenneth S. Korach, PhD

CONTENTS

1. INTRODUCTION

Estrogens are known to have significant direct and indirect effects on prostate gland development and homeostasis, and have been long suspected in playing a role in the etiology of prostatic diseases. These effects are mediated through estrogen receptors (ERs) ERα and ERβ, which are differentially expressed in prostatic stromal and epithelial cells, respectively, and whose expression changes over time and with disease progression. This chapter will review the evidence for a role of estrogens and specific ERs in prostate growth, differentiation, and carcinogenesis as well as discuss potential therapeutic strategies for growth regulation via these pathways.

From: *Current Clinical Oncology: Prostate Cancer:*
Signaling Networks, Genetics, and New Treatment Strategies
Edited by: R. G. Pestell and M. T. Nevalainen © Humana Press, Totowa, NJ

2. ESTROGENS IN THE MALE AND EFFECTS ON THE PROSTATE GLAND

Although low levels of circulating estrogens are present throughout life in males, there are two time periods, *in utero* development and aging, when males are exposed to relatively higher levels of circulating estradiol which have been shown to impact the prostate gland. In addition, estradiol may be produced locally within the prostate via conversion of testosterone by aromatase expressed within the prostate stroma *(1,2)*. Importantly, prostatic aromatase expression and promoter usage have been shown to shift in prostate cancer, which may contribute significantly to increased intraprostatic estrogen levels during disease progression *(3)*.

During the third trimester of *in utero* development in humans, rising maternal estradiol levels and declining fetal androgen production result in an increased estrogen/testosterone (E/T) ratio. This relative increase in estradiol has been shown to directly stimulate extensive squamous metaplasia within the developing prostatic epithelium, which regresses rapidly after birth when estrogen levels drop precipitously *(4–6)*. Although the natural role for estrogens during prostatic development is unclear, it has been proposed that excessive estrogenization during prostatic development may contribute to the high incidence of benign prostatic hyperplasia (BPH) and prostatic carcinoma currently observed in the aging male population *(7,8)*. African-American men have a twofold increased risk of prostatic carcinoma as compared with their Caucasian counterparts, and it has been suggested that this is related, in part, to elevated levels of maternal estrogens during early gestation in this population *(9,10)*. Indicators of pregnancy estrogen levels such as length of gestation, pre-eclampsia, and jaundice indicate a significant correlation between elevated estrogen levels and prostate cancer risk *(11,12)*. Furthermore, maternal exposure to diethylstilbestrol (DES), a potent synthetic estrogen agonist, during pregnancy was found to result in more extensive prostatic squamous metaplasia in male offspring than observed with maternal estradiol alone *(13)*. Whereas prostatic metaplasia eventually resolved following DES withdrawal, ectasia and persistent distortion of ductal architecture remained *(14)*. This has lead to the postulation that men exposed prenatally to DES may be at increased risk of prostatic disease later in life, although this has not been borne out in limited population studies conducted to date *(15)*. However, extensive studies with rodent models predict marked abnormalities in the adult prostate, including increased susceptibility to adult-onset carcinogenesis following early estrogenic exposures *(7,16–18)*. Although use of DES during pregnancy was discontinued in the early 1970s, the recent realization that certain environmental chemicals have potent estrogenic activities *(19)* has lead to a renewed interest in evaluating the effects and roles of exogenous estrogens during prostatic development *(20)*.

As men age, the relative levels of free circulating estrogens increase providing a potential environment for estrogenic stimulation of the prostate gland. Bioavailable testosterone levels decline in the aging male due to decreased production by the testis and increased sex hormone-binding globulin (SHGB) levels, which combine to lower free circulating testosterone *(21,22)*. During this time, circulating levels of free estradiol remain constant in the aging male, most likely a function of an age-related increase in body weight and adipose cells which express high levels of aromatase and peripherally convert androgens to estrogens *(23)*. The net result is a significant increase in the E/T ratio allowing the balance between androgen and estrogen regulation of prostate growth to shift toward estrogen dominance. It has been proposed that increased estrogenic stimulation of the prostate in the aging male may lead to reactivation of growth and subsequent neoplastic transformation *(24,25)*.

Estrogen is known to have significant direct effects on the adult prostate gland and has long been suspected in playing a role in the etiology of prostatic disease *(26–28)*. In 1936, Dorothy Price and colleagues demonstrated that estrogen administration to adult rodents leads to hyperplasia, squamous metaplasia, and keratinization of the prostate epithelium *(29,30)*. Long-term exposure of adult rats to supraphysiologic but non-pharmacologic levels of estradiol and physiologic levels of testosterone leads to prostatic intraepithelial neoplasia (PIN) in the dorsolateral lobe of Noble rats, and this is used as a model for estrogen-induced adenocarcinoma of the prostate *(31,32)*. Estrogen-induced aberrations in prostate epithelial growth have been observed not only in rodent studies but in dogs, monkeys, and humans as well with results varying according to species and experimental conditions *(33,34)*. In addition to epithelial effects, estrogens induce a preferential stimulation of stromal cell proliferation in the prostate gland, particularly in dogs *(35)*. Consequently, combined administration of estrogens with androgens has been used to exper-imentally induce BPH in dogs *(33,36)*. In humans, Krieg demonstrated that whereas the estradiol:DHT ratio increases moderately within normal prostate epithelial and stromal cells upon aging, the increased ratio is massive within BPH tissue, which directly implicates estradiol in the disease process *(37)*.

Estrogen effects on the prostate gland may also be indirectly mediated through alterations in other serum hormones. Estrogens increase circulating prolactin (PRL) levels and some, but not all, of estrogenic effects have been attributed to PRL action on the prostate *(28,38–40)*. Furthermore, estrogens affect the hypothalamic–hypophyseal–testicular axis via negative feedback regulation (i.e., chemical castration), thereby blocking gonadotropin secretion and testicular steroidogenesis of androgen hormones. Such treatments were the basis for high-dose estrogen therapy of prostate cancer for several decades *(41,42)*. Despite these indirect effects, there is ample evidence through hormone-controlled studies and with *in vitro* approaches to clearly document that many of the estrogenic effects on the prostate are directly mediated through prostatic expression of ERs *(43,44)*.

3. PROSTATE STEROID RECEPTORS

Hormone action occurs through multiple cellular mechanisms primarily involving receptor proteins. These receptors are part of the family of nuclear receptors (NRs) that are found ubiquitous throughout the animal kingdom. These receptors fulfill a plethora of physiological functions and are integral to the development and maintenance of the prostate. Commonalities in receptor dimerization and DNA-binding properties divide members of NR family into distinct classes *(45)*. In the prostate, androgen receptors (ARs) and ERs are referred to as the sex steroid receptors and are defined as ligand-induced homodimers that bind to DNA half-sites organized as inverted repeats *(45)*. Using a classification system based on the cytochrome P450 family *(46)*, ERs (ERα and ERβ) are the sole members of the NR3A subgroup, ERα and ERβ being NR3A1 and NR3A2, respectively *(47)*. The AR is the fourth member of the NR3C subgroup *(47)*. A detailed description of the structure and multitude of functions of the ER and AR family members is beyond the scope of this chapter but may be found in several excellent reviews. Genes encoding the ERα, ERβ, and AR are large and exhibit a highly conserved structural organization, composed of eight coding exons, and vary in length from 40 kb (ERβ) to >140 kb (ERα) *(48–50)*. In each case, the N-terminal domain (NTD) of the receptor is usually encoded by a single exon, the two zinc-fingers of the DNA-binding domain (DBD) each encoded by separate exons *(2,3)*, and the ligand-binding domain (LBD) encoded by exons *(4–8)*.

3.1. Estrogen Receptor

The human *ESR1* (ERα) cDNA was first cloned in 1985 *(51)* and has since been isolated from over 20 additional species *(52)*, from humans to fish. The encoded ERα proteins are 595 and 599 amino acids in length in humans and mice, respectively, with an approximate molecular weight of 66 kDa (Fig. 1) *(51,53)*. Numerous naturally occurring variants of the *ESR1* mRNA in normal and neoplastic tissues of several species have been described, but the existence of corresponding proteins remains controversial *(54–56)*. Decreased *ESR1* expression has been linked to receptor-mediated actions of vitamin D *(57)* and increased intracellular cyclic adenosine monophosphate (cAMP) or mitogen-activated protein kinase (MAPK) activity *(58)*, which may play a part in prostate cancer. Increased methylation of the *ESR1* promoter has also been implicated in reduced ER levels, especially in tumorigenic tissues *(58)*.

Surprisingly, a second ER gene termed *ESR2* (ERβ) was discovered in 1996 from rats *(59)* and humans *(60)*, and has since also been cloned in almost 20

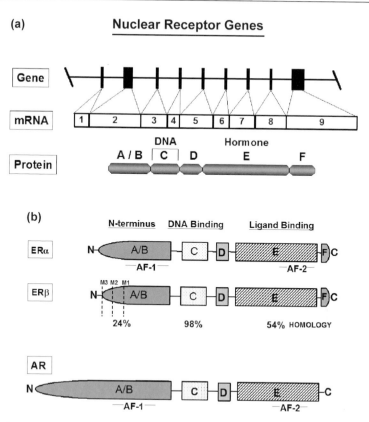

Fig. 1. (A) General diagrammatic structure of the genes encoding nuclear receptor (NR) proteins showing the 9 exons and their relative contributions to the different domain structure of the NRs. **(B)** Comparative domain structures of the androgen receptor (AR) and the alpha and beta estrogen receptors (ERs) N-terminal domain (NTD), DNA-binding domain (DBD), and ligand-binding domain (LBD). Homology listed is the percent comparison of ERα with ERβ showing the high homology in the DBD and poor homology in the NTD and LBD. M1-3 designated possible alternative ATG translational start sites in the ERβ mRNA.

species. Unlike the AR, or progesterone receptor (PR), ERα and ERβ are not isoforms but are encoded by separate genes possibly by gene duplication on different chromosomes, and therefore are distinct receptor forms. The *ESR2* genes of multiple species yield numerous transcripts that range from 1 to >9 kb in length, in contrast to the single predominant transcript of ~7 kb transcribed from the *ESR1* gene. Initial descriptions of human and rodent ERβ projected a protein of 485 amino acids. However, it is now apparent that translation of the *ESR2* mRNA initiates upstream of these original open reading frames and yields a receptor of 549 amino acids in rodents and 530 amino acids in humans, each with an approximate molecular weight of 60–63 kDa. Therefore, ERβ is slightly smaller than ERα and the majority of this difference lies within the N-terminus. A number of variant transcripts of the *ESR2* gene have been described; however, unlike ERα, there is growing evidence that some of these variants may co-exist with the wild-type form in certain tissues.

3.2. Androgen Receptor

The human *AR* cDNA was first cloned in 1988 *(61)* and has since been described in a number of species, including mouse *(62)*, rat *(63)*, rabbit *(64)*, monkey *(65)*, and fish *(66,67)*. The *AR* gene is located on the X chromosome and therefore genetic males possess only a single copy *(49,68,69)*. Two distinct start sites have been reported in the *AR* gene that are utilized to produce two isoforms, AR-B (110 kDa) and AR-A (87 kDa), which differ only in the N-terminus but exhibit subtle functional differences *(52)*. The *AR* gene has a unique feature compared with its sex steroid receptor counterparts by way of possessing polymorphic repeats of glutamine and glycine in the NTD, which have been linked to certain chronic diseases *(70)* but not prostate function.

3.3. Nuclear Receptor Structure

Common to all members of the NR family is a modular structure of domains, each of which harbors an autonomous function that is critical to total receptor action *(45,52,71)*. The sex steroid receptors are composed of five functional modules, an NTD (A/B), the DBD (C), a hinge region (D), and a ligand-binding domain (E) as encoded by nine exons of the NR genes (Fig. 1A). The ERs also possess a unique C-terminal F domain of unknown function. The functional autonomy of certain domains has been characterized by experimental studies in which corresponding domains were interchanged between NR family members without loss of function. Our understanding of the NR functional domains and their importance to overall receptor activity is largely derived from the *in vitro* study of mutant forms of the receptors and more recently from x-ray crystallography studies.

The NTD or A/B domains of the NR family members vary greatly in length and share little homology among the steroid receptors, although some structural features are conserved (Fig. 1B). Crystallography studies of the steroid receptor NTD have been largely unsuccessful because of the relatively unstructured nature of this portion of the receptor in aqueous solutions. However, evidence suggests that intramolecular interactions between the A/B and other receptor domains are likely to induce a more structured NTD, especially as it relates to the AR functionality *(69)*. The NTD of each of the sex steroid receptors harbors the transcriptional activation function-1 (AF-1) domain and specifies the cell- and promoter-specific activity of the receptor as well as a site for co-receptor protein interaction. Phosphorylation of the A/B domain is the most well characterized post-translational modification and occurs in the ER and AR receptors via the actions of multiple intracellular signaling pathways, including MAPK pathways, cAMP/protein kinase A (PKA) pathway, and cyclin-dependent kinases *(52,69–71)*.

The AR NTD is likely more important to overall AR transactivational activity than the C-terminal AF-2 domain (described on next page) *(69,70)*. Furthermore, the first 30 residues of the AR NTD are highly conserved and important for those interactions with the LBD that provide for agonist-induced stabilization of the receptor *(69)*. The greatest structural disparities between ERα and ERβ lie within the A/B domain, which is approximately 30 amino acids shorter in ERβ and exhibits only ~20% homology. This divergence likely accounts for the many functional differences that have been revealed from comparative studies of the two ER forms *(72–76)*. In general, ERβ tends to be a less effective transcriptional activator compared with ERα when *in vitro* *(60,72,73,77,78)*. Interestingly, certain antagonists (e.g., tamoxifen) that exhibit some agonist-like properties when bound to ERα exhibit no such agonist activity with ERβ *(72,76)*. Co-expression studies of ERα and ERβ *in vitro* show that ERβ attenuates the ERα activity. Whether this antagonism occurs *in vivo* to modulate biological activity of estrogen and ERα has yet to be well documented.

The C domain is termed the DNA-binding domain (or DBD), that portion of the receptor that specifically functions to recognize and bind to the *cis*-acting enhancer DNA sequences, or hormone response elements (HREs), which are located within the regulatory regions of target genes. It is the most highly conserved (55–80%) region among the NR family members *(45,79)*. A comparison of C domains among the human sex steroid receptor forms indicates 56% homology between the AR and ERα *(70)*. The C domains of ERα and ERβ are practically identical (>95% homology) in most species and are therefore expected to exhibit a similar affinity for the same HREs *(50,59,80)*. The functionality of the C domain is provided by a motif of two zinc-fingers, each composed of four cysteine residues that complex with a single Zn^{2+} ion, and encoded by separate exons. The first zinc-finger composes the P-box (proximal box) conferring specificity sequences on the receptor protein and also forming a "recognition helix" *(52)*. Amino acids of the second zinc-finger form the "D-box" (distal box) and are more specifically involved in spacer sequence and provide an interface for receptor dimerization *(52)*. P-box residues are identical among the AR, progesterone receptor (PR), and glucocorticoid receptor (GR); therefore, these receptors bind a common consensus HRE (or GRE). The consensus estrogen response element (ERE) bears the same arrangement but composed of a different half-site sequence *(52)*. In addition, ERα has been shown to induce gene expression when bound as a monomer to an ERE half-site in the vicinity of a GC-rich region or Sp1-binding sites *(81–83)*.

The D domain primarily serves to connect the more highly conserved C and E domains of the receptor (9). Commonly referred to as the "hinge" region, the D domain also harbors a nuclear localization signal that influences cellular compartmentalization of the receptor. The LBD, or E domain, of the steroid receptors is a highly structured, multifunctional region that primarily serves to specifically bind the hormone and provide for ligand-dependent transcriptional activity (52). An activation function-2 (AF-2) domain located in the C-terminus of the E domain mediates this latter function. The AF-2 domain is subject to post-translational receptor modifications (52) and is an especially strong activator of transcription in the ER but markedly weaker in the AR, where it is more involved in interactions with residues in the NTD (69). Also harbored within the E domain is a strong receptor dimerization interface. Although there is minimal homology in the primary sequence of the LBD for the NRs, comparative studies of the crystal structures of liganded and un-liganded LBDs indicate a highly conserved structural arrangement of 12 α-helices (52,84,85). Receptor binding to an agonist ligand leads to rearrangement of the LBD such that H11 is repositioned and H12 swings back to form a "lid" over the binding pocket. This agonist-induced repositioning of H12 leads to the formation of a hydrophobic cleft, which serves to recruit co-activators to the receptor complex to promote receptor transcriptional activity. By contrast, receptor antagonists are unable to induce a similar repositioning of H12, leading to a receptor formation that is incompatible with co-activator recruitment and therefore less likely to activate transcription.

The LBD of the AR in humans, rats, and mice is identical and provides for high-affinity binding of two endogenous androgens, testosterone and 5α-hydroxy-testosterone (DHT), the latter of which binds with much greater affinity (86). The LBDs of ERα and ERβ exhibit less than 60% homology but bind the endogenous hormone estradiol, estrone, and estriol with similar affinity (ERα = 0.1 nM; ERβ = 0.4 nM) (57). Both ER forms also bind the synthetic estrogen, DES, with relatively equal affinity (87). However, given the divergence in homology, it is not surprising that ERα and ERβ exhibit measurable differences in their affinity for other endogenous steroids and xenoestrogens (87–89). For example, ERβ tends to exhibit a stronger affinity for certain phytoestrogens (e.g., genestein and coumestrol) (87,88) as well as the endogenous androgen metabolite, 5α-androstane-3β,17β-diol (90). Numerous synthetic steroidal and non-steroidal ER antagonists have been developed over the years, but none have proven yet to be specific for one or the other ER subtype (91). However, more recent advances have allowed for the generation of ER-selective non-steroidal ligands (72,89,92) that exploit as well as illustrate differences between the LBDs of ERα and ERβ and provide for pharmacological tools to discern the overall function of each ER. Among the sex steroid receptors, the well-defined F domain is only found in ERs. This region is relatively unstructured and harbors little known function, although some data indicate a role in co-activator recruitment and receptor stability *(72,93)*.

4. RECEPTOR MECHANISMS OF ACTION

4.1. Ligand/HRE Dependent (Classic)

The "classic" model of steroid receptor action states that the receptor resides in the nucleus or cytoplasm but is sequestered in a multi-protein inhibitory complex in the absence of hormone; upon hormone binding, a conformational change occurs in the receptor, transforming it to an "activated" state that is now able to homodimerize, show increased phosphorylation, and bind to a HRE within target gene promoters. The ligand/HRE-bound receptor complex interacts with the general transcription apparatus either directly or indirectly via co-regulatory proteins to promote transcription of the target gene. This classic steroid receptor mechanism is dependent on the functions of both AF-1 and AF-2 domains of the receptor, which synergize via the recruitment of co-activator proteins, most notably the p160 family members. It is generally believed that the DNA-bound receptor/co-activator complex facilitates disruption of the chromatin and formation of a stable transcription pre-initiation complex. Depending on the cell and promoter context, the DNA-bound receptor complex may positively or negatively effect expression of the downstream target gene; the latter effect has been shown most clearly for AR only.

Several sex steroid regulated genes are regulated via the ligand/HRE-dependent mechanism; androgen/AR regulation of prostatic specific antigen, probasin, and keratinocyte growth factor are some examples *(69)*; and estradiol/ER regulation of several genes in prostate cancer have been reported *(20,94,95)*. These early findings have been corroborated, and the overall extent of ligand-dependent sex steroid receptor transactivation has been realized by the more recent use of differential display and microarray techniques to generate gene expression profiles in various cell types and tissues following hormonal stimulation. Similarly, differential display, microarray, and proteomics have been employed to identify a large number of androgen/AR-mediated *(96,97)* and estradiol/ER-mediated *(98–103)* genes.

When acting via a classic ERE-driven mechanism *in vitro*, ERα homodimers and ERα:ERβ heterodimers tend to be stronger activators of transcription compared with ERβ homodimers *(60,72,73,77,78)*. Corroborating *in vivo* evidence of differential regulation and heterodimer formation by the ER subtypes has been difficult to generate. However, a microarray study by Lindberg et al. found that ERβ generally inhibits ERα-mediated gene expression but, in the absence of ERα, ERβ can partially provide some estradiol-stimulated gene expression *(100)*.

4.2. Ligand Independent (Cross-Talk)

Following the initial description of membrane–NR coupling *(104)*, we now have ample evidence that the sex steroid receptors can be activated via intracellular second messenger and signaling pathways, allowing for the induction of sex steroid target genes in the absence of steroid ligand *(105,106)*. Polypeptide growth factors are able to activate ERα-mediated gene expression via MAPK activation of ERα in the absence of estradiol *(105)*. Similarly, interleukin-6 stimulation of cells leads to increased AR-mediated gene expression *(99,107,108)*. The intracellular signaling molecule cAMP, a common second messenger of G-protein-coupled receptors and an activator of the PKA pathway, can stimulate increased AR and ER-mediated transcription of target genes in the absence of steroid ligand *(109,110)*. Likewise, growth factors are able to mimic the effects of estradiol in the rodent uterus via estradiol-independent activation of ER, although such studies in the prostate are not as yet well described.

Ligand-independent activation of sex steroid receptors is believed to rely largely on the cellular kinase pathways that alter the phosphorylation state of the receptor and/or its associated proteins such as co-activators and/or heat-shock proteins. As in the classic, ligand-dependent mechanism described above, specific receptor domains are critical to ligand-independent activation as well. ER activation by peptide growth factor signaling pathways appears to be more dependent on AF-1 functions; whereas the effects of increased intracellular cAMP are postulated to depend on the AF-2 domain and do not require a functional AF-1 *(105)*.

4.3. HRE Independent (Tethered)

Ligand-activated steroid receptors can stimulate the expression of genes that lack a conspicuous HRE within their promoter. This has been especially demonstrated for ERα *(83,111)*. This mechanism of HRE-independent steroid receptor activation is postulated to involve a "tethering" of the ligand-activated receptor to transcription factors that are directly bound to DNA via their respective response elements *(83,111)*. Therefore, the sex steroid receptor acting in this fashion may be better defined as a co-regulator rather than a direct acting transcription factor. Estradiol/ERα regulation of several genes, including ovalbumin, collagenase, insulin-like growth factor-1, and cyclin D, is believed to occur via a tethering of the receptor to a DNA-bound AP-1 (Fos/Jun) complex with the gene promoter *(111)*. Details of this type of mechanism remain unclear but are postulated to involve a mediator component, e.g., p160, between ER and AP-1 versus direct interaction *(111)*. A similar HRE-independent mechanism of sex steroid receptor regulation has been documented for genes that possess a GC-rich region or Sp1-binding site within the promoter, upon which the actions of a bound Sp1 complex can be enhanced by ERα *(83,111)*. How much of this type of gene regulation is involved in prostate growth and stimulation is unknown at the present time. Further assessment will await development of animal models for testing the specific contribution of tethered gene regulation to prostate physiology and disease.

5. ERS IN THE PROSTATE GLAND

Original studies with ligand-binding assays, sucrose density gradients, and autoradiography were able to identify specific estrogen-binding sites in both epithelial and stromal cell fractions of the prostate gland in different species and these were assumed to be a single ER. With the discovery of ERβ in addition to ERα, the localization and relative contributions of each receptor type required reanalysis. Results demonstrated that for the most part, ERα and ERβ are expressed in different cellular compartments of the prostate gland. As discussed below, ERα localizes primarily to prostatic stromal cells (including smooth muscle cells), whereas ERβ is primarily expressed in prostate epithelium. This differential location as well as differential affinity of the two ERs for ligands, enhancers, and co-activators may explain the diverse biological functions of estrogens within the prostate gland and may also be exploited for regulation of prostate disease.

5.1. ERα

ERα is localized primarily to the stromal cells of the adult prostate gland in humans, dogs, monkeys, and rodents *(112–116)*. Immunohistochemical analysis reveals, however, that ERα expression is heterogeneous in stromal cells, i.e., only a portion of the cells are ERα positive whereas many remain ERα negative. The significance of this heterogeneous expression is not known. Studies in rodent prostate glands have shown relatively high stromal cell ERα expression during perinatal morphogenesis of the gland, which significantly declines thereafter suggesting a specific role for ERα in prostate development *(115,117,118)*. A decline in expression with puberty suggests that androgens may normally suppress ERα expression, a finding that has been borne out in direct studies *(119,120)*. In humans, ERα has been consistently observed in stromal cells during fetal development *(121)*. However, whereas one report restricts ERα protein to only stromal cells *(118)*, a recent report indicates the presence of ERα in fetal prostatic utricle and periurethral epithelium during mid-to-late gestation *(121)*. Importantly, squamous metaplasia, observed in all developing human prostates during the third trimester, is directly associated with epithelial ERα in the periurethral ducts and stromal ERα in the peripheral prostatic acini *(121)*.

It is believed that stromal proliferation, a hallmark response to estrogen treatment in most species, may be mediated through stromal ERα. In humans, there is evidence for an increased accumulation of estradiol in nuclei of stromal cells in BPH specimens *(122)*, suggesting that elevated ERα in stromal cells may be involved in the etiology of BPH. Of interest, epithelial cells in prostatic periurethral ducts have also been found to consistently express ERα in both normal and BPH tissue *(112)*. Because this is the prostatic region that forms BPH, it is possible that epithelial ERα in that specific region is involved. In rodents, periductal mesenchymal and smooth muscle cells, which are in close proximity to epithelial cells, express ERα whereas interductal fibroblasts are ERα negative *(115)*. The close proximity of ERα-positive stromal cells to epithelial cells allows for paracrine effects of estrogens on prostate epithelium. Indeed, work with ERα knockout (αERKO) mice demonstrated that estradiol-induced squamous metaplasia in adult prostates is mediated through ERα *(123)*. Similarly, using αERKO and βERKO mice, neonatal estrogenization of the prostate, which includes stromal as well as epithelial alterations, was shown to be mediated through stromal ERα *(124)*. It is noteworthy, however, that deletion of ERα in transgenic mice did not produce a marked phenotype in the prostate suggesting that ERα's role in the prostate gland is not necessary for normal growth and function *(125,126)*.

Recent reports have shown that in prostatic carcinoma, the ERα gene is methylated leading to silencing of this gene, loss of ERα transcription and ERα protein *(127,128)*. Although comparative data with normal prostate specimens were not available, it was noted that the incidence of ERα gene methylation and silencing increased with progression of prostatic disease from BPH to low-grade and to high-grade cancer. Thus it was proposed that ERα may have a tumor suppressor role in the prostate gland and loss of its expression may be an early event in prostatic disease. Interestingly, ERα expression has been observed in some prostate cancer cell lines *(128)* as well as in hormone refractory and metastatic lesions suggesting its re-emergence as cancer progresses *(129)* although this has not been consistently seen in all studies *(130)*. It is also noteworthy that prostate cancer risk has been associated with genetic polymorphisms in the ERα gene particularly within Japanese and African-American populations implicating a potential causal relationship between ERα-mediated estrogenic action and prostate cancer *(131–133)*.

5.2. ERβ

ERβ was originally cloned from a rat prostate cDNA library; thus it was not surprising to find that the rat prostate expressed this receptor at levels comparable with those found in other high-expressing reproductive organs such as the ovary, endometrium, and testis *(50,59)*. In the rat and murine prostate, ERβ mRNA and ERβ protein are primarily localized to differentiated luminal epithelial cells, which may preclude formation of ERα:ERβ heterodimers in this organ *(124,134,135)*. Expression of ERβ is low at birth, increases as epithelial cells cytodifferentiate, and reaches maximal expression with onset of secretory capacity at puberty, which suggests a role for ERβ in the differentiated function of the prostate *(134)*. In the adult, a proximal to distal gradient of ERβ mRNA signal intensity is observed with lowest levels proximally and increased expression distally, which may contribute to ductal heterogeneity and differential functions along the ductal length *(134)*. Unlike ERα, androgens have been shown to up-regulate expression of ERβ in the rodent prostate gland *(136,137)*. Studies with estrogens, however, do not indicate autoregulation of ERβ.

It is noteworthy that the developmental pattern for ERβ in the human prostate differs markedly from the rodent. As early as fetal week 7, ERβ is expressed throughout the urogenital sinus epithelium and stroma *(121)*. This strong expression is maintained in most epithelial and stromal cells throughout gestation, particularly in the active phase of branching morphogenesis during the second trimester suggesting the involvement of ERβ and estrogens in this process *(121)*. Although this pattern is maintained postnatally for several months, ERβ expression declines thereafter with a noticeable decrease in adluminal cells at puberty *(118)*. In the adult human prostate, ERβ expression is low relative to testicular expression, again showing a divergence from the rodent prostate gland *(50,60)*. Reports vary on ERβ localization in the human prostate which may be a function of antibodies used in immunohistochemical assays. Whereas some have shown that ERβ is expressed by basal epithelial cells with lower stromal cell expression *(116,130)*, others have shown high expression of ERβ in both basal and luminal epithelial cells of the adult human prostate *(138,139)*. In normal and tumorigenic human prostate epithelial cell lines, ERβ is expressed at high levels whereas ERα is typically absent *(128,140)*. In response to *in vitro* estrogen exposure, estrogen-regulated genes (progesterone receptor, pS2) are activated again pointing toward a role for ERβ in the differentiated function of the prostatic cell. Together, these findings indicate that, in the prostate epithelium, ERβ may be the key mediator of estrogen-induced events. The prostate gland also expresses ERβ isoform variants that have been shown to act as constitutive activators, transcription enhancers, or dominant negative regulators of estrogen action which further complicate estrogenic action within this gland *(141,142)*.

5.3. Putative Roles of ERβ

Although ERβ is the predominant ER expressed in the prostate gland, its role has not yet been clearly established. As stated above, indirect evidence exists for a role of ERβ in the differentiated state of the prostate epithelium. A recent study using βERKO mice showed a shift in basal, intermediate, and luminal epithelial cell markers in the prostate toward a less differentiated gland that supports this purported role *(143)*. As a counterpart to the hypothesis that ERβ plays a role in epithelial differentiation, it has also been suggested that ERβ has an antiproliferative role in the prostate and participates as a brake for androgenic stimulation of prostate growth *(144)*. Indirect evidence for this role was suggested by the hyperplastic and dysplastic adult prostate epithelium with reduced ERβ expression following neonatal estrogen exposure *(134)*. Direct studies for a role of ERβ in prostate proliferation using βERKO mice have yielded conflicting results. Although some studies show epithelial hyperplasia with increased BrdU labeling in the βERKO prostates *(144–146)*, this was not supported in subsequent studies using the same mice *(124,147)* or different βERKO models *(148)*.

Studies in our laboratory suggest that ERβ may play an immunomodulatory role in the prostate gland. βERKO and wild-type mice were aged to 6 or 12 months and a blinded histologic analysis by two independent investigators was performed for the entire glandular complex using serial sections (>100 sections/prostate complex). Aging associated changes were noted in both genotypes, which consisted of reduced secretions, flattened epithelium, and degenerated apoptotic cells within ductal lumens. Proliferative and apoptotic scores were not different for the two genotypes. However, a feature observed in βERKO but not in wild-type prostates was abundant-to-massive lymphoid aggregates that were at times associated with reactive (proliferative) epithelium consistent with injury-repair cycles (Fig. 2). Immune T-cell infiltration was blindly scored as 0 (absent), 1 (rare), 2 (focal), 3 (abundant), and 4 (massive) for the two genotypes. The percent of wild-type mice in each category was 50, 33, 17, 0, and 0%, respectively, indicating that rare or focal T-cell infiltration was present in 50% of animals whereas the remaining 50% of wild-type mice had no inflammatory cells present. In marked contrast, T-cell infiltration scores for the βERKO mice were 11, 11, 33, 22, and 22%, respectively, indicating that the vast majority of mice possessed prostatic inflammatory cells with 44% of cases presenting with abundant or massive immune cell infiltration (Fig. 2). Thus, we propose that ERβ may normally play an immunoprotective role in the prostate gland perhaps limiting tissue damage or modulating expression of stimulus for immune cell infiltration. Estrogens are widely known to affect the development and regulation of the immune system and have been shown to exert potent anti-inflammatory effects *(149,150)*. The present data suggest that the anti-inflammatory effects at the level of the prostate gland may be mediated through ERβ. This hypothesis is supported by a recent study that demonstrated that an ERβ-selective ligand was able to prevent inflammatory bowel disease in a rat model *(151)*. Such an approach may hold promise for treatment of prostatitis, which is the most prevalent of prostatic diseases. Furthermore, there is increasing evidence for a link between chronic prostatic inflammation and

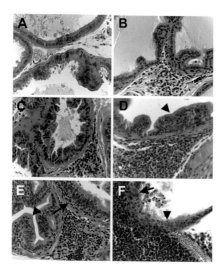

Fig. 2. Inflammatory cell infiltration in the ventral (**A–C, D, F**) and dorsal (**E**) prostate lobes of wild-type (**A**) and ERβ knockout (βERKO) (**B–F**) mice at 1 year of age. Wild-type prostates (**A**) were largely free of lymphoid infiltration. By contrast, focal (**B** and **C**), abundant (**D** and **E**) and massive (**F**) stromal lymphocytic infiltrate were routinely observed in the βERKO prostates. As infiltration became abundant, evidence for diapedesis in the epithelium was observed (**E** and **F**, arrows). In several instances, lymphoid aggregates were associated with reactive, proliferative epithelium in the immediate vicinity (**D–F**, arrowheads). Magnification, ×40.

prostate cancer etiology *(152)*, suggesting a potential relationship to prostatic ERβ in this regard. ERβ has also been proposed to play a role in the anti-oxidant pathway. ERβ can bind the electrophile/antioxidant response element (EpRE) and is a more potent activator at the EpRE element than ERα *(153)*. Thus, ERβ is capable of inducing genes that encode chemoprotective detox-ification enzymes [quinone reductase, glutathione S-transferase (GST)] and may play an active role in protecting prostate epithelial cells from carcinogens by detoxifying electrophiles. This is particularly relevant because GST-pi has been shown to be critical in protecting against prostate cancer through genome damage initiated by inflammatory cells and carcinogens *(154)*.

5.4. ERβ in Prostate Cancer

Dynamic changes in ERβ expression have been observed during the progression of prostate cancer, which suggests that estrogen action through ERβ may play an important role in prostate carcinogenesis, metastasis, and perhaps, androgen independence. There are several reports in the literature that describe ERβ expression in prostate cancer. Although there are a few conflicting reports *(155)*, most concur that ERβ expression declines in localized PCa with increasing grade from PIN through low to high Gleason scores *(130,139,156,157)*. This pattern fits with the hypothesis that ERβ plays an antiproliferative and pro-differentiation function in the prostate. Loss of its expression would permit unregulated proliferation and de-differentiation of prostatic epithelium. In support of this, a recent study using adenoviral vectors found that ERβ expression in prostate cell lines inhibited growth and invasiveness of cells suggesting that loss of ERβ in higher grade tumors allows for unregulated growth and eventual metastasis *(158)*. However, counter to this concept, ERβ expression re-emerges as prostate cancer metastasizes to distant sites with 100% of osseous and non-osseous metastatic cells expressing ERβ to varying degrees *(138,159)*. Loss of ERβ mRNA and protein with increasing tumor grade in organ-confined disease has been shown by Ho and colleagues to be epigenetically regulated by progressive hypermethylation of two 5′ flanking CpG islands due to methylation spreading from the untranslated first exonic CpG island to the promoter CpG island with methylation of the promotor CpG island causing transcriptional silencing *(160)*. Furthermore, as the cells metastasize, methylation at these two CpG islands is lost, which allows for high gene expression at metastatic sites. If ERβ is considered antiproliferative, it is unclear how high ERβ expression in metastatic disease permits uncontrolled proliferation with metastatic spread. Because one study reported that localized prostate cancers that retain ERβ in the primary tumor were associated with a higher rate of recurrence *(157)*, it is possible that expression of ERβ in metastatic cells in not due to demethylation in the tumor microenvironment and re-emergence of gene expression per se but rather is a function of a selective advantage of ERβ-retaining cells to metastasize. Thus, it is currently unclear whether ERβ function has an antiprolferative role in prostate cancer or whether it promotes growth and metastasis. Whatever the case, the important feature is the strong ERβ expression in metastatic, androgen-independent prostate cancer. This suggests that the metastatic cells are targets of estrogen action and thus may be potential targets for therapeutic interventions with antiestrogenic agents or more effectively ERβ-selective antagonists.

5.5. Selective ER Modulators, Aromatase Inhibitors, and Phytoestrogens as Therapeutic Agents in the Treatment of Prostate Cancer

Because of the above considerations concerning the effects of estrogens on prostate growth and carcinogenesis, and the expression of ERs in prostate stromal and epithelial cells, the use of estrogen/antiestrogen therapy may have efficacy for the treatment of prostate cancer. Thus many estrogens, antiestrogens, phytoestrogens, and selective ER modulators (SERMs) including DES, 2-methoxyestradiol, Raloxifene, ICI182,780, Trioxifene, and Torimifene have been shown to affect prostate tumor growth through various mechanisms *(24,161–167)*. Whereas some of the initial clinical trials using the SERMs tamoxifene and Torimifene proved to be unremarkable for the treatment of prostate disease *(168,169)*, more recent studies indicate that the Torimifene may in fact effectively block development and progression of clinical prostate cancer, and further clinical trials are underway *(166,170)*. It is unclear at present whether these agents act primarily through antagonism of ERα, ERβ, or both. Future studies with ERα- and ERβ-specific ligands *(151)* as well as the development of new specific SERMs may provide insight into the specific prostatic ER that is the most effective target for therapeutic use.

Another possible site for treatment of prostate cancer is through inhibition of estradiol production using aromatase inhibitors *(171,172)*. Recent work has shown high expression of aromatase in prostate cancer with alternate promotor utilization that suggests that intraprostatic production of estradiol may contribute to progression of this disease *(3)*. However, clinical trials with aromatase inhibitors have not shown the current agents, letrozole or anastrozole, to be effective in limiting prostate cancer progression *(173,174)*. Whether this is related to the timing of the intervention or the lack of efficacy of this approach remains unclear at present.

In addition to pharmaceutical approaches, there is considerable evidence that phytoestrogens may modulate prostate growth, which has been the basis for herbal supplements and dietary modulation for the treatment of abnormal prostate growth. Genestein *(175,176)*, resveratrol *(177)*, and soy *(178)* have all been shown to have beneficial effects, and consumption of these products has been inversely correlated with prostate cancer risk *(178–180)*. High prostatic expression of ERβ may explain why phytoestrogens (genestein, coumestrol) are beneficial to prostate health because these compounds have been found to bind to ERβ with an affinity up to 10 times higher than for ERα. However, clinical trials evaluating the efficacy of these approaches have been limited. Thus, caution in their usage must be issued until their safety and effectiveness has been demonstrated.

6. CONCLUSIONS

It is generally agreed that the prostate gland is an androgen target tissue and highly responsive to testosterone and DHT. Developmental dependence of the prostate on androgen and AR activity is also well documented both experimentally and clinically. Androgens' role in prostate disease including BPH and prostate cancer, and the therapeutic developments for successful treatment of these conditions using androgen antagonists are clearly established. On the other hand, a considerable amount of evidence now exists for an important role of estrogens in both normal prostate gland development and homeostasis.

Research continues to identify the different specific roles of androgens and estrogens and their mimics in prostate functionality. Additionally, development of prostatic diseases including BPH and prostate cancer has also been linked to estrogen exposures. Experimental studies have provided some evidence for the roles of the different NRs in the disease process. Therefore, such findings may provide new avenues or alternative treatments of prostate cancer with novel therapies directed at ERs or estrogen metabolism. Once the individual role of the hormones and their receptors are determined, a treatment approach could be developed in conjunction with androgen therapy or independent of androgen manipulation. Furthermore, manipulation of estrogen signaling in the prostate gland may be useful as a potential chemopreventive therapy.

REFERENCES

1. Matzkin H and Soloway MS (1992) Immunohistochemical evidence of the existence and localization of aromatase in human prostatic tissues. *Prostate* 21:309–314.
2. Risbridger GP, Bianco JJ, Ellem SJ and McPherson SJ (2003) Estrogens and prostate cancer. *Endocr Relat Cancer* 10:187–191.
3. Ellem SJ, Schmitt JF, Pedersen JS, Frydenberg M and Risbridger GP (2004) Local aromatase expression in human prostate is altered in malignancy. *J Clin Endocrinol Metab* 89:2431–2441.
4. Zondek T and Zondek LH (1975) The fetal and neonatal prostate. In: Goland M (ed) *Normal and Abnormal Growth of the Prostate*. Thomas C. Thomas, Springfield, IL, pp. 5–28.
5. Zondek T, Mansfield MD, Attree SL and Zondek LH (1986) Hormone levels in the fetal and neonatal prostate. *Acta Endocrinol (Copenh)* 112:447–456.
6. Wernet N, Kern L, Heitz PH, Bonkhoff H, Goebbels IR, Seitz G, Inniger R, Remberger K and Dhom G (1990) Morphological and immunohistochemical investigations of the utriculus prostaticus from the fetal period up to adulthood. *Prostate* 17:19–30.
7. Rajfer J and Coffey DS (1978) Sex steroid imprinting of the immature prostate. *Invest Urol* 16:186–190.
8. Santti R, Newbold RR, Makela S, Pylkkanen L and McLachlan JA (1994) Developmental estrogenization and prostatic neoplasia. *Prostate* 24:67–78.
9. Henderson BE, Bernstein L, Ross RK, Depue RH and Judd HL (1988) The early in utero oestrogen and testosterone environment of blacks and whites: potential effects on male offspring. *Br J Cancer* 57:216–218.
10. Platz EA and Giovannucci E (2004) The epidemiology of sex steroid hormones and their signaling and metabolic pathways in the etiology of prostate cancer. *J Steroid Biochem Mol Biol* 92:237–253.
11. Ekbom A, Wuu J, Adami HO, Lu CM, Lagiou P, Trichopoulos D and Hsieh C (2000) Duration of gestation and prostate cancer risk in offspring. *Cancer Epidemiol Biomarkers Prev* 9:221–223.
12. Ekbom A, Hsieh CC, Lipworth L, Wolk A, Ponten J, Adami HO and Trichopoulos D (1996) Perinatal characteristics in relation to incidence of and mortality from prostate cancer. *Br Med J* 313:337–341.
13. Driscoll SG and Taylor SH (1980) Effects of prenatal maternal estrogen on the male urogenital system. *Obstet Gynecol* 56:537–542.
14. Yonemura CY, Cunha GR, Sugimura Y and Mee SL (1995) Temporal and spatial factors in diethylstilbestrol-induced squamous metaplasia in the developing human prostate. II. Persistent changes after removal of diethylstilbestrol. *Acta Anat* 153:1–11.

15. Giusti RM, Iwamoto K and Hatch EE (1995) Diethylstilbestrol revisited: a review of the long-term health effects. *Ann Intern Med* 122:778–788.

16. Arai Y, Mori T, Suzuki Y and Bern HA (1983) Long-term effects of perinatal exposure to sex steroids and diethylstilbestrol on the reproductive system of male mammals. In: Bourne GHaD JF (ed) *International Review of Cytology*. Academic Press, Inc., New York, vol 84:235–268.

17. Prins GS, Birch L, Habermann H, Chang WY, Tebeau C, Putz O and Bieberich C (2001) Influence of neonatal estrogens on rat prostate development. *Reprod Fertil Dev* 13:241–252.

18. Huang L, Pu Y, Alam S, Birch L and Prins GS (2004) Estrogenic regulation of signaling pathways and homeobox genes during rat prostate development. *J Androl* 25:330–337.

19. McLachlan JA (2001) Environmental signaling: what embryos and evolution teach us about endocrine disrupting chemicals. *Endocr Rev* 22:319–341.

20. Ho SM, Wang WY, Belmonte J and Prins GS (2006) Developmental exposure estradiol and bisphenol A (BPA) increases susceptibility to prostate carcinogenesis and epigenetically regulates phosphodiesterase type 4 variant (PDE4D4) in the rat prostate. *Cancer Res* 66:5624–5632.

21. Vermeulen A (1976) Testicular hormonal secretion and aging in mammals. In: Grayhack J, Wilson J and Scherbenske M (ed) *Benign Prostatic Hyperplasia II*. DHEW Publication, NIH, Washington, DC, vol II:177–181.

22. Kaufman JM and Vermeulen A (2005) The decline of androgen levels in elderly men and its clinical and therapeutic implications. *Endocr Rev* 26:833–875.

23. Vermeulen A, Kaufman JM, Goemaere S and van Pottelberg I (2002) Estradiol in elderly men. *Aging Male* 5:98–102.

24. Ho SM (2004) Estrogens and antiestrogens: key mediators of prostate carcinogenesis and new therapeutic candidates. *J Cell Biochem* 91:491–503.

25. King KJ, Nicholson HD and Assinder SJ (2006) Effect of increasing ratio of estrogen: androgen on proliferation of normal and human prostate stromal and epithelial cells, and the malignant cell line LNCaP. *Prostate* 66:105–114.

26. Thomas JA and Keenan EJ (1994) Effects of estrogen on the prostate. *J Androl* 15:97–99.

27. Castagnetta LAM and Carruba G (1995) Human prostate cancer: a direct role for oestrogens. *Ciba Found Symp* 191:269–286.

28. Harkonen PL and Makela S (2004) Role of estrogens in the development of prostate cancer. *Steroid Biochem Mol Biol* 92:297–305.

29. Price D (1936) Normal development of the prostate and seminal vesicles of the rat with a study of experimental postnatal modifications. *Am J Anat* 60:79–127.

30. Price D (1963) Comparative aspects of development and structure in the prostate. In: Vollmer EP (ed) *Biology of the Prostate and Related Tissues*. National Cancer Institute, Washington, DC, vol 12:1–27.

31. Leav I, Merk FB, Kwan PWL and Ho SM (1989) Androgen-supported estrogen-enhanced epithelial proliferation in the prostates of intact Noble rats. *Prostate* 15:23–40.

32. Leav I, Ho S, Ofner P, Merk F, Kwan P and Damassa D (1988) Biochemical alterations in sex hormone-induced hyperplasia and dysplasia of the dorsolateral prostates of Noble rats. *J Natl Cancer Inst* 80:1045–1053.

33. Coffey D and Walsh P (1990) Clinical and experimental studies of benign prostatic hyperplasia. *Urol Clin North Am* 17:461–475.

34. El Etreby M and Habenicht U-F (1987) Experimental induction of stromal hyperplasia in prostate of castrated dogs treated with 4-androstene-3, 17-dione. In: Rodgers C, Coffey D, Cunha G, Grayhack J, Hinman F and Horton R (ed) *Benign Prostatic Hyperplasia II*. U.S. Department of Health and Human Services, NIH, vol II:303–310.

35. Walsh P and Wilson J (1976) The induction of prostatic hypertrophy in the dog with androstanediol. *J Clin Invest* 57:1093–1097.

36. DeKlerk D, Coffey D, Weing L, McDermott I, Reiener W, Robinson C, Scott W, Strandberg J, Talalay P, Walsh P, Wheaton L and Zirkin B (1979) Comparison of spontaneous and experimentally induced canine prostatic hyperplasia. *J Clin Invest* 64:842–849.

37. Krieg M, Klotzl G, Kaufmann J and Voigt K (1981) Stroma of human benign prostatic hyperplasia: preferential tissue for androgen metabolism and oestrogen binding. *Acta Endocrinol (Copenh)* 96:422–432.

38. Lee C, Prins GS, Henneberry MO and Grayhack JT (1981) Effect of estradiol on the rat prostate in the presence and absence of testosterone and pituitary. *J Androl* 2:293–299.

39. Gilleran JP, Putz O, De Jong M, De Jong S, Birch L, Pu Y, Huang L and Prins GS (2003) The role of prolactin in the prostatic inflammatory response to neonatal estrogen. *Endocrinology* 144:2046–2054.

40. Lane K, Leav I, Ziar J, Bridges R, Rand W and Ho SM (1997) Suppression of testosterone and estradiol-17beta-induced dysplasia in the dorsolateral prostate of Noble rats by bromocriptine. *Carcinogenesis* 18:1505–1520.

41. Huggins C and Hodges CF (1941) Studies on prostatic cancer. I. The effect of castration, of estrogen, and of androgen injection on serum phosphatases in metastatic carcinoma of the prostate. *Cancer Res* 1:293–297.

42. Griffiths K (2000) Estrogens and prostatic disease. *Prostate* 45:87–100.

43. Nevalainen MT, Harkonen PL, Valve EM, Ping W, Nurmi M and Martikainen PM (1993) Hormone regulation of human prostate in organ culture. *Cancer Res* 53:5199–5207.

44. Ellem SJ, Risbridger GP (2007) Treating prostate cancer: A rationale for targeting local oestrogens. *Nature Rev* 7:621–627.

45. Mangelsdorf DJ, Thummel C, Beato M, Herrlich P, Schutz G, Umesono K, Blumberg B, Kastner P, Mark M, Chambon P and Evans RM (1995) The nuclear receptor superfamily: the second decade. *Cell* 83:835–839.

46. Nebert DW, Adesnik M, Coon MJ, Estabrook RW, Gonzalez FJ, Guengerich FP, Gunsalus IC, Johnson EF, Kemper B, Levin W et al. (1987) The P450 gene superfamily: recommended nomenclature. *DNA* 6:1–11.

47. Committee NRN (1999) A unified nomenclature system for the nuclear receptor superfamily. *Cell* 97:161–163.

48. Ponglikitmongkol M, Green S and Chambon P (1988) Genomic organization of the human oestrogen receptor gene. *EMBO J* 7:3385–3388.

49. Kuiper GGJM, Faber PW, Van Rooij HCJ, Van der Korput JAGM, Ris-Stalpers C, Klaassen P, Trapman J and Brinkmann AO (1989) Structural organization of the human androgen receptor gene. *J Mol Endocrinol* 2:R1–R4.

50. Enmark E, Pelto-Huikko M, Grandien K, Lagercrantz S, Lagercrantz J, Fried G, Nordenskjold M and Gustafsson J (1997) Human estrogen receptor b -gene structure, chromosomal localization, and expression pattern. *J Clin Endocrinol Metab* 82:4258–4265.

51. Walter P, Green S, Greene G, Krust A, Bornert J-M, Jeltsch J-M, Staub A, Jensen E, Scrace G, Waterfield M and Chambon P (1985) Cloning of the human estrogen receptor cDNA. *Proc Natl Acad Sci USA* 82:7889–7893.

52. Laudet V and Gronemeyer H (2002) *The Nuclear Receptor: Factsbook.* Academic Press, San Diego.

53. White R, Lees JA, Needham M, Ham J and Parker M (1987) Structural organization and expression of the mouse estrogen receptor. *Mol Endocrinol* 1:735–744.

54. Murphy LC, Dotzlaw H, Leygue E, Douglas D, Coutts A and Watson PH (1997) Estrogen receptor variants and mutations. *J Steroid Biochem Mol Biol* 62:363–372.

55. Miksicek RJ (1994) Steroid receptor variants and their potential role in cancer. *Semin Cancer Biol* 5:369–379.

56. Sluyser M (1995) Mutations in the estrogen receptor gene. *Hum Mutat* 6:97–103.
57. Stoica A, Saceda M, Fakhro A, Solomon HB, Fenster BD and Martin MB (1999) Regulation of estrogen receptor-alpha gene expression by 1, 25-dihydroxyvitamin D in MCF-7 cells. *J Cell Biochem* 75:640–651.
58. Pinzone JJ, Stevenson H, Strobl JS and Berg PE (2004) Molecular and cellular determinants of estrogen receptor alpha expression. *Mol Cell Biol* 24:4605–4612.
59. Kuiper GGJM, Enmak E, Pelto-Huikko M, Nilsson S and Gustaffson JA (1996) Cloning of a novel estrogen receptor expressed in rat prostate and ovary. *Proc Natl Acad Sci USA* 93:5925–5930.
60. Mosselman S, Polman J and Dijkema R (1996) ER B: identification and characterization of a novel human estrogen receptor. *FEBS Lett* 392:49–53.
61. Lubahn D, Joseph D, Sullivan P, Willard H, French F and Wilson E (1988) Cloning of human androgen receptor complementary DNA and localization to the X chromosome. *Science* 240:327–330.
62. He WW, Fischer LM, Sun S, Bilhartz DL, Zhu X, Young CYF, Kelley DB and Tindall DJ (1990) Molecular cloning of androgen receptors from divergent species with a polymerase chain reaction technique: complete cDNA sequence of the mouse androgen receptor and isolation of androgen recepter cDNA probes from dog, guinea pig adn clawed frog. *Biochem Biophys Res Comm* 171:697–704.
63. Tan J, Joseph DR, Quarmby VE, Lubahn DB, Sar M, French FS and Wilson EM (1988) The rat androgen receptor: primary structure, autoregulation of its messenger ribonucleic acid, and immunocytochemical localization of the receptor protein. *Mol Endocrinol* 2:1276–1285.
64. Krongrad A, Wilson JD and McPhaul MJ (1995) Cloning and partial sequence of the rabbit androgen receptor: expression in fetal urogenital tissues. *J Androl* 16:209–212.
65. Choong CS, Kemppainen JA and Wilson EM (1998) Evolution of the primate androgen receptor: a structural basis for disease. *J Mol Evol* 47:334–342.
66. Takeo J and Yamashita S (1999) Two distinct isoforms of cDNA encoding rainbow trout androgen receptors. *J Biol Chem* 274:5674–5680.
67. Touhata K, Kinoshita M, Tokuda Y, Toyohara H, Sakaguchi M, Yokoyama Y and Yamashita S (1999) Sequence and expression of a cDNA encoding the red seabream androgen receptor. *Biochim Biophys Acta* 1450:481–485.
68. Lubahn DB, Brown TR, Simental JA, Higgs HN, Migeon CJ, Wilson EM and French FS (1989) Sequence of the intron/exon junctions of the coding region of the human androgen receptor gene and identification of a point mutation in a family with complete androgen insensitivity. *Proc Natl Acad Sci USA* 86:9534–9538.
69. McEwan IJ (2004) Molecular mechanisms of androgen receptor-mediated gene regulation: structure-function analysis of the AF-1 domain. *Endocr Relat Cancer* 11:281–293.
70. Gelmann EP (2002) Molecular biology of the androgen receptor. *J Clin Oncol* 20:3001–3015.
71. Aranda A and Pascual A (2001) Nuclear hormone receptors and gene expression. *Physiol Rev* 81:1269–1304.
72. Katzenellenbogen BS, Montano MM, Ediger TR, Sun J, Ekena K, Lazennec G, Martini PG, McInerney EM, Delage-Mourroux R, Weis K and Katzenellenbogen JA (2000) Estrogen receptors: selective ligands, partners, and distinctive pharmacology. *Recent Prog Horm Res* 55:163–193.
73. Tremblay GB, Tremblay A, Copeland NG, Gilbert DJ, Jenkins NA, Labrie F and Giguere V (1997) Cloning, chromosomal localization, and functional analysis of the murine estrogen receptor beta. *Mol Endocrinol* 11(3):353–365.
74. Delaunay F, Pettersson K, Tujague M and Gustafsson JA (2000) Functional differences between the amino-terminal domains of estrogen receptors alpha and beta. *Mol Pharmacol* 58:584–590.

75. Tremblay A, Tremblay GB, Labrie F and Giguere V (1999) Ligand-independent recruitment of SRC-1 to estrogen receptor beta through phosphorylation of activation function AF-1. *Mol Cell Biol* 3:513–519.

76. Watanabe T, Inoue S, Ogawa S, Ishii Y, Hiroi H, Ikeda K, Orimo A and Muramatsu M (1997) Agonist effect of tamoxifen is dependent on cell type, ERE-promoter context, and estrogen receptor subtype: functional difference between estrogen receptors alpha and beta. *Biochem Biophys Res Comm* 236:140–145.

77. Cowley SM and Parker MG (1999) A comparison of transcriptional activation by ER alpha and ER beta. *J Steroid Biochem Mol Biol* 69:165–175.

78. Cowley SM, Hoare S, Mosselman S and Parker MG (1997) Estrogen receptors alpha and beta form heterodimers on DNA. *J Biol Chem* 272:19858–19862.

79. Tsai M-J and O'Malley BW (1994) Molecular mechanisms of action of steroid/thyroid receptor superfamily members. *Annu Rev Biochem* 63:451–486.

80. Kuiper GGJM and Gustafsson J-A (1997) The novel estrogen receptor-b subtype: potential role in the cell- and promoter-specific actions of estrogens and anti-estrogens. *FEBS Lett* 410:87–90.

81. O'Lone R, Frith MC, Karlsson EK and Hansen U (2004) Genomic targets of nuclear estrogen receptors. *Mol Endocrinol* 18:1859–1875.

82. Kim K, Thu N, Saville B and Safe S (2003) Domains of estrogen receptor alpha (ERalpha) required for ERalpha/Sp1-mediated activation of GC-rich promoters by estrogens and antiestrogens in breast cancer cells. *Mol Endocrinol* 17:804–817.

83. Safe S and Kim K (2004) Nuclear receptor-mediated transactivation through interaction with Sp proteins. *Prog Nucleic Acid Res Mol Biol* 77:1–36.

84. Matias PM, Donner P, Coelho R, Thomaz M, Peixoto C, Macedo S, Otto N, Joschko S, Scholz P, Wegg A, Basler S, Schafer M, Egner U and Carrondo MA (2000) Structural evidence for ligand specificity in the binding domain of the human androgen receptor. Implications for pathogenic gene mutations. *J Biol Chem* 275:26164–26171.

85. Sack JS, Kish KF, Wang C, Attar RM, Kiefer SE, An Y, Wu GY, Scheffler JE, Salvati ME, Krystek SR Jr, Weinmann R and Einspahr HM (2001) Crystallographic structures of the ligand-binding domains of the androgen receptor and its T877A mutant complexed with the natural agonist dihydrotestosterone. *Proc Natl Acad Sci USA* 98:4904–4909.

86. Hiipakka RA and Liao S (1998) Molecular mechanisms of androgen action. *Trends Endocrinol Metab* 9:317–324.

87. Kuiper GGJM, Carlsson B, Grandien K, Enmark E, Haggblad J, Nilsson S and Gustafsson JA (1997) Comparison of the ligand binding specificity and transcript tissue distribution of estrogen receptors alpha and beta. *Endocrinology* 138:863–870.

88. Kuiper GGJM, Lemmen JG, Carlsson B, Corton JC, Safe SH, van der Saag PT, van der Burg B and Gustaffsson J-A (1998) Interaction of estrogenic chemicals and phytoestrogens with estrogen receptor beta. *Endocrinology* 139:4252–4263.

89. Harris HA, Bapat AR, Gonder DS and Frail DE (2002) The ligand binding profiles of estrogen receptors alpha and beta are species dependent. *Steroids* 67:379–384.

90. Weihua Z, Lathe R, Warner M and Gustafsson JA (2002) An endocrine pathway in the prostate, ERbeta, AR, 5alpha-androstane-3beta,17beta-diol, and CYP7B1, regulates prostate growth. *Proc Natl Acad Sci USA* 99:13589–13594.

91. Macgregor JI and Jordan VG (1998) Basic guide to the mechanisms of antiestrogen action. *Pharm Rev* 50:151–196.

92. Harris HA, Katzenellenbogen JA and Katzenellenbogen BS (2002) Characterization of the biological roles of the estrogen receptors, ERalpha and ERbeta, in estrogen target tissues in vivo through the use of an ERalpha-selective ligand. *Endocrinology* 143:4172–4177.

93. Montano MM, Muller V, Trobaugh A and Katzenellenbogen BS (1995) The carboxy-terminal F domain of the human estrogen receptor: role in the transcriptional activity of

the receptor and the effectiveness of antiestrogens as estrogen antagonists. *Mol Endocrinol* 9:814–825.

94. Fan S, Meng Q, Auborn K, Carter T and Rosen EM (2006) BRCA1 and BRCA2 as molecular targets for phytochemicals indole-3-carbinol and genistein in breast and prostate cancer cells. *Br J Cancer* 94(3):407–426.

95. Cao F, Jin TF and Zhou YF (2006) Inhibitory effect of isoflavones on prostate cancer cells and PTEN gene. *Biomed Environ Sci* 19(1):35–41.

96. Eder IE, Haag P, Basik M, Mousses S, Bektic J, Bartsch G and Klocker H (2003) Gene expression changes following androgen receptor elimination in LNCaP prostate cancer cells. *Mol Carcinog* 37:181–191.

97. Jiang F and Wang Z (2003) Identification of androgen-responsive genes in the rat ventral prostate by complementary deoxyribonucleic acid subtraction and microarray. *Endocrinology* 144:1257–1265.

98. Hewitt SC, Deroo BJ, Hansen K, Collins J, Grissom S, Afshari CA and Korach KS (2003) Estrogen receptor-dependent genomic responses in the uterus mirror the biphasic physiological response to estrogen. *Mol Endocrinol* 17:2070–2083.

99. Chen T, Wang LH and Farrar WL (2000) Interleukin 6 activates androgen receptor-mediated gene expression through a signal transducer and activator of transcription 3-dependent pathway in LNCaP prostate cancer cells. *Cancer Res* 60:2132–2135.

100. Lindberg MK, Moverare S, Skrtic S, Gao H, Dahlman-Wright K, Gustafsson JA and Ohlsson C (2003) Estrogen receptor (ER)-beta reduces ERalpha-regulated gene transcription, supporting a "ying yang" relationship between ERalpha and ERbeta in mice. *Mol Endocrinol* 17:203–208.

101. Fujimoto N, Igarashi K, Kanno J, Honda H and Inoue T (2004) Identification of estrogen-responsive genes in the GH3 cell line by cDNA microarray analysis. *J Steroid Biochem Mol Biol* 91:121–129.

102. Kato N, Shibutani M, Takagi H, Uneyama C, Lee KY, Takigami S, Mashima K and Hirose M (2004) Gene expression profile in the livers of rats orally administered ethinylestradiol for 28 days using a microarray technique. *Toxicology* 200:179–192.

103. Terasaka S, Aita Y, Inoue A, Hayashi S, Nishigaki M, Aoyagi K, Sasaki H, Wada-Kiyama Y, Sakuma Y, Akaba S, Tanaka J, Sone H, Yonemoto J, Tanji M and Kiyama R (2004) Using a customized DNA microarray for expression profiling of the estrogen-responsive genes to evaluate estrogen activity among natural estrogens and industrial chemicals. *Environ Health Perspect* 112:773–781.

104. Ignar-Trowbridge DM, Nelson KG, Bidwell MC, Curtis SW, Washburn TF, McLachlan JA and Korach KS (1992) Coupling of dual signaling pathways: epidermal growth factor action involves the estrogen receptor. *Proc Natl Acad Sci USA* 89:4658–4662.

105. Coleman KM and Smith CL (2001) Intracellular signaling pathways: nongenomic actions of estrogens and ligand-independent activation of estrogen receptors. *Front Biosci* 6:D1379–1391.

106. Weigel NL and Zhang Y (1998) Ligand-independent activation of steroid hormone receptors. *J Mol Med* 76:469–479.

107. Ueda T, Bruchovsky N and Sadar MD (2002) Activation of the androgen receptor N-terminal domain by interleukin-6 via MAPK and STAT3 signal transduction pathways. *J Biol Chem* 277:7076–7085.

108. Culig Z (2004) Androgen receptor cross-talk with cell signalling pathways. *Growth Factors* 22:179–184.

109. Sadar MD (1999) Androgen-independent induction of prostate-specific antigen gene expression via cross-talk between the androgen receptor and protein kinase A signal transduction pathways. *J Biol Chem* 274:7777–7783.

110. Nazareth LV and Weigel NL (1996) Activation of the human androgen receptor through a protein kinase A signaling pathway. *J Biol Chem* 271:19900–19907.
111. Kushner PJ, Agard DA, Greene GL, Scanlan TS, Shiau AK, Uht RM and Webb P (2000) Estrogen receptor pathways to AP-1. *J Steroid Biochem Mol Biol* 74:311–317.
112. Schulze H and Claus S (1990) Histological localization of estrogen receptors in normal and diseased human prostates by immunocytochemistry. *Prostate* 16:331–343.
113. Schulze H and Barrack E (1987) Immunocytochemical localization of estrogen receptors in the normal male and female canine urinary tract and prostate. *Endocrinology* 121:1773–1783.
114. Brenner RM, West N and McClellena M (1990) Estrogen and progestin receptors in the reproductive tract of male and female primates. *Biol Reprod* 42:11–19.
115. Prins GS and Birch L (1997) Neonatal estrogen exposure up-regulates estrogen receptor expression in the developing and adult rat prostate lobes. *Endocrinology* 138:1801–1809.
116. Tsurusaki T, Aoki D, Kanetake H, Inoue S, Muramatsu M, Hishikawa Y and Koji T (2003) Zone-dependent expression of estrogen receptors alpha and beta in human benign prostatic hyperplasia. *J Clin Endocrinol Metab* 88:1333–1340.
117. Tilley WD, Horsfall DJ, Skinner JM, Henderson DW and Marshall VR (1989) Effect of pubertal development on estrogen receptor levels and stromal morphology in the guinea pig prostate. *Prostate* 15:195–210.
118. Adams JY, Leav I, Lau KM, Ho SM and Pflueger SM (2002) Expression of estrogen receptor beta in the fetal, neonatal, and prepubertal human prostate. *Prostate* 52:69–81.
119. Kruithof-Dekker IG, Tetu B, Janssen PJ and Van der Kwast TH (1996) Elevated estrogen receptor expression in human prostatic stromal cells by androgen ablation therapy. *J Urol* 156:1194–1197.
120. Tilley WD, Horsfall DJ, McGee MA, Alderman JE and Marshall VR (1987) Effects of ageing and hormonal manipulations on the level of oestrogen receptors in the guinea-pig prostate. *J Endocrinol* 112:139–144.
121. Shapiro E, Huang H, Masch RJ, McFadden DE, Wilson EL and Wu XR (2005) Immunolocalization of estrogen receptor alpha and beta in human fetal prostate. *J Urol* 174:2051–2053.
122. Kozak I, Bartsch W, Krieg M and Voight K (1982) Nuclei of stroma: site of highest estrogen concentration in human benign prostatic hyperplasia. *Prostate* 3:433–438.
123. Risbridger GP, Wang H, Young P, Takeshi K, Wong YZ, Lubahn D, Gustafsson JA and Cunha G (2001) Evidence that epithelial and mesenchymal estrogen receptor-a mediates effects of estrogen on prostatic epithelium. *Dev Biol* 229:432–444.
124. Prins GS, Birch L, Couse JF, Choi I, Katzenellenbogen B and Korach KS (2001) Estrogen imprinting of the developing prostate gland in mediated through stromal estrogen receptor a: studies with aERKO and bERKO mice. *Cancer Res* 61:6089–6097.
125. Eddy EM, Washburn TF, Bunch DO, Goulding EH, Gladen BC, Lubahn DB and Korach KS (1996) Targeted disruption of the estrogen receptor gene in male mice causes alteration of spermatogenesis and infertility. *Endocrinology* 137:4796–4805.
126. Donaldson KM, Tong SY, Washburn T, Lubahn DB, Eddy EM, Hutson JM and Korach KS (1996) Morphometric study of the gubernaculum in male estrogen receptor mutant mice. *J Androl* 17:91–95.
127. Li L, Chui R, Nakajima K, Oh BR, Au HC and Dahiya R (2000) Frequent methylation of estrogen receptor in prostate cancer: correlation with tumor progression. *Cancer Res* 60:702–706.
128. Lau KM, LaSprina M, Long J and Ho SM (2000) Expression of estrogen receptor (ER)-a and ER-b in normal and malignant prostatic epithelial cells: regulation by methylation and involvement in growth regulation. *Cancer Res* 60:702–706.
129. Bonkhoff H, Fixemer T, Hunsicker I and Remberger K (1999) Estrogen receptor in prostate cancer and premalignant lesions. *Am J Pathol* 155:641–647.

130. Leav I, Lau KM, Adams J, NcNeal J, Taplin M, Wang J, Singh H and Ho SM (2001) Comparative studies of the estrogen receptors beta and alpha and the androgen receptor in normal human prostate gland, dysplasia, and in primary and metastatic carcinoma. *Am J Pathol* 1:79–92.

131. Hernandez J, Balic I, Johnson-Pais TL, Higgins BA, Torkko KC, Thompson I and Leach RJ (2006) Association between estrogen receptor alpha gene polymorphism and risk of prostate cancer in black men. *J Urol* 175:523–527.

132. Tanaka Y, Sasaki M, Kaneuchi M, Shiina H, Igawa M and Dahiya R (2003) Polymorphisms of estrogen receptor alpha in prostate cancer. *Mol Carcinog* 37:202–208.

133. Modugno F, Weissfeld JL, Trump DL, Zmuda JM, Shea P, Cauley JA and Ferrell RE (2001) Allelic variants of aromatase and the androgen and estrogen receptors: toward a multigenic model of prostate cancer risk. *Clin Cancer Res* 10:3092–3106.

134. Prins GS, Marmer M, Woodham C, Chang WY, Kuiper G, Gustafsson JA and Birch L (1998) Estrogen receptor-b messenger ribonucleic acid ontogeny in the prostate of normal and neonatally estrogenized rats. *Endocrinology* 139:874–883.

135. Makela S, Strauss L, Kuiper G, Valve EM, Salmi S, Santti R and Gustasson JA (2000) Differential expression of estrogen receptors alpha and beta in adult rat accessory sex glands and lower urinary tract. *Mol Cell Endocrinol* 164:109–116.

136. Chang W and Prins G (1999) Estrogen receptor-b: implications for the prostate gland. *Prostate* 40:115–124.

137. Asano K, Maruyama S, Usui T and Fujimoto N (2003) Regulation of estrogen receptor alpha and beta expression by testosterone in the rat prostate gland. *Endocr J* 50:281–287.

138. Fixemer T, Remberger K and Bonkoff H (2003) Differential expression of estrogen receptor beta in human prostate tissue, premalignant changes and in primary, metastaic and recurrent prostatic adenocarcinoma. *Prostate* 54:79–87.

139. Fujimura T, Takahashi S, Urano T, Ogawa S, Ouchi Y, Kitamura T, Muramatsu M and Inoue S (2001) Differential expression of estrogen receptor beta (ERbeta) and its C-terminal truncated splice variant ERbetacx as prognostic predictors in human prostatic cancer. *Biochem Biophys Res Commun* 289:692–699.

140. Hanstein B, Liu H, Yancisin M and Brown M (1999) Functional analysis of a novel estrogen receptor beta isoform. *Mol Endocrinol* 13:129–137.

141. Maruyama K, Endoh H, Sasaki-Iwaoka H, Kanou H, Shimaya E, Hashimoto S, Kato S and Kawashima H (1998) A novel isoform of rat estrogen receptor beta with 18 amino acid insertion in the ligand binding domain as a putative dominant negative regular of estrogen action. *Biochem Biophys Res Commun* 246:142–147.

142. Petersen D, Tkalcevic G, Koza-Taylor P, Turi T and Brown T (1998) Identification of estrogen receptor beta 2: a functional varient of estrogen receptor beta expressed in normal rat tissues. *Endocrinology* 139:1082–1092.

143. Imamov O, Morani A, Shim GJ, Omoto Y, Thulin-Andersson C, Warner M and Gustafsson JA (2004) Estrogen receptor beta regulates epithelial cellular differentiation in the mouse ventral prostate. *Proc Natl Acad Sci USA* 101:9375–9380.

144. Weihua Z, Makela S, Andersson LC, Salmi S, Saji S, Webster JI, Jensen EV, Nilsson S, Warner M and Gustafsson JA (2001) A role for estrogen receptor beta in the regulation of growth of the ventral prostate. *Proc Natl Acad Sci USA* 98:6330–6335.

145. Krege JH, Hodgin JB, Couse JF, Enmark E, Warner M, Mhler JF, Sar M, Korach KS, Gustafsson JA and Smithies O (1998) Generation and reproductive phenotypes of mice lacking estrogen receptor beta. *Proc Natl Acad Sci USA* 95:15677–15682.

146. Weihua Z, Warner M and Gustaffson J (2002) Estrogen receptor beta in the prostate. *Mol Cell Biol* 193:1–5.

147. Couse JF, Curtis-Hewitt S and Korach KS (2000) Receptor null mice reveal contrasting roles for estrogen receptor alpha and beta in reproductive tissues. *J Steroid Biochem Mol Biol* 74:287–296.

148. Dupont S, Krust A, Gansmuller A, Dierich A, Chambon P and Mark M 200 Effect of single and compound knockouts of estrogen receptors alpha (ERalpha) and beta (ERbeta) on mouse reproductive phenotypes. *Development* 127:4277–4291.

149. McMurray RW (2001) Estrogen, prolactin, and autoimmunity: actions and interactions. *Int Immunopharm* 1:995–1008.

150. Yellayi S, Teuscher C, Woods JA, Welsh TH Jr, Tung KS, Nakai M, Rosenfeld CS, Lubahn DB and Cooke PS (2000) Normal development of thymus in male and female mice requires estrogen/estrogen receptor-alpha signaling pathway. *Endocrine* 12:207–213.

151. Harris HA, Albert LM, Leathurby Y, Malamas MS, Mewshaw RE, Miller CP, Kharode YP, Marzolf J, Komm BS, Winneker RC, Frail DE, Henderson RA, Zhu Y and Keith JC (2003) Evaluation of an estrogen receptor-beta agonist in animal models of human disease. *Endocrinology* 144:4241–4249.

152. DeMarzo AM, Marchi VL, Epstein JI and Nelson WG (1999) Proliferative inflammatory atrophy of the prostate implications for prostatic carcinogenesis. *Am J Pathol* 155:1985–1992.

153. Montano M, Jaiswal A and Katzenellenbogen B (1998) Transcriptional regulation of the human quinone reductase gene by antiestrogen-liganded estrogen receptor-a and estrogen receptor-b. *Fertil Steril* 273:25443–25449.

154. Nelson WG, DeWeese TL and DeMarzo AM (2002) The diet, prostate inflammation, and the development of prostate cancer. *Cancer Metastasis Rev* 21:3–16.

155. Torlakovic E, Lilleby W, Torlakovic G, Fossa SD and Chibbar R (2002) Prostate carcinoma expression of estrogen receptor-beta as detected by PPG5/10 antibody has positive association with primary Gleason grade and Gleason score. *Hum Pathol* 33:646–651.

156. Pasquali D, Rossi V, Esposito D, Abbondanza C, Puca GA, Bellastella A and Sinisi AA (2001) Loss of estrogen receptor beta expression in malignant human prostate cells in primary cultures and in prostate cancer tissues. *J Clin Endocrinol Metab* 86:2051–2055.

157. Horvath LG, Henshall SM, Lee CS, Head DR, Quinn DI, Makela S, Delprado W, Golovsky D, Brenner PC, O'Neill G, Kooner R, Stricker PD, Grygiel JJ, Gustafsson JA and Sutherland RL (2001) Frequent loss of estrogen receptor-beta expression in prostate cancer. *Cancer Res* 61:5331–5335.

158. Cheng J, Lee EJ, Madison LD and Lazennec G (2004) Expression of estrogen receptor beta in prostate carcinoma cells inhibits invasion and proliferation and triggers apoptosis. *FEBS Lett* 566:169–172.

159. Lai JS, Brown LG, True LD, Hawley SJ, Etzioni RB, Higano CS, Ho SM, Vessella RL and Corey E (2004) Metastases of prostate cancer express estrogen receptor-beta. *Urology* 64:814–820.

160. Zhu X, Leav I, Leung Y, Wu M, Liu Q, Gao Y, McNeal JE and Ho SM (2004) Dynamic regulation of estrogen receptor-beta expression by DNA methylation during prostate cancer development and metastasis. *Am J Pathol* 164:1883–1886.

161. Garcia GE, Wisniewski HG, Lucia MS, Arevalo N, Slaga TJ, Kraft SL, Strange R and Kumar AP (2006) 2-Methoxyestradiol inhibits prostate tumor development in transgenic adenocarcinoma of mouse prostate: role of tumor necrosis factor-alpha-stimulated gene 6. *Clin Cancer Res* 12:980–988.

162. Schulz P, Bauer HW, Brade WP, Keller A and Fittler F (1988) Evaluation of the cytotoxic activity of diethylstilbestrol and its mono- and diphosphate towards prostatic carcinoma cells. *Cancer Res* 48(10):2867–2870.

163. Kim IY, Kim BC, Seong DH, Lee DK, Seo JM, Hong YJ, Kim HT, Morton RA and Kim SJ (2002) Raloxifene, a mixed estrogen agonist/antagonist, induces apoptosis in androgen-independent human prostate cancer cell lines. *Cancer Res* 62:5365–5369.

164. Neubauer BL, McNulty AM, Chedid M, Chen K, Goode RL, Johnson MA, Jones CD, Krishnan V, Lynch R, Osborne HE and Graff JR (2003) The selective estrogen receptor modulator trioxifene (LY133314) inhibits metastasis and extends survival in the PAIII rat prostatic carcinoma model. *Cancer Res* 63:6056–6062.

165. Corey E, Quinn JE, Emond MJ, Buhler KR, Brown LG and Vessella RL (2002) Inhibition of androgen-independent growth of prostate cancer xenografts by 17beta-estradiol. *Clin Cancer Res* 8:1003–1007.

166. Raghow S, Hooshdraran MZ, Katiyar S and Steiner MS (2002) Toremifene prevents prostate cancer in the transgenic adenocarcinoma of mouse prostate model. *Cancer Res* 62:1370–1376.

167. Guerini V, Sau D, Scaccianoce E, Rusmini P, Ciana P, Maggi A, Martini PG, Katzenellenbogen BS, Martini L, Motta M and Poletti A (2005) The androgen derivative 5alpha-androstane-3beta,17beta-diol inhibits prostate cancer cell migration through activation of the estrogen receptor beta subtype. *Cancer Res* 65:5445–5453.

168. Bergan RC, Reed E, Myers CE, Headlee D, Brawley O, Cho HK, Figg WD, Tompkins A, Linehan WM, Kohler D, Steinberg SM and Blagosklonny MV (1999) Phase II study of high-dose tamoxifen in patients with hormone-refractory prostate cancer. *Clin Cancer Res* 5:2366–2373.

169. Stein S, Zoltick B, Peacock T, Holroyde C, Haller D, Armstead B, Malkowicz SB and Vaughn DJ (2001) Phase II trial of toremifene in androgen-independent prostate cancer: a Penn cancer clinical trials group trial. *Am J Clin Oncol* 24:283–285.

170. Steiner MS and Pound CR (2003) Phase IIA clinical trials to test the efficacy and safety of Toremifene in men with high-grade prostatic intraepithelial neoplasia. *Clin Prostate Cancer* 2:32–33.

171. Ellem SJ and Risbridger GP (2006) Aromatase and prostate cancer. *Minerva Endocrinol* 31:1–12.

172. Narashimamurthy J, Rao AR and Sastry GN (2004) Aromatase inhibitors: a new paradigm in breast cancer treatment. *Curr Med Chem Anticancer Agents* 4:523–534.

173. Santen RJ, Petroni GR, Fisch MJ, Meyers CE, Theodorescu D, Cohen RB (2001) Use of the aromatase inhibitor anastrozole in the treatment of advanced prostate carcinnoma. *Cancer* 92:2095–2101.

174. Smith MR, Kaufman D, George D, Oh WK, Kazanis M, Manola J, Kantoff PW (2002) Selective aromatatse inhibiton for patients with androgen-independent prostate carcinoma. *Cancer* 95:1864–1868.

175. Wang J, Eltoum IE and Lamartiniere CA (2002) Dietary genistein suppresses chemically induced prostate cancer in Lobund-Wistar rats. *Cancer Lett* 186(1):11–18.

176. Mentor-Marcel R, Lamartiniere CA, Eltoum A, Greenberg NM and Elgavish A (2005) Dietary genistein improves survival and reduces expression of osteopontin in the prostate of transgenic mice with prostatic adenocarcinoma (TRAMP). *J Nutr* 135:989–995.

177. Jones SB, DePrimo SE, Whitfield ML and Brooks JD (2005) Resveratrol-induced gene expression profiles in human prostate cancer cells. *Cancer Epidemiol Biomarkers Prev* 14:596–604.

178. Lee MM, Gomez SL, Chang JS, Wey M, Wang RT and Hsing AW (2003) Soy and isoflavone consumption in relation to prostate cancer risk in China. *Cancer Epidemiol Biomarkers Prev* 12:665–668.

179. Ozasa K, Nakao M, Watanabe Y, Hayashi K, Miki T, Mikami K, Mori M, Sakauchi F, Washio M, Ito Y, Suzuki K, Wakai K and Tamakoshi A (2004) Serum phytoestrogens and prostate cancer risk in a nested case-control study among Japanese men. *Cancer Sci* 95:65–71.

180. Manson MM, Farmer PB, Gescher A and Steward WP (2005) Innovative agents in cancer prevention. *Recent Results Cancer Res* 166:257–275.

10 Hypoxia-Inducible Factor 1 in the Angiogenesis of Prostate Cancer

Jonathan W. Simons, MD

CONTENTS

INTRODUCTION
HIF: FUNCTION AND BIOLOGY
REGULATION OF HIF-1
HIF-1 AND TRANSLATIONAL RESEARCH IN PROSTATE CANCER
HIF AS THERAPEUTIC TARGET IN PROSTATE CANCER
SUMMARY
ACKNOWLEDGMENTS
REFERENCES

1. INTRODUCTION

Lethal clones of prostate cancer express multiple genes involved in driving angiogenesis. A Google Scholar search of "angiogenesis and prostate cancer" yields over 1000 citations. Nevertheless, until recently the signaling pathways, critical epigenetic factors, and genetic alterations were not well defined for human prostate cancer angiogenesis. Hypoxia-inducible factor 1 (HIF-1) is an important transcription factor associated with the lethality of human prostate cancer and many other cancers *(1)*. HIF-1 regulates multiple genes, including vascular endothelial growth factor (VEGF) in angiogenesis *(1–3)*. Furthermore, HIF-1 also appears to be involved in the overexpression of genes implicated in osseous metastasis *(1)*. As the signaling pathways for HIF-1 have been identified, HIF-1 has become a candidate therapeutic target

From: *Current Clinical Oncology: Prostate Cancer:
Signaling Networks, Genetics, and New Treatment Strategies*
Edited by: R. G. Pestell and M. T. Nevalainen © Humana Press, Totowa, NJ

in prostate cancer; pursuit of HIF-1 inhibitors has the attractive aspect that they may centrally inhibit angiogenesis, as well as other survival functions of HIF-1.

2. HIF: FUNCTION AND BIOLOGY

To survive, eukaryotic cells need to endure metabolic stresses and manage satisfactory O_2 levels; they respond to hypoxia with adaptive changes in mRNA and protein expression *(4)*. Mammals are sustained by multiple complex physiological pathways to maintain adequate O_2 delivery to all tissues. Cells have systems of gene expression that respond to the stress of hypoxia. Adaptive reactions to acute oxygen tension reductions are modulated by protein phosphorylation or redox state changes, but chronic hypoxia—akin to the metabolism of tumors in vivo—causes cellular alterations in gene expression throughout the genome *(4)*.

In seminal work studying the regulation of erythropoietin and glycolytic enzymes under hypoxia, Semenza and colleagues discovered that the transcriptional regulator hypoxia-inducible factor 1 (HIF-1), is an essential mediator of O_2 homeostasis in chronic hypoxia *(6,7)*. HIF-1β is a dimer as a transcription factor, and is composed of HIF-1α and HIF-1β [also known as the aryl hydrocarbon nuclear translocator (ARNT)] subunits *(3)*. The two subunits form heterodimers, and the amino-terminal half of each subunit contains basic helix-loop-helix (bHLH) and PER–ARNT–SIM homology (PAS) domains (Fig. 1). The bHLH domain mediates dimerization; the basic domain binds DNA *(7)*. Interestingly, the bHLH–PAS proteins represent a relatively small family of bHLH proteins that are only found in multicellular metazoan species *(3)*. HIF, like angiogenesis itself, is central to multicellular development and organogenesis. Mice homozygous for a loss-of-function mutation in the gene encoding HIF-1α or HIF-1β die as embryos with major vascular defects primarily involving the embryonic and extraembryonic circulation, respectively *(4)*.

Hereditary tumor suppressor gene research has helped unmask the importance of HIF-1 in tumor cell survival and metastasis. Renal carcinoma cell lines, which lack expression of the von Hippel–Lindau (VHL) tumor suppressor protein, profoundly express HIF-1 under even nonhypoxic conditions, and O_2-regulated downregulation is restored in cells that have been transfected with a wild-type VHL cDNA expression vector *(8,9)*. The VHL gene functions as a component of a ubiquitin-protein ligase, and constitutive expression of HIF-1α in kidney cancer is due to a lack of ubiquitination under nonhypoxic conditions (Fig. 2). HIF-1 is regulated post-transcriptionally in normoxia by ubiquitination and interaction with the von Hippel-Lindau (VHL) tumor suppressor protein (pVHL), then degraded by the 26S proteasome. After HIF-1α is hydroxylated at the proline residues 402 and 577 it is recognized by pVHL, a member of

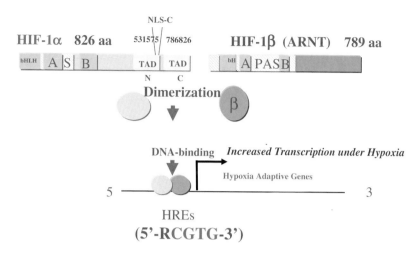

Fig. 1. *The HIF-1 Complex Structure.* HIF-1α is an 826-amino-acid polypeptide, whereas HIF-1β is expressed as a 774- or 789-amino-acid polypeptide as a result of alternative splicing. Basic helix-loop-helix (bHLH) and PER–aryl hydrocarbon nuclear translocator (ARNT)–SIM (PAS) domains are required for heterodimerization that is needed for DNA binding. Additional regulatory domains of HIF-1α include N- and C-terminal nuclear localization signals (NLS-N and NLS-C, respectively), the proline–serine–threonine-rich protein stabilization domain (PSTD), amino- and carboxy-terminal transactivation domains (TAD-N and TAD-C, respectively), and the transcriptional inhibitory domain (ID). Transcriptional coactivators CBP and p300 interact with TAD-C.

the E$_3$ ubiquitination complex. This critical regulatory process is mediated by prolyl hydroxylase domain containing (PHD) proteins. The PHDs require O$_2$ as a cofactor for activity and have been intensively studied *(8)*. HIF-1 biological activity in cancer is determined by the presence or absence of the HIF-1α subunit in the nucleus. The regulation of HIF-1α expression and activity *in vivo* occurs at multiple levels, including mRNA expression, translation, nuclear localization, and post-translational modification *(7)*.

HIF-1α dimerizes with HIF-1β and the heterodimer binds to DNA at sites represented by the consensus sequence 5′-RCGTG-3′ to activate transcription of target genes *(7)*. The 5′-RCGTG-3′-binding site is present within a hypoxia response element (HRE), a *cis*-acting transcriptional regulatory sequence that can be located within 5′-flanking, 3′-flanking, or intervening sequences of genes that are activated by HIF-1 *(7)*.

Normally, biochemical detection of intracellular O$_2$ levels directly is mediated by the PHD enzymes which then regulate HIF-1α protein levels. When cells are hypoxic, PHD activity is inhibited from the low availability of O$_2$. HIF-1α hydroxylation and pVHL association is lowered and steady-state HIF-1α protein levels rise (Fig. 2). HIF-1α protein with its nuclear

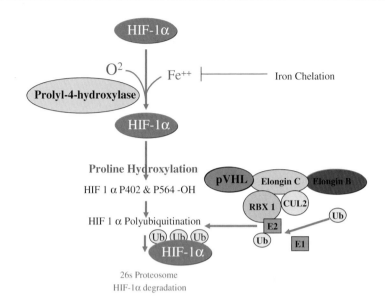

Fig. 2. *HIF-1α Protein Level Control Pathway.* Hypoxia is signaled by prolyl hydroxylase domains (PHDs), which lack oxygen, which in turn release HIF-1α and redirects HIF-1α from the proteasome toward the signaling pathway to the nucleus. This pathway is a complex multi-responsive series of different phosphorylations originating from K-ras to either PI3 kinase/AKT or RAF/MEK pathway. In turn, HIF-1α dimerizes with HIF1β/hydrocarbon nuclear translocator (ARNT), translocates to the nucleus, and interacts possibly with signal and transducer of transcription 3 (STAT3) and p300/CBP to initiate an increase in vascular endothelial growth factor (VEGF) transcription.

localization sequence is then translocated to the nucleus; in this process it heterodimerizes with HIF-1β, and the HIF-1 complex binds to hypoxic responsive elements (HREs) upstream of hypoxic-regulated genes, where it functions as a transcription factor. HIF-2α is another structurally related HIF protein, also termed (EPAS1, HRF, HLF, MOP2), has recently been implicated in the hypoxic response of endometrial carcinoma and other cancers *(7)*. A small group exists of HIF subunit genes exists in the genome. The slightly different HIF-αs may have different functions in different tissues. They have been discovered within the human and mouse genomes by database searches for cDNA sequences encoding structurally similar proteins. The HIF-1α, which has been studied the most in cancer, was cloned directly from its functional activity in erythropoietin *(6,7)*. HIF-1α, -2α, and -3α each heterodimerize with ARNT (HIF-1β), ARNT2, or ARNT3. HIF-1α and ARNT (HIF-1β) mRNA are expressed in most, if not all, human and rodent tissues *(8,9)*. In contrast, HIF-2α, HIF-3α, ARNT2, and ARNT3 have a narrower set of tissues and cell lineages of expression. To date, the best evidence suggests that HIF-1α plays a very central and widespread role by signaling the existence of hypoxia to the transcriptional machinery in the nucleus. Comparative analyses of HIF-1α and HIF-2α knockout mice suggest HIF-2α and HIF-3α play more limited or specialized roles in O_2 adaptive responses. Both HIF-1α and HIF-2 at the level of protein are expressed in human prostate cancer compared to normal prostate cells immediately adjacent to the tumor (7). Both HIF-1 and HIF-2 have structural homology, and both are regulated by oxygen tension through the PHDs (7). HIF-2 also regulates hypoxia-inducible genes through binding to HREs. The functional differences between the two HIF-α proteins, their tissue distribution, and roles in human tumor biology are still being defined. The HIF-1 heterodimers associate with the transcription coactivator CBP, bind to a, hypoxia response element (HRE) promoter, and enhances transcription of the downstream target genes (3). The number of HRE upregulated transcripts activated by HIF-1 identified in metastatic tumors continues to increase, and includes genes whose protein products are involved in angiogenesis, energy metabolism, erythropoiesis, cell proliferation, apoptosis, stromal growth and remodeling, vascular remodeling, and vasomotor responses (7). At least 20 genes upregulated by HIF-1 are known to be involved in the tumor biology of invasion, angiogenesis, cell survival, and metastasis of prostate cancer (7).

Human cancers have the plasticity to adapt to unfavorable metabolic environments, to generate a vasculature, and withstand hypoxic environments in both primary and metastatic sites. Hypoxia is defined as a loss of oxygen in tissues and is widespread in solid tumors (epithelial or mesenchymal stem cell origin) due to the tumors ability to outgrow the existing vasculature. Oxygen tension in normal tissues has a mean of approximately 7% oxygen; in human tumors, the mean oxygen tension averages about 1.5% oxygen (3,10–12). Eppendorf partial oxygen pressure histograph measurements have demonstrated the existence of hypoxia in patient's prostate cancers in situ.

Tumor cells must survive by epigenetically adapting to the low pO_2 in their microenvironments, or by increasing vascularization, or both. Many gene products are implicated and have been studied individually in tumor neo-angiogenesis. One of the most investigated and "drugable" targets of the hypoxic response is Vascular Endothelial Growth Factor (VEGF), which is secreted by hypoxic tumor cells. In addition to increased vascularization, hypoxia initiates multiple cellular responses such as activation of proto-oncogenes (14,15), increase in glucose transport (14), induction of glycolytic enzymes (7,16), and induction of various apoptotic related genes (17). It is estimated that perhaps as much as 1% of the genome is HIF-1 hypoxia regulated (3). Tumor hypoxia tends to result in poorer prognosis at diagnosis (10) in several types of cancer because these hypoxic adaptations make the tumors more difficult to treat, and confer increased resistance to chemotherapy and radiotherapy (3,10).

3. REGULATION OF HIF-1

In cancers, HIF-1 is biomodulated and upregulated in addition to hypoxia by oncogene activation ligand–receptor interactions, and tumor suppressor gene mutations and deletions *(7)*. One of the most recently identified pathways modulating and influencing HIF-1 regulation is the Ras–ERK pathway *(18)*. Several laboratories have recently shown that Ras affects VEGF expression through HIF-1 *(3,16)*. This interaction is mediated through the tyrosine kinase signaling, Raf/MEK1/ERK, a pathway shared with Akt-1 *(19)*.

Another oncogene that has been shown to induce VEGF through HIF-1 signaling is the protein tyrosine kinase c-Src, and/or its downstream mediator phosphatidyl-inositol 3-kinase (PI3K) *(20–23)*. A rapid increase in the Src activity has been seen in both tumor and normal cells under hypoxia. More recently, signal and transducer of transcription 3 (STAT3) activation has also been shown to upregulate VEGF transcription through HIF-1 *(21,23,24)*. In addition, both HIF-1 and STAT3 bind the transcriptional coactivator CBP/p300 in the initiation of HIF regulated genes, suggesting that simultaneous occupancy of the VEGF promoter may occur and be part of a single large transcriptional complex *(25–28)*. HIF-1 physically associates with STAT3, CBP/p300, and Redox effector factor 1/apurinic/apyrimidinic endonuclease (Ref-1/Ape) *(24)*. Several investigators have demonstrated that optimal transcriptional control of the VEGF promoter requires binding of both HIF-1 and STAT3, and these factors are part of a large transcriptional complex coordinated in part by the coactivators CBP/p300 and Ref-1/APE *(24)*.

Not only is HIF-1 activated by oncogenes in normoxic conditions, but the loss of tumor suppressor gene function can upregulate HIF-1 as well. Mutations and allelic loss of the *PTEN* tumor suppressor gene in prostate cancer and other tumors can activate HIF-1α through increased downstream signaling from Akt-1 *(3,26)*. The loss of PTEN repression of the PI3-kinase pathway has sparked considerable interest in this pathway relative to HIF-1 expression and activation with regards to activation of early angiogenesis during early prostate tumorigenesis *(29)*. Studies have shown that PTEN-deficient prostate cancer cells display a higher HIF-1 activation response to hypoxia, suggesting that this may play a role in the biological aggressiveness of a tumor. Given the finding that PTEN loss in prostate cancer is involved in metastatic prostate cancer, a major emphasis is being placed on the role of PTEN$^{-/+}$ and PTEN$^{-/-}$ genotypes in prostate cancer and downstream activation of HIF dependent genes in the establishment of metastatic sites *(29)*.

Despite being a transcription factor rather than an enzyme, HIF-1α is potentially a "drugable" downstream target based on PTEN studies *(2,3,30)*. Inhibitors of mTOR function like rapamycin, CCI 779, and Rad 001 decrease HIF-1 protein levels in both normoxic and hypoxic cells by reducing protein synthesis *(2,7,29)*. These findings suggest in the right genetic context—where tumors are dependent on Tor function for survival—clinical trials now underway against prostate cancer using "rapalogues" would in part be evaluating the efficacy of the inhibition of HIF-1-activated downstream genes in tumors, and in effect be antiangiogenic.

4. HIF-1 AND TRANSLATIONAL RESEARCH IN PROSTATE CANCER

We originally observed HIF-1 was overexpressed in primary prostate cancers, compared with normal prostate epithelium *(1)*. HIF-1 overexpression was observed in prostate cancer bone metastases as well *(1)*. Hypoxic regions exist in human prostate carcinoma, and increasing levels of hypoxia are associated with higher clinical stages. In a clinical observation of high-grade prostate intraepithelial neoplasia (PIN) lesions, with PIN lesions considered the precursor for the majority of invasive prostate adenocarcinomas, we found an increase in HIF-1α relative to the respective normal epithelium, stromal cells, and benign prostatic hyperplasia (BPH) *(4)*. Upregulation of HIF-1α is associated with high-grade PIN, and thus HIF-1α may be a future biomarker for pre-malignant lesions of the prostate *(4,29)*. One possible explanation for these high levels of HIF-1 in prostate cancer is the possible amplification of the *HIF-1α* gene; however, this was found to be infrequent by Saramaki and colleagues, who showed an amplification of the HIF-1 locus only in the PC-3 prostate cancer cell line and not in clinical prostate tumors *(31)*. Fu et al. also suggest that high levels of HIF-1α expression are not due to an increase in stabilization resulting from a high frequency rate of mutations in human prostate cancers in the oxygen degradation domain (ODD) region of HIF-1α *(27)*. Data taken together at this time suggest that HIF-1 is being upregulated not as itself a target of amplification or mutations, but rather an important downstream "effector" of epigenetic or genetic alterations the early transformed prostate epithelial cell is programmed with.

HIF-1 overexpression in very early prostate tumorigenesis has been identified in transgenic mouse models of prostate cancer *(28)*. HIF-1 overexpression appears to be a very early event in prostate cancer pathogenesis in these systems. In transgenic mice expressing human Akt-1 in the ventral prostate, there is an mTOR-dependent survival signal that activates HIF-1 in early tumors. Transcriptome profiling showed that HIF-1/HRE targets, including genes encoding most glycolytic enzymes, constituted the dominant transcriptional response to Akt activation and mTOR inhibition *(28)*.

Androgens working through the androgen receptor (AR) have been discovered to modulate HIF-1 protein levels in human prostate cancer *(32–34)*.

We tested the hypothesis that HIF-1 overexpression may not just be epigenetic from intratumoral hypoxia but is affected by androgen receptor (AR) and androgens. We tested as well the hypothesis that the anti-angiogenic effects of anti-androgens in androgen-responsive prostate cancer (PCA) cells occurs from blocking HIF-1 transcriptional pathway. Mabjeesh and colleagues found that *in vitro* dihydrotestosterone (DHT) stimulates HIF-1 protein expression, HIF-1 transcriptional activity, and VEGF production in AR-positive LNCaP cells; conversely the anti-androgen flutamide reduced these effects on both HIF and downstream VEGF *(34)*. Androgen-induction of HIF-protein expression and function are regulated in part through an autocrine loop mechanism. Increased secretion of EGF is androgen-mediated in prostate cancer *in vitro*; EGF then acts through the EGF receptor the PI3K/Akt/Tor pathway in prostate cancer cells to increase HIF-1 levels *(34)*. Thus, the efficacy of anti-androgen therapy, a mainstay of prostate cancer endocrine therapy for 60 years, may be based in part on downregulation of VEGF by suppression of HIF-1 *(35)*.

Dihydrotestosterone (DHT) has also been shown to increase the stabilization of HIF-1α and EGF mRNAs through the AU-rich elements (AREs) within the 3' untranslated regions (UTRs) of these transcripts. DHT can act post-transcriptionally as well as post-translationally to modulate HIF-1 activity *(3)*. Recently, Boddy et al. showed that the expression of HIF-1 and HIF-2 were significantly correlated with AR expression in human tumors, supporting the previously cited *in vitro* data *(35)*. A significant area for inquiry is how androgen refractory prostate cancer develops resistance pathways from blockade of androgen receptor action. An additional area of focus will be testing new anti-androgen agents in clinical trials from Bristol Meyers Squibb, Cougar Biotechnology, and Medivation Biotechnology, for example, and their potency compared to flutamide and casodex in the clinic for repressing Dihydrotestosterone-Anderogen Receptor (DHT–AR) crosstalk that upregulates both HIF-1 and VEGF.

5. HIF AS THERAPEUTIC TARGET IN PROSTATE CANCER

Transcription factors are difficult targets for anti-cancer drug discovery and development, as they are not enzymes for the identification of active site inhibitors. Given the overexpression of HIF-1 is common in so many tumor types, the role of HIF-1 with tumor angiogenesis, and that HIF-1 is not overexpressed in normal cells, many approaches have been taken to identify agents that reduce HIF-1 protein levels *(3,33)*. High-throughput HIF-1 reporter gene assays have been used to start to screen diverse chemical and natural product libraries *(32)*.

Other groups have sought disruption of HIF-1-signaling upstream of it by attacking Ras-related proteins. Blum and colleagues showed that the

Ras-inhibitor trans-farnesylthiosalicylic acid (FTS) exhibits profound anti-oncogenic effects in U87 GBM cells. FTS inhibited active Ras and its signaling to phosphatidylinosital-3-kinase (PI3K) and Akt pathway; hence, HIF-1α was lowered and downstream HIF regulated genes were reduced in expression accordingly *(36)*.

Another "upstream approach" to inhibit HIF-1 downstream utilizes the anti-epidermal growth factor receptor monoclonal antibody cetuximab (C225; Erbitux), which is approved for the treatment of metastatic colorectal cancer. This monoclonal antibody inhibits tumor cell VEGF secretion *in vitro* and *in vivo* as a consequence of treatment. Luwor and colleagues showed that cetuximab reduces HIF-1 in epidermoid carcinoma cells under both normoxic and hypoxic conditions. This inhibition is through the Ras pathway and confirms that VEGF secretion can be modulated by signal transduction inhibition of HIF-1 protein translation *(37)*.

Novel HIF-1 functional inhibitors are also in medical oncology clinical trials of breast and prostate cancer. Of these, 2-methoxyestradiol (2ME2) is particularly interesting. 2ME2 is an estradiol metabolite with significant anti-proliferative and anti-angiogenic activity *independent* of estrogen receptor status. It is currently in clinical trials (Panzem) and inhibits tumor growth

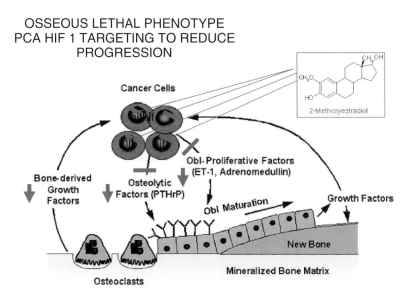

Fig. 3. *HIF-1 Inhibition as a Strategy to Reduce Bone Metastatic Progression.* HIF-1 antagonists such as 2-methoxyestradiol (2ME2), docetaxel, and antiandrogens may inhibit the secretion of key factors that are secreted by tumor cells that activate osteoclasts and osteoblasts involved in the progression of prostate cancer in the bone.

and angiogenesis at concentrations that efficiently disrupt tumor microtubules (MTs) in vivo. 2ME2 downregulates HIF-1 at the post-transcriptional level and inhibits HIF-1-induced transcriptional activation of VEGF expression *(31)*. Inhibition of HIF-1 occurs downstream of the 2ME2/tubulin interaction, as disruption of interphase MTs is required for HIF-1 downregulation *(46)*. These data establish 2ME2 as a small molecule inhibitor of HIF-1, and provide a mechanistic link between the disruption of the MT cytoskeleton with drugs like taxol and taxotere and inhibition of angiogenesis *(49)*. The demonstration by Giannakakou and colleagues at Emory that taxotere (docetaxel) treatment inhibits HIF-1 and VEGF creates an important hypothesis: is the modest but real increase in survival in hormone refractory prostate cancer patients treated with docetaxel due in part to HIF-1 inhibition at the sites of metastases? Guise and Simons and others are currently assessing inhibition of prostate cancer osseous metastases models by targeting HIF-1 directly (Fig. 3).

6. SUMMARY

HIF-1 has emerged as an important transcription factor in human cancers and specifically prostate cancer tumor biology. Mouse transgenic models have validated translational research from human tumor banks that HIF-1 is associated with the angiogenic switch in early tumorigenesis. In prostate cancer, upregulation is early, making HIF-1α a logical target for chemoprevention strategies in patients at higher genetic risk of breast and prostate cancer. Agents like 2ME2 as well other small molecules under study are centering on HIF as a target for new approaches to inhibiting angiogenesis in prostate cancer. HIF-1 and HIF-2 upregulation is impacted by multiple factors that are epigenetic and genetic. One very important question is whether HIF-1 upregulation in the $PTEN^{-/+}$ and $PTEN^{-/-}$ genotypes of clinical prostate cancer is so clinically significant that it actually helps the metastatic seeding of circulating clones in the bone and other tissues. These questions may also be ascertainable with the advent of detection of circulating tumor cells. Androgens upregulate HIF-1 in prostate cells through androgen-regulated autocrine secretion and then receptor tyrosine kinase receptor/PIP3K/AKT-1/mTor signaling. HIF is at the crossroads for important signals in lethal clone survival, and important inroads to slow progression or elimination of lethal clones are projected to come from new strategies that eliminate the ability of HIF to rescue cancer cells from death and the metabolic stresses they face.

ACKNOWLEDGMENTS

This work is supported by the Emory-Georgia Tech NCI Center for Cancer Nanotechnology Excellence. Graphics provided by Seraphim Graphics.

REFERENCES

1. Zhong H, De Marzo AM, Laughner E, Lim M, Hilton DA, Zagzag D, Buechler P, Isaacs WB, Semenza GL and Simons JW. Overexpression of hypoxia-inducible factor 1 alpha in common human cancers and their metastases. *Cancer Res* 59:5830–5835, 1999.
2. Zhong H, Chiles K, Feldser D, Laughner E, Hanrahan C, Georgescu MM, Simons JW and Semenza GL. Modulation of hypoxia-inducible factor 1 alpha expression by the epidermal growth factor/phosphatidylinositol 3-kinase/PTEN/AKT/FRAP pathway in human prostate cancer cells: implications for tumor angiogenesis and therapeutics. *Cancer Res* 60: 1541–1545, 2000.
3. Semenza GL. HIF-1 and tumor progression: pathophysiology and therapeutics. *Trends Mol Med* 8:S62–67, 2002.
4. Semenza GL. HIF-1, O(2), and the 3 PHDs: how animal cells signal hypoxia to the nucleus. *Cell* 107:1–3, 2001.
5. Zhang H, Semenza GL, Simons JW and De Marzo AM. Up-regulation of hypoxia-inducible factor 1 alpha is an early event in prostate carcinogenesis. *Cancer Detect Prev* 28:88–93, 2004.
6. Semenza GL, Jiang BH, Leung SW, Passantino R, Concordet JP, Maire P and Giallongo A. Hypoxia response elements in the aldolase A, enolase 1, and lactate dehydrogenase A gene promoters contain essential binding sites for hypoxia-inducible factor 1. *J Biol Chem* 271:32529–32537, 1996.
7. Semenza GL. Targeting HIF-1 for cancer therapy. *Nat Rev Cancer* 3:721–732, 2003.
8. Jaakkola P, Mole DR, Tian YM, Wilson MI, Gielbert J, Gaskell SJ, Kriegsheim A, Hebestreit HF, Mukherji M and Schofield CJ. Targeting of HIF-alpha to the von Hippel-Lindau ubiquitylation complex by O_2-regulated prolyl hydroxylation. *Science* 292:468–472, 2001.
9. Kondo K and Kaelin WG, Jr. The von Hippel-Lindau tumor suppressor gene. *Exp Cell Res* 264:117–125, 2001.
10. Vaupel P, Kelleher DK and Hockel M. Oxygen status of malignant tumors: pathogenesis of hypoxia and significance for tumor therapy. *Semin Oncol* 28:29–35, 2001.
11. Vaupel P. The role of hypoxia-induced factors in tumor progression. *Oncologist* 9 Suppl 5:10–17, 2004.
12. Chun YS, Lee KH, Choi E, Bae SY, Yeo EJ, Huang LE, Kim MS and Park JW. Phorbol ester stimulates the nonhypoxic induction of a novel hypoxia-inducible factor 1 alpha isoform: implications for tumor promotion. *Cancer Res* 63:8700–8707, 2003.
13. Gleadle JM and Ratcliffe PJ. Induction of hypoxia-inducible factor-1, erthropoietin, vascular endothelial growth factor, and glucose transporter-1 by hypoxia: evidence against a regulatory role for Src kinase. *Blood* 89:503–509, 1997.
14. Airley R, Loncaster J, Davidson S, Bromley M, Roberts S, Patterson A, Hunter R, Stratford I and West C. Glucose transporter glut-1 expression correlates with tumor hypoxia and predicts metastasis-free survival in advanced carcinoma of the cervix. *Clin Cancer Res* 7:928–934, 2001.
15. Price BD and Calderwood SK. Gadd45 and Gadd153 messenger RNA levels are increased during hypoxia and after exposure of cells to agents which elevate the levels of the glucose-regulated proteins. *Cancer Res* 52:3814–3817, 1992.
16. Lim JH, Lee ES, You HJ, Lee JW, Park JW and Chun YS. Ras-dependent induction of HIF-1 alpha785 via the Raf/MEK/ERK pathway: a novel mechanism of Ras-mediated tumor promotion. *Oncogene* 23:9427–9431, 2004.
17. Alvarez-Tejado M, Naranjo-Suarez S, Jimenez C, Carrera AC, Landazuri MO and del Peso L. Hypoxia induces the activation of the phosphatidylinositol 3-kinase/Akt cell survival pathway in PC12 cells: protective role in apoptosis. *J Biol Chem* 276:22368–22374, 2001.

18. Mukhopadhyay D, Tsiokas L, Zhou XM, Foster D, Brugge JS and Sukhatme VP. Hypoxic induction of human vascular endothelial growth factor expression through c-Src activation. *Nature* 375:577–581, 1995.

19. Zhong XS, Zheng JZ, Reed E and Jiang BH. SU5416 inhibited VEGF and HIF-1 alpha expression through the PI3K/AKT/p70S6K1 signaling pathway. *Biochem Biophys Res Commun* 324:471–480, 2004.

20. Niu G, Wright KL, Huang M, Song L, Haura E, Turkson J, Zhang S, Wang T, Sinibaldi D, Coppola D et al. Constitutive Stat3 activity up-regulates VEGF expression and tumor angiogenesis. *Oncogene* 21:2000–2008, 2002.

21. Wei D, Le X, Zheng L, Wang L, Frey JA, Gao AC, Peng Z, Huang S, Xiong HQ and Abbruzzese JL. Stat3 activation regulates the expression of vascular endothelial growth factor and human pancreatic cancer angiogenesis and metastasis. *Oncogene* 22:319–329, 2003.

22. Arany Z, Huang LE, Eckner R, Bhattacharya S, Jiang C, Goldberg MA, Bunn HF and Livingston DM. An essential role for p300/CBP in the cellular response to hypoxia. *Proc Natl Acad Sci USA* 93:12969–12973, 1996.

23. Schuringa JJ, Schepers H, Vellenga E and Kruijer W. Ser727-dependent transcriptional activation by association of p300 with STAT3 upon IL-6 stimulation. *FEBS Lett* 495:71–76, 2001.

24. Gray MJ, Zhang J, Ellis LM, Semenza GL, Evans DB, Watowich SS and Gallick GE. HIF-1 alpha, STAT3, CBP/p300 and Ref-1/APE are components of a transcriptional complex that regulates Src-dependent hypoxia-induced expression of VEGF in pancreatic and prostate carcinomas. *Oncogene* 24:3110–3120, 2005.

25. Jung JE, Lee HG, Cho IH, Chung DH, Yoon SH, Yang YM, Lee JW, Choi S, Park JW and Ye SK. STAT3 is a potential modulator of HIF-1-mediated VEGF expression in human renal carcinoma cells. *FASEB J* 174–178, 2005.

26. Zundel W, Schindler C, Haas-Kogan D, Koong A, Kaper F, Chen E, Gottschalk, Johnson RS, Jefferson AB et al. Loss of PTEN facilitates HIF-1-mediated gene expression. *Genes Dev* 14:391–396, 2000.

27. Fu XS, Choi E, Bubley GJ and Balk SP. Identification of hypoxia-inducible factor-1 alpha (HIF-1 alpha) polymorphism as a mutation in prostate cancer that prevents normoxia-induced degradation. *Prostate* 63:215–221, 2005.

28. Majumder PK, Febbo PG, Bikoff R, Berger R, Xue Q, McMahon LM, Manola J, Brugarolas J, McDonnell TJ and Golub TR. mTOR inhibition reverses Akt-dependent prostate intraepithelial neoplasia through regulation of apoptotic and HIF-1-dependent pathways. *Nat Med* 10:594–601, 2004.

29. Carroll VA and Ashcroft M. Targeting the molecular basis for tumour hypoxia. *Expert Rev Mol Med* 7:1–16, 2005.

30. Mabjeesh NJ, Escuin D, LaVallee TM, Pribluda VS, Swartz GM, Johnson MS, Willard MT, Zhong H, Simons JW and Giannakakou P. 2ME2 inhibits tumor growth and angiogenesis by disrupting microtubules and dysregulating HIF. *Cancer Cell* 3:363–375, 2003b.

31. Saramaki OR, Savinainen KJ, Nupponen NN, Bratt O and Visakorpi T. Amplification of hypoxia-inducible factor 1 alpha gene in prostate cancer. *Cancer Genet Cytogenet* 128: 31–34, 2001.

32. Tan C, de Noronha RG, Roecker AJ, Pyrzynska B, Khwaja F, Zhang Z, Zhang H, Teng Q, Nicholson AC and Giannakakou P. Identification of a novel small-molecule inhibitor of the hypoxia-inducible factor 1 pathway. *Cancer Res* 65:605–612, 2005.

33. Sheflin LG, Zou AP and Spaulding SW. Androgens regulate the binding of endogenous HuR to the AU-rich 3'UTRs of HIF-1 alpha and EGF mRNA. *Biochem Biophys Res Commun* 322:644–651, 2004.

34. Boddy JL, Fox SB, Han C, Campo L, Turley H, Kanga S, Malone PR and Harris AL. The androgen receptor is significantly associated with vascular endothelial growth factor and hypoxia sensing via hypoxia-inducible factors HIF-1a, HIF-2a, and the prolyl hydroxylases in human prostate cancer. *Clin Cancer Res* 11:7658–7663, 2005.

35. Mabjeesh NJ, Willard MT, Frederickson CE, Zhong H and Simons JW. Androgens stimulate hypoxia-inducible factor 1 activation via autocrine loop of tyrosine kinase receptor/phosphatidylinositol 3´-kinase/protein kinase B in prostate cancer cells. *Clin Cancer Res* 9(7):2416–2425, 2003 Jul.

36. Blum R, Jacob-Hirsch J, Amariglio N, Rechavi G and Kloog Y. Ras inhibition in glioblastoma down-regulates hypoxia-inducible factor-1 alpha, causing glycolysis shutdown and cell death. *Cancer Res* 65:999–1006, 2005.

37. Luwor RB, Lu Y, Li X, Mendelsohn J and Fan Z. The antiepidermal growth factor receptor monoclonal antibody cetuximab/C225 reduces hypoxia-inducible factor-1 alpha, leading to transcriptional inhibition of vascular endothelial growth factor expression. *Oncogene* 24:4433–4441, 2005.

38. Escuin D, Kline ER and Giannakakou P. Both microtubule-stabilizing and microtubule-destabilizing drugs inhibit hypoxia-inducible factor-1 alpha accumulation and activity by disrupting microtubule function. *Cancer Res* 65:9021–9028, 2005.

11 Signal Transduction by the Ras–MAP Kinase Pathway in Prostate Cancer Progression

Daniel Gioeli, PhD, Sarah Kraus, PhD, and Michael J. Weber, PhD

1. INTRODUCTION

Prostate cancer (PC) initially appears as an androgen-dependent disease but progresses to one that is refractory to hormone ablation therapy. Substantial evidence suggests that autocrine and paracrine growth factor loops fuel PC progression to hormone independence: *(1)* overexpression of numerous growth factors and their cognate receptors correlates with PC progression; *(2)* ectopic overexpression of some receptors can drive hormone-independent growth; and *(3)* neuroendocrine (NE)-like cells that produce growth-regulatory

From: *Current Clinical Oncology: Prostate Cancer:*
Signaling Networks, Genetics, and New Treatment Strategies
Edited by: R. G. Pestell and M. T. Nevalainen © Humana Press, Totowa, NJ

neuropeptides are associated with aggressive, advanced PCs. The Ras–mitogen-activated protein (MAP) kinase (MAPK) pathway is a signaling pathway that is utilized by virtually all the receptor systems known to be upregulated in PC, and it is likely that this signaling pathway plays a role in the progression of at least a subset of advanced cancers. However, the mechanisms by which this pathway modulates PC growth, survival, and hormone independence are unclear. Importantly, androgen-independent PC almost always retains expression of the androgen receptor (AR), which often is mutated and overexpressed, and the AR continues to play an important role in cancer cell growth at this stage of disease. Therefore, a key to understanding hormone-independent PC is to determine the mechanism(s) by which the AR can function even in the absence of physiologic levels of circulating androgen. In this review, we examine the organization and regulation of the Ras–MAPK pathway, the role this pathway might play in progression to hormone independence, and the ways that this pathway might be exploited therapeutically in the treatment of PC.

2. PROGRESSION TO ANDROGEN INDEPENDENCE

When PC initially presents in the clinic, the tumor is dependent on androgen for growth *(1)*, and androgen ablation is commonly used as front-line therapy for metastatic disease. However, therapeutic success is often temporary, and the disease almost invariably recurs. Such cancers are called "androgen independent" or "hormone refractory" because they no longer respond to hormone ablation. However, in general the tumors still depend on the AR. Thus, there is considerable interest in determining the mechanism(s) by which the AR becomes activated in advanced disease.

It is worth noting that "androgen independent" or "hormone refractory" PCs often are not fully independent of androgen, but have become sensitive to very low levels of androgen *(2–5)*. Clinically, these cancers may appear "androgen independent." However, because hormone ablation therapies do not eliminate all traces of androgen at the molecular level, these cancers still depend on androgen and on the AR. This is a source of semantic confusion, because the term "androgen independent" is used in the literature, regardless of whether the cells are completely or only partially androgen independent. In this review, the term "androgen independent" is used, with the understanding that the cells may actually be hypersensitive to androgen rather than being completely independent of the steroid.

Even in advanced, androgen-independent disease, the AR is retained, and substantial evidence supports its critical role in the continued growth and survival of PC cells *(6–9)*. Work by Tindall and colleagues used an AR hammerhead ribozyme to disrupt AR expression in LNCaP PC cells; knockdown of AR expression resulted in a dramatic inhibition of cell growth

(6). Work from Culig and colleagues used antisense oligonucleotides to downregulate AR expression both in tissue culture cells and in PC xenografts *(8,9)*. Collectively, these studies suggest that although advanced PC may be functionally independent of androgen, it is not independent of the AR. The AR-dependent regulatory mechanisms are subverted, not bypassed.

How is the AR activated in advanced disease? Current evidence points to at least three mechanisms that are likely synergistic: (1) increases in expression levels of the AR and its co-activators; (2) mutations in the AR that broaden its selectivity for agonists; (3) modulation of the activity state of the AR and its co-regulators via post-translational modifications that are triggered by signal transduction pathways.

Late-stage androgen-independent PCs often overexpress the AR despite the near-absence of circulating androgens *(10,11)*. Overexpression is often associated with gene amplification (30%) *(11,12)*. An increase in AR expression was the only consistent change detected in a genome-wide survey comparing androgen-dependent and androgen-independent PCs *(3)*. Chen et al. used microarray gene profiling of isogenic PC xenografts and found that a two to fivefold increase in AR mRNA was the only gene expression change consistently associated with androgen-independent disease. This increase in expression hypersensitized the AR to low levels of ligand. The authors also showed that the increase in AR levels was both necessary and sufficient to drive PC progression to androgen independence. Surprisingly, the increase in AR levels converted AR antagonists to agonists. Anti-androgens such as Casodex function by inducing an AR conformation that recruits co-repressors rather than co-activators. The conversion of antagonists to agonists in cells overexpressing the AR was associated with changes in the recruitment of co-activators and co-repressors at AR target promoters *(3)*. Thus, AR expression level can be a determinant both of "androgen independence" and of resistance to anti-androgens. It is not clear whether this conversion is solely and directly a consequence of increased AR expression, or whether concordant changes in other regulatory networks, such as signaling pathways, have occurred.

The importance of AR expression level is supported by other work examining AR protein stability in various PC cell lines *(13)*. The AR protein was two to four times more stable in recurrent PC cell lines compared with androgen-dependent lines. A high level of AR expression and stability was associated with increased sensitivity to hormone. Furthermore, overexpression of transcriptional co-activators often accompanies PC progression, and this facilitates the activity of the AR *(2–15)*.

Frequently, the AR is mutated in advanced PCs (10–40%), which often results in an AR that can be activated by non-androgen ligands *(7,13,16–18)*. In these cases, the AR functions even in the face of androgen ablation by utilizing alternative ligands. Cells expressing such mutated receptors have a selective growth and survival advantage under conditions of androgen ablation.

In addition to changes in AR expression and mutations as described above, the AR can be activated functionally in response to signal transduction from growth factors *(4,19,20)*. The mechanisms driving cancer progression that are based on growth factor signaling are not mutually exclusive with the mechanisms based on AR modification, and indeed are likely to be mutually reinforcing. A therapy that targets any aspect of the AR compensatory mechanisms could effectively treat androgen-dependent and androgen-independent PC. The rest of this chapter focuses on the mechanisms by which growth factor signaling alters AR function, because this is a mechanism that is likely to play a role in at least half of advanced PCs, is subject to therapeutic intervention, and provides an opportunity to understand how growth factor and Ras signaling can integrate with the functions of nuclear receptors.

3. GROWTH FACTORS IN PROSTATE CANCER PROGRESSION

The prostate gland requires both androgens and polypeptide growth factors for proliferation, differentiation, and maintenance of function *(21)*. Androgen action on stromal cells leads to the secretion of peptide growth factors— andromedins *(22)*. Andromedins diffuse to the epithelial cell compartment and regulate proliferation and survival of basal and secretory luminal epithelial cells. Androgen action on epithelial cells stimulates the transcription of genes encoding prostate-specific differentiation factors. In fact, androgen action in epithelial cells suppresses the growth stimulatory effects of andromedins and promotes the differentiated phenotype. However, during prostate tumorigenesis this system is dysregulated, allowing for growth stimulatory interactions to occur between androgens and growth factors—interactions that are the opposite of those seen during prostate development and maintenance. Stimulation of PC cells with growth factors can decrease the requirement for androgen, and the expression of these growth factors and receptors increases as PC progresses *(19, 23–25)*. Thus, the AR plays a paradoxical role in the prostate, being essential for normal differentiation and maintenance, but subsequently essential for driving malignant behavior.

Increases in autocrine and paracrine growth factor loops are among the most commonly reported changes correlated with PC progression from a localized and androgen-dependent to a disseminated and androgen-independent disease. In advanced PC, epidermal growth factor (EGF), transforming growth factor-α (TGF-α), keratinocyte growth factor (KGF), basic fibroblast growth factor (bFGF), and insulin-like growth factor (IGF-I) as well as their cognate receptors are all reported to be overexpressed *(17,24,26)*. Additionally, there appears to be a paracrine relationship between the EGF receptor (EGFR) and TGF-α in primary androgen-dependent prostatic tumors; the tumor cells express the EGFR and the surrounding stromal cells express TGF-α *(25)*. However, in androgen-independent metastases, the prostate tumor cells co-express the EGFR and TGF-α, consistent with autocrine regulation. A similar observation has been made with bFGF expression in PC cell lines. The androgen-dependent PC cell line, LNCaP, does not produce bFGF and requires co-inoculation of bone or prostatic fibroblasts which express large amounts of bFGF, or matrigel for effective tumor formation in athymic mice *(27–29)*. By contrast, the androgen-independent PC cell lines, DU145 and PC-3, produce bFGF and readily form tumors in athymic mice. These PC cell lines also require autocrine production of IGF-I for growth *(30)*. That transgenic mice expressing KGF under the control of the mouse mammary tumor virus (MMTV) promoter develop prostatic hyperplasia further suggests that chronic exposure to growth factor can dysregulate prostate cell growth *(31)*.

Previous studies have demonstrated that polypeptide growth factor signal transduction pathways can stimulate AR activation, suggesting that the increase in growth factor and receptor expression could be causal in PC progression to androgen independence. Growth factor stimulation has been reported to render androgen-response element (ARE)-driven promoters hypersensitive to, or "independent" of, androgen *(17,19,20,24,26,32–35)*. Culig et al. investigated the effects of growth factors on stimulation of AR-mediated transcription *(26)*. DU145 cells, a PC cell line that expresses neither AR nor prostate-specific antigen (PSA), were co-transfected with an expression vector encoding the AR and chloramphenicol acetyltransferases (CAT) reporter constructs driven by either a synthetic ARE or the PSA promoter. IGF-I was able to stimulate reporter gene expression to levels comparable with those induced by the synthetic androgen R1881. This observation was independent of the promoter used. EGF and KGF were also able to induce reporter gene expression, but only in experiments using the ARE-driven reporter construct. Growth factor-induced reporter gene expression was dependent on co-transfection of the AR expression construct and was blocked by the AR antagonist Casodex. In this same study, activation of endogenous AR by IGF-I in LNCaP cells was demonstrated using PSA production as a marker. Again, Casodex blocked the effect of IGF-I on PSA production.

Olli Janne and colleagues did similar studies using CV-1 and HeLa cells *(4)*. Activation of an ARE-driven CAT reporter construct was induced by EGF in the absence of androgen when AR was co-expressed in these cell lines. However, when the MMTV promoter was used to drive expression of a CAT reporter, EGF (and IGF-I) stimulation was dependent on the presence of androgen. Unlike Culig et al., these investigators failed to detect ligand-independent activation of the AR. However, they did see stimulation of androgen-dependent activity in response to growth factors. Thus, the ability and degree of ligand-independent activation of the AR appears dependent on the promoter and cell type. Collectively, these experiments suggest that growth factor signaling can regulate androgen responsive genes by a mechanism that is AR dependent.

Consistent with this, forced overexpression of HER2/neu in androgen-dependent PC cells could drive androgen-independent growth (14,36). After the initial observation that androgen-independent sublines of LAPC4 cells expressed elevated levels of HER2/neu, Craft et al. generated LAPC4 cell lines overexpressing HER2/neu (37). These cell lines displayed androgen-independent growth and activated the AR pathway in the absence of ligand. Importantly, the HER2/neu expressing cells synergized with low levels of androgen to activate PSA transcription and growth. Yeh et al. overexpressed HER2/neu in LNCaP cells and demonstrated that activation of the pathway induced AR transcription and androgen-independent growth (14). Additionally, activation of HER2/neu and HER3 by heregulins increased AR transactivation and PC cell growth in vivo (38).

Studies have also shown that the small molecule dual EGFR/HER2 inhibitor, PKI-166, can inhibit the growth of PC xenografts (39). Additionally, use of a monoclonal antibody, 2C4, that sterically hinders HER2 heterodimerization inhibited in vitro and in vivo PC cell growth (40). Inhibition of HER2/neu with the kinase inhibitor, GW572016, or an inhibiting antibody impaired AR transcriptional activity and androgen-dependent cell growth (41). However, examination of clinical specimens has provided conflicting results—depending on the study, either no effect or an increase in HER2/neu levels was reported (42–45).

Small molecule inhibitors and siRNA were used by Sawyers and colleagues to study the relative roles of ErbB receptor tyrosine kinases in PC progression (46). PKI-166 inhibited AR transcriptional activity, protein stability, DNA binding, and phosphorylation on Serine 81. Use of EGFR selective small molecule inhibitors, EGFR negative cells, and siRNA demonstrated that the PKI-166-mediated anti-androgen effects were due to inhibition of HER2, not EGFR. This work demonstrates that signals emanating from HER2/ErbB3 heterodimers regulate AR activity in LNCaP and LAPC4 cell lines.

Although most emphasis has been placed on tyrosine kinase receptors, there is evidence that G-protein-coupled receptors (GPCRs) can also play an important role in modulating AR activity. These receptors often regulate intracellular levels of cAMP and hence of protein kinase A (PKA). Nazareth and Weigel reported that the AR could be activated by a PKA activator in the absence of androgen (33). This activation can be blocked by a PKA inhibitor peptide and the AR antagonists Casodex and flutamide, indicating that the activation effect was due to PKA and dependent on AR. Furthermore, Sadar found that treatment of LNCaP cells with PKA activators resulted in a dose- and time-dependent increase in PSA mRNA levels that was blocked by Casodex *(47)*. NE-like cells, which often occur in aggressive forms of PC, secrete neuropeptides that can increase intracellular cAMP and hence activate PKA *(48)*. As described below, changes in cAMP/PKA signaling can synergize with Ras signaling in the context of PC.

Collectively, the above-mentioned studies suggest that growth factor receptor signals can activate or sensitize the AR to reduced levels of ligand. The hypothesis is that the activation/sensitization is mediated by tyrosine kinase receptors and their downstream signaling effectors through regulated changes in phosphorylation of the AR or of AR-associated proteins. The diversity of these changes in autocrine and paracrine signaling predicts that, at least in the context of PC, attempts to utilize a single receptor/ligand pair as a therapeutic target will not generally be effective. To identify optimal targets for therapy, it will be necessary to identify the downstream signaling intermediates that are shared by these diverse receptors and ligands.

4. RAS SIGNALING

4.1. The Ras Family

The large Ras superfamily contains over 150 members involved in diverse cellular activities including growth, apoptosis, cell shape, and intracellular transport *(49)*. They are small (21 kDa) GTP-binding proteins that are activated when bound to GTP and inactive when bound to GDP. These states are regulated by the balance between the intrinsic GTPase activity of the proteins, their interactions with inactivating proteins that accelerate their GTPase activity [GTPase-activating proteins (GAPs)], and with activating proteins that regulate the exchange of GDP for GTP [GTP exchange factors (GEFs)]. Thus, they can function as both molecular switches and timers. The GAPs and GEFs provide the major mechanisms for Ras regulation and for linking Ras proteins to other signaling systems. For example, there are GEFs (such as SOS) that activate Ras proteins in response to tyrosine kinase receptors, such as members of the HER family; and there are GEFs (such as EPACs *(50,51)*) that can activate Ras proteins in response to changes in cAMP.

The Ras superfamily contains four founding members that constitute the Ras subfamily—H-Ras, K-Ras4A, KRas4B, and N-Ras. Genetic and biochemical evidences indicate that these isoforms have different intracellular functions and locations, but the "division of labor" between them is not fully explored. The founding members of the Ras subfamily, H-Ras and K-Ras, were discovered as oncogenes and most of the related proteins also have oncogenic activity when overexpressed in an activated form in the appropriate cell background. Most of our knowledge about the biochemistry of Ras signaling is based on analysis of H-Ras; however, K-Ras is the isoform that is most frequently mutated in human cancers *(52,53)*. We detect K-Ras and N-Ras in PC cell lines, but very little H-Ras (unpublished observation).

Virtually all of the growth factor receptors upregulated in PC activate Ras or Ras-related proteins for a portion of their signal transduction activity. Ras mutations are infrequent in PC *(54)*, but this is consistent with our hypothesis that wild-type Ras is chronically activated by autocrine and paracrine growth factor stimulation in PC, thus creating an "ecosystem" where there is little selective advantage for growth or survival of cells with mutationally activated Ras. Because Ras signaling represents a convergence point for numerous, diverse extracellular signals, Ras and its effectors may be appropriate targets for therapeutic intervention.

4.2. Ras Effectors

Ras is a multi-effector signaling molecule that has been shown to engage multiple signaling pathways (Fig. 1) *(55)* Ras effectors are defined as proteins that: (1) preferentially bind to the GTP-bound form of Ras; (2) have activity modulated by Ras; and (3) transduce the biological activity of Ras. Repasky et al. *(55)* list 25 distinct effectors, regulating 10 different pathways. It is

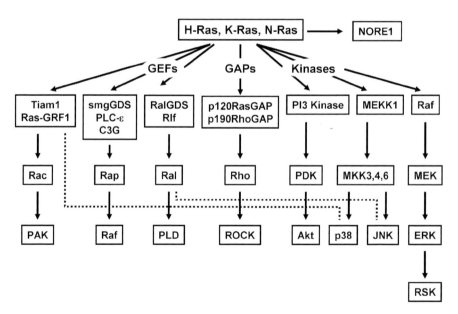

Fig. 1. Ras family members can be regulated by multiple activators [GTP exchange factors (GEFs)] and deactivators [GTPase-activating proteins (GAPs)], many of which are utilized by several members of the Ras superfamily. In addition, Ras proteins utilize multiple direct effectors. Some of these effectors also interact with other members of the Ras family. Collectively, these overlapping and redundant interactions create a signaling network in which information moves laterally as well as vertically.

clear that the regulatory effects of Ras activation vary with cellular context, depending critically on the expression of various effector molecules.

The best-characterized Ras effectors are the Raf serine/threonine kinases, A-Raf, B-Raf, and Raf-1, which regulate the MAPK kinase (MEK)1/2 and extracellular regulated kinase (ERK)1/2 kinase signaling cascade. Most studies have been performed with Raf-1, which is the founding member of the family. However, the identification of activating B-Raf mutations in melanoma and other cancers is stimulating heightened interest in this isoform. B-Raf was originally viewed as a "neural" form of Raf; however, we find its expression in PC cell lines to be comparable with neural cells *(56)*.

Another well-characterized Ras effector is PI3 kinase (PI3K), which has a role in both cell proliferation and survival signaling. The main activity of PI3K is to convert the phosphoinositide PIP_2 to PIP_3, which in turn facilitates Akt activation. There is a well-established role of PI3K signaling in PC progression *(57,58)* and the loss of PTEN expression in late-stage PCs underscores the role of PI3K signaling in the disease *(59–61)* These issues are discussed in detail in the accompanying chapter by Tindall and colleagues.

Recently, members of the RASSF gene family that potentially act as tumor suppressors have been identified as candidate Ras effectors *(62–64)*. Loss of expression of NORE1 and RASSF1, members of the RASSF gene family, has been observed in a variety of cancers *(65–67)*. The interaction of Ras with NORE1 has been shown to regulate apoptosis *(68)*. Thus, it seems very likely that the ability of Ras to trigger either growth or apoptosis depends on the balance of interactions between pro-growth, pro-survival, and pro-death effectors *(57,69)*.

The signaling activity of Ras GTPases occur not only through engagement of direct effectors but also by the recruitment of other GTPases, especially other members of the Ras sub-family (e.g., Rap) and members of the Rho subfamily (e.g., RhoA, Rac1, cdc42—which control cell shape and migration). This "hierarchical networking" between different Ras isoforms is controlled in part by interactions with GEFs, GAPs, and downstream effectors *(70–75)* For instance, RalGEFs are important in Ras-mediated transformation. RalGEFs, such as RalGDS, link Ras signaling to the activation of the small GTPases RalA and RalB. In human cells, the Ras effector loop mutant that preferentially activates RalGDS was able to transform cells rendered transformation sensitive *(76)* and knockout of RalGDS inhibits tumorigenesis *(77)*.

5. MAP KINASE SIGNALING

5.1. Architecture of the MAP Kinase Pathways

There are several MAPK pathways in mammalian cells. The best characterized are the original MAPK or ERK pathway, and the so-called "stress kinase" pathways that lead to activation of ERK homologs of the p38 and

Jun Kinase (JNK) families. Most information related to PC concerns the ERK pathway. As with all MAPK pathways, the ERK pathway has an evolutionarily-conserved architecture consisting of a three-kinase cascade: MAPK/ERK, which is activated by phosphorylation on threonine and tyrosine by a dual-specificity MAPK kinase (MEK), which in turn is activated by phosphorylation by a MEK kinase (Raf). Raf is in turn activated by a complex mechanism that can involve one or more Ras-dependent phosphorylations.

There are several isoforms for each member of the MAPK pathway. There are two ERKs, ERK1 and ERK2, and two MEKs, MEK1 and MEK2, all of which are widely expressed in tissues. Like Ras itself, MAPK functions as an information-neutral "switch" that modulates diverse biological outcomes dependent on context. For example, it can play a role in regulating cell growth, morphogenesis, migration, and/or gene expression, and its effects can be stimulatory or inhibitory. Little is known about how ERK1 and ERK2 differ in regulation and function. However, mice with an ERK1 knockout are viable and display a mild defect in T-cell development, whereas ERK2 knockout mice are embryonic lethal *(78,79)*.

There are three Raf isoforms that differ in expression, regulation, and functions: Raf-1, A-Raf, and B-Raf. The Raf-1 isoform has been most extensively studied and can act both as an activator of MAPK signaling and as an inhibitor of apoptosis, at least in part by its regulation of the kinase MST2 *(80)* and by interaction directly with mitochondria and the machinery of apoptosis *(81,82)*. It is unproven whether B-Raf has comparable anti-apoptotic activity, but it clearly is oncogenic: activating mutations in B-Raf occur in 60% of melanomas *(83)* and this renders the cells highly sensitive to inhibitors of Raf and MEK, such as Sorafenib and PD325901 *(84,85)*. B-Raf is highly expressed in prostate cell lines; however, activating B-Raf mutations have not been reported in PC as of this writing.

5.2. Integration of Signaling Pathways

Most information about signaling pathways is derived from studies of acute stimulation with a single agonist. However, signaling *in vivo* occurs in a context of multiple inputs from diverse agonists and other cells. Understanding the ways in which signaling pathways function within signaling networks is very incomplete, but will be important for rational selection of drugs that target signaling molecules.

A hallmark of malignant transformation of epithelial and fibroblastic cells is the ability to grow in suspension, without a matrix—"anchorage independent growth." Recent research on integrin signaling has shed light on important aspects of this process. Anchorage to matrix molecules such as fibronectin and collagen engages integrins, which are the cognate receptors for these matrix molecules. Engagement of integrins activates a signaling pathway that includes the kinases Src, Focal Adhesion Kinase (FAK), and p21 Activated kinase (PAK). PAK1 phosphorylates Raf-1 and MEK1 and this enhances the formation of signaling complexes of Raf-1, MEK1, and ERK (Fig. 2) *(86–88)* B-Raf and MEK2 lack the homologous phosphorylation sites. Thus, it is likely that Raf-1 and MEK1 serve important and specific roles in integrating cellular responses to growth factors, adhesion, and cell–cell interactions. Because independence from integrin engagement is a fundamental alteration in many solid cancers (e.g., anchorage independence), the specific roles of Raf-1 and MEK1 in fostering uncontrolled growth and metastasis are an area ripe for investigation.

The fact that the prostate displays high B-Raf expression levels has important consequences for the regulation of MAPK pathway by PI3K and PKA pathways (Fig. 3). Raf-1 is subject to negative regulation by Akt and by PKA *(89–91)* In cells that depend primarily on Raf-1, elevation of PI3K/Akt signaling or of cAMP/PKA signaling will inhibit MAPK signaling. However, B-Raf does not have analogous PKA phosphorylation sites, and Akt phosphorylation has been reported to have no effect on B-Raf *(92)*. Therefore, in PC cells that have ample expression of B-Raf, MAPK signaling can be activated even when Akt and

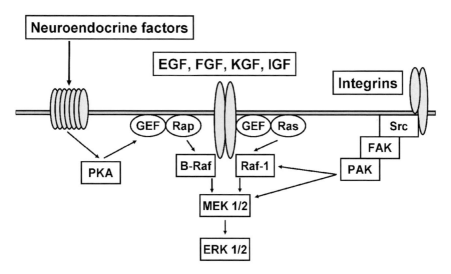

Fig. 2. Neuroendocrine-like cells produce neuropeptides and growth factors that can interact with G-coupled receptors in prostate epithelial and cancer cells, leading to activation of protein kinase A (PKA). This can turn on the mitogen-activated protein kinase (MAPK) pathway via GTP exchange factors (GEFs) that activate Raps, or can sensitize the pathway to classic growth factors such as epidermal growth factor (EGF). In addition, adhesion signaling via integrins monitors cell–cell and cell–matrix interactions controlling the activation levels of the FAK, Src, and PAK protein kinases. PAK phosphorylates both Raf-1 and MEK1, facilitating the assembly of a functional MAPK signaling complex.

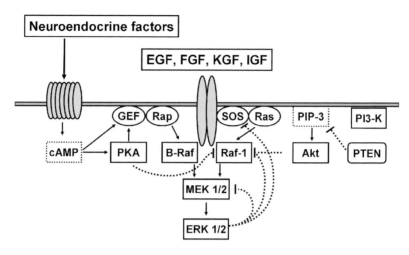

Fig. 3. Physiologic signaling is negatively regulated in amplitude and in duration by feedback loops. These limit Raf-1 activity by protein kinase A (PKA) and PI3 kinase signaling by PTEN. Active ERK phosphorylates SOS, Raf-1, and MEK. In addition, ERK regulates the expression of mitogen-activated protein (MAP) kinase (MAPK) phosphatases (not shown). In oncogenic signaling, these negative controls are defective.

PKA are also activated. There are two important physiologic circumstances in which this happens: (1) PC often displays deletion of the lipid phosphatase PTEN, which results in persistent activation of the PI3K/Akt pathway; (2) many advanced and aggressive prostate tumors contain NE-like cells, which produce neuropeptides that elevate cAMP/PKA signaling.

Although PKA signaling inhibits Raf-1-dependent MAPK signaling, it actually enhances MAPK signaling in the B-Raf-rich environment of PC. B-Raf can be activated not only by Ras but by the close homolog Rap-1. Rap-1 can be activated by GEFs (EPACs) that in turn are responsive to cAMP and Ca++ *(50,51)*. The synergistic interaction of these pathways has been demonstrated in LNCaP cells, where PKA agonists were shown to enhance the sensitivity of MAPK activation by EGF *(56)*. Rap1 was shown to be necessary for this synergism, since it could be blocked by a dominant-negative Rap1 mutant (N17).

Interestingly, MAPK is able to phosphorylate and inhibit the activity of a phosphodiesterase that inactivates cAMP, the long form of PDE4, and activates the short form of PDE4 *(93)*. This provides a potential mechanism for either negative or positive feedback loops between MAPK and PKA signaling. Overall, the combination of PKA-elevating NE cells with MAPK-activating growth factors potentially can enhance MAPK activation and the aggressive phenotypes that depend on the MAPK pathway.

5.3. Scaffold Proteins in Ras MAP Kinase Signaling

Scaffold proteins are the structural elements of signaling pathways. They may or may not have enzymatic activity, but their scaffolding role is to assemble multiple elements of a signaling pathway to ensure its appropriate specificity and intracellular localization. Several proteins have been reported as potential MAPK scaffolds based on their ability to bind multiple kinases *(94)*, including kinase suppressor of Ras (KSR), MEK-partner 1 (MP1), MAPK organizer 1 (MORG1), and β-arrestins for the ERK cascade, beta-arrestin 2 and JNK-interacting proteins (JIP 1/2/3) for the JNK pathway, and Sprouty, which appears to serve as a negative feedback scaffold. These proteins likely contribute to the physiological regulation of MAPK modules in mammals and are a potential and relatively unexplored mechanism for deregulation of signaling in cancer.

5.3.1. KINASE SUPPRESSOR OF RAS (KSR)

KSR was identified in a genetic screen as a positive regulator of Ras/MAPK signaling *(95)*. Members of the KSR family display structural similarity to proteins of the Raf family but apparently lack kinase activity. KSR was found to localize MEK with activated Raf-1 at the plasma membrane and provide a docking platform for ERK, thereby facilitating the sequential phosphorylation events required for ERK activation *(96)*. Other proteins that have been reported to interact with KSR include Hsp70, Hsp90, p50^{cdc37}, G-protein beta-gamma, and 14-3-3 *(94)*. Nguyen and co-workers *(97)* demonstrated in a Ras-dependent tumor model that loss of KSR expression could attenuate tumor growth. This is consistent with the role of KSR as a positive regulator of Ras MAPK signaling and is in agreement with the original genetic screens in which loss-of-function KSR mutants were able to suppress the effects of constitutively active transgenes *(98)*.

5.3.2. β-ARRESTINS

β-Arrestins function as GPCR transducers and MAPK scaffolds *(99)* in addition to their well-known role in the desensitization and termination of GPCR signaling. Binding of β-arrestin not only prevents the receptor from activating its G-protein but can also target it for internalization via clathrin-coated vesicles and can mediate signaling to proteins regulating endocytosis. β-Arrestins may also recruit other molecules to the receptor. Thus, β-arrestin 1 was shown to recruit c-Src to the β-adrenergic receptor, and β-arrestin 2 can recruit a JNK-signaling module that consists of ASK1, MKK4, and JNK3 *(100)*. β-Arrestin 2 contains a MAPK docking site that conforms to the D-domain consensus sequence and selectively binds JNK3 *(101)*. Recently, it was shown that β-arrestin mediates the activation of ERK by the GnRH receptor *(102)* and may serve as a scaffold for the ERK cascade. GnRH receptors are expressed in the pituitary and in several extra-pituitary sites, and these may mediate direct inhibition of proliferation and apoptosis of hormone-dependent tumors, mainly PC *(103)*.

5.3.3. Receptor for Activated C-Kinase 1 (RACK-1)

Receptor for activated c-kinase 1 (RACK1) is a member of the WD40 family of proteins characterized by highly conserved internal WD40 repeats (Trp-Asp). RACK1 was initially identified as an anchoring protein for activated PKC, with PKCβ II seemingly being the preferred binding partner *(104)*. Later reports showed that RACK1 is able to interact with several signaling molecules including c-Src, integrin β-subunit, phosphodiesterase PDE4D5, STAT1, IGF-I receptor, and others *(105)*. Therefore, RACK1 has been suggested to play a central role as a scaffold protein in multiple signaling pathways.

RACK1 was identified as an AR-interacting protein by using yeast two-hybrid screening and minimal domain mapping of RACK1–AR interactions *(106)*. Results from this study showed that RACK1 facilitates ligand-independent AR nuclear translocation upon activation of PKC by indolactam V. Prolonged exposure to indolactam V inhibited mRNA PSA expression in LNCaP cells and reduced PKC-induced AR recruitment to the AR-responsive PSA promoter. Thus, RACK1 may function as a scaffold for the association and modification of AR by PKC, enabling translocation of AR to the nucleus but rendering AR unable to activate transcription of its target genes.

Recent findings from our laboratory have identified RACK1, along with another WD40 protein, MORG1 *(107)* as a binding partner of the ERK pathway. Similar to MORG1, RACK1 specifically associates with several components of the ERK module, including Raf-1, MEK, and ERK, and enhances their functional interaction in cells (Vomastek T., unpublished observation). RACK1 also serves as an adaptor for PKC-mediated JNK activation and was identified as a novel ATF2-interacting protein in a yeast two-hybrid screen by using ATF2 as bait *(108)*. Inhibition of RACK1 expression by siRNA attenuates JNK activation, sensitizes melanoma cells to UV-induced apoptosis, and reduces their tumorigenicity in nude mice. These data support a role for RACK1 in the activation of JNK by physiological stimuli that activate PKC.

In PC, studies from our laboratory have provided evidence that RACK1 modulates the tyrosine phosphorylation of AR and its interaction with c-Src. Moreover, downregulation of RACK1 by siRNA inhibits growth and stimulates PSA gene expression (Kraus et al. submitted for publication). Our results suggest that RACK1 mediates the crosstalk of AR with c-Src, and facilitates the tyrosine phosphorylation of AR, which may be associated with its transcriptional activation. Thus, RACK1 may represent a critical link between different classes of signal-transducing molecules, such as the AR, c-Src, and possibly other signaling components including the MAPK pathway.

5.3.4. CAVEOLIN-1

Caveolin-1, a 21- to 24-kDa integral membrane protein, is a major component of caveolae membrane structures involved in endocytosis and has been implicated as a principal scaffold for the organization of diverse cytoplasmic signaling complexes (109). Thus, caveolins interact with GPCRs, growth factor receptors, eNos, c-Src-family tyrosine kinases, MAPKs, the signaling adaptor proteins Shc and GRB2, and more (110). The majority of these interactions are mediated through the caveolin-scaffolding domain, which appears to inhibit the downstream activation signaling of some of these proteins by holding them in the inactive conformation until activation by an appropriate stimulus. Previous studies have shown that targeted downregulation of caveolin-1 expression in 3T3 cells harboring caveolin-1 antisense results in hyperactivation of the ERK pathway as well as loss of anchorage-dependent cell growth (111). Moreover, downregulation of caveolin-1 can induce tumor formation in immunodeficient mice and is sufficient to mediate cell transformation.

Recent findings correlate caveolin-1 expression with androgen sensitivity in murine PC (112) Furthermore, immunohistochemical staining of patient specimens suggests that caveolin expression may be an independent predictor of PC progression (113). Using a mammalian two-hybrid assay system, Lu and co-workers (114) have shown that caveolin-1 directly interacts with the ligand-binding domain of AR and found that overexpression of caveolin-1 potentiates ligand-dependent AR activation. Conversely, downregulation of caveolin-1 by a caveolin-1 antisense expression construct can downregulate ligand-dependent AR activation.

5.3.5. SPROUTY

Sprouty was discovered in Drosophila as a negative regulator of receptor tyrosine kinase signaling through the ERK pathway (115). There are four Sprouty proteins and three related Spred proteins in mammals. The Sprouty proteins have been reported to interact with numerous members of the MAPK pathway, including the adaptor Grb2, caveolin, the phosphatase Shp2, Raf-1, B-Raf, and the E3-ubiquitin ligase Cbl, among others, but a coherent model for its mechanism of action has not been established. Epistasis experiments in Drosophila, designed to determine where in the signaling pathway Sprouty acts, have been inconclusive, as is often the case with scaffold proteins, because they interact at multiple places in a pathway. Sprouty expression is increased in response to ERK signaling, so it appears to be part of a negative feedback loop. Thus, one would predict that in settings where ERK signaling is chronically activated, Sprouty expression would be silenced or reduced. This is the case for PC, breast cancer, and hepatocellular carcinoma (116). Because chronic activation of a signaling pathway is a likely pre-requisite for "oncogene addiction," it will be important to identify the ways that negative feedback of the pathways is disrupted in cancer.

5.3.6. JNK-Interacting Proteins (JIPs)

JIPs include JIP1 and the related protein JIP2 (also known as IB1 and IB2, respectively). JIP1 was initially identified by a yeast two-hybrid screen using JNK as bait *(117)*. The JIP proteins can interact with various signaling components, including members of both the JNK and p38MAPK modules, the MAPK phosphatase MKP7, p190 RhoGEF, Rac exchange factors such as Tiam1 and RasGRF1, and others.

Several studies have demonstrated that JIP1 can regulate the biological actions of the JNK-signaling pathway. Overexpression of JIP1 inhibits the normal signal transduction pathway, probably by sequestering signaling molecules *(118)*. JIP1 caused cytoplasmic retention of JNK and inhibition of JNK-regulated gene expression. In addition, JIP1 expression suppressed the effects of JNK on cellular proliferation, including transformation by the *Bcr-Abl* oncogene *(118)*. Other reports have indicated that overexpression of JIP1 could block JNK-mediated cell death in various cellular models. For example, it was shown that overexpressing the JNK-binding domain (JBD) of JIP1 prevented neuronal apoptosis induced by NGF deprivation *(119)*. However, the evidence that JIP1 acts as an inhibitor may be a result of an artificial situation generated from overexpression of JIP and sequestration of signaling component into incomplete and inactive complexes. Some studies have shown that JIP1 can enhance JNK activation by the MLK–MKK7 pathway *(117,120,121)*. When JIP1 is overexpressed with components of the JNK pathway, JIP1 enhances JNK activation by MKK7 but not by MKK4. Interestingly, a mutation in the JIP1 gene was recently identified and associated with accelerated cell death of pancreatic β-cells and secondary diabetes in humans *(122)*. In certain cellular contexts, JIP1 may act as an Akt1 scaffold, which promotes Akt1 activation. Thus, JIP1 expression can also exert signaling effects that are independent of JNK activity *(123)*.

In PC cells, activation of the JNK-ignaling pathway has been reported to be involved in the response to apoptosis mediated by 12-O-Tetradecanoyl-phorbol-13-acetate (TPA), thapsigargin *(124)* and 4-hydroxyphenyl retinamide (4-HPR), a retinoic acid analog known to induce apoptosis in several cell types *(56)*. A recent study showed that JIP1 significantly attenuated TPA-mediated growth arrest and apoptosis in LNCaP cells stably expressing the JBD of JIP1 *(125)*, thereby supporting a role for JIP1 in the regulation of JNK activity in these cells.

JIP1 expression is mostly restricted to the endocrine pancreas and to the brain. Recently, the presence of JIP1 was described in rat prostate epithelium as well as in LNCaP cells *(126)*. Moreover, JIP1 levels were increased in LNCaP during NE differentiation. This regulated expression of JIP1 was secondary to a loss of the neuronal transcriptional repressor neuron restrictive silencing factor (NRSF/REST) function that is known to repress JIP1. Collectively, these data indicate that JIP1 plays a significant role in acquisition of the NE phenotype in LNCaP cells, which is generally associated with tumor progression and androgen independence.

Although many important details need to be further investigated, it is becoming evident that MAPK scaffold proteins may play an important role in the progression to androgen independency in some human PCs.

5.4. Activation of MAP Kinase Pathways in Prostate Cancer

To test whether Ras might be activated during PC progression in patients, we examined 82 paraffin thin sections from primary and metastatic prostate tumor specimens with an activation-state-specific phospho-ERK antibody *(114,127)*. Activation of ERK in this case was used as a surrogate for Ras activation, because it is not possible to directly measure Ras activity in these samples. Non-neoplastic prostate tissue showed little or no staining with activated ERK antiserum. In prostate tumors, the level of activated ERK increased with increasing Gleason score and tumor stage. ERK activation was seen both in locally advanced disease and in metastases. Moreover, tumor samples from two patients showed no activation of ERK before androgen ablation therapy; however, following androgen ablation treatment, high levels of activated ERK were detected in the recurrent tumors. Finally, we found that in the hormone-dependent CWR22 PC xenograft, although the tumor regresses after castration, its recurrence correlated with upregulation of phospho-ERK (unpublished observation). Thus, our results show a strong correlation between activation of the ERK pathway and PC progression in various settings. These findings have been corroborated by others *(128–130)* Price and colleagues observed an increase in active ERK in prostate tumors when compared with normal tissue *(129)* and Royuela et al. observed an increase in active ERK in both benign prostatic hyperplasia (BPH) and PC, with the highest levels of active MAPK in the PC specimens *(130)*. Burger et al. also observed active ERK in over half of tumors examined; however, they did not observe a correlation between MAPK activity and clinical or histopathological characteristics *(128)*.

In contrast to the above studies, two reports do not find this correlation. Ugare et al. report that in the TRAMP mouse model ERK activation occurs in prostatic intraepithelial neoplasia (PIN), but then declines as the disease progresses *(131)*. It is possible that this reflects an idiosyncrasy of the TRAMP model. More disturbing, Malik et al. have reported a *decrease* in phospho-ERK staining in tumor samples *(132)*. Because of the lability of phospho-epitopes, it is easier to suspect false-negative results than false-positive results. Although Malik et al. were able to detect phospho-Akt in the same samples and also could detect phospho-ERK in normal, BPH, and PIN samples, it is possible that the tissue and intracellular distribution of various phosphatases could generate these differences. Alternatively, it is also possible that different antibodies recognize phospho-ERK differentially depending on intracellular context. In any event, it is important to understand the underlying causes of these different results if phospho-ERK is to be used as a diagnostic or prognostic tool.

Another mechanism for activating the MAPK pathway in PC is suggested by the work of Fu et al. *(133)*, who found that a decreased expression of the Raf inhibitor RKIP was a valid prognostic indicator for PC progression in human tissues. Thus, a MAPK pathway inhibitor behaves like a tumor suppressor, consistent with the concept that this pathway plays a functional role in driving PC.

Mechanistic studies *in vitro* and in mouse models increase our confidence in the correlation between PC progression and ERK activation. Pioneering studies by Gelmann and collaborators showed that expression of an activated v-Ha-Ras in androgen-dependent LNCaP cells enabled LNCaP cells to grow in the absence of androgen *(134))*. We extended those results and showed that expression of activated Ras mutants that activated the ERK pathway rendered LNCaP cells hypersensitive to androgen, as measured by growth *in vitro* and in xenografts. More recent studies from our laboratory suggest that Ras stimulates activation of androgen-responsive gene expression through Raf and ERK both in an AR-dependent and androgen-independent manner (unpublished observation). Effector domain mutants of Ras were co-transfected into LNCaP cells with a PSA or ARE promoter-driven luciferase reporter construct. The Ras effector domain mutant selective for Raf signaling stimulated luciferase expression, whereas those mutants selective for either RalGDS or PI3K did not. Co-transfection of activated MEK (a downstream effector of Raf) also stimulated expression of reporters driven either by the PSA promoter-enhancer or by synthetic AREs (unpublished observation). Therefore, ERK signaling appears to regulate androgen-responsive gene expression.

The *necessity* of Ras signaling in progression has been shown in at least one model: expression of a dominant-negative N17 Ha-Ras actually can restore androgen dependence to an androgen-independent cell line. C4-2 cells were derived by Leland Chung and colleagues from LNCaP cells by serial passage in castrated mice *(135)*. C4-2 cells demonstrate decreased androgen dependence of growth both *in vitro* and *in vivo*, increased tumorigenicity *in vivo* and the ability to grow in soft agarose (anchorage independence) compared with the parental LNCaP cells. Importantly, this model retains expression of the AR, as do most cases of advanced PC. Expression of the dominant-negative Ras under the control of a tetracycline-inducible promoter in C4-2 PC cells restored androgen-dependence to the androgen-independent C4-2 cells *(136)*. When implanted in nude mice, the C4-2 derivatives continued to grow after castration or when dominant-negative N17-Ras was induced with Doxycycline. However, the tumors regressed, in most cases completely, when the mice were castrated and treated with Doxycycline to induce N17Ras.

In summary, the overexpression of growth factors and receptors utilizing Ras signaling, and the activation of MAPK, correlates with PC progression. Additionally, experimental models such as CWR22 display activated ERK, and expression of activated Ras makes LNCaP cells less dependent on androgen. Moreover, expression of dominant-negative Ras restores androgen dependence to C4-2 cells. Thus, our findings and those previously published clearly implicate ERK signaling in progression to androgen independence.

Stress kinase signaling has been implicated in various cellular processes including oncogenesis and tumor suppression *(137)*. MKK4 has been reported to function as a suppressor of tumorigenesis or metastasis in various cell types *(138,139)*. Rinker-Schaeffer and colleagues demonstrated that MKK4 could suppress AT6.1 rat PC metastasis *in vivo (140)*. More significantly, the same research group demonstrated a loss of MKK4 expression in advanced prostate tumors from patients *(141)*. An immunohistochemical and loss of heterozygosity analysis of human prostate tumor material showed a consistent downregulation of MKK4 expression in advanced prostate tumors and loss of heterozygosity (LOH) within the MKK4 locus in 31% of prostate tumors examined. Loss of stress kinase signaling components in advanced PC was also observed in the TRAMP model; activated forms of p38 are reduced or absent in both late-stage adenocarcinomas and metastatic deposits *(131)*. In contrast, Ricote et al. *(142)* have shown increased expression of MKK6 and activation of the p38 pathway in PC, including enhanced phosphorylation and nuclear localization of Elk-1, a transcription factor that is a target of both stress kinase and ERK signaling. Activation of stress kinases in PC could conceivably be triggered by the surgical and pathological procedures used; however, the reported differences in activation state between normal and tumor tissue argue strongly for (at the least) a difference in sensitivity to activation associated with malignant progression.

As in the case of ERK activation, mechanistic studies will be required to bolster these correlations and to reveal whether this pathway would be an appropriate therapeutic target. We have shown that decreasing MKK4 and MKK6 expression with siRNA can increase AR transactivation of PSA. Thus, one implication of the decrease in MKK4 expression in advanced PC is that it may hypersensitize the AR to androgen and thereby promote the acquisition of androgen-independent disease. It remains to be determined whether the loss of MKK4 (or MKK6) expression correlates with the appearance of androgen-independent PC, or whether these kinases can drive progression to androgen independence.

Both p38 and JNK inhibitors have been tested for growth inhibition of PC cells *(143,144)*. Interestingly, JNK inhibition induced apoptosis in multiple PC cell lines in vitro and slowed DU145 xenograft growth. Given the multitude of cell processes regulated by stress kinases, it is not surprising that there are data to suggest stress kinase signaling may provide both a tumor suppressor and oncogenic role in PC *(137)*. These observations are likely because of multiple factors including cell type, strength and duration of signal, method of signal modulation, and cell signaling environment. Moreover, it is possible that the role of stress kinase signaling may change as PC progresses. Consistent with this, well-differentiated TRAMP tumors showed elevated levels of p38 activity, whereas p38 activity is decreased or absent in late stage adenocarcinomas and metastatic deposits *(131)*.

6. INTERSECTION OF KINASE SIGNALING WITH AR FUNCTION

6.1. AR Phosphorylation

Much of the literature referenced above suggests that kinase cascades regulate AR function in part by activating the AR in the absence of ligand or sensitizing the AR to reduced levels of ligand. These functional alterations could occur by direct post-translational modifications of the AR and/or by modifications of AR co-regulatory molecules and/or chromatin.

There is clear evidence that phosphorylation of the estrogen receptor and progesterone receptor modifies their activities in a physiologically significant way. However, the link between AR phosphorylation and functional regulation is less clear. Candidate phosphorylation sites on the AR have been proposed based on *in vitro* phosphorylation reactions and/or by identifying kinase consensus sites and then mutagenizing them. Sites so identified include serines 81, 94, 213, 515, 650, and 791 *(14,145–148)*. [All AR amino acid numbers in this chapter are based on NCB Accession number AAA51729 *(149,150)*.] However, these determinations, although a useful first step, are not definitive because *in vitro* kinase reactions are often not selective, and mutagenesis can alter the phosphorylations on sites distinct from the ones mutagenized. Ser 308 was directly identified as a phosphorylation site in Baculovirus-overexpressed AR using mass spectrometry *(151)*. This was the first site identified in living cells either by mass spectrometry or by *in vivo* metabolic labeling. We identified seven major sites of AR phosphorylation in living cells, using both metabolic labeling and mass spectrometry *(152)*. One site is constitutively phosphorylated, six sites are regulated in response to androgen, and one of these, S650, becomes phosphorylated in response to a number of non-steroid agonists, including EGF, PMA, forskolin, and anisomycin *(152)*. Phosphorylation on S650 regulates nuclear export and is regulated by stress kinase signaling *(153)*. We also have found that when one androgen-induced phosphorylation site, S308, is mutated to Alanine, the AR gives a heightened transcriptional response to steroid, as measured by a reporter assay *(152)*. The functional significance of the other phosphorylations is thus far unknown.

Although ERK is capable of phosphorylating AR *in vitro* on S515 (14), we did not detect phosphorylation on this residue in living cells *(152)*. Moreover, the peak of ERK activation following growth factor stimulation occurs in about 10–15 min, whereas the AR phosphorylations we observed occur more slowly, peaking after one or more hours following agonist stimulation. Addition of a MEK inhibitor did not substantially alter the pattern of AR phosphorylations. However, this does not necessarily mean that the AR is not an *in vivo* ERK substrate. It is possible that the S515 phosphorylation occurs under conditions we did not investigate or that the stoichiometry of phosphorylation is low. A low-stoichiometry phosphorylation can be highly significant—it might, for example, be transitory, yet regulate a key aspect of receptor function. Thus, it is not resolved whether the AR is a direct substrate for MAPK, but the weight of evidence suggests that the AR, in contrast to the ER *(154)*, is not directly phosphorylated by ERK pathway kinases.

The studies by Wen et al. and Lin et al. show AR is an Akt substrate *in vitro (147,148)* Lin et al. show phosphorylation of exogenous AR in COS-1 cells stimulated with IGF-1. However, both studies examined only overall AR phosphorylation of wild-type and Ser-to-Ala mutants, not the phosphorylation of individual residues. Recently, the use of phospho-specific antibodies has strongly suggested AR is phosphorylated on Ser213 *(155,156)*.

Interestingly, a role for Akt in regulating the HER2 modulation of AR function was recently excluded *(157)*. Reconstitution studies with Akt failed to rescue the effects of PKI-166 on AR activity. Although this study did not directly examine any of the putative Akt phosphorylation sites on the AR, the lack of any functional evidence suggests that Akt is not involved in regulating AR function under the conditions reported.

Recent observations have demonstrated that the AR becomes phosphorylated on tyrosine, in response both to steroid and to growth factors (Kraus et al. submitted; and Qiu, submitted). This phosphorylation is transitory (peaking at 15 min after stimulation) and thus was not detected in our studies *(152)* that determined steady-state AR phosphorylations and phosphorylations which were more long-lasting. Src is the kinase likely responsible for this phosphorylation, and the activity of the AR in regulating both growth and transcription appears to be altered.

6.2. AR Partners

Transcriptional co-regulators are frequently overexpressed in advanced PC, facilitating AR activity *(2,14,15)*. A direct role for co-regulatory proteins in PC has been implicated by a number of studies. Knockout of SRC-1 in mice results in defective growth of the prostate *(158)*. Additionally, SRC-1 and TIF2/GRIP1 are overexpressed in recurrent PCs *(2)*. Overexpression of TIF2/GRIP1, ARA55, or ARA70 increases the transcriptional activity of AR in response to low-affinity ligands (e.g., DHEA, androstenedione, estradiol) or to low concentrations of DHT *(14,159)*. It is possible that phosphorylation of these co-activators provides an alternative to overexpression as a mechanism for regulating AR. Consistent with this, p300 mediates IL6 activation of AR, and overexpression of p300 can overcome the ability of MEK-inhibition to block the IL-6 simulated transactivation *(160)*. Several co-regulators have been identified as targets of signaling pathways, including the MAPK pathway (e.g., SRC-1, CBP, p300, AIB1). Rowan et al. comprehensively mapped the SRC-1 phosphorylation sites and showed that ERK can phosphorylate SRC-1 *in vitro* on T1179 and S1185 *(161)*. Furthermore, EGF potentiated ligand activation of the progesterone receptor by SRC-1. A subsequent study from Rowan et al. demonstrated that PKA and ERK signaling directly regulates SRC-1 phosphorylation and activity *(162)*. Similarly, Gregory et al. showed that in an androgen-independent PC cell line, EGF increased AR transcriptional activity through MAPK-dependent increases in GRIP1 phosphorylation *(163)*.

Kinase cascades other than the MAPK cascade also might modify Ar co-regulators, in a way that could complement Ras and MAPK signaling. For example, as described above, PC containing NE-like cells is expected to have enhanced signaling via cAMP and PKA. CBP was first described as the partner for the cAMP-regulated transcription factor CREB *(164)*. Moreover, it also is a phosphoprotein and is subject to phosphorylation by PKC, CaM Kinase, and others *(15,34,159,161,162,165–175)*.

In addition to chromatin-remodeling proteins, the AR also interacts physically and functionally with other transcription factors including c-Jun and Ets family members *(176–183)*. Chung and colleagues *(184)* have mapped the PSA promoter to determine the regions that are responsible for the differential basal gene expression between LNCaP and C4-2. They identified both the ARE in the enhancer and a site with similarity to SP-1 family sites, near the promoter. These data are consistent with the concept that transcription factors that can directly bind DNA could be involved in progression to decreased androgen dependence. Particularly noteworthy are the recent reports that gene fusions involving Ets-family transcription factor genes play an important and early role in PC *(133,183)*, as Ets transcription factors are often targets of MAPK signaling. Thus, one could imagine that these factors could be activated either by gene fusion or by post-translational modification.

AR and co-activators are also regulated by other post-translational modifications, such as sumoylation and methylation as well as acetylation *(171,185,186)*. This chapter focused on phosphorylation because our goal is to understand the intersection between Ras signaling—which activates kinase cascades—and the AR. However, it is possible that the targets of phosphory-lation could be regulators of these other processes.

7. IMPLICATIONS FOR THERAPY

Pathways that result in AR-dependent androgen-independent growth and survival of PCs present an attractive target for therapeutic intervention. In this review, we presented evidence that growth factors, receptors, Ras and MAPK signaling, the AR, and AR co-regulatory molecules all participate in the progression process. However, because each of these individually can contribute to androgen independence, and because for each there are redundant pathways that would circumvent any individual blockade, it seems highly probable that interventions will need to be combinatorial.

Several inhibitors of growth factor receptors are in development, especially inhibitors of members of the EGFR family. Use of growth factor receptors as therapeutic targets is complicated by the functional redundancy of many receptor types. This is further complicated by the well-established but widely ignored observation that kinase-dead EGFRs are capable of intracellular signaling, apparently by dimerization with other receptors or kinases (187,188). Thus, it is not certain that an essential target (as determined with knockout or dominant-negative methodologies) would be a useful target for a small molecule catalytic inhibitor.

Intracellular signaling may provide effective targets, because, although redundancy is common, some functional nodes where signaling pathways converge have been identified. The Ras–MAPK pathway represents one of those sites of regulatory convergence (189). It is widely believed that the downside of targeting intracellular signaling is that the same regulatory modules are used in multiple functions, and thus that drugs that inhibit these pathways might display widespread mechanism-induced toxicities. However, Sorafenib, a small molecule that inhibits Raf (in addition to other kinases) is proving effective and well tolerated in various settings and is being approved by the FDA for renal cell carcinoma (190–192). MEK inhibitors from Pfizer and Astra-Zeneca are under development and also appear to be well tolerated despite the centrality of the MAPK pathway (193,194). The sensitivity of tumor cells to these inhibitors varies depending on which pathway is the essential driver of tumor cell growth. Thus, melanoma cells carrying B-Raf mutations and dependent on Raf and MAPK signaling are hypersensitive to these inhibitors of the MAPK pathway, whereas cells dependent on other pathways (including Ras) may show less sensitivity (84). It remains to be seen whether tumor cells driven by pathways that are activated by overexpression rather than mutation will be hypersensitive. However, even if they are only partially sensitive, it could be possible to identify other inhibitors that will render these cells as sensitive to MAPK inhibition as if they were mutationally activated. For example, we find that dominant-negative N17-Ras has little effect on xenograft growth by itself, but is cytotoxic when combined with androgen ablation. Ablation of the AR itself (in addition to just the androgen ligand) could provide even more effectiveness. Androgen ablation strategies include not only siRNA and antisense approaches but also exploitation of the fact that the AR is a client protein of the chaperone Hsp90 *(195)*. Moreover, we also find that MEK inhibition is cytostatic rather than cytotoxic, even when combined with androgen ablation, suggesting that a third type of intervention will be needed to achieve cytotoxicity.

Synthetic lethality with MEK inhibitors might be achieved by identifying pathways inhibited by dominant-negative N17-Ras that are not blocked by MEK inhibitors. These could include the PI3K, Ral, and RASSF1 pathways. However, other pathways also play an important role in prostate development and are likely to be important in PC as well, such as the Hedgehog and wnt pathways. Work from Beachey and colleagues has provided compelling evidence for the importance of Hedgehog pathway signaling in PC, demonstrating regression of xenografts in mice treated with the Hedgehog inhibitor cyclopamine *(196)*. Synergies between MEK inhibition and inhibition of these other pathways are attractive opportunities for examination.

ACKNOWLEDGMENTS

This work was supported by grants from DOD (W81XWH-04-1-0112), and National Institutes of Health, National Cancer Institute (P01 CA76465, R01 CA105402).

REFERENCES

1. Perez, C. A., Fair, W. R., Ihde, D. C. and Labrie, F. (1985) *Cancer. Principles and Practice of Oncology.* In: DeVita, V. T., Hellman, S. and Rosenberg, S. A. (ed). J.B. Lippincott Company, Philadelphia, pp 929–64.
2. Gregory, C. W., He, B., Johnson, R. T., et al. (2001) A mechanism for androgen receptor-mediated prostate cancer recurrence after androgen deprivation therapy. *Cancer Res* **61**, 4315–9.
3. Chen, C. D., Welsbie, D. S., Tran, C., et al. (2004) Molecular determinants of resistance to antiandrogen therapy. *Nat Med* **10**, 33–9.
4. Reinikainen, P., Palvimo, J. J. and Janne, O. A. (1996) Effects of mitogens on androgen receptor-mediated transactivation. *Endocrinology* **137**, 4351–7.
5. Weber, M. J. and Gioeli, D. (2004) Ras signaling in prostate cancer progression. *J Cell Biochem* **91**, 13–25.
6. Zegarra-Moro, O. L., Schmidt, L. J., Huang, H. and Tindall, D. J. (2002) Disruption of androgen receptor function inhibits proliferation of androgen-refractory prostate cancer cells. *Cancer Res.* **62**, 1008–13.

7. Culig, Z., Klocker, H., Bartsch, G. and Hobisch, A. (2001) Androgen receptor mutations in carcinoma of the prostate: significance for endocrine therapy. *Am J Pharmacogenomics* **1**, 241–9.
8. Eder, I. E., Hoffmann, J., Rogatsch, H., et al. (2002) Inhibition of LNCaP prostate tumor growth in vivo by an antisense oligonucleotide directed against the human androgen receptor. *Cancer Gene Ther* **9**, 117–25.
9. Eder, I. E., Culig, Z., Ramoner, R., et al. (2000) Inhibition of LncaP prostate cancer cells by means of androgen receptor antisense oligonucleotides. *Cancer Gene Ther* **7**, 997–1007.
10. Hobisch, A., Culig, Z., Radmayr, C., Bartsch, G., Klocker, H. and Hittmair, A. (1995) Distant metastases from prostatic carcinoma express androgen receptor protein. *Cancer Res* **55**, 3068–72.
11. Linja, M. J., Savinainen, K. J., Saramaki, O. R., Tammela, T. L., Vessella, R. L. and Visakorpi, T. (2001) Amplification and over expression of androgen receptor gene in hormone-refractory prostate cancer. *Cancer Res* **61**, 3550–5.
12. Brown, R. S., Edwards, J., Dogan, A., et al. (2002) Amplification of the androgen receptor gene in bone metastases from hormone-refractory prostate cancer. *J Pathol* **198**, 237–44.
13. Gregory, C. W., Johnson, R. T., Jr., Mohler, J. L., French, F. S. and Wilson, E. M. (2001) Androgen receptor stabilization in recurrent prostate cancer is associated with hypersensitivity to low androgen. *Cancer Res* **61**, 2892–8.
14. Yeh, S., Kang, H. Y., Miyamoto, H., et al. (1999) Differential induction of androgen receptor transactivation by different androgen receptor co-activators in human prostate cancer DU145 cells. *Endocrine* **11**, 195–202.
15. Comuzzi, B., Lambrinidis, L., Rogatsch, H., et al. (2003) The transcriptional co-activator cAMP response element-binding protein-binding protein is expressed in prostate cancer and enhances androgen- and anti-androgen-induced androgen receptor function. *Am J Pathol* **162**, 233–41.
16. Bentel, J. M. and Tilley, W. D. (1996) Androgen receptors in prostate cancer. *J Endocrinol* **151**, 1–11.
17. Culig, Z., Klocker, H., Bartsch, G. and Hobisch, A. (2002) Androgen receptors in prostate cancer. *Endocr Relat Cancer* **9**, 155–70.
18. Isaacs, W., De Marzo, A. and Nelson, W. G. (2002) Focus on prostate cancer. *Cancer Cell* **2**, 113–6.
19. Culig, Z., Bartsch, G. and Hobisch, A. (2002) Interleukin-6 regulates androgen receptor activity and prostate cancer cell growth. *Mol Cell Endocrinol* **197**, 231–8.
20. Ikonen, T., Palvimo, J. J., Kallio, P. J., Reinikainen, P. and Janne, O. A. (1994) Stimulation of androgen-regulated transactivation by modulators of protein phosphorylation. *Endocrinology* **135**, 1359–66.
21. Culig, Z., Hobisch, A., Cronaur, M. V., Radmayr, C., Hittmair, A., Zhang, J., Thurnher, M., Bartsch, G. and Klocker, H. (1996) Regulation of prostatic growth and function by peptide growth factors. *The Prostate* **28**, 392–405.
22. Litvinov, I. V., De Marzo, A. M. and Isaacs, J. T. (2003) Is the Achilles' heel for prostate cancer therapy a gain of function in androgen receptor signaling? *J Clin Endocrinol Metab* **88**, 2972–82.
23. Godoy-Tundidor, S., Hobisch, A., Pfeil, K., Bartsch, G. and Culig, Z. (2002) Acquisition of agonistic properties of nonsteroidal antiandrogens after treatment with oncostatin M in prostate cancer cells. *Clin Cancer Res* **8**, 2356–61.
24. Culig, Z., Hobisch, A., Bartsch, G. and Klocker, H. (2000) Androgen receptor–an update of mechanisms of action in prostate cancer. *Urol Res* **28**, 211–9.
25. Scher, H. I., Sarkis, A., Reuter, V., et al. (1995) Changing pattern of expression of the epidermal growth factor receptor and transforming growth factor alpha in the progression of prostatic neoplasms. *Clin Cancer Res* **1**, 545–50.

26. Culig, Z., Hobisch, A., Cronauer, M. V., et al. (1994) Androgen receptor activation in prostatic tumor cell lines by insulin-like growth factor-I, keratinocyte growth factor, and epidermal growth factor. *Cancer Res* **54**, 5474–8.

27. Thalmann, G. N., Anezinis, P. E., Chang, S. M., et al. (1994) Androgen-independent cancer progression and bone metastasis in the LNCaP model of human prostate cancer. *Cancer Res* **54**, 2577–81.

28. Nakamoto, T., Chang, C. S., Li, A. K. and Chodak, G. W. (1992) Basic fibroblast growth factor in human prostate cancer cells. *Cancer Res* **52**, 571–7.

29. Gleave, M. E., Hsieh, J. T., von Eschenbach, A. C. and Chung, L. W. (1992) Prostate and bone fibroblasts induce human prostate cancer growth in vivo: implications for bidirectional tumor-stromal cell interaction in prostate carcinoma growth and metastasis. *J Urol* **147**, 1151–9.

30. Pietrzkowski, Z., Mulholland, G., Gomella, L., Jameson, B. A., Wernicke, D. and Baserga, R. (1993) Inhibition of growth of prostatic cancer cell lines by peptide analogues of insulin-like growth factor 1. *Cancer Res* **53**, 1102–6.

31. Kitsberg, D. I. and Leder, P. (1996) Keratinocyte growth factor induces mammary and prostatic hyperplasia and mammary adenocarcinoma in transgenic mice. *Oncogene* **13**, 2507–15.

32. Steiner, H., Godoy-Tundidor, S., Rogatsch, H., et al. (2003) Accelerated in vivo growth of prostate tumors that up-regulate interleukin-6 is associated with reduced retinoblastoma protein expression and activation of the mitogen-activated protein kinase pathway. *Am J Pathol* **162**, 655–63.

33. Nazareth, L. V. and Weigel, N. L. (1996) Activation of the human androgen receptor through a protein kinase A signaling pathway. *J Biol Chem* **271**, 19900–7.

34. Weigel, N. L. (1996) Steroid hormone receptors and their regulation by phosphorylation. *Biochem J* **319 (Pt 3)**, 657–67.

35. Weigel, N. L. and Zhang, Y. (1998) Ligand-independent activation of steroid hormone receptors. *J Mol Med* **76**, 469–79.

36. Craft, N., Shostak, Y., Carey, M. and Sawyers, C. L. (1999) A mechanism for hormone-independent prostate cancer through modulation of androgen receptor signaling by the HER-2/neu tyrosine kinase. *Nat Med* **5**, 280–5.

37. Abreu-Martin, M. T., Chari, A., Palladino, A. A., Craft, N. A. and Sawyers, C. L. (1999) Mitogen-activated protein kinase kinase kinase 1 activates androgen receptor-dependent transcription and apoptosis in prostate cancer. *Mol Cell Biol* **19**, 5143–54.

38. Gregory, C. W., Whang, Y. E., McCall, W., et al. (2005) Heregulin-induced activation of HER2 and HER3 increases androgen receptor transactivation and CWR-R1 human recurrent prostate cancer cell growth. *Clin Cancer Res* **11**, 1704–12.

39. Mellinghoff, I. K., Tran, C. and Sawyers, C. L. (2002) Growth inhibitory effects of the dual ErbB1/ErbB2 tyrosine kinase inhibitor PKI-166 on human prostate cancer xenografts. *Cancer Res* **62**, 5254–9.

40. Agus, D. B., Akita, R. W., Fox, W. D., et al. (2002) Targeting ligand-activated ErbB2 signaling inhibits breast and prostate tumor growth. *Cancer Cell* **2**, 127–37.

41. Liu, Y., Majumder, S., McCall, W., et al. (2005) Inhibition of HER-2/neu kinase impairs androgen receptor recruitment to the androgen responsive enhancer. *Cancer Res* **65**, 3404–9.

42. Savinainen, K. J., Saramaki, O. R., Linja, M. J., et al. (2002) Expression and gene copy number analysis of ERBB2 oncogene in prostate cancer. *Am J Pathol* **160**, 339–45.

43. Calvo, B. F., Levine, A. M., Marcos, M., et al. (2003) Human epidermal receptor-2 expression in prostate cancer. *Clin Cancer Res* **9**, 1087–1097.

44. Osman, I., Scher, H. I., Drobnjak, M., et al. (2001) HER-2/neu (p185neu) protein expression in the natural or treated history of prostate cancer. *Clin Cancer Res.* **7**, 2643–7.

45. Signoretti, S., Montironi, R., Manola, J., et al. (2000) Her-2-neu expression and progression toward androgen independence in human prostate cancer. *J Natl Cancer Inst* **92**, 1918–25.

46. Mellinghoff, I. K., Vivanco, I., Kwon, A., Tran, C., Wongvipat, J. and Sawyers, C. L. (2004) HER2/neu kinase-dependent modulation of androgen receptor function through effects on DNA binding and stability. *Cancer Cell* **6**, 517–27.

47. Sadar, M. D. (1999) Androgen-independent induction of prostate-specific antigen gene expression via cross-talk between the androgen receptor and protein kinase A signal transduction pathways. *J Biol Chem* **274**, 7777–83.

48. Abrahamsson, P. A. (1999) Neuroendocrine cells in tumor growth of the prostate. *Endocr Relat Cancer* **6**(4), 503–19.

49. Wennerberg, K., Rossman, K. L. and Der, C. J. (2005) The Ras superfamily at a glance. *J Cell Sci* **118**, 843–6.

50. Bos, J. L., de, B. K., Enserink, J., et al. (2003) The role of Rap1 in integrin-mediated cell adhesion. *Biochem Soc Trans* **31**, 83–6.

51. Hattori, M. and Minato, N. (2003) Rap1 GTPase: functions, regulation, and malignancy. *J Biochem (Tokyo)* **134**, 479–84.

52. Bos, J. L. (1989) ras oncogenes in human cancer: a review. *Cancer Res* **49**, 4682–9.

53. Lowy, D. R. and Willumsen, B. M. (1993) Function and regulation of ras. *Annu Rev Biochem* **62**, 851–91.

54. Carter, B. S., Epstein, J. I. and Isaacs, W. B. (1990) ras gene mutations in human prostate cancer. *Cancer Res* **50**, 6830–2.

55. Repasky, G. A., Chenette, E. J. and Der, C. J. (2004) Renewing the conspiracy theory debate: does Raf function alone to mediate Ras oncogenesis? *Trends Cell Biol* **14**, 639–47.

56. Chen, T., Cho, R. W., Stork, P. J. and Weber, M. J. (1999) Elevation of cyclic adenosine 3′,5′-monophosphate potentiates activation of mitogen-activated protein kinase by growth factors in LNCaP prostate cancer cells. *Cancer Res* **59**, 213–8.

57. Carson, J. P., Kulik, G. and Weber, M. J. (1999) Antiapoptotic signaling in LNCaP prostate cancer cells: a survival signaling pathway independent of phosphatidylinositol 3′-kinase and Akt/protein kinase B. *Cancer Res* **59**, 1449–53.

58. Murillo, H., Huang, H., Schmidt, L. J., Smith, D. I. and Tindall, D. J. (2001) Role of PI3K signaling in survival and progression of LNCaP prostate cancer cells to the androgen refractory state. *Endocrinology* **142**, 4795–805.

59. Cairns, P., Okami, K., Halachmi, S., et al. (1997) Frequent inactivation of PTEN/MMAC1 in primary prostate cancer. *Cancer Res* **57**, 4997–5000.

60. McMenamin, M. E., Soung, P., Perera, S., Kaplan, I., Loda, M. and Sellers, W. R. (1999) Loss of PTEN expression in paraffin-embedded primary prostate cancer correlates with high Gleason score and advanced stage. *Cancer Res* **59**, 4291–6.

61. Whang, Y. E., Wu, X., Suzuki, H., et al. (1998) Inactivation of the tumor suppressor PTEN/MMAC1 in advanced human prostate cancer through loss of expression. *Proc Natl Acad Sci USA* **95**, 5246–50.

62. Vos, M. D., Ellis, C. A., Bell, A., Birrer, M. J. and Clark, G. J. (2000) Ras uses the novel tumor suppressor RASSF1 as an effector to mediate apoptosis. *J Biol Chem.* **275**, 35669–72.

63. Vos, M. D., Ellis, C. A., Elam, C., Ulku, A. S., Taylor, B. J. and Clark, G. J. (2003) RASSF2 is a novel K-Ras-specific effector and potential tumor suppressor. *J Biol Chem* **278**, 28045–51.

64. Tommasi, S., Dammann, R., Jin, S. G., Zhang, X. F., Avruch, J. and Pfeifer, G. P. (2002) RASSF3 and NORE1: identification and cloning of two human homologues of the putative tumor suppressor gene RASSF1. *Oncogene* **21**, 2713–20.

65. Dreijerink, K., Braga, E., Kuzmin, I., et al. (2001) The candidate tumor suppressor gene, RASSF1A, from human chromosome 3p21.3 is involved in kidney tumorigenesis. *Proc Natl Acad Sci USA* **98**, 7504–9.

66. Dammann, R., Li, C., Yoon, J. H., Chin, P. L., Bates, S. and Pfeifer, G. P. (2000) Epigenetic inactivation of a RAS association domain family protein from the lung tumour suppressor locus 3p21.3. *Nat Genet* **25**, 315–9.

67. Hesson, L., Dallol, A., Minna, J. D., Maher, E. R. and Latif, F. (2003) NORE1A, a homologue of RASSF1A tumour suppressor gene is inactivated in human cancers. *Oncogene* **22**, 947–54.

68. Khokhlatchev, A., Rabizadeh, S., Xavier, R., et al. (2002) Identification of a novel Ras-regulated proapoptotic pathway. *Curr Biol* **12**, 253–65.

69. Feig, L. A., Urano, T. and Cantor, S. (1996) Evidence for a Ras/Ral signaling cascade. *Trends Biochem Sci* **21**, 438–41.

70. Ehrhardt, A., Ehrhardt, G. R., Guo, X. and Schrader, J. W. (2002) Ras and relatives—job sharing and networking keep an old family together. *Exp Hematol* **30**, 1089–1106.

71. Innocenti, M., Tenca, P., Frittoli, E., et al. (2002) Mechanisms through which Sos-1 co-ordinates the activation of Ras and Rac. *J Cell Biol* **156**, 125–36.

72. Walsh, A. B. and Bar-Sagi, D. (2001) Differential activation of the Rac pathway by Ha-Ras and K-Ras. *J Biol Chem* **276**, 15609–615.

73. Ward, Y., Wang, W., Woodhouse, E., Linnoila, I., Liotta, L. and Kelly, K. (2001) Signal pathways which promote invasion and metastasis: critical and distinct contributions of extracellular signal-regulated kinase and Ral-specific guanine exchange factor pathways. *Mol Cell Biol* **21**, 5958–69.

74. Weijzen, S., Rizzo, P., Braid, M., et al. (2002) Activation of Notch-1 signaling maintains the neoplastic phenotype in human Ras-transformed cells. *Nat Med* **8**, 979–86.

75. Weston, C. R. and Davis, R. J. (2002) The JNK signal transduction pathway. *Curr Opin Genet Dev* **12**, 14–21.

76. Hamad, N. M., Elconin, J. H., Karnoub, A. E., et al. (2002) Distinct requirements for Ras oncogenesis in human versus mouse cells. *Genes Dev* **16**, 2045–2057.

77. Gonzalez-Garcia, A., Pritchard, C. A., Paterson, H. F., Mavria, G., Stamp, G. and Marshall, C. J. (2005) RalGDS is required for tumor formation in a model of skin carcinogenesis. *Cancer Cell* **7**, 219–26.

78. Pages, G., Guerin, S., Grall, D., et al. (1999) Defective thymocyte maturation in p44 MAP kinase (Erk 1) knockout mice. *Science* **286**, 1374–7.

79. Yao, Y., Li, W., Wu, J., et al. (2003) Extracellular signal-regulated kinase 2 is necessary for mesoderm differentiation. *Proc Natl Acad Sci USA* **100**, 12759–64.

80. O'Neill, E. and Kolch, W. (2005) Taming the Hippo: Raf-1 controls apoptosis by suppressing MST2/Hippo. *Cell Cycle* **4**, 365–7.

81. Jin, S., Zhuo, Y., Guo, W. and Field, J. (2005) p21-activated Kinase 1 (Pak1)-dependent phosphorylation of Raf-1 regulates its mitochondrial localization, phosphorylation of BAD, and Bcl-2 association. *J Biol Chem* **280**, 24698–705.

82. Matsumoto, S., Miyagishi, M., Akashi, H., Nagai, R. and Taira, K. (2005) Analysis of double-stranded RNA-induced apoptosis pathways using interferon-response noninducible small interfering RNA expression vector library. *J Biol Chem* **280**, 25687–96.

83. Davies, H., Bignell, G. R., Cox, C., et al. (2002) Mutations of the BRAF gene in human cancer. *Nature* **417**, 949–54.

84. Solit, D. B., Garraway, L. A., Pratilas, C. A., et al. (2006) BRAF mutation predicts sensitivity to MEK inhibition. *Nature* **439**, 358–62.

85. Sharma, A., Trivedi, N. R., Zimmerman, M. A., Tuveson, D. A., Smith, C. D. and Robertson, G. P. (2005) Mutant V599EB-Raf regulates growth and vascular development of malignant melanoma tumors. *Cancer Res* **65**, 2412–21.

86. Eblen, S. T., Slack-Davis, J. K., Tarcsafalvi, A., Parsons, J. T., Weber, M. J. and Catling, A. D. (2004) Mitogen-activated protein kinase feedback phosphorylation regulates

MEK1 complex formation and activation during cellular adhesion. *Mol Cell Biol* **24**, 2308–317.

87. Slack-Davis, J. K., Eblen, S. T., Zecevic, M., et al. (2003) PAK1 phosphorylation of MEK1 regulates fibronectin-stimulated MAPK activation. *J Cell Biol* **162**, 281–91.

88. Slack-Davis, J. K. and Parsons, J. T. (2004) Emerging views of integrin signaling: implications for prostate cancer. *J Cell Biochem* **91**, 41–46.

89. Wu, J., Dent, P., Jelinek, T., Wolfman, A., Weber, M. J. and Sturgill, T. W. (1993) Inhibition of the EGF-activated MAP kinase signaling pathway by adenosine 3´,5´-monophosphate. *Science* **262**, 1065–9.

90. Dhillon, A. S., Meikle, S., Peyssonnaux, C., et al. (2003) A Raf-1 mutant that dissociates MEK/extracellular signal-regulated kinase activation from malignant transformation and differentiation but not proliferation. *Mol Cell Biol* **23**, 1983–93.

91. Mischak, H., Seitz, T., Janosch, P., et al. (1996) Negative regulation of Raf-1 by phosphorylation of serine 621. *Mol Cell Biol* **16**, 5409–18.

92. Ikenoue, T., Kanai, F., Hikiba, Y., et al. (2005) Functional consequences of mutations in a putative Akt phosphorylation motif of B-raf in human cancers. *Mol Carcinog* **43**, 59–63.

93. Dumaz, N. and Marais, R. (2005) Integrating signals between cAMP and the RAS/RAF/MEK/ERK signalling pathways. Based on the anniversary prize of the Gesellschaft fur Biochemie und Molekularbiologie Lecture delivered on 5 July 2003 at the Special FEBS Meeting in Brussels. *FEBS J* **272**, 3491–504.

94. Morrison, D. K. and Davis, R. J. (2003) Regulation of MAP kinase signaling modules by scaffold proteins in mammals. *Annu Rev Cell Dev Biol* **19**, 91–118.

95. Kornfeld, K., Hom, D. B. and Horvitz, H. R. (1995) The ksr-1 gene encodes a novel protein kinase involved in Ras-mediated signaling in C. elegans. *Cell* **83**, 903–13.

96. Cacace, A. M., Michaud, N. R., Therrien, M., et al. (1999) Identification of constitutive and ras-inducible phosphorylation sites of KSR: implications for 14–3-3 binding, mitogen-activated protein kinase binding, and KSR overexpression. *Mol Cell Biol* **19**, 229–40.

97. Nguyen, A., Burack, W. R., Stock, J. L., et al. (2002) Kinase suppressor of Ras (KSR) is a scaffold which facilitates mitogen-activated protein kinase activation in vivo. *Mol Cell Biol* **22**, 3035–45.

98. Ohmachi, M., Rocheleau, C. E., Church, D., Lambie, E., Schedl, T. and Sundaram, M. V. (2002) C. elegans ksr-1 and ksr-2 have both unique and redundant functions and are required for MPK-1 ERK phosphorylation. *Curr Biol* **12**, 427–33.

99. Luttrell, L. M. and Lefkowitz, R. J. (2002) The role of beta-arrestins in the termination and transduction of G-protein-coupled receptor signals. *J Cell Sci* **115**, 455–65.

100. Miller, W. E., McDonald, P. H., Cai, S. F., Field, M. E., Davis, R. J. and Lefkowitz, R. J. (2001) Identification of a motif in the carboxyl terminus of beta -arrestin2 responsible for activation of JNK3. *J Biol Chem* **276**, 27770–7.

101. Enslen, H. and Davis, R. J. (2001) Regulation of MAP kinases by docking domains. *Biol Cell* **93**, 5–14.

102. Caunt, C. J., Finch, A. R., Sedgley, K. R., Oakley, L., Luttrell, L. M. and McArdle, C. A. (2006) Arrestin-mediated ERK activation by gonadotropin-releasing hormone receptors: receptor-specific activation mechanisms and compartmentalization. *J Biol Chem* **281**, 2701–10.

103. Kraus, S., Naor, Z. and Seger, R. (2006) Gonadotropin-releasing hormone in apoptosis of prostate cancer cells. *Cancer Lett* **234**, 109–23.

104. Ron, D., Chen, C. H., Caldwell, J., Jamieson, L., Orr, E. and Mochly-Rosen, D. (1994) Cloning of an intracellular receptor for protein kinase C: a homolog of the beta subunit of G proteins. *Proc Natl Acad Sci USA* **91**, 839–43.

105. McCahill, A., Warwicker, J., Bolger, G. B., Houslay, M. D. and Yarwood, S. J. (2002) The RACK1 scaffold protein: a dynamic cog in cell response mechanisms. *Mol Pharmacol* **62**, 1261–73.

106. Rigas, A. C., Ozanne, D. M., Neal, D. E. and Robson, C. N. (2003) The scaffolding protein RACK1 interacts with androgen receptor and promotes cross-talk through a protein kinase C signaling pathway. *J Biol Chem* **278**, 46087–6093.

107. Vomastek, T., Schaeffer, H. J., Tarcsafalvi, A., Smolkin, M. E., Bissonette, E. A. and Weber, M. J. (2004) Modular construction of a signaling scaffold: MORG1 interacts with components of the ERK cascade and links ERK signaling to specific agonists. *Proc Natl Acad Sci USA* **101**, 6981–86.

108. Lopez-Bergami, P., Habelhah, H., Bhoumik, A., Zhang, W., Wang, L. H. and Ronai, Z. (2005) RACK1 mediates activation of JNK by protein kinase C [corrected]. *Mol Cell* **19**, 309–20.

109. Okamoto, T., Schlegel, A., Scherer, P. E. and Lisanti, M. P. (1998) Caveolins, a family of scaffolding proteins for organizing "preassembled signaling complexes" at the plasma membrane. *J Biol Chem* **273**, 5419–22.

110. Williams, T. M. and Lisanti, M. P. (2004) The Caveolin genes: from cell biology to medicine. *Ann Med* **36**, 584–95.

111. Galbiati, F., Volonte, D., Engelman, J. A., et al. (1998) Targeted downregulation of caveolin-1 is sufficient to drive cell transformation and hyperactivate the p42/44 MAP kinase cascade. *EMBO J* **17**, 6633–48.

112. Nasu, Y., Timme, T. L., Yang, G., et al. (1998) Suppression of caveolin expression induces androgen sensitivity in metastatic androgen-insensitive mouse prostate cancer cells. *Nat Med* **4**, 1062–4.

113. Yang, G., Truong, L. D., Wheeler, T. M. and Thompson, T. C. (1999) Caveolin-1 expression in clinically confined human prostate cancer: a novel prognostic marker. *Cancer Res* **59**, 5719–23.

114. Lu, M. L., Schneider, M. C., Zheng, Y., Zhang, X. and Richie, J. P. (2001) Caveolin-1 interacts with androgen receptor. A positive modulator of androgen receptor mediated transactivation *J Biol Chem* **276** (16), 13442–51. Epub 2001 Jan 18.

115. Mason, J. M., Morrison, D. J., Basson, M. A. and Licht, J. D. (2006) Sprouty proteins: multifaceted negative-feedback regulators of receptor tyrosine kinase signaling. *Trends Cell Biol* **16**, 45–54.

116. Lo, T. L., Fong, C. W., Yusoff, P., et al. (2006) Sprouty and cancer: the first terms report. *Cancer Res.* **66** (4), 2048–58.

117. Whitmarsh, A. J., Cavanagh, J., Tournier, C., Yasuda, J. and Davis, R. J. (1998) A mammalian scaffold complex that selectively mediates MAP kinase activation. *Science* **281**, 1671–4.

118. Dickens, M., Rogers, J. S., Cavanagh, J., et al. (1997) A cytoplasmic inhibitor of the JNK signal transduction pathway. *Science* **277**, 693–6.

119. Harding, T. C., Xue, L., Bienemann, A., et al. (2001) Inhibition of JNK by over expression of the JNL binding domain of JIP-1 prevents apoptosis in sympathetic neurons. *J Biol Chem* **276**, 4531–4.

120. Yasuda, J., Whitmarsh, A. J., Cavanagh, J., Sharma, M. and Davis, R. J. (1999) The JIP group of mitogen-activated protein kinase scaffold proteins. *Mol Cell Biol* **19**, 7245–54.

121. Kelkar, N., Gupta, S., Dickens, M. and Davis, R. J. (2000) Interaction of a mitogen-activated protein kinase signaling module with the neuronal protein JIP3. *Mol Cell Biol* **20**, 1030–43.

122. Waeber, G., Delplanque, J., Bonny, C., et al. (2000) The gene MAPK8IP1, encoding islet-brain-1, is a candidate for type 2 diabetes. *Nat Genet* **24**, 291–5.

123. Kim, A. H., Sasaki, T. and Chao, M. V. (2003) JNK-interacting protein 1 promotes Akt1 activation. *J Biol Chem* **278**, 29830–6.

124. Engedal, N., Korkmaz, C. G. and Saatcioglu, F. (2002) C-Jun N-terminal kinase is required for phorbol ester- and thapsigargin-induced apoptosis in the androgen responsive prostate cancer cell line LNCaP. *Oncogene* **21**, 1017–27.

125. Ikezoe, T., Yang, Y., Taguchi, H. and Koeffler, H. P. (2004) JNK interacting protein 1 (JIP-1) protects LNCaP prostate cancer cells from growth arrest and apoptosis mediated by 12–0-tetradecanoylphorbol-13-acetate (TPA). *Br J Cancer* **90**, 2017–24.

126. Tawadros, T., Martin, D., Abderrahmani, A., Leisinger, H. J., Waeber, G. and Haefliger, J. A. (2005) IB1/JIP-1 controls JNK activation and increased during prostatic LNCaP cells neuroendocrine differentiation. *Cell Signal* **17**, 929–39.

127. Gioeli, D., Mandell, J. W., Petroni, G. R., Frierson, H. F., Jr. and Weber, M. J. (1999) Activation of mitogen-activated protein kinase associated with prostate cancer progression. *Cancer Res* **59**, 279–84.

128. Burger, M., Denzinger, S., Hammerschmied, C., et al. (2006) Mitogen-activated protein kinase signaling is activated in prostate tumors but not mediated by B-RAF mutations. *Eur Urol*, Nov 50(5), 1102–9.

129. Price, D. T., Della Rocca, G., Guo, C., Ballo, M. S., Schwinn, D. A. and Luttrell, L. M. (1999) Activation of extracellular signal-regulated kinase in human prostate cancer. *J Urol* **162** (4), 1537–42.

130. Royuela, M., Arenas, M. I., Bethencourt, F. R., Sanchez-Chapado, M., Fraile, B. and Paniagua, R. (2002) Regulation of proliferation/apoptosis equilibrium by mitogen-activated protein kinases in normal, hyperplastic, and carcinomatous human prostate. *Hum Pathol* **33**, 299–306.

131. Uzgare, A. R., Kaplan, P. J. and Greenberg, N. M. (2003) Differential expression and/or activation of P38MAPK, erk1/2, and jnk during the initiation and progression of prostate cancer. *Prostate* **55**, 128–39.

132. Malik, S. N., Brattain, M., Ghosh, P. M., et al. (2002) Immunohistochemical demonstration of phospho-Akt in high Gleason grade prostate cancer. *Clin Cancer Res* **8**, 1168–71.

133. Fu, Z., Kitagawa, Y., Shen, R., Shah, R., Mehra, R., Rhodes, D., Keller, P. J., Mizokami, A., Dunn, R., Chinnaiyan, A. M., Yao, Z. and Keller, E. T. (2006) Metastasis suppressor gene Raf kinase inhibitor protein (RKIP) is a novel prognostic marker in prostate cancer. *Prostate* **66**(3), 248–56.

134. Voeller, H. J., Wilding, G. and Gelmann, E. P. (1991) v-rasH expression confers hormone-independent in vitro growth to LNCaP prostate carcinoma cells. *Mol Endocrinol* **5**, 209–16.

135. Thalmann, G. N., Sikes, R. A., Wu, T. T., et al. (2000) LNCaP progression model of human prostate cancer: androgen-independence and osseous metastasis. *Prostate* **44**, 91–103.

136. Bakin, R. E., Gioeli, D., Bissonette, E. A. and Weber, M. J. (2003) Attenuation of Ras signaling restores androgen sensitivity to hormone-refractory C4–2 prostate cancer cells. *Cancer Res* **63**, 1975–80.

137. Engelberg, D. (2004) Stress-activated protein kinases-tumor suppressors or tumor initiators? *Semin Cancer Biol* **14**, 271–82.

138. Cazillis, M., Bringuier, A. F., Delautier, D., et al. (2004) Disruption of MKK4 signaling reveals its tumor-suppressor role in embryonic stem cells. *Oncogene* **23**, 4735–44.

139. Teng, D. H., Perry, W. L., 3rd, Hogan, J. K., et al. (1997) Human mitogen-activated protein kinase kinase 4 as a candidate tumor suppressor. *Cancer Res* **57**, 4177–82.

140. Yoshida, B. A., Dubauskas, Z., Chekmareva, M. A., Christiano, T. R., Stadler, W. M. and Rinker-Schaeffer, C. W. (1999) Mitogen-activated protein kinase kinase 4/stress-activated protein/Erk kinase 1 (MKK4/SEK1), a prostate cancer metastasis suppressor gene encoded by human chromosome 17. *Cancer Res* **59**, 5483–7.

141. Kim, H. L., Vander Griend, D. J., Yang, X., et al. (2001) Mitogen-activated protein kinase kinase 4 metastasis suppressor gene expression is inversely related to histological pattern in advancing human prostatic cancers. *Cancer Res* **61**, 2833–7.

142. Ricote, M., Garcia-Tunon, I., Bethencourt, F., et al. (2006) The p38 transduction pathway in prostatic neoplasia. *J Pathol* **208**, 401–7.

143. Ennis, B. W., Fultz, K. E., Smith, K. A., et al. (2005) Inhibition of tumor growth, angiogenesis, and tumor cell proliferation by a small molecule inhibitor of c-Jun N-terminal kinase. *J Pharmacol Exp Ther* **313**, 325–32.

144. Uzgare, A. R. and Isaacs, J. T. (2004) Enhanced redundancy in Akt and mitogen-activated protein kinase-induced survival of malignant versus normal prostate epithelial cells. *Cancer Res* **64**, 6190–9.

145. Zhou, Z. X., Kemppainen, J. A. and Wilson, E. M. (1995) Identification of three proline-directed phosphorylation sites in the human androgen receptor. *Mol Endocrinol* **9**, 605–15.

146. Jenster, G., de Ruiter, P. E., van der Korput, H. A., Kuiper, G. G., Trapman, J. and Brinkmann, A. O. (1994) Changes in the abundance of androgen receptor isotypes: effects of ligand treatment, glutamine-stretch variation, and mutation of putative phosphorylation sites. *Biochemistry* **33**, 14064–4072.

147. Lin, H. K., Yeh, S., Kang, H. Y. and Chang, C. (2001) Akt suppresses androgen-induced apoptosis by phosphorylating and inhibiting androgen receptor. *Proc Natl Acad Sci USA* **98**, 7200–5.

148. Wen, Y., Hu, M. C., Makino, K., et al. (2000) HER-2/neu promotes androgen-independent survival and growth of prostate cancer cells through the Akt pathway. *Cancer Res* **60**, 6841–5.

149. Lubahn, D. B., Joseph, D. R., Sar, M., et al. (1988) The human androgen receptor: complementary deoxyribonucleic acid cloning, sequence analysis and gene expression in prostate. *Mol Endocrinol* **2**, 1265–75.

150. Lubahn, D. B., Joseph, D. R., Sullivan, P. M., Willard, H. F., French, F. S. and Wilson, E. M. (1988) Cloning of human androgen receptor complementary DNA and localization to the X chromosome. *Science* **240**, 327–30.

151. Zhu, Z., Becklin, R. R., Desiderio, D. M. and Dalton, J. T. (2001) Identification of a novel phosphorylation site in human androgen receptor by mass spectrometry. *Biochem Biophys Res Commun* **284**, 836–44.

152. Gioeli, D., Ficarro, S. B., Kwiek, J. J., et al. (2002) Androgen receptor phosphorylation. Regulation and identification of the phosphorylation sites. *J Biol Chem* **277**, 29304–314.

153. Gioeli, D., Black, B. E., Gordon, V., et al. (2006) Stress kinase signaling regulates androgen receptor phosphorylation, transcription, and localization. *Mol Endocrinol* **20**, 503–515.

154. Lannigan, D. A. (2003) Estrogen receptor phosphorylation. *Steroids* **68**, 1–9.

155. Lin, H. K., Hu, Y. C., Yang, L., et al. (2003) Suppression versus induction of androgen receptor functions by the phosphatidylinositol 3-kinase/Akt pathway in prostate cancer LNCaP cells with different passage numbers. *J Biol Chem* **278**, 50902–7.

156. Taneja, S. S., Ha, S., Swenson, N. K., et al. (2005) Cell-specific regulation of androgen receptor phosphorylation in vivo. *J Biol Chem* **280**, 40916–24.

157. Mellinhoff, I. K., Vivanco, I., Kwon, A., Tran, C., Wongvipat, J. and Sawyers, C. L. (2004) HER2/new kinase-dependent modulation of androgen receptor function through effects on DNA binding and stability. *Cancer Cell* **6**, 1–11.

158. Xu, J., Qiu, Y., DeMayo, F. J., Tsai, S. Y., Tsai, M. J. and O'Malley, B. W. (1998) Partial hormone resistance in mice with disruption of the steroid receptor coactivator-1 (SRC-1) gene. *Science* **279**, 1922–5.

159. Heinlein, C. A. and Chang, C. (2002) Androgen receptor (AR) coregulators: an overview. *Endocr Rev* **23**, 175–200.

160. Debes, J. D., Schmidt, L. J., Huang, H. and Tindall, D. J. (2002) p300 mediates androgen-independent transactivation of the androgen receptor by interleukin 6. *Cancer Res* **62**, 5632–6.

161. Rowan, B. G., Weigel, N. L. and O'Malley, B. W. (2000) Phosphorylation of steroid receptor coactivator-1. Identification of the phosphorylation sites and phosphorylation through the mitogen- activated protein kinase pathway. *J Biol Chem* **275**, 4475–83.

162. Rowan, B. G., Garrison, N., Weigel, N. L. and O'Malley, B. W. (2000) 8-Bromo-cyclic AMP induces phosphorylation of two sites in SRC-1 that facilitate ligand-independent activation of the chicken progesterone receptor and are critical for functional co-operation between SRC-1 and CREB binding protein. *Mol Cell Biol* **20**, 8720–30.

163. Gregory, C. W., Fei, X., Ponguta, L. A., et al. (2004) Epidermal growth factor increases coactivation of the androgen receptor in recurrent prostate cancer. *J Biol Chem* **279**, 7119–30.

164. Chrivia, J. C., Kwok, R. P., Lamb, N., Hagiwara, M., Montminy, M. R. and Goodman, R. H. (1993) Phosphorylated CREB binds specifically to the nuclear protein CBP. *Nature* **365**, 855–9.

165. Goodman, R. H. and Smolik, S. (2000) CBP/p300 in cell growth, transformation, and development. *Genes Dev* **14**, 1553–77.

166. Aarnisalo, P., Palvimo, J. J. and Janne, O. A. (1998) CREB-binding protein in androgen receptor-mediated signaling. *Proc Natl Acad Sci USA* **95**, 2122–7.

167. Chadee, D. N., Hendzel, M. J., Tylipski, C. P., et al. (1999) Increased Ser-10 phosphorylation of histone H3 in mitogen-stimulated and oncogene-transformed mouse fibroblasts. *J Biol Chem* **274**, 24914–20.

168. Featherstone, M. (2002) Coactivators in transcription initiation: here are your orders. *Curr Opin Genet Dev* **12**, 149–55.

169. Font, d. M. and Brown, M. (2000) AIB1 is a conduit for kinase-mediated growth factor signaling to the estrogen receptor. *Mol Cell Biol* **20**, 5041–7.

170. Fronsdal, K., Engedal, N., Slagsvold, T. and Saatcioglu, F. (1998) CREB binding protein is a coactivator for the androgen receptor and mediates cross-talk with AP-1. *J Biol Chem* **273**, 31853–9.

171. Gaughan, L., Logan, I. R., Cook, S., Neal, D. E. and Robson, C. N. (2002) Tip60 and histone deacetylase 1 regulate androgen receptor activity through changes to the acetylation status of the receptor. *J BiolChem* **277**, 25904–913.

172. Gnanapragasam, V. J., Leung, H. Y., Pulimood, A. S., Neal, D. E. and Robson, C. N. (2001) Expression of RAC 3, a steroid hormone receptor co-activator in prostate cancer. *Br J Cancer* **85**, 1928–36.

173. McKenna, N. J. and O'Malley, B. W. (2002) Minireview: nuclear receptor coactivators—an update. *Endocrinology* **143**, 2461–5.

174. McKenna, N. J. and O'Malley, B. W. (2002) Combinatorial control of gene expression by nuclear receptors and coregulators. *Cell* **108**, 465–74.

175. Wu, D., Foreman, T. L., Gregory, C. W., et al. (2002) Protein kinase cepsilon has the potential to advance the recurrence of human prostate cancer. *Cancer Res* **62**, 2423–9.

176. Bubulya, A., Wise, S. C., Shen, X. Q., Burmeister, L. A. and Shemshedini, L. (1996) c-Jun can mediate androgen receptor-induced transactivation. *J Biol Chem* **271**, 24583–9.

177. Bubulya, A., Chen, S. Y., Fisher, C. J., Zheng, Z., Shen, X. Q. and Shemshedini, L. (2001) c-Jun potentiates the functional interaction between the amino and carboxyl termini of the androgen receptor. *J Biol Chem* **276**, 44704–711.

178. Bubulya, A., Zhou, X. F., Shen, X. Q., Fisher, C. J. and Shemshedini, L. (2000) c-Jun targets amino terminus of androgen receptor in regulating androgen-responsive transcription. *Endocrine* **13**, 55–62.

179. Tillman, K., Oberfield, J. L., Shen, X. Q., Bubulya, A. and Shemshedini, L. (1998) c-Fos dimerization with c-Jun represses c-Jun enhancement of androgen receptor transactivation. *Endocrine* **9**, 193–200.

180. Wise, S. C., Burmeister, L. A., Zhou, X. F., et al. (1998) Identification of domains of c-Jun mediating androgen receptor transactivation. *Oncogene* **16**, 2001–2009.

181. Edwards, J., Krishna, N. S., Mukherjee, R. and Bartlett, J. M. (2004) The role of c-Jun and c-Fos expression in androgen-independent prostate cancer. *J Pathol* **204**, 153–8.

182. Oettgen, P., Finger, E., Sun, Z., et al. (2000) PDEF, a novel prostate epithelium-specific ets transcription factor, interacts with the androgen receptor and activates prostate-specific antigen gene expression. *J Biol Chem* **275**, 1216–25.

183. Lobaccaro, J. M., Poujol, N., Terouanne, B., et al. (1999) Transcriptional interferences between normal or mutant androgen receptors and the activator protein 1—dissection of the androgen receptor functional domains. *Endocrinology* **140**, 350–57.

184. Yeung, F., Li, X., Ellett, J., Trapman, J., Kao, C. and Chung, L. W. (2000) Regions of prostate-specific antigen (PSA) promoter confer androgen-independent expression of PSA in prostate cancer cells. *J Biol Chem* **275**, 40846–55.

185. Poukka, H., Karvonen, U., Janne, O. A. and Palvimo, J. J. (2000) Covalent modification of the androgen receptor by small ubiquitin-like modifier 1 (SUMO-1). *Proc Natl Acad Sci USA* **97**, 14145–50.

186. Stallcup, M. R. (2001) Role of protein methylation in chromatin remodeling and transcriptional regulation. *Oncogene* **20**, 3014–3020.

187. Wright, J. D., Reuter, C. W. and Weber, M. J. (1995) An incomplete program of cellular tyrosine phosphorylations induced by kinase-defective epidermal growth factor receptors. *J Biol Chem* **270**, 12085–2093.

188. Coker, K. J., Staros, J. V. and Guyer, C. A. (1994) A kinase-negative epidermal growth factor receptor that retains the capacity to stimulate DNA synthesis. *Proc Natl Acad Sci USA* **91**, 6967–71.

189. Boldt, S. and Kolch, W. (2004) Targeting MAPK signalling: Prometheus' fire or Pandora's box? *Curr Pharm Des* **10**, 1885–905.

190. Ratain, M. J., Eisen, T., Stadler, W. M., et al. (2006) Phase II placebo-controlled randomized discontinuation trial of Sorafenib in patients with metastatic renal cell carcinoma. *J Clin Oncol* **24** (16):2505–12.

191. Staehler, M., Rohrmann, K., Haseke, N., Stief, C. G. and Siebels, M. (2005) Targeted agents for the treatment of advanced renal cell carcinoma. *Curr Drug Targets* **6**, 835–46.

192. Stadler, W. M. (2005) Targeted agents for the treatment of advanced renal cell carcinoma. *Cancer* **104**, 2323–33.

193. Rinehart, J., Adjei, A. A., Lorusso, P. M., et al. (2004) Multicenter phase II study of the oral MEK inhibitor, CI-1040, in patients with advanced non-small-cell lung, breast, colon, and pancreatic cancer. *J Clin Oncol* **22**, 4456–62.

194. Lorusso, P. M., Adjei, A. A., Varterasian, M., et al. (2005) Phase I and pharmacodynamic study of the oral MEK inhibitor CI-1040 in patients with advanced malignancies. *J Clin Oncol* **23**, 5281–93.

195. Chen, L., Meng, S., Wang, H., et al. (2005) Chemical ablation of androgen receptor in prostate cancer cells by the histone deacetylase inhibitor LAQ824. *Mol Cancer Ther* **4**, 1311–9.

196. Karhadkar, S. S., Bova, G. S., Abdallah, N., et al. (2004) Hedgehog signalling in prostate regeneration, neoplasia and metastasis. *Nature* **431**, 707–12.

12 Transcription Factors STAT5 and STAT3

Survival Factors for Prostate Cancer Cells

Zoran Culig, MD,
Richard G. Pestell, MD, PhD,
and Marja T. Nevalainen, MD, PhD

CONTENTS

1. INTRODUCTION

The specific mechanisms underlying growth promotion of prostate cancer cells need to be determined for identification of new therapeutic target molecules for treatment development for prostate cancer. Protein kinase

From: *Current Clinical Oncology: Prostate Cancer:*
Signaling Networks, Genetics, and New Treatment Strategies
Edited by: R. G. Pestell and M. T. Nevalainen © Humana Press, Totowa, NJ

signaling pathways are one key area of interest. Protein kinase signaling pathways, often activated by local growth factors, may provide the critical growth signals for premalignant lesions to progress to clinical prostate cancer and organ-confined primary prostate cancer to progress to hormone-refractory disseminated disease. Specifically, protein kinase signaling pathways may have predominantly a permissive role for proliferation and survival of prostate cancer cell clones in which the initial genetic lesions have already occurred.

This chapter will review and discuss *signal transducer and activator of transcription* (STAT)5 and STAT3 signaling pathways in the regulation of proliferation and survival of prostate cancer cells. STAT5 is the major signal transducer downstream of prolactin (Prl) receptors and Jak2 in human prostate cancer cells (1,2), whereas STAT3 is known as the key mediator of interleukin (IL)-6 effects (3–5). Other potential molecular mechanisms and factors underlying activation of these two key survival signaling pathways in prostate cancer will be reviewed. Moreover, STAT5 and STAT3 as therapeutic target proteins for prostate cancer will be discussed.

2. STAT FAMILY OF TRANSCRIPTION FACTORS

2.1. Structure and Function of STAT Proteins

The STAT family of transcription factors has seven members: STAT1, STAT2, STAT3, STAT4, STAT5a, STAT5b, and STAT6, which all are encoded by separate genes. The STAT proteins range in size from 750 to 900 amino acids (90–115 kDa). STAT5 proteins are divided into five structurally and functionally conserved domains (Fig. 1). The N-terminal domain (amino acids 1–125) is well conserved and is involved in stabilizing interactions between two STAT dimers to form tetramers that will bind to adjacent non-consensus STAT-binding sites in DNA (6–9). Tetramer formation strengthens STAT–DNA interaction at adjacent sites (6) and is needed for maximal transcriptional activation of weak promoters. Next to the N-terminal domain is the coiled-coil domain (amino acids 135–315), which consists of a four-helix bundle (10,11). This domain facilitates multiple protein–protein interactions (10,11) that are important for transcriptional regulation. The DNA-binding domain (DBD; amino acids 320–480) mediates direct binding of STATs to DNA and recognizes members of the GAS family of enhancers. The stability of DNA binding is modified by the adjacent linker domain (amino acids 480–575), which is important for confirming the appropriate structure of the DNA-binding motif (12,13). The most highly conserved domain of STAT proteins is the SH2 domain (amino acids 575–680), which mediates both receptor-specific recruitment and STAT dimerization (14). Specifically, dimerization requires the binding of a phosphorylated tyrosine residue of one STAT subunit to the SH2 domain of the other subunit (13–15). Homodimers or heterodimers, but not monomers, are competent to bind DNA. Finally, the carboxy terminus carries

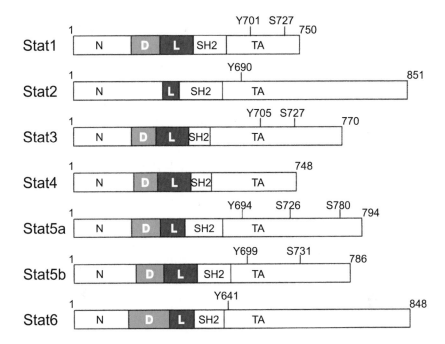

Fig. 1. The major structural features and the phosphorylation sites of mammalian STAT proteins. STAT proteins share an overall general structure that is organized into functional modular domains: N-terminal domain (N), DNA-binding domain (D), linker domain (L), SH2-domain (SH2), and transactivation domain (TA). All STAT molecules have a highly conserved tyrosine phosphorylation site (Y) at or around residue 700. Tyrosine phosphorylation follows ligand-induced activation and is required for dimerization and nuclear translocation. Phosphoserine (S) at or around residue 725 is present in activated STATs 1, 3, 5a, and 5b and is involved in transcriptional regulation. This figure has been reprinted from Chatterjee-Kishore et al. *(270)* with the permission from Elsevier.

a transactivation domain (TAD), which varies considerably in both length and sequence between different STAT family members. The TAD binds critical co-activators and is therefore directly involved in facilitating the initiation of transcription *(13,16)*. The nuclear localization signal is located in the DBD and N-terminal domain in some of the STATs *(17–21)*, whereas the export of STATs from the nucleus has been reported to depend on residues in the coiled-coiled domain and also in the DBD *(19,22)*.

STAT5 comprises two distinct, but highly homologous isoforms, STAT5a (94 kDa) and STAT5b (92 kDa) (Fig. 1), which are encoded by separate genes *(23,24)*. STAT5a has 20 amino acids that are unique in its C-terminal sequence, whereas 8 amino acids in the C-terminus are specific to STAT5b. Furthermore, STAT5b has a five-residue abbreviation of the STAT5a phosphotyrosyl tail segment between the SH2 domain and TAD (Table 1).

Table 1
Peptide Sequence Similarities Between STAT5a and STAT5b

Description	Positions(s)	Similarity (%)
N-domain	1–144	91.0
Coiled-coil domain	145–330	98.7
DNA-binding domain	331–496	97.6
Linker domain	497–592	97.9
SH2 domain	593–685	93.5
Phosphotyrosyl segment	686–701/686–706	87.5
Transactivation domain	702–794/707–786	82.5
Entire coding sequence		94.6

STAT transcription factors are implicated in a wide variety of biologic events. Five of the STATs have a narrow activation profile: STAT1 is activated by interferons (IFNs) and IL-6 *(25,26)*; STAT2 is activated only by IFN-α *(27)*; STAT3 is activated by IL-6 family proteins *(3–5)*; STAT4 in lymphocytes is activated by IL-12 and IFN-α *(28,29)*; and STAT6 is activated only by IL-4 and IL-13 *(30,31)*. By contrast, STAT5a and STAT5b are activated by multiple different ligands such as Prl, growth hormone (GH), erythropoietin (EPO), and thrombopoietin (TPO) *(32)*.

Gene-targeting studies have provided important insight into the distinct roles of STATs in mediating biological responses (Table 2). STAT1, STAT2, *(25–27)* STAT4, and STAT6 *(27,30,31)* are critically involved in immune responses (Table 2). STAT3 knockouts die before gastrulation *(33)*, and conditional gene-targeting studies have revealed a critical role of STAT3 in multiple tissues such as skin, thymus, mammary gland, liver, white blood cells, and neuronal tissue *(13)*. Specifically, hair cycle and wound-healing processes *(34)*, survival of T cells *(34,35)*, acute-phase response of liver cells *(36)*, and appropriate inflammatory response of macrophages and neutrophils are severely compromised in STAT3 mutant mice *(37)*. Moreover, apoptosis of mammary epithelial cells after lactation is delayed *(38)*, whereas survival of neuronal cells is impaired *(39)* in conditional STAT3 mutant mice, indicating both positive and negative tissue-specific effects of STAT3 on apoptotic cell death.

Importantly, the phenotypes of STAT5a/b knockout mice are not fully functionally redundant. Specifically, STAT5a-null female mice are defective in Prl-dependent mammary gland development *(40)* and STAT5a-null male mice exhibit defects in prostate epithelium *(41)*, which will be discussed in more detail below. By contrast, STAT5b-null mice fail to respond effectively to GH *(42,43)* and have severe anemia suggesting defective hematopoiesis due to impaired response to hematopoietins *(44)*. Lymphoid development

Table 2
Functions of STAT Proteins as Revealed by Gene-Targeting in Mice[a]

STAT protein	Phenotype of null mice	Activating ligands
STAT1	Impaired responses to interferons (IFNs)	IFNs, IL-6
STAT2	Impaired responses to IFNs	IFNs
STAT3	Embryonic lethality; multiple defects in adult tissues including impaired cell survival (both positive and negative) and impaired response to pathogens	IL-6 family
STAT4	Impaired $T_H 1$ differentiation owing to loss of IL-12 responsiveness	IL-12, IFN-α
STAT5a	Impaired mammary gland development owing to loss of prolactin (Prl) responsiveness	Prl, GH, EPO, TPO, G-CSF, Leptin, OSM, LIF
STAT5b	Impaired growth owing to loss of growth hormone responsiveness	GH, Prl, EPO, TPO, G-CSF, Leptin, OSM, LIF
STAT6	Impaired $T_H 1$ differentiation owing to loss of IL-4 responsiveness	IL-4, IL-13

[a]Reviewed in refs. *(13)*, *(15)*, and *(16)*. IL, interleukin; $T_H 1$, T helper 1 cell; GH, growth hormone; EPO, erythropoietin; TPO, thrombopoietin; OSM, oncostatin M; LIF, leukemia inhibitory factor; G-CSF, granulocyte-colony stimulating factor.

and differentiation are severely impaired in STAT5a/b double-knockout mice. Moreover, absence of STAT5a/b also abrogates T-cell receptor γ-rearrangement and peripheral CD8[+] T-cell survival *(45)*. No reports of prostate phenotypes of STAT3, STAT5b, or STAT5a/b double-knockout mice currently exist, which are important unresolved issues.

Considerable evidence has accumulated suggesting STAT3 and STAT5 contribution to growth of human prostate cancer. By contrast, little is currently known about involvement of STAT1, STAT2, STAT4, and STAT6 in differentiation or growth of prostate cancer. Activated STAT3 has been linked also to head and neck cancers, mammary cancers, skin cancers, and hematological malignancies *(46–49)*, whereas STAT5a/b has been mostly associated with initiation *(50–53)* and differentiation *(54–56)* of breast cancer, growth of hematological malignancies *(46,57,58)*, and head and neck cancers *(59–63)*.

2.2. Genetic Location of STAT Transcription Factors

The STAT proteins are thought to have risen from a single gene, and the distribution of STAT genes into three genetic loci suggests that the initial STAT gene underwent several consecutive duplications. Specifically, the members of STAT family are genetically localized to three chromosomal regions *(64)*. The genes encoding STAT3, STAT5a, and STAT5b map to human chromosome 17 (bands q11-1 to q22); STATs 1 and 4 map to human chromosome 2 (bands q12 to q33); and STATs 2 and 6 map to human chromosome 12 (bands q13 to q14-1). Additional variety in STAT action in cells is produced by differential splicing. The extent of variable splicing has not been widely explored *(15)*.

2.3. The Jak–STAT Signaling Cascade

Most STAT-activating cytokine receptors do not have tyrosine kinase activity, which is provided by receptor-associated cytoplasmic proteins from the Janus kinase (JAK) family *(12,13,15,16,65)*. There are four JAK proteins in mammalian cells, JAK1, JAK2, JAK3, and TYK2 *(13,15,16,66)*. They range in size from 120 to 130 kDa and, except for Jak3, Jaks are ubiquitously expressed *(67)*. Jak proteins share seven regions of high homology. Jak homology domain 1 (JH1), which is the kinase domain, resides near the carboxyl terminus of the protein. Immediately upstream of JH1 is the JH2 pseudokinase domain, which resembles the JH1 domain and has a negative regulatory function. The amino-terminal JH domains, JH3–JH7, constitute a FERM (four-point-one, ezrin, radixin, moesin) domain and mediate association with cytokine receptors *(12,13,15,16,66,68)*. The predominant Jak that associates with STAT5a/b is Jak2, whereas Jak1 is the most important tyrosine kinase upstream of STAT3 *(12)*.

Ligand-induced receptor dimerization brings two JAK2 molecules close to each other allowing the JAKs to activate each other by phosphorylating specific receptor tyrosine motifs. STATs and other signaling molecules that recognize these motifs, typically through their SH2 domains, are recruited to these docking sites and are then themselves activated by JAK-dependent tyrosine phosphorylation of a conserved tyrosine residue in the C-terminus. Phosphorylation of a C-terminal tyrosine residue activates STAT3 (Y705) and STAT5 (5a, Y694; 5b Y699). Once active, STATs dimerize through a phosphotyrosine–SH2 domain interaction *(10,11)* (Fig. 2). Diverse protein kinases phosphorylate STATs on serine residues, allowing additional cellular signaling pathways to potentiate the primary STAT-activating stimulus *(69)*. Activation of STAT3 is supplemented by phosphorylation of a specific serine residue of STAT3 (S727), whereas the corresponding serine phosphorylation of STAT5 (5a, S725; 5b, S730) might have a negative role in transcriptional activity of STAT5 *(70–72)*.

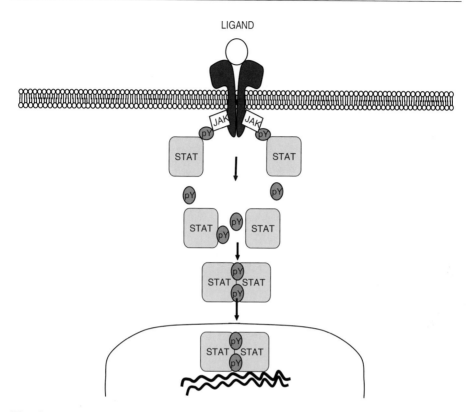

Fig. 2. Jak–STAT signaling pathway. Binding of the ligands to the cytokine receptor results in activation of receptor-associated kinases of Janus kinase tyrosine kinase family. These tyrosine kinases subsequently phosphorylate the cytoplasmic tails of the receptors to provide docking sites for STAT proteins. STAT proteins become substrate for tyrosine phosphorylation, and phosphorylated STAT proteins dimerize and translocate to nucleus to bind to the promoter regions of the target genes.

STAT dimers are transported from the cytoplasm to the nucleus to bind to the promoters of their target genes to regulate transcription. Recent studies support the concept that non-phosphorylated STAT proteins shuttle freely between the cytoplasm and nucleus in the absence of cytokine activation *(73–76)*. While non-phosphorylated STATs cycle between the cytoplasm and nucleus, dimerization of STAT proteins has been suggested to preclude carrier-free nuclear import of STATs and results in switch to carrier-dependent translocation *(73)*. Most STAT dimers recognize an 8- to 10-bp inverted repeat DNA element with a consensus sequence of TTCC (C or G) GGAA, which is referred to as a GAS element (γ-IFN activation sequence) *(77)*.

2.4. Negative Regulators of STAT Signaling

A number of different mechanisms regulate the duration and magnitude of STAT activation at the cytoplasmic and nuclear levels. These mechanisms include the actions of cytoplasmic and nuclear phosphatases, the interaction of inhibitory proteins, and covalent modifications of STAT proteins that result in targeted degradation of the active STAT protein or inhibition of STAT transcription.

Both cytoplasmic and/or nuclear phosphatases inactivate STAT proteins. The protein tyrosine phosphatase (PTP), SHP-2, directly interacts with STATs in the cytoplasm (78,79). Other known phosphatases inactivating STATs are cytosolic PTP, PTP1B, and a nuclear phosphatase, TC-PTP (80,81). Protein inhibitors of activated STAT (PIAS) proteins inhibit STAT protein activation by direct association to STATs. The mammalian PIAS family members include PIAS1, PIAS3, PIASx and PIASy, and alternative splicing variants of PIASx (82,83). DNA binding of STAT1 and STAT3 is selectively inhibited by PIAS1 and PIAS3, respectively (82,84).

A third mechanism for down-regulation of signaling by STAT proteins involves cytokine-inducible suppressors of cytokine signaling (SOCS) proteins (85). There are eight members of the SOCS family, including CIS (cytokine-inducible SH2 domain protein) and SOCS1–7. These proteins are all structurally related: they possess a central SH2 domain and a conserved C-terminal motif, termed the SOCS box (85). Individual SOCS proteins have several different strategies to attenuate STAT signaling. These include direct binding to JAKs to inhibit their activity, binding to receptors to block the binding of JAKs, and competing for STAT binding to activated receptors (85).

3. REGULATION OF GROWTH AND DIFFERENTIATION OF NORMAL AND MALIGNANT PROSTATE EPITHELIUM BY STAT5 SIGNALING PATHWAY

3.1. Transcription Factor STAT5a/b Regulates the Viability of Prostate Cancer Cells

The background and rationale for the studies leading to identification of transcription factor STAT5a/b as a crucial protein for survival of prostate cancer cells were based on the identification of STAT5a and STAT5b as the key signaling proteins that mediate the effects of Prl in normal and malignant prostate tissue (1,2,86). Prl, in turn, has been shown to be a powerful mitogen and survival factor for prostate epithelium in numerous studies (86–121). Importantly, Prl is produced as autocrine growth factor in both normal and malignant prostate epithelial cells (2,87,88,122).

The first step in examining STAT5a/b in the regulation of function of the prostate gland was carried out using STAT5a knockout mice. The prostate phenotype of STAT5a-null mice was the primary interest to analyze, because studies of Prl signaling in the mammary gland showed that STAT5a is the principal mediator of Prl effects in breast tissue *(40,123)*. The prostate acinar epithelium was defective in STAT5a knockout mice. Specifically, the epithelial defect was characterized by acinar cyst formation, local disorganization and shedding of the epithelial cells to the glandular lumini *(41)*. Affected acini were typically filled with desquamated granular epithelial cells that had become embedded in dense coagulated secretory material inside the broken acini *(41)*. The defective prostate tissue architecture in STAT5a-deficient mice was not associated with increased prostate size or morphological hallmarks of epithelial hyperplasia, which suggested that STAT5a is rather involved in growth promotion of prostate gland.

Analysis of the prostate phenotypes of STAT5b or STAT5ab double-knockout mice will be important in future studies. This is because the prostate phenotype of STAT5a-null mice was likely to be compromised by the redundant function of other STAT family proteins, particularly STAT5b *(124,125)*. An example of such redundant function of STATs is in mammary glands of STAT5a-null mice in which STAT5a is critical for normal lactogenesis but STAT5b can compensate for its absence after multiple pregnancies *(124,125)*. Similarly, STAT5b might have compensated for the lack of STAT5a in the prostates of STAT5a-deficient mice. In addition, because the lack of STAT5a in this study was due to a germ-line mutation of STAT5a gene, a long period of time was allowed for functional compensation by other STATs to occur throughout the development of the mice. Even more specific results on the significance of STAT5a/b in growth regulation of normal prostate epithelium would be achieved by conditional prostate-specific targeting of STAT5a/STAT5b/STAT5a/b.

On the basis of the prostate phenotype of STAT5a-null mice, it became important to determine the consequences of blocking the action of STAT5a and STAT5b in human prostate cancer cells. A dominant-negative (DN) mutant of STAT5a, which inhibits both STAT5a and STAT5b, was created by deletion of the TAD of STAT5a. This mutant was cloned to an adenoviral expression vector to achieve high expression levels of DNSTAT5a/b in human prostate cancer cells and, thus, effective inhibition of STAT5a/b *(126)*. The results were interesting: inhibition of STAT5a/b in all STAT5-expressing prostate cancer cell lines (CWR22Rv1, LNCap, DU 145) induced massive apoptosis within 2–4 days as determined by several different assays such as cell morphology, metabolic cell viability assays, DNA fragmentation, and activation of Caspases 3 and 9 *(126)*. These initial data showing a critical involvement of STAT5a/b in the regulation of viability of prostate cancer cells were later confirmed by studies by Kazansky et al., in the TRAMP mouse prostate cancer model *(127)*. Specifically, inhibition of STAT5a and STAT5b in TRAMP mouse prostate tumor cells by inducible expression of a carboxy-terminally truncated STAT5b mutant, which inhibits both STAT5a and STAT5b, decreased growth in soft agar and tumor formation in nude mice. Future studies need to determine the individual roles of STAT5a and STAT5b as survival factors for prostate cancer. In addition, identification of the molecular mechanisms underlying rapid death of prostate cancer cells when STAT5a/b is inhibited may reveal additional therapeutic target proteins for prostate cancer and is therefore important.

3.2. Active STAT5a/b is Abundant in High-Grade Prostate Cancer

When clinical human prostate tissue sections were analyzed for active STAT5a/b, it became clear that STAT5a/b is constitutively activated in human prostate cancer cells, but not in the epithelium of adjacent normal prostate glands. The first analysis was done in a material from 40 human prostate cancer specimens of various histological grades *(126)*. The next study focused on determining the distribution of active STAT5a/b in clinical prostate cancers of different histological grades. The results of this study including 114 prostate cancer samples showed that activation of STAT5a/b was strongly ($p <$ 0.001) associated with high-grade human prostate cancer *(2)*. The preferential expression of active STAT5a/b in high-grade prostate cancers was confirmed by the third study of 357 clinical prostate cancer specimens, which was performed using tissue microarrays *(128)*. In this independent set of prostate cancers, active Stat5a/b was associated with high Gleason grades ($p = 0.03$). Collectively, these data indicated that active STAT5a/b is particularly abundant in high-grade prostate cancer.

3.3. Activation of STAT5a/b in Primary Prostate Cancer Predicts Poor Clinical Outcome

The concept of involvement of STAT5a/b in clinical progression of prostate cancer was further supported by evidence of predictive role of STAT5a/b of early recurrence of prostate cancer *(128)*. Specifically, active STAT5a/b in a primary prostate tumor predicted early recurrence of prostate cancer after the initial treatment *(128)*. Most importantly, if only prostate cancers of intermediate Gleason grades *(3,4)* were analyzed, active STAT5a/b still remained an independent prognostic marker of early disease recurrence *(128)*. This result suggested that the presence of active STAT5a/b in primary prostate cancer of intermediate histological grade is associated with progressive disease.

The distribution of active STAT5a versus STAT5b in prostate cancers of different histological grades should be determined, and the prognostic value of STAT5a versus STAT5b assessed. The optimal material for such a study would be prostate cancer specimens from patients who have not received adjuvant therapies besides radical prostatectomy, because the different adjuvant therapies might affect the activation of STAT5a/b. Because active STAT5 in primary prostate cancer predicted early disease recurrence which is often hormone-refractory prostate cancer, and because STAT5a/b promotes growth of prostate cancer cells, the contribution of STAT5a/b to androgen-independent growth of prostate cancer needs to be evaluated. Specifically, the involvement of STAT5a/b in androgen-independent growth of prostate cancer should be investigated from two different directions: *(1)* STAT5a/b as an androgen receptor (AR)-independent survival factor for prostate cancer cells during androgen ablation, *(2)* STAT5a/b as a survival factor acting through enhancing the transcriptional activity of AR in androgen-deprived prostate cancer. On the basis of the finding that active STAT5 predicts poor clinical outcome of prostate cancer, stimulation of migration, motility, and invasion of human prostate cancer cells by STAT5a/b should be evaluated as well.

3.4. Pathways Leading to Activation of STAT5a/b in Prostate Cancer Cells

Prl is one of the predominant peptide factors currently known to activate Jak2–STAT5a/b in normal and malignant prostate epithelium *(1,2)*. Prl promotes proliferation and survival of prostate cells, and Prl is produced locally by normal prostate epithelium and by prostate cancer *(86–90,115–117)*.

3.4.1. PROLACTIN IS LOCALLY EXPRESSED IN NORMAL AND MALIGNANT PROSTATE EPITHELIUM

Prl is a polypeptide hormone that belongs to Prl/GH/placental lactogen (PL) family. Prl is similar to GH (16% amino acid identity) and PL (13% amino acid identity) in its primary structure *(129)*. The signal peptide of 30 amino acids is removed from the amino-terminal end of Prl prohormone by proteolytic cleavage *(130,131)* producing a secretory product with a molecular weight of approximately 23,000 23KD *(132,133)*. Prl is encoded by a single gene containing six exons in humans and is located on chromosome 6 (6p22.2–p21.3) *(134,135)*. Transcription of the gene is driven by two tissue-specific promoters, a proximal promoter that is used in the pituitary gland and a distal promoter that is used in extrapituitary cells and tissues, including decidua, myometrium, and lymphoid cells *(136–138)*. A non-coding exon 1a is only expressed in extrapituitary tissues and has a transcriptional start site 5.8 kb upstream of the pituitary start site *(136)*. In extrapituitary sites, exon 1a is spliced to the first pituitary exon 1b, generating a transcript that is approximately 150 bp larger than the pituitary counterpart *(139)*, differing only in the 5′-untranslated region. The downstream promoter that directs transcription in pituitary lactotrophs is under control of the POU-homeodomain transcription factor Pit-1. Transcriptional control of the distal, non-pituitary start site includes two consensus binding sites for CCAAT/enhancer-binding proteins (C/EBP) in human decidual cells *(140)*.

The Prl protein and gene are expressed at a high level in normal and malignant prostate epithelial cells *(2,87,88,122)*. First, Prl protein synthesis in rodent and human prostate epithelium was shown by immunohistochemistry *(87,88)*, and Prl protein was localized to the secretory granules of prostate epithelial cells by immuno-electron-microscopy *(88)*. These results suggested a secretory route of Prl protein from the prostate epithelial cells. Typically, Prl-producing cells were single epithelial cells dispersed throughout the epithelium or clusters of epithelial cells distributed along the prostate ducts *(87,88)* resembling neuroendocrine cells *(141)*. Expression of the Prl gene was detected by *in situ* hybridization and reverse transcription (RT)-PCR in normal prostate *(87,88)*, and later, by Xu et al. *(122)* using RT-PCR in DU-145 human prostate cancer cell line grown as xenograft tumors in nude mice. Studies focused on clinical human prostate cancer specimens showed that Prl protein expression is associated with high histological grade of human prostate cancer *(2)*. Autocrine Prl in prostate cancer may be the factor responsible for constitutive activation of STAT5a/b in human prostate cancer and, therefore, would also possibly serve as a therapeutic target protein for prostate cancer. Future studies should determine the promoter of Prl gene that drives Prl gene expression in normal and malignant prostate cells. This will allow identification of actors regulating Prl gene expression in prostate cancer.

3.4.2. PROLACTIN RECEPTORS ARE EXPRESSED IN NORMAL AND MALIGNANT PROSTATE TISSUE

Prl exerts its biological actions via membrane-bound receptors *(142,143)*. The human Prl-receptor (PrlR) gene is located on chromosome 5 (p13–14) and contains 11 exons *(144,145)*. The PrlR protein is a glycosylated, single-pass transmembrane protein with the N-terminus in the extracellular space. The PrlR gene in both rodents and humans gives rise to three PrlR isoforms, a long, an intermediate, and a short form *(144,145)*. The different PrlR isoforms have strictly identical extracellular (ligand-binding) domains and differ only by the length of their cytoplasmic tail *(146)*. There are no known enzymatic motifs intrinsic to the cytoplasmic domain of PrlRs. In both rat prostate and human prostate, specific Prl-binding sites have been demonstrated in a number of studies over the years *(92,93,101,147–151)*. After the cDNA encoding the PrlR was cloned *(152)*, the expression of both the short (42–44 kDa) and long (82–95 kDa) PrlR isoforms were identified in normal rat and human prostate at protein and mRNA levels *(86,87)*. Specifically, PrlR proteins in prostate are primarily expressed in epithelial cells, in which the major immunohistochemical reaction is localized to the apical surface of the cells *(86,87)*. This localization would allow direct stimulation of the epithelium by autocrine/paracrine Prl. A study by Leav et al. *(93)* demonstrated PrlR expression in clinical human prostate cancer specimens. In addition to human prostate cancer samples, PrlRs are expressed in human prostate cancer cell lines DU-145, LnCaP, PC-3, and CWR22Rv1 *(122,153)*.

3.4.3. PROLACTIN PROMOTES GROWTH OF NORMAL AND MALIGNANT PROSTATE EPITHELIUM

Involvement of Prl in development and growth of normal prostate and prostate neoplasia has been established and continues to be a subject of active research. In fetal human prostate, receptors for Prl are expressed at a high level *(93)*, and Prl induces growth of the developing prostate gland in rodents *(94,154,155)*. Targeted disruption of the Prl gene in mice caused significant reduction in prostate size *(95)*, whereas aromatase knockout mice, which are deficient of estrogens but have high serum testosterone and Prl levels, had enlarged prostates *(96)*. In mature prostate gland, Prl regulates accumulation and secretion of citrate, which are physiological functions of prostate. Specifically, Prl stimulates metabolic events related to citrate production and zinc uptake of prostate epithelium *(92,99,100,102,156–160)*.

Transgenic mice overexpressing Prl developed hyperplastic prostate enlargement *(115)*. Importantly, the development of the hyperplasia was later shown not to be dependent on serum androgen levels *(117)*. Moreover, induction of preneoplastic lesions in prostate epithelium by chemical carcinogens was enhanced by hyperprolactinemia in rats *(109)*, and hyperprolactinemia-induced increases in weight, nucleic acid content, and protein content of prostate have been reported in a number of studies *(101,109,112,113,161–164)*. Surprisingly, no specific prostate phenotype was found in mice deficient of PrlRs *(165)*. However, it is important to note that the lack of the expected phenotype in the gene deletion models may be due to the redundant regulatory mechanisms that become active during organ development.

Shown by organ cultures of normal and malignant prostate tissues, Prl directly stimulates proliferation of prostate epithelial cells *(87,89)* and inhibits apoptosis *(90)*. Direct growth-promoting effects of Prl have been shown also for human prostate cancer cells *(118,153)*. The demonstration that Prl directly affects growth of prostate cells is important because a number of preceding *in vivo* studies suggested Prl action on prostate epithelium only through enhancement of androgen action *(100,119–121,161,164,166,167)* and through secondary endocrinological changes *(168)*.

Autocrine Prl–PrlR–Jak2–STAT5a/b signaling pathway potentially provides prostate cancer cells with the ability to survive in a growth environment lacking androgens. Interestingly, during estrogen, antiestrogen, antiandrogen, or GnRH analog therapy for prostate cancer, Prl levels can increase and are predictive of poor prognosis *(169)*. Also, locally produced Prl by prostate epithelium *(87,88)* may be the major source of Prl that is available for prostate cancer cells, and may also explain the limited success of adjuvant therapies of prostate cancer with inhibitors of pituitary Prl secretion *(170,171)*.

3.5. Other Potential Pathways Leading to Activation of STAT5a/b in Prostate Cancer

Mechanisms that are responsible for constitutive activation of STAT5a/b in prostate cancer other than autocrine Prl secretion are virtually unknown. In addition to locally produced Prl, constitutive activation of STAT5 in malignant prostate epithelium might be due to amplification of STAT5a/b genes. Specifically, chromosome 17q, where STAT5a and STAT5b are located, is frequently altered in both hereditary and incidental prostate cancer *(172–182)*. Chromosome 17q showed allelic imbalance in prostate cancer in eight studies *(183–190)*, and gains in chromosome 17q were detected in five studies *(185–188,190)*. Moreover, three large studies linked a prostate cancer susceptibility gene to chromosome 17q (17q21) *(191–193)* suggesting involvement of genes in this region in inherited form of prostate cancer. Moreover, STAT5a/b might be activated in prostate cancer cells by activating mutations of Jak2 *(194–196)* or by tyrosine kinases such as Src *(47,197)*, Bcr-Abl *(198)*, or Tel-Jak *(199)*. In addition, decrease of STAT5a/b phosphatases or inhibitory proteins of STAT5a/b (PIAS) may result in constitutively active STAT5a/b in malignant prostate epithelium. GH, which is one of the principal activators of STAT5b in a number of tissues, might be as well involved in activating STAT5a/b in malignant prostate epithelium *(200–205)*. However, no evidence of direct effects of GH on stimulation of growth of prostate cancer cells currently exists. Finally, there are no reports on contribution of other local growth factors such as epidermal growth factor (EGF) *(206–208)* and fibroblast growth factor (FGF) *(172,209)* on STAT5a/b activation in prostate cancer cells.

4. REGULATION OF PROSTATE CANCER CELL VIABILITY BY INTERLEUKIN-6–STAT3 SIGNALING PATHWAY

4.1. Regulation of Prostate Cancer Cell Growth by STAT3

The oncogenic role of STAT3 has been well documented in various tumors, most frequently in multiple myeloma (for a more detailed review, see ref. *173*). Nuclear translocation and activation of STAT3 has been observed in areas of chronic inflammatory disease such as colitis or viral hepatitis. Investigation of early lesions that occur during early prostate carcinogenesis may therefore uncover novel signal transduction mechanisms that include activation of STAT3 by other growth factors as well as interactions between STAT3 and other signaling proteins *(173)*.

The contribution of Stat3 to growth of established prostate cancer is more complex. Specifically, it is unclear to some extent whether STAT3 promotes or inhibits growth of prostate cancer cells. Different methodological approaches for investigating the role of STAT3 in prostate cancer may have contributed to the controversy. Gao's group has established a model based on stable transfection of IL-6 cDNA in LNCaP cells *(174)*. It was shown that there was an increased phosphorylation of STAT3 and mitogen-activated protein kinase (MAPK) associated with acquisition of a growth advantage in LNCaP cells stably expressing IL-6. Moreover, LNCaP cells that overexpressed IL-6 gained the ability to proliferate in an androgen-independent manner. Because the major problem in prostate cancer is development of resistance to therapy with androgen ablation or hormonal antagonists, the LNCaP subline generated by stable transfection of IL-6 cDNA is an excellent model for studies of tumor progression toward therapy resistance. STAT3 activation in prostate cancer was studied by other researchers in non-transfected cells after stimulation with IL-6. Findings similar to those of Gao's group were reported by some but not all investigators *(175,176)*. Specifically, IL-6 caused growth retardation of LNCaP cells and this was associated with induction of STAT3 phosphorylation. These controversial results have not been clarified yet. In another study, LNCaP cells stably transfected with a DN STAT3 showed a proliferative response *(177)*. By contrast, two studies have implicated STAT3 in the promotion of proliferation and inhibition of apoptosis in the DU 145 cell line *(178,179)*. These studies, using an antisense approach, showed that STAT3 acts as a positive growth factor required for cell survival. Moreover, analysis of JAK/STAT3 activation in a LNCaP derivative generated after chronic treatment with IL-6 (LNCaP–IL-6$^+$) and in CWR22Rv1 cells derived from the relapsed xenograft CWR22R revealed lack of STAT3 phosphorylation after stimulation with IL-6 *(180,181)*. Taken together, these data from several laboratories indicate that there is a complex and incompletely understood relationship between IL-6, activation of the JAK/STAT3 pathway, and cell growth in prostate cancer. Development of new experimental models and additional studies are needed to obtain more information on mechanisms involved in the regulation of STAT3 phosphorylation and prostate cancer cell growth and survival.

STAT3 phosphorylation has been investigated by several researchers in clinical prostate cancer by immunohistochemistry. Compared with benign tissue, nuclear expression of phosphorylated STAT3 was increased in cancer *(179,182,210)*. However, activation of STAT3 in primary prostate cancer was not associated with clinical outcome in a study including 357 primary prostate cancers *(128)* with 30-year clinical follow-up data (Nevalainen, unpublished results). In the same material, active STAT5a/b in primary prostate cancer predicted poor clinical outcome *(128)*. A study in breast cancer showed that patients who presented with increased nuclear expression of activated STAT3 had a survival benefit *(211)*.

4.2. Pathways Leading to Activation of STAT3 in Prostate Cancer Cells: Interleukin-6

The major activator of STAT3 in prostate cancer is IL-6 *(3–5)*. IL-6 is a pleiotropic cytokine of molecular weight between 21 and 28 kDa. IL-6 is composed of four long helices and an additional mini helix. One long loop is joining A and B, another one C and D helices, and there is a short loop between B and C helices *(212)*. IL-6 helices are straight whereas the A helix of the IL-6-related cytokines oncostatin M (OSM) and leukemia inhibitory factor (LIF) are kinked *(213)*. This structural finding has implications on differential recruitment of cytokine receptors and will be discussed in more detail below.

4.2.1. IL-6 PRODUCTION IN PROSTATE CELLS

There are several positive and negative factors that determine expression of IL-6. IL-6 gene expression is up-regulated by nuclear factor NF-κB, prostaglandin E2, activating protein-1, and the homeobox gene GBX2. By contrast, vitamin D, retinoblastoma (Rb), and steroid receptors are known as IL-6 repressors. NF-κB is a major positive regulator of IL-6, and its expression in prostate cancer is higher than in normal prostate tissue *(214)*. Increased levels of cyclooxygenase-2 and prostaglandin E2 may be responsible for up-regulation of IL-6 in chronic inflammation areas. Contribution of NF-κB and members of the activating protein-1 complex JunD and Fra-1 to increased expression of IL-6 has been demonstrated in prostate cancer cells *(215)*. Mutations of the binding sites for those factors in the IL-6 gene promoter were shown to lead to a significant down-regulation of IL-6. The GBX2 expression is also higher in malignant prostate tissue compared with benign prostate tissue and may therefore contribute to IL-6 elevation in prostate cancer *(216)*.

High IL-6 levels were measured in supernatants from PC-3 and DU 145 cells that are AR negative *(217)*. Studies carried out in osteoblasts have demonstrated that steroids may decrease the expression of IL-6 and its receptor *(218,219)*. Lack of AR expression may be a possible explanation for high IL-6 expression in PC-3 and DU 145 cells and, conversely, for the absence of IL-6 in the LNCaP cell line, which is AR positive. An LNCaP subline, which was generated in the presence of exogenous IL-6, expresses IL-6 mRNA and protein, in contrast to parental cells that do not produce IL-6 *(220)*. IL-6 is expressed in human prostate tissue expression has been shown by immunohistochemistry In benign tissue, IL-6 immunoreactivity was confined to basal cells, whereas glandular cells were weakly positive *(221)*. By contrast, IL-6 levels were high in the supernatants obtained from prostate stromal cells, which suggests a possible paracrine growth-regulatory loop in prostate tissue. Co-localization studies are needed to clarify the relationship between IL-6 and AR expression in clinical human prostate cancer.

4.2.2. IL-6 RECEPTORS IN PROSTATE CANCER

The IL-6 receptor (IL-6R) consists of two subunits: gp80 or IL-6Rα, which is cytokine-specific, and gp130, which is shared by IL-6 and related cytokines. IL-6 binds first to IL-6Rα *(222)*, which then recruits the signal-transducing subunit gp130. The association of gp130 with IL-6 and IL-6Rα leads to the formation of the high-affinity IL-6R complex, to the homodimerization of two gp130 subunits, and to signal transduction. Signaling through gp130 can be enhanced by a soluble receptor form that is generated by limited proteolysis of the membrane-bound receptor or translation of alternatively spliced RNA *(223,224)*. The responsiveness of a given cell type to IL-6 is mainly determined by expression of α-subunits or the presence of soluble receptors. IL-6Rs are phosphorylated rapidly at tyrosine residues by JaK1 and JaK2, and this step is followed by nuclear translocation and phosphorylation of STAT3 *(225)*.

IL-6R subunits are ubiquitously expressed in benign and malignant prostate tissue *(221)*. In addition, expression of IL-6 and its receptor has been quantified in benign and malignant tissue extracts, and the expression levels are increased in early stages of prostate carcinogenesis, suggesting that IL-6 regulates cellular events during early stages of malignant transformation *(226)*. Moreover, high IL-6 levels are measured in sera from patients with metastatic disease. In conclusion, prostate cancer cells are a source of IL-6 that acts through its respective receptors.

4.2.3. REGULATION OF PROSTATE CANCER CELL GROWTH BY IL-6

Clinical observations suggest that there is an association between serum IL-6 and prostate cancer morbidity *(217,227)*. Serum levels of IL-6 higher than 7 pg/ml were considered a bad prognostic factor in prostate cancer *(228)*. This initial work served as the rationale for experimental studies on IL-6 signaling in prostate cancer. However, the results on the effects of IL-6 on growth and survival of prostate cancer cells have been controversial *(203,204,229)*. This could be in part explained by activation of several different signaling pathways by IL-6 in prostate cancer. Specifically, treatment with IL-6 could cause either exclusive or non-exclusive phosphorylation of intracellular cascades of JaK/STAT, p42/p44 or p38 MAPK, or phosphatidylinositol 3-kinase (PI3K) *(230)*. Under certain experimental conditions, observed growth stimulation might be caused by interaction between the IL-6R and Her-2 that leads to activation of the MAPK signaling *(231)*. Besides affecting cell growth and apoptosis, IL-6 is implicated in a reversible regulation of neuroendocrine phenotype of prostate cancer *(232–234)*. LNCaP cells show typical processes

and elongations after treatment with IL-6 or compounds that increase intracellular cAMP or when cultured in steroid-depleted culture conditions. Morphological changes were paralleled with expression of neuroendocrine markers
chromogranin A and neuron-specific enolase or peptides such as bombesin or
serotonin that influence growth of adjacent epithelial cells. Neuroendocrine
differentiation was also induced in PC-3 cells by stable transfection with constitutively active STAT3 (235). Presence of high levels of neuropeptides in sera
of prostate cancer patients is in most cases associated with advanced prostate
cancer grade or stage.

To mimic conditions in prostate cancer patients, LNCaP cells were treated
with IL-6 for several months (220). In the newly generated LNCaP subline,
exogenous IL-6 did not cause growth inhibition. Instead, LNCaP–IL-6$^+$ cells
exhibited growth advantage in vitro and in vivo when grown as xenograft
tumors in nude mice. Two independent experiments demonstrated that LNCaP–
IL-6$^+$ tumor volumes were significantly higher than tumors of the parental
LNCaP cells. Moreover, cyclin-dependent kinases that govern cell cycle
progression were expressed at a higher level in LNCaP–IL-6$^+$ cells compared
with parental LNCaP cells (236), whereas the tumor suppressors p27 and pRb
were undetectable. IL-6 acts as an autocrine growth factor through the p42/p44
MAPK in LNCaP–IL-6$^+$ cells, in contrast to paracrine inhibition observed in
the parental cells. Increased tumorigenicity in vivo was also recently reported
for the LNCaP derivative generated after co-culture with stromal cells that
produce IL-6. Specifically, IL-6 caused indirect effects through up-regulation
of vascular endothelial growth factor (VEGF) and its receptor VEGFR-2 in
LNCaP–IL-6$^+$ (237). The presence of VEGF and VEGFR-2 in prostate cancer
cells and partial inhibition of proliferation of these cells by anti-VEGFR-2
antibodies suggested that VEGF autocrine loop was developed in conditions
in which IL-6 levels were increased. IL-6 has been shown to act as a positive
growth factor also in primary cultures of prostate cells and in a cell line
derived from prostate epithelial neoplasia (176,237). The autocrine IL-6 loop
is also present in DU 145 and PC-3 cells (238). In PC-3 cells, endogenous IL-6
promoted cell survival through phosphorylation of PI3K (175). In summary,
most studies performed with different prostate cancer models suggest that IL-6
is a potent growth factor in prostate cancer.

4.3. Other Potential Pathways Leading to Activation of STAT3 in Prostate Cancer

IL-6-related cytokines such as IL-11, OSM, LIF, or ciliary neurotrophic
growth factor may activate STAT3 signaling pathway in prostate cancer cells.
The IL-11 receptor levels are increased in prostate cancer, and this could in part
explain constitutive activation of STAT3 observed in prostate cancers (182). In
addition, the OSMR is expressed in several different prostate cancer cell lines,

and the highest OSMR mRNA levels were measured in DU 145 cells *(239)*. OSM was able to signal through the OSMR–gp130 or LIFR–gp130 complex without a specific α-subunit. The name "oncostatin" is, however, misleading in prostate cancer because it is a growth-stimulatory factor in contrast to its function in breast cancer *(240)*. In contrast to DU 145 cells, growth stimulation of androgen-independent CWR22Rv1 human prostate cancer cells by either IL-6 or OSM was not associated with activation of STAT3 *(181)*. Instead, there was increased phosphorylation of the p38 MAPK and PI3K in CWR22Rv1 cells. An immunohistochemical study indicated that the OSM autocrine/paracrine loop exists in prostate cancers with increased expression of the OSMR in prostate cancers of higher Gleason grades *(241)*. By contrast, LIFR was detectable in prostate cancers of lower Gleason grades.

4.4. Functional Interaction of STAT3 Signaling with Androgen Receptor in Prostate Cancer

AR expression continues during androgen deprivation therapy of prostate cancer. Depending on how androgen action is inhibited in the prostate cancer patients, the AR may be activated by non-androgenic steroids, adrenal androgens, and hormonal antagonists *(242)*. Another mechanism of AR activation in hormone-refractory prostate cancer is the membrane receptors with intrinsic kinase activity that leads to activation of intracellular kinases and may result in a ligand-independent activation of the AR. Sawyers' group has demonstrated that ligand-independent activation of the AR by Her-2/neu is associated with tumor progression *in vivo* *(243)*. On the basis of these experimental results, novel experimental treatments with an aim to down-regulate AR expression were proposed. A potential limitation of this type of therapy is the inability to distinguish between AR–cofactor interactions responsible for growth versus differentiation. In addition, long-term consequences of AR depletion will need to be evaluated.

Because IL-6 expression is elevated in prostate cancer, it was reasoned that this pluripotent cytokine may also trigger ligand-independent activation of the AR. IL-6-induced activation of the AR was first demonstrated by Hobisch and associates and confirmed by other researchers *(244–248)*. Treatment of DU 145 cells with IL-6 causes a ligand-independent activation of the AR. Similarly, IL-6 induced prostate-specific antigen (PSA) expression in the absence of androgens in LNCap cells *(244)*. The sequences of the AR required for interaction with IL-6 are located in the N-terminal region of the AR receptor. Activation of transcriptional activity of AR by IL-6 occurs through recruitment of the co-activators p300 and SRC-1. Two studies demonstrated increased expression of SRC-1 in therapy-resistant prostate cancer. Another approach for therapy intervention in prostate cancer may be based on blocking interactions between specific co-activators and the AR. Importantly, the co-activator p300 was demonstrated to be able to induce expression of AR target genes in the late passages of LNCaP–IL-6$^+$ cells that do not express the AR *(249)*. Collectively, these observations are of importance for better understanding of the molecular mechanisms underlying the expression of AR-target genes in prostate cancer tissues from patients who failed endocrine treatment.

However, according to the report by Coetzee's group, the expression of PSA was diminished by IL-6, thus suggesting a negative regulation of the AR by IL-6 *(250)*. These controversial findings might be partly explained by differential activation of the PI3K pathway by IL-6 in LNCaP sublines or LNCaP cells of different passage numbers *(230)*. In addition to IL-6, OSM was reported to activate the AR in DU 145 cells *(239)*. In contrast to IL-6-induced AR activation, non-steroidal antiandrogens were not able to inhibit OSM-induced AR activity. Several experimental approaches demonstrated that there is a physical interaction between STAT3 and the N-terminal region of the AR that is IL-6 dependent *(251,252)*. Consistent with these findings, treatment of cells with the DN STAT3 down-regulated AR activation. Thus, STAT3 is one of the numerous co-activators of the AR.

5. TRANSCRIPTION FACTORS STAT5 AND STAT3 AS THERAPEUTIC TARGET MOLECULES FOR PROSTATE CANCER

In those prostate cancers in which STAT5a/b or Stat3 promotes growth, Prl–STAT5a/b and IL-6–Stat3 signaling pathways can be inhibited by several different approaches. In addition to locally delivered antisense Oligodeoxy nucleotides (ODNs) or siRNAs, inhibition of STAT5a/b or STAT3 can be achieved by small-molecule inhibitors for STAT5a/b/STAT3. Specifically, STAT action provides multiple levels for rational drug design: dimerization of STATs can be inhibited by targeting the SH2 domain, transactivation of STATs can be inhibited by targeting the C-terminal TAD, and DNA-binding can be blocked by targeting the DBD of STATs. A number small-molecule inhibitors for STAT3 have already been developed *(253–258)*. Moreover, PrlR, which activates STAT5a/b in prostate cancer cells, provides an additional molecular target for pharmacological inhibition of STAT5a/b signaling pathway in prostate cancer. This may be achieved by a specific PrlR antagonist that inhibits PrlR dimerization, developed by Goffin's laboratory *(259)* or by phosphory-lated Prl *(122)*. The specific mechanism of action of phosphorylated Prl is at present not entirely clear and may involve induction of the expression of DN PrlRs *(260)*. A chimeric anti-IL-6 antibody CNTO 328 has been developed and shown to be effective in inhibiting growth of PC-3 prostate cancer xenograft tumors in nude mice *(261,262)*. Reduction in IL-6 levels may be achieved also by inhibition of NF-κB activity by the proteasome inhibitor PS-341. Targeting the IL-6-STAT3 signaling pathway may be also achieved by the IL-6 superantagonist SANT7, a compound that prevents signal transduction by abrogating the complex of the gp80/gp130/IL-6. This drug has been tested in myeloma models and its efficacy *in vivo* was potentiated by combination with dexamethasone *(263)*. Finally, to overcome adverse effects of systemic drugs, novel approaches for prostate cancer-specific delivery of pharmacological agents are under development in various laboratories *(264–269)*. These specific delivery methods will potentially be applicable for specific pharmacological inhibition of STAT5a/b/STAT3 in prostate cancer cells.

ACKNOWLEDGMENTS

We thank Ms. Jacqueline Lutz for her help in designing the graphs. This work was supported by grants (M.T.N.) from American Cancer Society (RSG-04-196-01-MGO), DOD Department of Defense Prostate Cancer Research Program (W81XWH-05-01-0062 and W81XWH-06-01-0076), American Institute for Cancer Research (03B115-Rev), and NIH (NCI) (1RO1CA113580-01A1).

REFERENCES

1. Ahonen TJ, Harkonen PL, Rui H, Nevalainen MT (2002) PRL signal transduction in the epithelial compartment of rat prostate maintained as long-term organ cultures in vitro. *Endocrinology* 143:228–38.
2. Li H, Ahonen TJ, Alanen K, et al. (2004) Activation of signal transducer and activator of transcription 5 in human prostate cancer is associated with high histological grade. *Cancer Res* 64:4774–82.
3. Darnell JE, Jr., Kerr IM, Stark GR (1994) Jak-STAT pathways and transcriptional activation in response to IFNs and other extracellular signaling proteins. *Science* 264:1415–21.
4. Stahl N, Boulton TG, Farruggella T, et al. (1994) Association and activation of Jak-Tyk kinases by CNTF-LIF-OSM-IL-6 beta receptor components. *Science* 263:92–5.
5. Lutticken C, Wegenka UM, Yuan J, et al. (1994) Association of transcription factor APRF and protein kinase Jak1 with the interleukin-6 signal transducer gp130. *Science* 263:89–92.
6. Vinkemeier U, Cohen SL, Moarefi I, Chait BT, Kuriyan J, Darnell JE, Jr. (1996) DNA binding of in vitro activated Stat1 alpha, Stat1 beta and truncated Stat1: interaction between NH2-terminal domains stabilizes binding of two dimers to tandem DNA sites. *EMBO J* 15:5616–26.
7. Horvath CM, Wen Z, Darnell JE, Jr. (1995) A STAT protein domain that determines DNA sequence recognition suggests a novel DNA-binding domain. *Genes Dev* 9:984–94.
8. John S, Vinkemeier U, Soldaini E, Darnell JE, Jr., Leonard WJ (1999) The significance of tetramerization in promoter recruitment by Stat5. *Mol Cell Biol* 19:1910–8.
9. Meyer WK, Reichenbach P, Schindler U, Soldaini E, Nabholz M (1997) Interaction of STAT5 dimers on two low affinity binding sites mediates interleukin 2 (IL-2) stimulation of IL-2 receptor alpha gene transcription. *J Biol Chem* 272:31821–8.
10. Chen X, Vinkemeier U, Zhao Y, Jeruzalmi D, Darnell JE, Jr., Kuriyan J (1998) Crystal structure of a tyrosine phosphorylated STAT-1 dimer bound to DNA. *Cell* 93:827–39.

11. Becker S, Groner B, Muller CW (1998) Three-dimensional structure of the Stat3beta homodimer bound to DNA. *Nature* 394:145–51.

12. Schindler CW (2002) Series introduction. JAK-STAT signaling in human disease. *J Clin Invest* 109:1133–7.

13. Levy DE, Darnell JE, Jr. (2002) Stats: transcriptional control and biological impact. *Nat Rev Mol Cell Biol* 3:651–62.

14. Kisseleva T, Bhattacharya S, Braunstein J, Schindler CW (2002) Signaling through the JAK/STAT pathway, recent advances and future challenges. *Gene* 285:1–24.

15. Ihle JN (2001) The Stat family in cytokine signaling. *Curr Opin Cell Biol* 13:211–7.

16. Darnell JE, Jr. (1997) STATs and gene regulation. *Science* 277:1630–5.

17. Vinkemeier U, Moarefi I, Darnell JE, Jr., Kuriyan J (1998) Structure of the amino-terminal protein interaction domain of STAT-4. *Science* 279:1048–52.

18. Meyer T, Gavenis K, Vinkemeier U (2002) Cell type-specific and tyrosine phosphorylation-independent nuclear presence of STAT1 and STAT3. *Exp Cell Res* 272:45–55.

19. McBride KM, McDonald C, Reich NC (2000) Nuclear export signal located within theDNA-binding domain of the STAT1transcription factor. *EMBO J* 19:6196–206.

20. McBride KM, Banninger G, McDonald C, Reich NC (2002) Regulated nuclear import of the STAT1 transcription factor by direct binding of importin-alpha. *EMBO J* 21:1754–63.

21. Melen K, Kinnunen L, Julkunen I (2001) Arginine/lysine-rich structural element is involved in interferon-induced nuclear import of STATs. *J Biol Chem* 276:16447–55.

22. Begitt A, Meyer T, van Rossum M, Vinkemeier U (2000) Nucleocytoplasmic translocation of Stat1 is regulated by a leucine-rich export signal in the coiled-coil domain. *Proc Natl Acad Sci USA* 97:10418–23.

23. Liu JX, Mietz J, Modi WS, John S, Leonard WJ (1996) Cloning of human Stat5B. Reconstitution of interleukin-2-induced Stat5A and Stat5B DNA binding activity in COS-7 cells. *J Biol Chem* 271:10738–44.

24. Liu X, Robinson GW, Gouilleux F, Groner B, Hennighausen L (1995) Cloning and expression of Stat5 and an additional homologue (Stat5b) involved in prolactin signal transduction in mouse mammary tissue. *Proc Natl Acad Sci USA* 92:8831–5.

25. Meraz MA, White JM, Sheehan KC, et al. (1996) Targeted disruption of the Stat1 gene in mice reveals unexpected physiologic specificity in the JAK-STAT signaling pathway. *Cell* 84:431–42.

26. Durbin JE, Hackenmiller R, Simon MC, Levy DE (1996) Targeted disruption of the mouse Stat1 gene results in compromised innate immunity to viral disease. *Cell* 84:443–50.

27. Park C, Li S, Cha E, Schindler C (2000) Immune response in Stat2 knockout mice. *Immunity* 13:795–804.

28. Thierfelder WE, van Deursen JM, Yamamoto K, et al. (1996) Requirement for Stat4 in interleukin-12-mediated responses of natural killer and T cells. *Nature* 382:171–4.

29. Kaplan MH, Sun YL, Hoey T, Grusby MJ (1996) Impaired IL-12 responses and enhanced development of Th2 cells in Stat4-deficient mice. *Nature* 382:174–7.

30. Shimoda K, van Deursen J, Sangster MY, et al. (1996) Lack of IL-4-induced Th2 response and IgE class switching in mice with disrupted Stat6 gene. *Nature* 380:630–3.

31. Takeda K, Tanaka T, Shi W, et al. (1996) Essential role of Stat6 in IL-4 signalling. *Nature* 380:627–30.

32. Grimley PM, Dong F, Rui H (1999) Stat5a and Stat5b: fraternal twins of signal transduction and transcriptional activation. *Cytokine Growth Factor Rev* 10:131–57.

33. Takeda K, Noguchi K, Shi W, et al. (1997) Targeted disruption of the mouse Stat3 gene leads to early embryonic lethality. *Proc Natl Acad Sci USA* 94:3801–4.

34. Sano S, Itami S, Takeda K, et al. (1999) Keratinocyte-specific ablation of Stat3 exhibits impaired skin remodeling, but does not affect skin morphogenesis. *EMBO J* 18:4657–68.

35. Stein PL, Vogel H, Soriano P (1994) Combined deficiencies of Src, Fyn, and Yes tyrosine kinases in mutant mice. *Genes Dev* 8:1999–2007.

36. Alonzi T, Maritano D, Gorgoni B, Rizzuto G, Libert C, Poli V (2001) Essential role of STAT3 in the control of the acute-phase response as revealed by inducible gene inactivation [correction of activation] in the liver. *Mol Cell Biol* 21:1621–32.

37. Takeda K, Clausen BE, Kaisho T, et al. (1999) Enhanced Th1 activity and development of chronic enterocolitis in mice devoid of Stat3 in macrophages and neutrophils. *Immunity* 10:39–49.

38. Chapman RS, Lourenco P, Tonner E, et al. (2000) The role of Stat3 in apoptosis and mammary gland involution. Conditional deletion of Stat3. *Adv Exp Med Biol* 480:129–38.

39. Alonzi T, Middleton G, Wyatt S, et al. (2001) Role of STAT3 and PI 3-kinase/Akt in mediating the survival actions of cytokines on sensory neurons. *Mol Cell Neurosci* 18:270–82.

40. Liu X, Robinson GW, Wagner KU, Garrett L, Wynshaw-Boris A, Hennighausen L (1997) Stat5a is mandatory for adult mammary gland development and lactogenesis. *Genes Dev* 11:179–86.

41. Nevalainen MT, Ahonen TJ, Yamashita H, et al. (2000) Epithelial defect in prostates of Stat5a-null mice. *Lab Invest* 80:993–1006.

42. Teglund S, McKay C, Schuetz E, et al. (1998) Stat5a and Stat5b proteins have essential and nonessential, or redundant roles in cytokine responses. *Cell* 93:841–50.

43. Udy GB, Towers RP, Snell RG, et al. (1997) Requirement of STAT5b for sexual dimorphism of body growth rates and liver gene expression. *Proc Natl Acad Sci USA* 94:7239–44.

44. Socolovsky M, Fallon AE, Wang S, Brugnara C, Lodish HF (1999) Fetal anemia and apoptosis of red cell progenitors in Stat5a-/-5b-/- mice: a direct role for Stat5 in Bcl-X(L) induction. *Cell* 98:181–91.

45. Yao Z, Cui Y, Watford WT, et al. (2006) Stat5a/b are essential for normal lymphoid development and differentiation. *Proc Natl Acad Sci USA* 103:1000–1005.

46. Bowman T, Garcia R, Turkson J, Jove R (2000) STATs in oncogenesis. Oncogene 19:2474–88.

47. Silva CM (2004) Role of STATs as downstream signal transducers in Src family kinase-mediated tumorigenesis. *Oncogene* 23:8017–23.

48. Pedranzini L, Leitch A, Bromberg J (2004) Stat3 is required for the development of skin cancer. *J Clin Invest* 114:619–22.

49. Levy DE, Lee CK (2002) What does Stat3 do? *J Clin Invest* 109:1143–8.

50. Miyoshi K, Shillingford JM, Smith GH, et al. (2001) Signal transducer and activator of transcription (Stat) 5 controls the proliferation and differentiation of mammary alveolar epithelium. *J Cell Biol* 155:531–42.

51. Humphreys RC, Hennighausen L (1999) Signal transducer and activator of transcription 5a influences mammary epithelial cell survival and tumorigenesis. *Cell Growth Differ* 10:685–94.

52. Ren S, Cai HR, Li M, Furth PA (2002) Loss of Stat5a delays mammary cancer progression in a mouse model. *Oncogene* 21:4335–9.

53. Iavnilovitch E, Groner B, Barash I (2002) Overexpression and forced activation of stat5 in mammary gland of transgenic mice promotes cellular proliferation, enhances differentiation, and delays postlactational apoptosis. *Mol Cancer Res* 1:32–47.

54. Cotarla I, Ren S, Zhang Y, Gehan E, Singh B, Furth PA (2004) Stat5a is tyrosine phosphorylated and nuclear localized in a high proportion of human breast cancers. *Int J Cancer* 108:665–71.

55. Nevalainen MT, Xie J, Torhorst J, et al. (2004) Signal transducer and activator of transcription-5 activation and breast cancer prognosis. *J Clin Oncol* 22:2053–60.

56. Sultan AS, Xie J, LeBaron MJ, Ealley EL, Nevalainen MT, Rui H (2005) Stat5 promotes homotypic adhesion and inhibits invasive characteristics of human breast cancer cells. *Oncogene* 24:746–60.

57. Benekli M, Baer MR, Baumann H, Wetzler M (2003) Signal transducer and activator of transcription proteins in leukemias. *Blood* 101:2940–54.

58. Mitchell TJ, John S (2005) Signal transducer and activator of transcription (STAT) signalling and T-cell lymphomas. *Immunology* 114:301–12.

59. Lai SY, Childs EE, Xi S, et al. (2005) Erythropoietin-mediated activation of JAK-STAT signaling contributes to cellular invasion in head and neck squamous cell carcinoma. *Oncogene* 24:4442–9.

60. Hsiao JR, Jin YT, Tsai ST, Shiau AL, Wu CL, Su WC (2003) Constitutive activation of STAT3 and STAT5 is present in the majority of nasopharyngeal carcinoma and correlates with better prognosis. *Br J Cancer* 89:344–9.

61. Xi S, Zhang Q, Dyer KF, et al. (2003) Src kinases mediate STAT growth pathways in squamous cell carcinoma of the head and neck. *J Biol Chem* 278:31574–83.

62. Xi S, Zhang Q, Gooding WE, Smithgall TE, Grandis JR (2003) Constitutive activation of Stat5b contributes to carcinogenesis in vivo. *Cancer Res* 63:6763–71.

63. Leong PL, Xi S, Drenning SD, et al. (2002) Differential function of STAT5 isoforms in head and neck cancer growth control. *Oncogene* 21:2846–53.

64. Copeland NG, Gilbert DJ, Schindler C, et al. (1995) Distribution of the mammalian Stat gene family in mouse chromosomes. *Genomics* 29:225–8.

65. Ihle JN (1996) STATs: signal transducers and activators of transcription. *Cell* 84:331–4.

66. Aaronson DS, Horvath CM (2002) A road map for those who don't know JAK-STAT. *Science* 296:1653–5.

67. Leonard WJ, O'Shea JJ (1998) Jaks and STATs: biological implications. *Annu Rev Immunol* 16:293–322.

68. Schindler C, Darnell JE, Jr. (1995) Transcriptional responses to polypeptide ligands: the JAK-STAT pathway. *Annu Rev Biochem* 64:621–51.

69. Decker T, Kovarik P (2000) Serine phosphorylation of STATs. *Oncogene* 19:2628–37.

70. Kirken RA, Malabarba MG, Xu J, et al. (1997) Prolactin stimulates serine/tyrosine phosphorylation and formation of heterocomplexes of multiple Stat5 isoforms in Nb2 lymphocytes. *J Biol Chem* 272:14098–103.

71. Kirken RA, Malabarba MG, Xu J, et al. (1997) Two discrete regions of interleukin-2 (IL2) receptor beta independently mediate IL2 activation of a PD98059/rapamycin/wortmannin-insensitive Stat5a/b serine kinase. *J Biol Chem* 272:15459–65.

72. Yamashita H, Nevalainen MT, Xu J, et al. (2001) Role of serine phosphorylation of Stat5a in prolactin-stimulated beta- casein gene expression. *Mol Cell Endocrinol* 183:151–63.

73. Vinkemeier U (2004) Getting the message across, STAT! Design principles of a molecular signaling circuit. *J Cell Biol* 167:197–201.

74. Meyer T, Vinkemeier U (2004) Nucleocytoplasmic shuttling of STAT transcription factors. *Eur J Biochem* 271:4606–12.

75. Meyer T, Begitt A, Lodige I, van Rossum M, Vinkemeier U (2002) Constitutive and IFN-gamma-induced nuclear import of STAT1 proceed through independent pathways. *EMBO J* 21:344–54.

76. Marg A, Shan Y, Meyer T, Meissner T, Brandenburg M, Vinkemeier U (2004) Nucleo-cytoplasmic shuttling by nucleoporins Nup153 and Nup214 and CRM1-dependent nuclear export control the subcellular distribution of latent Stat1. *J Cell Biol* 165:823–33.

77. Decker T, Lew DJ, Mirkovitch J, Darnell JE, Jr. (1991) Cytoplasmic activation of GAF, an IFN-gamma-regulated DNA-binding factor. *EMBO J* 10:927–32.

78. Chen Y, Wen R, Yang S, et al. (2003) Identification of Shp-2 as a Stat5A phosphatase. *J Biol Chem* 278:16520–7.

79. Chughtai N, Schimchowitsch S, Lebrun JJ, Ali S (2002) Prolactin induces SHP-2 association with Stat5, nuclear translocation, and binding to the beta-casein gene promoter in mammary cells. *J Biol Chem* 277:31107–14.

80. Aoki N, Matsuda T (2000) A cytosolic protein-tyrosine phosphatase PTP1B specifically dephosphorylates and deactivates prolactin-activated STAT5a and STAT5b. *J Biol Chem* 275:39718–26.

81. Aoki N, Matsuda T (2002) A nuclear protein tyrosine phosphatase TC-PTP is a potential negative regulator of the PRL-mediated signaling pathway: dephosphorylation and deactivation of signal transducer and activator of transcription 5a and 5b by TC-PTP in nucleus. *Mol Endocrinol* 16:58–69.

82. Chung CD, Liao J, Liu B, et al. (1997) Specific inhibition of Stat3 signal transduction by PIAS3. *Science* 278:1803–5.

83. Schmidt D, Muller S (2003) PIAS/SUMO: new partners in transcriptional regulation. *Cell Mol Life Sci* 60:2561–74.

84. Rogers RS, Inselman A, Handel MA, Matunis MJ (2004) SUMO modified proteins localize to the XY body of pachytene spermatocytes. *Chromosoma* 113:233–43.

85. Alexander WS, Hilton DJ (2004) The role of suppressors of cytokine signaling (SOCS) proteins in regulation of the immune response. *Annu Rev Immunol* 22:503–29.

86. Nevalainen MT, Valve EM, Ingleton PM, Harkonen PL (1996) Expression and hormone regulation of prolactin receptors in rat dorsal and lateral prostate. *Endocrinology* 137: 3078–88.

87. Nevalainen MT, Valve EM, Ingleton PM, Nurmi M, Martikainen PM, Harkonen PL (1997) Prolactin and prolactin receptors are expressed and functioning in human prostate. *J Clin Invest* 99:618–27.

88. Nevalainen MT, Valve EM, Ahonen T, Yagi A, Paranko J, Harkonen PL (1997) Androgen-dependent expression of prolactin in rat prostate epithelium in vivo and in organ culture. *Faseb J* 11:1297–307.

89. Nevalainen MT, Valve EM, Makela SI, Blauer M, Tuohimaa PJ, Harkonen PL (1991) Estrogen and prolactin regulation of rat dorsal and lateral prostate in organ culture. *Endocrinology* 129:612–22.

90. Ahonen TJ, Harkonen PL, Laine J, Rui H, Martikainen PM, Nevalainen MT (1999) Prolactin is a survival factor for androgen-deprived rat dorsal and lateral prostate epithelium in organ culture. *Endocrinology* 140:5412–21.

91. Rui H (1987) *Effects and Mechanisms of Action of Prolactin on the Prostate: Studies Based on the Rat.* University of Oslo, Oslo, Norway.

92. Rui H, Purvis K (1987) Independent control of citrate production and ornithine decarboxylase by prolactin in the lateral lobe of the rat prostate. *Mol Cell Endocrinol* 52:91–5.

93. Leav I, Merk FB, Lee KF, et al. (1999) Prolactin receptor expression in the developing human prostate and in hyperplastic, dysplastic, and neoplastic lesions. *Am J Pathol* 154: 863–70.

94. Hostetter MW, Piacsek BE (1977) The effect of prolactin deficiency during sexual maturation in the male rat. *Biol Reprod* 17:574–7.

95. Steger RW, Chandrashekar V, Zhao W, Bartke A, Horseman ND (1998) Neuroendocrine and reproductive functions in male mice with targeted disruption of the prolactin gene. *Endocrinology* 139:3691–5.

96. McPherson SJ, Wang H, Jones ME, et al. (2001) Elevated androgens and prolactin in aromatase-deficient mice cause enlargement, but not malignancy, of the prostate gland. *Endocrinology* 142:2458–67.

97. Grayhack J, Lebowitz J (1967) Effect of prolactin on citric acid of lateral lobe of prostate of Sprague-Dawley rat. *Invest Urol* 5:87–94.

98. Grayhack JT (1963) Pituitary factors influencing growth of the prostate. *Natl Cancer Inst Monogr* 12:159–99.

99. Grayhack JT, Lebowitz JM (1967) Effect of prolactin on citric acid of lateral lobe of prostate of Sprague-Dawley rat. *Invest Urol* 5:87–94.

100. Walvoord DJ, Resnick MI, Grayhack JT (1976) Effect of testosterone, dihydrotestosterone, estradiol, and prolactin on the weight and citric acid content of the lateral lobe of the rat prostate. *Invest Urol* 14:60–5.

101. Rui H, Purvis K (1987) Prolactin selectively stimulates ornithine decarboxylase in the lateral lobe of the rat prostate. *Mol Cell Endocrinol* 50:89–97.

102. Rui H, Haug E, Mevag B, Thomassen Y, Purvis K (1985) Short-term effects of prolactin on prostatic function in rats with lisuride-induced hypoprolactinemia. *J Reprod Fert* 75:421–32.

103. Franklin R, Costello L (1992) Prolactin stimulates transcription of mitochondrial aspartate aminotransferase in prostate epithelial cells. *Mol Cell Endocrinol* 90:27–32.

104. Franklin RB, Costello LC (1990) Prolactin directly stimulates citrate production and mitochondrial aspartate aminotransferase of prostate epithelial cells. *Prostate* 17:13–8.

105. Franklin RB, Ekiko DB, Costello LC (1992) Prolactin stimulates transcription of aspartate aminotransferase in prostate cells. *Mol Cell Endocrinol* 90:27–32.

106. Franklin RB, Zou J, Gorski E, Yang YH, Costello LC (1997) Prolactin regulation of mitochondrial aspartate aminotransferase and protein kinase C in human prostate cancer cells. *Mol Cell Endocrinol* 127:19–25.

107. Costello LC, Liu Y, Franklin RB (1996) Testosterone and prolactin stimulation of mitochondrial aconitase in pig prostate epithelial cells. *Urology* 48:654–9.

108. Costello LC, Liu Y, Franklin RB (1995) Prolactin specifically increases pyruvate dehydrogenase E1 alpha in rat lateral prostate epithelial cells. *Prostate* 26:189–93.

109. Nakamura A, Shirai T, Ogawa K, et al. (1990) Promoting action of prolactin released from a grafted transplantable pituitary tumor (MtT/F84) on rat prostate carcinogenesis. *Cancer Lett* 53:151–7.

110. Thomas JA, Manandhar MSP (1977) Effects of prolactin on the dorsolateral lobe of the rat prostate gland. *Invest Urol* 14:398–9.

111. Prins GS (1987) Prolactin influence on cytosol and nuclear androgen receptors in the ventral, dorsal, and lateral lobes of the rat prostate. *Endocrinology* 120:1457–64.

112. Prins GS, Lee C (1982) Influence of prolactin-producing pituitary grafts on the in vivo uptake, distribution, and disappearance of [3H]testosterone and [3H]dihydrotestosterone by the rat prostate lobes. *Endocrinology* 110:920–5.

113. Sissom JF, Eigenbrodt ML, Porter JC (1988) Anti-growth action on mouse mammary and prostate glands of a monoclonal antibody to prolactin receptor. *Am J Pathol* 133:589–95.

114. Schacht MJ, Niederberger CS, Garnett JE, Sensibar JA, Lee C, Grayhack JT (1992) A local direct effect of pituitary graft on growth of the lateral prostate in rats. *Prostate* 20:51–8.

115. Wennbo H, Kindblom J, Isaksson OG, Tornell J (1997) Transgenic mice overexpressing the prolactin gene develop dramatic enlargement of the prostate gland. *Endocrinology* 138:4410–5.

116. Kindblom J, Dillner K, Sahlin L, et al. (2003) Prostate hyperplasia in a transgenic mouse with prostate-specific expression of prolactin. *Endocrinology* 144:2269–78.

117. Kindblom J, Dillner K, Ling C, Tornell J, Wennbo H (2002) Progressive prostate hyperplasia in adult prolactin transgenic mice is not dependent on elevated serum androgen levels. *Prostate* 53:24–33.

118. Janssen T, Darro F, Petein M, et al. (1996) In vitro characterization of prolactin-induced effects on proliferation in the neoplastic LNCaP, DU-145, and PC3 models of the human prostate. *Cancer* 77:144–9.

119. Coert A, Nievelstein H, Kloosterboer HJ, Loonen P, van deer Vies J (1985) Effects of hyperprolactinemia on the accessory sexual organs of the male rat. *Prostate* 6:269–276.

120. Jones R, Riding PR, Parker MG (1983) Effects of prolactin on testosterone-induced growth and protein synthesis in rat accessory sex glands. *J Endocrinol* 96:407–16.

121. Holland JM, Lee C (1980) Effects of pituitary grafts on testosterone stimulated growth of rat prostate. *Biol Reprod* 22:351–5.

122. Xu X, Kreye E, Kuo CB, Walker AM (2001) A molecular mimic of phosphorylated prolactin markedly reduced tumor incidence and size when DU-145 human prostate cancer cells were grown in nude mice. *Cancer Res* 61:6098–104.

123. Teglund S, McKay C, Schuetz E, et al. (1998) Stat5a and Stat5b proteins have essential and nonessential, or redundant, roles in cytokine responses. *Cell* 93:841–50.

124. Liu X, Gallego MI, Smith GH, Robinson GW, Hennighausen L (1998) Functional release of Stat5a-null mammary tissue through the activation of compensating signals including Stat5b. *Cell Growth Differ* 9:795–803.

125. Nevalainen MT, Xie J, Bubendorf L, Wagner KU, Rui H (2002) Basal activation of transcription factor signal transducer and activator of transcription (Stat5) in nonpregnant mouse and human breast epithelium. *Mol Endocrinol* 16:1108–24.

126. Ahonen TJ, Xie J, LeBaron MJ, et al. (2003) Inhibition of transcription factor Stat5 induces cell death of human prostate cancer cells. *J Biol Chem* 278:27287–92.

127. Kazansky AV, Spencer DM, Greenberg NM (2003) Activation of signal transducer and activator of transcription 5 is required for progression of autochthonous prostate cancer: evidence from the transgenic adenocarcinoma of the mouse prostate system. *Cancer Res* 63:8757–62.

128. Li H, Zhang Y, Glass A, et al. (2005) Activation of signal transducer and activator of transcription-5 in prostate cancer predicts early recurrence. *Clin Cancer Res* 11:5863–8.

129. Niall HD, Hogan ML, Sauer R, Rosenblum IY, Greenwood FC (1971) Sequences of pituitary and placental lactogenic and growth hormones: evolution from a primordial peptide by gene reduplication. *Proc Natl Acad Sci USA* 68:866–70.

130. Maurer RA, Gorski J, McKean DJ (1977) Partial amino acid sequence of rat pre-prolactin. *Biochem J* 161:189–92.

131. Lingappa VR, Devillers-Thiery A, Blobel G (1977) Nascent prehormones are intermediates in the biosynthesis of authentic bovine pituitary growth hormone and prolactin. *Proc Natl Acad Sci USA* 74:2432–6.

132. Li CH, Dixon JS, Lo TB, Schmidt KD, Pankov YA (1970) Studies on pituitary lactogenic hormone. XXX. The primary structure of the sheep hormone. *Arch Biochem Biophys* 141:705–37.

133. Shome B, Parlow AF (1977) Human pituitary prolactin (hPRL): the entire linear amino acid sequence. *J Clin Endocrinol Metab* 45:1112–5.

134. Owerbach D, Rutter WJ, Cooke NE, Martial JA, Shows TB (1981) The prolactin gene is located on chromosome 6 in humans. *Science* 212:815–6.

135. Evans AM, Petersen JW, Sekhon GS, DeMars R (1989) Mapping of prolactin and tumor necrosis factor-beta genes on human chromosome 6p using lymphoblastoid cell deletion mutants. *Somat Cell Mol Genet* 15:203–13.

136. Berwaer M, Martial JA, Davis JR (1994) Characterization of an up-stream promoter directing extrapituitary expression of the human prolactin gene. *Mol Endocrinol* 8:635–42.

137. DiMattia GE, Gellersen B, Duckworth ML, Friesen HG (1990) Human prolactin gene expression. The use of an alternative noncoding exon in decidua and the IM-9-P3 lymphoblast cell line. *J Biol Chem* 265:16412–21.

138. Gellersen B, Kempf R, Telgman R, DiMattia GE (1994) Nonpituitary human prolactin gene transcription is independent of Pit-1 and differentially controlled in lymphocytes and in endometrial stroma. *Mol Endocrinol* 8:356–373.

139. Gellersen B, DiMattia GE, Friesen HG, Bohnet HG (1989) Prolactin (PRL) mRNA from human decidua differs from pituitary PRL mRNA but resembles the IM-9-P3 lymphoblast PRL transcript. *Mol Cell Endocrinol* 64:127–30.

140. Pohnke Y, Kempf R, Gellersen B (1999) CCAAT/enhancer-binding proteins are mediators in the protein kinase A- dependent activation of the decidual prolactin promoter. *J Biol Chem* 274:24808–18.
141. Vashchenko N, Abrahamsson PA (2005) Neuroendocrine differentiation in prostate cancer: implications for new treatment modalities. *Eur Urol* 47:147–55.
142. Boutin JM, Jolicoeur C, Okamura H, et al. (1988) Cloning and expression of the rat prolactin receptor, a member of the growth hormone/prolactin receptor gene family. *Cell* 53:69–77.
143. Kelly PA, Djiane J, Postel-Vinay MC, Edery M (1991) The prolactin/growth hormone receptor family. *Endocr Rev* 12:235–51.
144. Arden KC, Boutin JM, Djiane J, Kelly PA, Cavenee WK (1990) The receptors for prolactin and growth hormone are localized in the same region of human chromosome 5. *Cytogenet Cell Genet* 53:161–5.
145. Hu ZZ, Meng J, Dufau ML (2001) Isolation and characterization of two novel forms of the human prolactin receptor generated by alternative splicing of a newly identified exon 11. *J Biol Chem* 276(44):41086–94.
146. Kelly PA, Ali S, Rozakis M, et al. (1993) The growth hormone/prolactin receptor family. *Recent Prog Horm Res* 48:123–64.
147. Dave JR, Krieg RJ, Jr., Witorsch RJ (1985) Modulation of prolactin binding sites in vitro by membrane fluidizers. Effects on male prostatic and female hepatic membranes in alcohol-fed rats. *Biochim Biophys Acta* 816:313–20.
148. Blankenstein MA, Bolt-de Vries J, van Aubel OG, van Steenbrugge GJ (1988) Hormone receptors in human prostate cancer. *Scand J Urol Nephrol Suppl* 107:39–45.
149. Leake A, Chisholm GD, Habib FK (1983) Characterization of the prolactin receptor in human prostate. *J Endocrinol* 99:321–8.
150. Rui H, Purvis K (1988) Hormonal control of prostate function. *Scand J Urol Nephrol Suppl* 107:32–8.
151. Rui H, Torjesen PA, Jacobsen H, Purvis K (1985) Testicular and glandular contributions to the prolactin pool in human semen. *Arch Androl* 15:129–36.
152. Boutin J, Jolicoeur C, Okamura H, et al. (1988) Cloning and expression of the rat prolactin receptor, a member of the growth hormone-prolactin receptor gene family. *Cell* 53:69–77.
153. Melck D, De Petrocellis L, Orlando P, et al. (2000) Suppression of nerve growth factor Trk receptors and prolactin receptors by endocannabinoids leads to inhibition of human breast and prostate cancer cell proliferation. *Endocrinology* 141:118–26.
154. Negro-Vilar A, Saad WA, McCann SM (1977) Evidence for a role of prolactin in prostate and seminal vesicle in immature male rats. *Endocrinology* 100:729–737.
155. Perez-Villamil B, Bordiu E, Puente-Cueva M (1992) Involvement of physiological prolactin levels in growth and prolactin receptor content of prostate glands and testes in developing male rats. *J Endocrinol* 132:449–59.
156. Grayhack JT (1965) Effect of testosterone-estradiol administration on citric acid and fructose content of the rat prostate. *Endocrinology* 77:1068–74.
157. Franklin RB, Costello LC (1990) Prolactin directly stimulates citrate production and mitochondrial aspartate aminotransferase in prostate epithelial cells. *Prostate* 17:13–18.
158. Costello LC, Franklin RB (1997) Citrate metabolism of normal and malignant prostate epithelial cells. *Urology* 50:3–12.
159. Costello LC, Franklin RB (1994) Effect of prolactin on the prostate. *Prostate* 24:162–6.
160. Costello LC, Liu Y, Zou J, Franklin RB (2000) Mitochondrial aconitase gene expression is regulated by testosterone and prolactin in prostate epithelial cells. *Prostate* 42:196–202.
161. Thomas JA, Manandhar MS (1977) Effects of prolactin on the dorsolateral lobe of the rat prostate gland. *Invest Urol* 14:398–9.
162. Prins GS, Lee C (1983) Biphasic response of the rat lateral prostate to increasing levels of serum prolactin. *Biol Reprod* 29:938–45.

163. Johnson MP, Thompson SA, Lubaroff DM (1985) Differential effects of prolactin on rat dorsolateral prostate and R3327 prostatic tumor sublines. *J Urol* 133:1112–20.

164. Schacht MJ, Niederberger CS, Garnett JE, Sensibar JA, Lee C, Grayhack JT (1992) A local direct effect of pituitary grafts on growth of the lateral prostate in rats. *Prostate* 20:51–58.

165. Ormandy CJ, Camus A, Barra J, et al. (1997) Null mutation of the prolactin receptor gene produces multiple reproductive defects in the mouse. *Genes Dev* 11:167–78.

166. Keenan EJ, Klase PA, Thomas JA (1981) Effects of prolactin on DNA synthesis and growth of the accessory sex organs in male mice. *Endocrinology* 109:170–5.

167. Prins GS (1987) Prolactin influence cytosol and nuclear androgen receptors in the ventral, dorsal, and lateral lobes of rat prostate. *Endocrinology* 120:1457–64.

168. Adams JB (1985) Control of secretion and the function of C19-delta 5-steroids of the human adrenal gland. *Mol Cell Endocrinol* 41:1–17.

169. Matzkin H, Kaver I, Lewyshon O, Ayalon D, Braf Z (1988) The role of increased prolactin levels under GnRH analogue treatment in advanced prostatic carcinoma. *Cancer* 61: 2187–91.

170. Horti J, Figg WD, Weinberger B, Kohler D, Sartor O (1998) A phase II study of bromocriptine in patients with androgen-independent prostate cancer. *Oncol Rep* 5:893–6.

171. Rana A, Habib FK, Halliday P, et al. (1995) A case for synchronous reduction of testicular androgen, adrenal androgen and prolactin for the treatment of advanced carcinoma of the prostate. *Eur J Cancer* 31A:871–5.

172. Sahni M, Ambrosetti DC, Mansukhani A, Gertner R, Levy D, Basilico C (1999) FGF signaling inhibits chondrocyte proliferation and regulates bone development through the STAT-1 pathway. *Genes Dev* 13:1361–6.

173. Hodge DR, Hurt EM, Farrar WL (2005) The role of IL-6 and STAT3 in inflammation and cancer. *Eur J Cancer* 41:2502–512.

174. Lee SO, Lou W, Hou M, de Miguel F, Gerber L, Gao AC (2003) Interleukin-6 promotes androgen-independent growth in LNCaP human prostate cancer cells. *Clin Cancer Res* 9:370–76.

175. Chung TD, Yu JJ, Kong TA, Spiotto MT, Lin JM (2000) Interleukin-6 activates phosphatidylinositol-3 kinase, which inhibits apoptosis in human prostate cancer cell lines. *Prostate* 42:1–7.

176. Giri D, Ozen M, Ittmann M (2001) Interleukin-6 is an autocrine growth factor in human prostate cancer. *Am J Pathol* 159:2159–65.

177. Spiotto MT, Chung TD (2000) STAT3 mediates IL-6-induced growth inhibition in the human prostate cancer cell line LNCaP. *Prostate* 42:88–98.

178. Barton BE, Karras JG, Murphy TF, Barton A, Huang H (2004) Signal transducer and activator of transcription (STAT3) activation in prostate cancer: direct STAT3 inhibition induces apoptosis in prostate cancer lines. *Mol Cancer Ther* 3:11–20.

179. Mora LB, Buettner R, Seigne J, et al. (2002) Constitutive activation of Stat3 in human prostate tumors and cell lines: direct inhibition of Stat3 signaling induces apoptosis of prostate cancer cells. *Cancer Res* 62:6659–66.

180. Steiner H, Godoy-Tundidor S, Rogatsch H, et al. (2003) Accelerated in vivo growth of prostate tumors that up-regulate interleukin-6 is associated with reduced retinoblastoma protein expression and activation of the mitogen-activated protein kinase pathway. *Am J Pathol* 162:655–63.

181. Godoy-Tundidor S, Cavarretta IT, Fuchs D, et al. (2005) Interleukin-6 and oncostatin M stimulation of proliferation of prostate cancer 22Rv1 cells through the signaling pathways of p38 mitogen-activated protein kinase and phosphatidylinositol 3-kinase. *Prostate* 64: 209–216.

182. Campbell CL, Jiang Z, Savarese DM, Savarese TM (2001) Increased expression of the interleukin-11 receptor and evidence of STAT3 activation in prostate carcinoma. *Am J Pathol* 158:25–32.

183. von Knobloch R, Konrad L, Barth PJ, et al. (2004) Genetic pathways and new progression markers for prostate cancer suggested by microsatellite allelotyping. *Clin Cancer Res* 10:1064–73.

184. Verhage BA, van Houwelingen K, Ruijter TE, Kiemeney LA, Schalken JA (2003) Allelic imbalance in hereditary and sporadic prostate cancer. *Prostate* 54:50–7.

185. Kasahara K, Taguchi T, Yamasaki I, Kamada M, Yuri K, Shuin T (2002) Detection of genetic alterations in advanced prostate cancer by comparative genomic hybridization. *Cancer Genet Cytogenet* 137:59–63.

186. Wolter H, Gottfried HW, Mattfeldt T (2002) Genetic changes in stage pT2N0 prostate cancer studied by comparative genomic hybridization. *BJU Int* 89:310–6.

187. Wolter H, Trijic D, Gottfried HW, Mattfeldt T (2002) Chromosomal changes in incidental prostatic carcinomas detected by comparative genomic hybridization. *Eur Urol* 41:328–34.

188. Alers JC, Rochat J, Krijtenburg PJ, et al. (2000) Identification of genetic markers for prostatic cancer progression. *Lab Invest* 80:931–42.

189. Latil A, Baron JC, Cussenot O, et al. (1994) Oncogene amplifications in early-stage human prostate carcinomas. *Int J Cancer* 59:637–8.

190. Bova GS, Isaacs WB (1996) Review of allelic loss and gain in prostate cancer. *World J Urol* 14:338–46.

191. Lange EM, Gillanders EM, Davis CC, et al. (2003) Genome-wide scan for prostate cancer susceptibility genes using families from the University of Michigan prostate cancer genetics project finds evidence for linkage on chromosome 17 near BRCA1. *Prostate* 57:326–34.

192. Gillanders EM, Xu J, Chang BL, et al. (2004) Combined genome-wide scan for prostate cancer susceptibility genes. *J Natl Cancer Inst* 96:1240–7.

193. Zuhlke KA, Madeoy JJ, Beebe-Dimmer J, et al. (2004) Truncating BRCA1 mutations are uncommon in a cohort of hereditary prostate cancer families with evidence of linkage to 17q markers. *Clin Cancer Res* 10:5975–80.

194. Baxter EJ, Scott LM, Campbell PJ, et al. (2005) Acquired mutation of the tyrosine kinase JAK2 in human myeloproliferative disorders. *Lancet* 365:1054–61.

195. James C, Ugo V, Le Couedic JP, et al. (2005) A unique clonal JAK2 mutation leading to constitutive signalling causes polycythaemia vera. *Nature* 434:1144–8.

196. Levine RL, Wadleigh M, Cools J, et al. (2005) Activating mutation in the tyrosine kinase JAK2 in polycythemia vera, essential thrombocythemia, and myeloid metaplasia with myelofibrosis. *Cancer Cell* 7:387–97.

197. Yu H, Jove R (2004) The STATs of cancer–new molecular targets come of age. *Nat Rev Cancer* 4:97–105.

198. de Groot RP, Raaijmakers JA, Lammers JW, Jove R, Koenderman L (1999) STAT5 activation by BCR-Abl contributes to transformation of K562 leukemia cells. *Blood* 94:1108–12.

199. Schwaller J, Parganas E, Wang D, et al. (2000) Stat5 is essential for the myelo- and lymphoproliferative disease induced by TEL/JAK2. *Mol Cell* 6:693–704.

200. Wang Z, Prins GS, Coschigano KT, et al. (2005) Disruption of growth hormone signaling retards early stages of prostate carcinogenesis in the C3(1)/tag mouse. *Endocrinology* 146(12):5188–96.

201. Weiss-Messer E, Merom O, Adi A, et al. (2004) Growth hormone (GH) receptors in prostate cancer: gene expression in human tissues and cell lines and characterization, GH signaling and androgen receptor regulation in LNCaP cells. *Mol Cell Endocrinol* 220:109–23.

202. Stangelberger A, Schally AV, Varga JL, et al. (2005) Inhibitory effect of antagonists of bombesin and growth hormone-releasing hormone on orthotopic and intraosseous growth and invasiveness of PC-3 human prostate cancer in nude mice. *Clin Cancer Res* 11:49–57.

203. Letsch M, Schally AV, Busto R, Bajo AM, Varga JL (2003) Growth hormone-releasing hormone (GHRH) antagonists inhibit the proliferation of androgen-dependent and-independent prostate cancers. *Proc Natl Acad Sci USA* 100:1250–5.

204. Halmos G, Schally AV, Czompoly T, Krupa M, Varga JL, Rekasi Z (2002) Expression of growth hormone-releasing hormone and its receptor splice variants in human prostate cancer. *J Clin Endocrinol Metab* 87:4707–14.
205. Chopin LK, Veveris-Lowe TL, Philipps AF, Herington AC (2002) Co-expression of GH and GHR isoforms in prostate cancer cell lines. *Growth Horm IGF Res* 12:126–36.
206. Boerner JL, Gibson MA, Fox EM, et al. (2005) Estrogen negatively regulates EGF-mediated STAT5 signaling in HER family receptor overexpressing breast cancer cells. *Mol Endocrinol* 19(11):2660–70.
207. Kloth MT, Catling AD, Silva CM (2002) Novel activation of STAT5b in response to epidermal growth factor. *J Biol Chem* 277:8693–701.
208. Kloth MT, Laughlin KK, Biscardi JS, Boerner JL, Parsons SJ, Silva CM (2003) STAT5b, a Mediator of Synergism between c-Src and the Epidermal Growth Factor Receptor. *J Biol Chem* 278:1671–9.
209. Deo DD, Axelrad TW, Robert EG, Marcheselli V, Bazan NG, Hunt JD (2002) Phosphorylation of STAT-3 in response to basic fibroblast growth factor occurs through a mechanism involving platelet-activating factor, JAK-2, and Src in human umbilical vein endothelial cells. Evidence for a dual kinase mechanism. *J Biol Chem* 277:21237–45.
210. Dhir R, Ni Z, Lou W, DeMiguel F, Grandis JR, Gao AC (2002) Stat3 activation in prostatic carcinomas. *Prostate* 51:241–6.
211. Dolled-Filhart M, Camp RL, Kowalski DP, Smith BL, Rimm DL (2003) Tissue microarray analysis of signal transducers and activators of transcription 3 (Stat3) and phospho-Stat3 (Tyr705) in node-negative breast cancer shows nuclear localization is associated with a better prognosis. *Clin Cancer Res* 9:594–600.
212. Somers W, Stahl M, Seehra JS (1997) 1.9 A crystal structure of interleukin-6: implications for a novel mode of receptor dimerization and signaling. *EMBO J* 16:989–97.
213. Heinrich PC, Behrmann I, Haan S, Hermanns HM, Muller-Newen, Schaper F (2003) Principles of interleukin-6 (IL-6)-type cytokine signalling and its regulation. *Biochem J* 374:1–20.
214. Ross JS, Kallakury BV, Sheehan CE, et al. (2004) Expression of nuclear factor-kappa B and I kappa B alpha proteins in prostatic adenocarcinomas: correlation of nuclear factor-kappa B immunoreactivity with disease recurrence. *Clin Cancer Res* 10:2466–72.
215. Zerbini LF, Wang Y, Cho JY, Libermann TA (2003) Constitutive activation of nuclear factor kappaB p50/p65 and Fra-1 and JunD is essential for deregulated interleukin 6 expression in prostate cancer. *Cancer Res* 63:2206–215.
216. Gao AC, Lou W, Isaacs JT (2000) Enhanced GBX2 expression stimulates growth of human prostate cancer cells via transcriptional regulation of the interleukin 6 gene. *Clin Cancer Res* 6:493–7.
217. Twillie DA, Eisenberger MA, Carducci MA, Hseih W-S, Kim WY, Simons JW (1995) Interleukin-6: a candidate mediator of human prostate cancer morbidity. *Urology* 45:542–9.
218. Bellido T, Jilka RL, Boyce BF, et al. (1995) Regulation of interleukin-6, osteoclastogenesis, and bone mass by androgens. *J Clin Invest* 95:2886–95.
219. Lin SC, Yamate T, Taguchi Y, et al. (1997) Regulation of the gp80 and gp130 subunits of the IL-6 receptor by sex steroids in the murine bone marrow. *J Clin Invest* 100:1980–90.
220. Hobisch A, Ramoner R, Fuchs D, et al. (2001) Prostate cancer cells (LNCaP) generated after long-term interleukin-6 treatment express interleukin-6 and acquire an interleukin-6-partially resistant phenotype. *Clin Cancer Res* 7:2941–8.
221. Hobisch A, Rogatsch H, Hittmair A, et al. (2000) Immunohistochemical localization of interleukin-6 and its receptor in benign, premalignant and malignant prostate tissue. *J Pathol* 191:239–44.
222. Yamasaki K, Taga T, Hirata Y, et al. (1988) Cloning and expression of the human interleukin-6 (BSF-2/IFN beta 2) receptor. *Science* 241:825–8.

223. Mackiewicz A, Rose-John S, Schooltink H, Laciak M, Gorny A, Heinrich PC (1992) Soluble human interleukin-6-receptor modulates interleukin-6-dependent N-glycosylation of alpha 1-protease inhibitor secreted by HepG2 cells. *FEBS Lett* 306:257–61.

224. Taga T, Hibi M, Hirata Y, et al. (1989) Interleukin-6 triggers the association of its receptor with a possible signal transducer, gp130. *Cell* 58:573–81.

225. Heinrich PC, Behrmann I, Muller-Newen G, Schaper F, Graeve L (1998) Interleukin-6-type cytokine signalling through the gp130/Jak/STAT pathway. *Biochem J* 334 (Pt 2):297–314.

226. Giri D, Ozen M, Ittmann M (2001) Interleukin-6 is an autocrine growth factor in human prostate cancer. *Am J Pathol* 159:2159–65.

227. Drachenberg DE, Elgamal AA, Rowbotham R, Peterson M, Murphy GP (1999) Circulating levels of interleukin-6 in patients with hormone refractory prostate cancer. *Prostate* 41: 127–33.

228. Nakashima J, Tachibana M, Horiguchi Y, et al. (2000) Serum interleukin-6 as a prognostic factor in patients with prostate cancer. *Clin Cancer Res* 6:2702–706.

229. Degeorges A, Tatoud R, Fauvel Lafeve F, et al. (1996) Stromal cells from human benign prostate hyperplasia produce a growth-inhibitory factor for LNCaP prostate cancer cells, identified as interleukin-6. *Int J Cancer* 68:207–14.

230. Yang L, Wang L, Lin HK, et al. (2003) Interleukin-6 differentially regulates androgen receptor transactivation via PI3K-Akt, STAT3, and MAPK, three distinct signal pathways in prostate cancer cells. *Biochem Biophys Res Commun* 305:462–9.

231. Qiu Y, Ravi L, Kung H-J (1998) Requirement of ErbB2 for signalling by interleukin-6 in prostate carcinoma cells. *Nature* 393:83–5.

232. Spiotto MT, Chung TD (2000) STAT3 mediates IL-6-induced neuroendocrine differentiation in prostate cancer cells. *Prostate* 42:186–95.

233. Deeble PD, Murphy DJ, Parsons SJ, Cox ME (2001) Interleukin-6 and cyclic-AMP-mediated signaling potentiates neuroendocrine differentiation of LNCaP prostate tumor cells. *Mol Cell Biol* 21:8471–82.

234. Cox ME, Deeble PD, Lakhani S, Parsons SJ (1999) Acquisition of neuroendocrine characteristics by prostate cancer is reversible: implications for prostate cancer progression. *Cancer Res* 59:3821–30.

235. Spiotto MT, Chung TD (2000) STAT3 mediates IL-6-induced neuroendocrine differentiation in prostate cancer cells. *Prostate* 42:186–95.

236. Steiner H, Godoy-Tundidor S, Rogatsch H, et al. (2003) Accelerated in vivo growth of prostate tumors that up-regulate interleukin-6 is associated with reduced retinoblastoma protein expression and activation of the mitogen-activated protein kinase pathway. *Am J Pathol* 162:655–63.

237. Steiner H, Berger AP, Godoy-Tundidor S, et al. (2004) Vascular endothelial growth factor autocrine loop is established in prostate cancer cells generated after prolonged treatment with interleukin-6. *Eur J Cancer* 40:1066–72.

238. Chung TD, Yu JJ, Spiotto MT, Bartkowski M, Simons JW (1999) Characterization of the role of IL-6 in the progression of prostate cancer. *Prostate* 38:199–207.

239. Godoy-Tundidor S, Hobisch A, Pfeil K, Bartsch G, Culig Z (2002) Acquisition of agonistic properties of nonsteroidal antiandrogens after treatment with oncostatin M in prostate cancer cells. *Clin Cancer Res* 8:2356–61.

240. Mori S, Murakami-Mori K, Bonavida B (1999) Oncostatin M (OM) promotes the growth of DU-145 human prostate cancer cells, but not PC-3 or LNCaP, through the signaling of the OM specific receptor. *Anticancer Res* 19:1011–1015.

241. Royuela M, Ricote M, Parsons MS, Garcia-Tunon I, Paniagua R, de Miguel MP (2004) Immunohistochemical analysis of the IL-6 family of cytokines and their receptors in benign, hyperplastic, and malignant human prostate. *J Pathol* 202:41–9.

242. Culig Z, Steiner H, Bartsch G, Hobisch A (2005) Mechanisms of endocrine therapy-responsive and-unresponsive prostate tumors. *Endocr Relat Cancer* 12:229–44.

243. Craft N, Shostak Y, Carey M, Sawyers CL (1999) A mechanism for hormone-independent prostate cancer through modulation of androgen receptor signaling by the HER-2/neu tyrosine kinase. *Nat Med* 5:280–85.

244. Hobisch A, Eder IE, Putz T, et al. (1998) Interleukin-6 regulates prostate-specific protein expression in prostate carcinoma cells by activation of the androgen receptor. *Cancer Res* 58:4640–45.

245. Lin DL, Whitney MC, Yao Z, Keller ET (2001) Interleukin-6 induces androgen responsiveness in prostate cancer cells through up-regulation of androgen receptor expression. *Clin Cancer Res* 7:1773–81.

246. Ueda T, Mawji NR, Bruchovsky N, Sadar MD (2002) Ligand-independent activation of the androgen receptor by interleukin-6 and the role of steroid receptor coactivator-1 in prostate cancer cells. *J Biol Chem* 277:38087–94.

247. Ueda T, Bruchovsky N, Sadar MD (2002) Activation of the androgen receptor N-terminal domain by interleukin-6 via MAPK and STAT3 signal transduction pathways. *J Biol Chem* 277:7076–85.

248. Debes JD, Schmidt LJ, Huang H, Tindall DJ (2002) p300 mediates interleukin-6-dependent transactivation of the androgen receptor. *Cancer Res* 62:5632–6.

249. Debes JD, Comuzzi B, Schmidt LJ, Dehm SM, Culig Z, Tindall DJ (2005) p300 regulates androgen receptor-independent expression of prostate-specific antigen in prostate cancer cells treated chronically with interleukin-6. *Cancer Res* 65:5965–73.

250. Jia L, Choong CS, Ricciardelli C, Kim J, Tilley WD, Coetzee GA (2004) Androgen receptor signaling: mechanism of interleukin-6 inhibition. *Cancer Res* 64:2619–26.

251. Chen T, Wang LH, Farrar WL (2000) Interleukin 6 activates androgen receptor-mediated gene expression through a signal transducer and activator of transcription 3-dependent pathway in LNCaP prostate cancer cells. *Cancer Res* 60:2132–5.

252. Junicho A, Matsuda T, Yamamoto T, et al. (2000) Protein inhibitor of activated STAT3 regulates androgen receptor signaling in prostate carcinoma cells. *Biochem Biophys Res Commun* 278:9–13.

253. Turkson J, Ryan D, Kim JS, et al. (2001) Phosphotyrosyl peptides block Stat3-mediated DNA binding activity, gene regulation, and cell transformation. *J Biol Chem* 276:45443–55.

254. Turkson J, Kim JS, Zhang S, et al. (2004) Novel peptidomimetic inhibitors of signal transducer and activator of transcription 3 dimerization and biological activity. *Mol Cancer Ther* 3:261–9.

255. Jing N, Sha W, Li Y, Xiong W, Tweardy DJ (2005) Rational drug design of G-quartet DNA as anti-cancer agents. *Curr Pharm Des* 11:2841–54.

256. Jing N, Zhu Q, Yuan P, Li Y, Mao L, Tweardy DJ (2006) Targeting signal transducer and activator of transcription 3 with G-quartet oligonucleotides: a potential novel therapy for head and neck cancer. *Mol Cancer Ther* 5:279–86.

257. Jing N, Li Y, Xiong W, Sha W, Jing L, Tweardy DJ (2004) G-quartet oligonucleotides: a new class of signal transducer and activator of transcription 3 inhibitors that suppresses growth of prostate and breast tumors through induction of apoptosis. *Cancer Res* 64:6603–9.

258. Song H, Wang R, Wang S, Lin J (2005) A low-molecular-weight compound discovered through virtual database screening inhibits Stat3 function in breast cancer cells. *Proc Natl Acad Sci USA* 102:4700–705.

259. Goffin V, Bernichtein S, Touraine P, Kelly PA (2005) Development and potential clinical uses of human prolactin receptor antagonists. *Endocr Rev* 26(3):400–22.

260. Wu W, Ginsburg E, Vonderhaar BK, Walker AM (2005) S179D prolactin increases vitamin D receptor and p21 through up-regulation of short 1b prolactin receptor in human prostate cancer cells. *Cancer Res* 65:7509–515.

261. Smith PC, Keller ET (2001) Anti-interleukin-6 monoclonal antibody induces regression of human prostate cancer xenografts in nude mice. *Prostate* 48:47–53.
262. Zaki MH, Nemeth JA, Trikha M (2004) CNTO328, a monoclonal antibody to IL-6, inhibits human tumor-induced cachexia in nude mice. *Int J Cancer* 111:592–5.
263. Tassone P, Neri P, Burger R, et al. (2005) Combination therapy with interleukin-6 receptor superantagonist Sant7 and dexamethasone induces antitumor effects in a novel SCID-hu In vivo model of human multiple myeloma. *Clin Cancer Res* 11:4251–8.
264. Isaacs JT (2005) New strategies for the medical treatment of prostate cancer. *BJU Int* 96 Suppl 2:35–40.
265. Denmeade SR, Isaacs JT (2005) The SERCA pump as a therapeutic target: making a "smart bomb" for prostate cancer. *Cancer Biol Ther* 4:14–22.
266. Denmeade SR, Jakobsen CM, Janssen S, et al. (2003) Prostate-specific antigen-activated thapsigargin prodrug as targeted therapy for prostate cancer. *J Natl Cancer Inst* 95: 990–1000.
267. Denmeade SR, Litvinov I, Sokoll LJ, Lilja H, Isaacs JT (2003) Prostate-specific antigen (PSA) protein does not affect growth of prostate cancer cells in vitro or prostate cancer xenografts in vivo. *Prostate* 56:45–53.
268. Mhaka A, Denmeade SR, Yao W, Isaacs JT, Khan SR (2002) A 5-fluorodeoxyuridine prodrug as targeted therapy for prostate cancer. *Bioorg Med Chem Lett* 12:2459–61.
269. Jakobsen CM, Denmeade SR, Isaacs JT, Gady A, Olsen CE, Christensen SB (2001) Design, synthesis, and pharmacological evaluation of thapsigargin analogues for targeting apoptosis to prostatic cancer cells. *J Med Chem* 44:4696–703.
270. Chatterjee-Kishore M, van den Akker F, Stark GR (2000) Association of STATs with relatives and friends. *Trends Cell Biol* 10:106–11.

13 Role of PI3K–Akt and PTEN in the Growth and Progression of Prostate Cancer

Haojie Huang, PhD
and Donald J. Tindall, PhD

CONTENTS

1. INTRODUCTION

Prostate cancer is the most commonly diagnosed cancer and the second leading cause of cancer death in American men. Prostate cancer is a very heterogeneous disease. Development and progression of this disease involve multiple genes and gene networks. This heterogeneity makes it very difficult to define any one individual therapeutic option. The treatment of prostate-confined

From: *Current Clinical Oncology: Prostate Cancer:*
Signaling Networks, Genetics, and New Treatment Strategies
Edited by: R. G. Pestell and M. T. Nevalainen © Humana Press, Totowa, NJ

cancers typically includes radical prostatectomy, external beam radiotherapy, or brachytherapy. Because most prostate cancers depend on androgens for growth and survival, androgen ablation therapy has been a standard treatment of advanced (metastatic) prostate cancer. However, the majority of prostate cancers evolve into androgen-refractory or androgen-depletion independent *(1)* disease, from which most patients eventually die.

It is conceivable that androgen-independent pathways may compensate for the mitogenic and/or anti-apoptotic impact of androgens on the growth and survival of androgen-depletion independent prostate cancer. Emerging evidence suggests that androgen-independent activation of the androgen receptor (AR) plays an essential role in proliferation and progression of prostate cancer under androgen-deprivation conditions *(2,3)*. Loss of heterozygosity (LOH) at 10q23 and/or deletions and mutations in the tumor suppressor gene *PTEN* are one of the most common somatic genetic alterations in prostate cancer *(4)*. Although it has been demonstrated that PTEN can function as a protein phosphatase, the tumor suppressor function of PTEN is primarily mediated by its lipid phosphatase activity that antagonizes the phosphatidylinositol 3-kinase (PI3K)/Akt pathway. Loss of PTEN results in a constitutive activation of Akt. Activated Akt, in turn, activates or inhibits a number of downstream targets, therefore favoring proliferation and survival of cancer cells.

2. PTEN AS A NEGATIVE REGULATOR OF THE PI3K/AKT PATHWAY

2.1. PI3K/Akt Function as an Oncogenic Pathway

PI3K possesses two subunits, the p85 regulatory subunit and the p110 catalytic subunit. Activation of receptor tyrosine kinases by mitogenic or survival factors results in the activation of PI3K. Activated PI3K catalyzes the phosphorylation of the inositol ring of phosphatidylinositol-4,5-phosphatase [PtdIns$(3,4,5)$P3] at the D3 position, which leads to the production of the intracellular second messenger PtdIns$(3,4,5)$P3 *(5)*. One of the critical downstream targets of PI3K is Akt/PKB, a serine/threonine protein kinase that plays an essential role in proliferation, apoptosis, and differentiation. Binding of growth factor receptors (GFRs) to their cognate ligands results in the activation of PI3K and production of PtdIns$(3,4,5)$P3 in the plasma membrane, where PtdIns$(3-5)$P recruits Akt and PDK1 to the plasma membrane via an interaction between the phosphoinositide and the pleckstrin homology (PH) domains of Akt or PDK1. Once recruited into the plasma membrane, Akt is phosphorylated and activated by PDK1 and possibly PDK2 at threonine 308 and serine 473 *(6–10)*. Akt promotes cell growth and survival via inactivation or activation of a number of tumor suppressor or oncogenic proteins such as FOXO proteins, BAD, TSC2, GSK3, MDM2, p21, and the AR *(11–17)*.

2.2. PTEN Acts Primarily as a Lipid Phosphatase for Tumor Suppression

The *PTEN* gene encodes a protein with intrinsic sequence similarity to protein phosphatase. *In vitro* studies have suggested that PTEN acts as a dual-specificity phosphatase with activities of both serine/threonine phosphatase and tyrosine phosphatase *(18)*. Surprisingly, PTEN also functions as a lipid phosphatase against a phosphate at the D3 position in PtdIns(3,4,5)P3 *(19)*. Importantly, mutated alleles of PTEN derived from human tumors and Cowden disease, an inherited genetic disease with increased susceptibility to cancer, often retain their protein phosphatase activity but lose their lipid phosphatase activity *(20,21)*. These findings suggest that the lipid phosphatase activity of PTEN, rather than its protein phosphatase activity, is important for its tumor suppressor function. As summarized in Table 1, the *PTEN* gene is deleted or mutated in a number of prostate cancer cell lines *(22,23)*. Because of loss of PTEN, Akt becomes constitutively active in PTEN-null prostate cancer cells *(24,25)*. Inhibition of Akt by treatment of cells with PI3K antagonists or transfection of cells with wild-type PTEN induces growth arrest or apoptosis in PTEN-null prostate cancer cells *(21,25,26)*. However, no such effect is observed when a mutated form of PTEN (PTEN-G129E), which lacks the lipid phosphatase activity, but retains protein phosphatase activity, is expressed *(25)*. These findings suggest that constitutively activated Akt is important for the growth and survival of these PTEN-null prostate cancer

Table 1
PTEN Status and Akt Activation in Prostate Cancer Cell Lines

Cell line	PTEN Del/Mut[a]	PTEN protein	Akt activation	References
LNCaP	Nonsense Mut	No	High	*(21,23)*
PC-3	Deletion	No	High	*(21,23)*
DU145	Met to Leu	High	Weak	*(21,23)*
CWR22rv1	ND[b]	High	Weak	([c])
LAPC-4	ND[b]	High	Weak	*(23,[c])*
LAPC-9	ND[b]	No	High	*(23)*
BPH-1	ND[b]	Low	Weak	([c])
RWPE-1	ND[b]	Low	Weak	([c])
WPE1-NA22	ND[b]	Moderate	Weak	([c])
WPE1-NB26	ND[b]	Low	Weak	([c])

[a]Del/Mut: deletion/mutation; [b]non-determined; [c]unpublished data from H.H. and D.J.T.

cells. Restoration of PTEN in PTEN-null prostate cancer cells also represses their growth in soft agar and in nude mice *(27)*. A number of downstream effector proteins have been identified in mediating the inhibitory effects of Akt on cell growth and survival. Inhibition of PI3K by transfection of *PTEN* or treating cells with chemical inhibitors such as wortmannin or LY294002 results in an increased expression of p27^{Kip1} and G1 arrest in prostate cancer cells *(26,28,29)*. The mechanisms underlying PTEN-mediated cell cycle arrest or apoptosis are quite complex. Activation of Akt or loss of PTEN can lead to upregulation of Skp2, an E3 ubiquitin ligase that targets p27^{Kip1} for proteasome degradation *(30)*. Elevated levels of p27^{Kip1} protein are caused, in part, by increased expression of p27^{Kip1} messenger RNA due to activation of FOXO transcription factors such as FOXO1 and FOXO3a in prostate cancer cells *(28,31)*. The pro-survival effect of PTEN loss has been linked to the over-expression of an anti-apoptotic protein Bcl-2. Bcl-2 is frequently overexpressed in the androgen-depletion independent prostate cancer *(32)*. Activation of the AR by physiological concentrations of androgens represses expression of Bcl-2 in prostate and breast cancer cells *(33,34)*. By contrast, depletion of androgens results in increased expression of Bcl-2 in androgen-depletion independent prostate cancer cells *in vitro (25,33)*. Therefore, androgen deprivation could provide one explanation for the overexpression of Bcl-2 in androgen-depletion independent prostate tumors. Another mechanism appears to involve the loss of PTEN. It has been demonstrated that overexpression of Bcl-2 is inversely correlated with loss of PTEN protein in advanced prostate tumors *(25)*. Indeed, ectopic expression of PTEN suppresses expression of Bcl-2 in prostate cancer cells, and requires the lipid phosphatase activity of PTEN and inhibition of Akt-mediated phosphorylation and activation of cAMP-responsive element-binding (CREB) protein *(25)*. Importantly, PTEN-mediated suppression of Bcl-2 plays an essential role in the chemosensitivity induced by PTEN in prostate cancer cells *(25)*.

2.3. Interaction Between PI3K/Akt and Androgen Signaling in Prostate Cancer

Androgens are critical for the growth and maintenance of prostate cells *(35)*. Androgens are also major risk factors for prostate cancer *(36,37)*. Besides their well-documented mitogenic effects, androgens also inhibit apoptosis in both normal and malignant prostatic epithelial cells *(38)*. Although the mechanism by which androgens function as survival factors in prostate cancer is largely unknown, progress has been made recently. Androgens protect prostate cancer cells from death induced by the cancer chemotherapeutic agent etoposide *(39)*. They also inhibit death of prostate cancer cells mediated by tumor necrosis factor α (TNFα) or Fas activation *(40)*. The PI3K/Akt pathway plays a critical role in the survival of PTEN-mutated prostate cancer cells. Inhibition of this survival pathway by PI3K inhibitors (LY294002 or wortmannin) or ectopic expression of PTEN induces apoptosis in the LNCaP cell line *(26,41,42)*. Importantly, these studies demonstrate that death of prostate cancer cells is blocked by pretreatment of cells with androgens. Logical targets for these effects include Akt and PTEN. However, neither Akt activity nor PTEN function is affected by androgens *(26,40)*. These data suggest that downstream effectors of the PTEN/Akt pathway may be targets for androgen action on proliferation and apoptosis in prostate cancer cells. Recently, it has been demonstrated that androgens through the AR can abrogate the transcriptional activity of FOXO1, thereby inhibiting its pro-apoptotic function in prostate cancer cells *(43)*. This effect can be blocked by the activation of the AR *(42,43)*. Mechanistically, androgens induce the proteolytic cleavage of FOXO1 and abolish the function of FOXO1 in prostate cancer cells. Importantly, this effect can be prevented by anti-androgens *(43)*.

3. DEREGULATION OF PTEN AND THE PI3K/AKT PATHWAY IN PROSTATE CANCER

3.1. Genetic Alterations in PTEN

The *PTEN* gene maps to the 10q23.3 locus *(22)*. Many prostate tumors exhibit a significant allelic loss, termed loss of heterozygosity (LOH) in chromosome 10q. Approximately 16% of localized prostate tumors have LOH in the PTEN locus, whereas advanced prostate tumors have over 60% *(44–46)*. Several studies have combined mutation screening with LOH analysis to determine the rate of biallelic loss of PTEN. Biallelic inactivation of PTEN in prostate cancer was found to range from 2 to 20% *(45,47)*. However, the rate of biallelic loss of PTEN increases to greater than 80% in metastatic prostate cancer. Monoallelic or biallelic loss of PTEN is also very prevalent in prostate cancer cell lines and xenografts, including PC-3, LNCaP, and LAPC9 *(24,48)*. The allelic loss of PTEN involves large deletions and small mutations. Unlike other tumor types, detailed mutation screening for PTEN has not been carried out systematically in prostate cancer. In the few studies done, missense and frameshift mutations have been detected, and all mutations described thus far result in a truncated protein *(49)*.

3.2. Epigenetic Inactivation of PTEN

Loss of function or inactivation of PTEN can also be attributed to epigenetic mechanisms that involve gene methylation and post-translational modifications. Among the epigenetic inactivation of tumor suppressor genes, promoter methylation is one potential mechanism. A reduction or loss of PTEN protein expression in the absence of corresponding mutations has been described *(23)*.

Treatment of prostate cancer cell lines with 5-azadeoxycytidine, whose incorporation into replicating DNA leads to hypomethylation, was able to restore the levels of PTEN protein in cells, suggesting that promoter methylation of the *PTEN* gene may be one mechanism of its inactivation in prostate tumors, although the details by which this may be accomplished have not been fully addressed. Moreover, oxidative stress may affect cell proliferation and survival. H_2O_2 induces an increase in phosphorylation of Akt *(50)*, although the exact mechanism is not clear. Interestingly, it has been demonstrated that H_2O_2 can cause inactivation of PTEN via the formation of a disulfide bond between cysteine 71 and cysteine 124, which is the active site of the lipid phosphatase activity of PTEN *(51)*. Furthermore, the function of PTEN is also regulated by the rate of protein turnover. The PTEN protein is composed of an N-terminal dual-specificity phosphatase domain and a C-terminal regulatory domain containing a C2 lipid-binding domain (for membrane insertion), two PEST domains (for protein stability), and a PDZ-binding motif (for protein–protein interaction) *(52)*. The function of PTEN protein is regulated by phosphorylation. The C-terminal domain of PTEN is rich in putative phosphorylation sites. Three amino acid residues, serine 380, threonine 383, and threonine 383, are the major sites of phosphorylation in PTEN *(53)*. Phosphorylation of these site leads to a decrease in the activity of PTEN *(53)*. CK2 appears to be the protein kinase responsible for the phosphorylation of PTEN *(54)*. Phosphorylation of the PTEN protein also affects its stability. Unphosphorylated PTEN, while having a higher functionality, is more sensitive to degradation via a proteasome-dependent pathway *(53)*.

3.3. Overexpression of Akt

The Akt family is composed of three members, including Akt1, Akt2, and Akt3. Akt1 and Akt2 are amplified in pancreatic, ovarian, and gastric tumors *(55,56)*. No amplification of these loci has been reported in prostate cancer. However, activation of Akt has been observed in prostate tumors by means of immunohistochemistry. Some studies have shown that Akt is phosphorylated at serine 473 in almost all prostatic intraepithelial neoplasia (PIN) lesions and invasive prostate tumors. By contrast, other studies have shown that the intensity of phosphorylation of Akt at serine 473 is correlated with high preoperative levels of prostate-specific antigen (PSA) in the serum *(57)* and is significantly higher with Gleason scores 8–10 than in PIN lesions *(58,59)*. However, whether the elevated phosphorylation of Akt at serine 473 is caused by loss of PTEN or increased activation of PI3K by upstream factors such as Her2/Neu amplification *(60)* or increased expression of insulin-like growth factor (IGF) *(61)* in prostate cancer is unknown.

3.4. Alterations in PI3K

In addition to the intrinsic activity of PI3K in regulating oncogenic pathways, the gene encoding the catalytic subunit of the type I PI3Kα (*PIK3CA* or p110) has been found to be amplified frequently in ovarian, breast, liver, and brain tumors *(62,63)*. The *PI3KCA* gene is also frequently mutated in these cancers. However, neither amplification nor activating mutations have been reported in prostate cancer to date.

4. FOXO PROTEINS: CRITICAL DOWNSTREAM TARGETS OF THE PI3K/PTEN/AKT PATHWAY

4.1. The FOXO Transcription Factors Function as Tumor Suppressors in Prostate Cancer

Mutations in the insulin receptor or PI3K result in an extended lifespan in *Caenorhabditis elegans*, and this is reversed when DAF-16, the *C. elegans* ortholog of FOXO, is mutated *(64,65)*. These findings suggest that FOXO factors play critical roles downstream of the PI3K/Akt pathway. Indeed, mammalian FOXO proteins are phosphorylated directly by Akt *(11,66–68)*. Akt phosphorylates FOXO proteins *in vitro* and *in vivo* and abrogates their cellular functions. FOXO transcription factors play important roles in regulating many cellular functions, including proliferation, cell survival, DNA damage, and oxidative stress. FOXO factors mediate G_1 cell cycle progression by upregulation of the cyclin-dependent kinase inhibitor p27^{KIP1} and downregulation of the D-type cyclins including D1 and D2 *(28,69–71)*. Moreover, these factors affect transition from M to G_1 of the cell cycle by directly regulating expression of mitotic genes such as *cyclin B* and *polo-like kinase (Plk) (72)*. Furthermore, FOXO transcription factors modulate the expression of several other genes that are involved in the cell cycle including *Wip1*, *EXT1*, and *cyclin G2 (73)*. Expression of the DNA damage response gene *gadd45-α* and expression of free-radical scavenger proteins such as catalase and superoxide dismutase are also regulated by FOXO transcription factors, suggesting important roles in surveillance of DNA damage and oxidative stress *(50,73,74)*. A number of pro-apoptotic proteins such as Fas ligand (FasL), Bim, TRAIL, the IGF-binding protein-1 (IGFBP1), NIP3, and legumain are transcriptionally regulated by members of the FOXO subfamily *(11,75,76)*. On the basis of these findings, it has been proposed that the FOXO transcription factors provide nodal points for diverse cell-signaling pathways, thereby modulating multiple cellular functions including differentiation, metabolism, proliferation, and survival *(77)*.

The cellular functions of the FOXO factors as demonstrated by the aforementioned in vitro studies have been investigated also *in vivo*. FOXO1a, FOXO3a, and FOXO4 have been disrupted in mice. FOXO1-null embryos die on embryonic day 10.5 from a major defect in vascular development, suggesting a crucial role of FOXO1 in angiogenesis *(78,79)*. FOXO1 possesses an intrinsic capability of inducing expression of several genes involved in blood vessel destabilization and remodeling such as angiopoietin-2, sema 3C, and slit-2 *(80)*, and this function can be inhibited by angiopoietin-1, an essential regulator of vascular development, in an Akt-dependent manner. In contrast to FOXO1, both FOXO3a- and FOXO4-null mice are viable and grossly indistinguishable from control littermates *(78,81)*. However, aged FOXO3a knockout female mice exhibit a distinctive ovarian phenotype of global follicular activation, oocyte death, and infertility. No histological changes in FOXO4-null mice have been identified *(78)*. Thus, these studies demonstrate that the physiological roles of the FOXO genes are functionally diverse in mammals.

4.2. Deregulation of FOXO Proteins During Tumorigenesis in Prostate Cancer

The potent cellular functions of the FOXO factors stipulate that these proteins are strictly regulated so that they exert their activities exclusively upon demand. A primary regulatory mechanism of FOXO proteins is mediated by Akt-dependent phosphorylation. Akt inhibits the tumor suppressor function of the FOXO factors via phosphorylation, which leads to nuclear exportation, together with the chaperone protein 14-3-3 *(11,66–68,82–84)*. The Akt-mediated nuclear exportation is suspended in cells under oxidative stress *(85)*. This effect appears to be mediated by Jun N-terminal kinase (JNK)-promoted phosphorylation of 14-3-3 *(86)*. Nuclear-to-cytoplasm trafficking of FOXO3a is also regulated by inflammatory factors such as TNFα. Activation of IκB kinase β (IKKβ) by TNFα results in the nuclear exportation and loss of function of FOXO3a *(87)*. In addition to excluding FOXO factors from the nucleus, Akt-dependent phosphorylation also plays a critical role in the proteasome degradation of FOXO1 and FOXO3a *(88–90)*. Ubiquitin-dependent proteasome degradation of FOXO1 is mediated by its interaction with Skp2, the substrate-binding component of the Skp1/Culin1/the F-box protein (SCF^{Skp2}) ubiquitin E3 ligase complex *(91)*. This effect of Skp2 requires Akt-specific phosphorylation of FOXO1 at serine 256. Moreover, activation of Akt due to loss of PTEN leads to upregulation of Skp2 *(30)*. Therefore, both Skp2 and Akt play central roles in destruction of the FOXO1 protein (Fig. 1). The potential for translating these findings into cancer therapy is underscored by the observation that Akt-mediated phosphorylation of FOXO proteins is reversible. PTEN expression or treatment with PI3K inhibitors results in a loss of Akt-dependent phosphorylation of FOXO1, thereby inducing cell death in PTEN-mutated prostate cancer cells. The proteasome inhibitor, Bortezomib (Velcade), is currently in clinical use for the treatment of multiple myeloma. A phase I clinical trial of Bortezomib in prostate cancer has been completed. However, the results

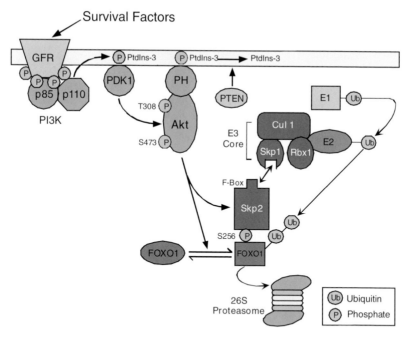

Fig. 1. Signaling pathways that regulate the ubiquitin-dependent degradation of FOXO1 protein. Binding of growth factor receptors (GFRs) to their cognate ligands, such as survival factors, results in the activation of phosphatidylinositol 3-kinase (PI3K) and phosphorylation of lipid phosphatidylinositol at the D3 position (PtdIns-3). This process in non-malignant cells is counterbalanced primarily by the lipid phosphatase function of the tumor suppressor gene PTEN. Recruitment of Akt and PDK1 kinases by their binding to the phosphatidylinositol triphosphate PI(3–5)P3 via the pleckstrin homology (PH) domain results in the phosphorylation of Akt. Activated Akt phosphorylates the FOXO1 protein at serine-256 and induces increased expression of Skp2, a key component of Skp1/Cullin1/ F-box protein (SCF) E3 ligase complex. Elevated levels of Skp2 and phosphorylation of FOXO1 at serine-256 are two key events that are required for the ubiquitination and degradation of FOXO1. Amplification of PI3K and Akt, loss of PTEN, and overexpression of Skp2 often seen in a wide range of tumor types may lead to the degradation and loss of tumor suppressor function of FOXO1. Image modified with permission from *ASBMB Today* May 2005: 10–11.

are less promising than that in multiple myeloma *(92)*. Thus, greater efficacy of Bortezomib may require a specific molecule or pathway for targeting. Given that Skp2 is overexpressed frequently in a number of human cancers including prostate cancer, breast cancer, lymphoma, small-cell lung cancer, oral squamous cell carcinoma, and colorectal carcinoma *(93–96)*, it would be worthwhile to investigate whether Bortezomib has effective anti-cancer activity in tumors that exhibit high levels of Skp2 and low levels of FOXO1.

Recent studies indicate that the FOXO factors are also regulated by protein acetylation/deacetylation. The acetylases CREB-binding protein (CBP) and p300 have been shown to be transcriptional coactivators of the FOXO factors in mammals and *C. elegans (97,98).* However, whether the effect of the acetylase activities of CBP and p300 is mediated by their specific acetylation of FOXO proteins or by their acetylation of histones as well as other transcription cofactors in general has not been determined. The silent information regulator-2 (Sir2), a protein deacetylase, extends the lifespan of the nematode in a manner that requires the FOXO transcription factor DAF-16 *(99).* Activation of the mammalian homolog of Sir-2, SIRT1, under stress counteracts the CBP and/or p300-mediated acetylation of FOXO1, FOXO3a, and FOXO4. This attenuates FOXO-induced apoptosis and potentiates FOXO-induced cell cycle arrest *(85,100,101).* CBP binds to and acetylates the positively charged lysine residues in the DNA-binding domain of FOXO1, thereby attenuating the ability of FOXO1 to bind to its cognate DNA sequence *(102).* By contrast, SIRT1 binds to and deacetylates FOXO1 at residues acetylated by CBP, thus increasing the expression of the anti-oxidative gene manganese superoxide dismutase *(MnSOD)* and the cell cycle inhibitor $p27^{KIP1}$ *(103).* Therefore, the interaction between FOXO and SIRT1 appears to tip FOXO-dependent responses away from apoptosis and toward stress resistance. Alterations in the expression of CBP and p300 are often observed in prostate cancer specimens *(104–106).* Therefore, it will be important to determine how the functions of FOXO proteins are affected by the expression of these acetylases in prostate cancer cells.

5. PROSTATE CANCER MOUSE MODELS WITH LOSS OF PTEN OR OVEREXPRESSION OF AKT

5.1. Conventional Knockout of PTEN in Mice

The concept that PTEN functions as a tumor suppressor gene in the prostate is further supported by findings in mouse studies. Conventional homozygous deletion of the *PTEN* gene results in developmental defects and embryonic lethality *(107,108).* By contrast, mice with heterozygous deletions of PTEN develop PIN with almost 100% penetrance *(107).* However, no progression to invasive prostate cancer was observed in these mice. Nonetheless, whether loss of one allele of the *PTEN* gene can promote formation of metastatic prostate cancer in aged mice is unclear because tumors develop in other organs such as intestines, mammary, thyroid, and endometrial and adrenal glands, which kill the mice before cancer is detected in the prostate. Moreover, mice with heterozygous PTEN in combination of homozygous deletion of $p27^{KIP1}$ develop prostate cancer within 3 months with 100% penetrance *(109).* However, contradictory results were reported in an independent study where deletion of both alleles of p27^{KIP1} diminished the tumor progression, even though deletion of one allele of p27^{KIP1} accelerated tumor formation in the PTEN heterozygous prostate *(110)*. Whether this discrepancy is due to the differences of mouse strains used in these studies is unknown. Other examples of a synergism between loss of PTEN and other genes in the development of prostate cancer from PIN lesions are found where one allele of PTEN is lost in conjunction with a deletion of Nkx3.1, Ink4a/p19arf, or TSC2 *(111–113)*.

5.2. Conditional Knockout of PTEN in the Mouse Prostate

The tumor suppressor function of PTEN is further suggested by conditional knockout of the *PTEN* gene in the mouse prostate. A mouse line with floxed alleles of PTEN was generated *(114)*. Prostate-specific deletion of PTEN was achieved by crossbreeding of PTEN-floxed mice with transgenic mice (PB4-Cre) expressing a Cre recombinase driven by a composite prostate-specific probasin promoter *(115)*. Mice lacking both alleles in the prostate developed PIN lesions much earlier than PTEN heterozygous mice, and importantly invasive and metastatic prostate cancer was documented in those mice *(116,117)*. Similar results were seen in mice where the PTEN gene was homozygously deleted by intercrossing the PTEN-floxed mice with transgenic mice of PSA- or mouse mammary tumor virus (MMTV)-driven Cre recombinase *(118,119)*. Thus, these studies suggest that PTEN plays a critical role in the initiation and progression of prostate cancer. They also support findings in human prostate cancer, although inactivation of PTEN is more often seen in metastatic prostate cancer than in primary prostate cancer.

5.3. Transgenic Expression of Akt in the Mouse Prostate

The importance of Akt activation in the development of prostate cancer has also been assessed directly in an Akt transgenic mouse model, where a myristoylated and therefore constitutively activated form of human Akt1 was specifically expressed in the mouse prostate using the rat probasin promoter *(120)*. Overexpression of Akt resulted in the development of histological features similar to the PIN lesion in the mouse prostate. However, the PIN phenotype in Akt transgenic mice did not progress to cancer with aging. Whether the phenotypic differences observed between Akt overexpression and PTEN deletion are due to the strain differences used by these different studies or reflect the biological differences between activation of Akt and loss of PTEN is unknown. Intriguingly, it has been shown recently that activation of Akt1 inhibits invasion in some cancer cells including breast cancer *(121)*. Whether loss of PTEN in the prostate leads to the activation of Akt-independent pathways that antagonizes the inhibitory effect of Akt on cancer invasion remains to be elucidated.

6. SUMMARY

Prostate cancer continues to be a major cause of cancer-related mortality in the USA. Genetic and biochemical studies demonstrate that loss of the tumor suppressor gene PTEN by means of mutations, deletions, and epigenetic mechanisms plays an essential role in the development and progression of prostate cancer. Activation of Akt due to inactivation of PTEN promotes growth and survival by inactivation of a number of tumor-suppressing molecules such as FOXO proteins, BAD, and $p27^{KIP1}$ or activation of a number of tumor-promoting molecules such as the AR and MDM2. All these findings provide promise for the discovery of options for the prevention, diagnosis, and treatment of prostate cancer.

ACKNOWLEDGMENTS

This work was supported by DK65236, DK60920, CA92956, CA15083, and the T.J. Martell Foundation.

REFERENCES

1. Roy-Burman, P., Tindall, D.J., Robins, D.M. et al. (2005) Androgens and prostate cancer: are the descriptors valid? *Cancer Biol. Ther.* **4**, 4–5.
2. Zegarra-Moro, O.L., Schmidt, L.J., Huang, H. and Tindall, D.J. (2002) Disruption of androgen receptor function inhibits proliferation of androgen-refractory prostate cancer cells. *Cancer Res.* **62**, 1008–13.
3. Chen, C.D., Welsbie, D.S., Tran, C. et al. (2004) Molecular determinants of resistance to antiandrogen therapy. *Nat. Med.* **10**, 33–9.
4. Deocampo, N.D., Huang, H. and Tindall, D.J. (2003) The role of PTEN in the progression and survival of prostate cancer. *Minerva Endocrinol.* **28**, 145–53.
5. Cantley, L.C. and Neel, B.G. (1999) New insights into tumor suppression: PTEN suppresses tumor formation by restraining the phosphoinositide 3-kinase/AKT pathway. *Proc. Natl. Acad. Sci. USA* **96**, 4240–5.
6. Downward, J. (1998) Mechanisms and consequences of activation of protein kinase B/Akt. *Curr. Opin. Cell Biol.* **10**, 262–7.
7. Toker, A. and Newton, A.C. (2000) Akt/protein kinase B is regulated by autophosphorylation at the hypothetical PDK-2 site. *J. Biol. Chem.* **275**, 8271–4.
8. Persad, S., Attwell, S., Gray, V. et al. (2001) Regulation of protein kinase B/Akt-serine 473 phosphorylation by integrin-linked kinase: critical roles for kinase activity and amino acids arginine 211 and serine 343. *J. Biol. Chem.* **276**, 27462–9.
9. Feng, J., Park, J., Cron, P., Hess, D. and Hemmings, B.A. (2004) Identification of a PKB/Akt hydrophobic motif Ser-473 kinase as DNA-dependent protein kinase. *J. Biol. Chem.* **279**, 41189–96.
10. Sarbassov, D.D., Guertin, D.A., Ali, S.M. and Sabatini, D.M. (2005) Phosphorylation and regulation of Akt/PKB by the rictor-mTOR complex. *Science* **307**, 1098–101.
11. Brunet, A., Bonni, A., Zigmond, M.J. et al. (1999) Akt promotes cell survival by phosphorylating and inhibiting a Forkhead transcription factor. *Cell* **96**, 857–68.
12. Datta, S.R., Dudek, H., Tao, X. et al. (1997) Akt phosphorylation of BAD couples survival signals to the cell-intrinsic death machinery. *Cell* **91**, 231–41.

13. Inoki, K., Li, Y., Zhu, T., Wu, J. and Guan, K.L. (2002) TSC2 is phosphorylated and inhibited by Akt and suppresses mTOR signalling. *Nat. Cell Biol.* **4**, 648–57.

14. Cross, D.A., Alessi, D.R., Cohen, P., Andjelkovich, M. and Hemmings, B.A. (1995) Inhibition of glycogen synthase kinase-3 by insulin mediated by protein kinase B. *Nature* **378**, 785–9.

15. Zhou, B.P., Liao, Y., Xia, W., Zou, Y., Spohn, B. and Hung, M.C. (2001) HER-2/neu induces p53 ubiquitination via Akt-mediated MDM2 phosphorylation. *Nat. Cell Biol.* **3**, 973–82.

16. Zhou, B.P., Liao, Y., Xia, W., Spohn, B., Lee, M.H. and Hung, M.C. (2001) Cytoplasmic localization of p21Cip1/WAF1 by Akt-induced phosphorylation in HER-2/neu-overexpressing cells. *Nat. Cell Biol.* **3**, 245–52.

17. Wen, Y., Hu, M.C., Makino, K. et al. (2000) HER-2/neu promotes androgen-independent survival and growth of prostate cancer cells through the Akt pathway. *Cancer Res.* **60**, 6841–5.

18. Myers, M.P., Stolarov, J.P., Eng, C. et al. (1997) P-TEN, the tumor suppressor from human chromosome 10q23, is a dual-specificity phosphatase. *Proc. Natl. Acad. Sci. USA* **94**, 9052–7.

19. Maehama, T. and Dixon, J.E. (1998) The tumor suppressor, PTEN/MMAC1, dephospho-rylates the lipid second messenger, phosphatidylinositol 3,4,5-trisphosphate. *J. Biol Chem.* **273**, 13375–8.

20. Myers, M.P., Pass, I., Batty, I.H. et al. (1998) The lipid phosphatase activity of PTEN is critical for its tumor supressor function. *Proc. Natl. Acad. Sci. USA* **95**, 13513–8.

21. Ramaswamy, S., Nakamura, N., Vazquez, F. et al. (1999) Regulation of G1 progression by the PTEN tumor suppressor protein is linked to inhibition of the phosphatidylinositol 3-kinase/Akt pathway. *Proc. Natl. Acad. Sci. USA* **96**, 2110–5.

22. Li, J., Yen, C., Liaw, D. et al. (1997) PTEN, a putative protein tyrosine phosphatase gene mutated in human brain, breast, and prostate cancer. *Science* **275**, 1943–7.

23. Whang, Y.E., Wu, X., Suzuki, H. et al. (1998) Inactivation of the tumor suppressor PTEN/MMAC1 in advanced human prostate cancer through loss of expression. *Proc. Natl. Acad. Sci. USA* **95**, 5246–50.

24. Wu, X., Senechal, K., Neshat, M.S., Whang, Y.E. and Sawyers, C.L. (1998) The PTEN/MMAC1 tumor suppressor phosphatase functions as a negative regulator of the phosphoinositide 3-kinase/Akt pathway. *Proc. Natl. Acad. Sci. USA* **95**, 15587–91.

25. Huang, H., Cheville, J.C., Pan, Y., Roche, P.C., Schmidt, L.J. and Tindall, D.J. (2001) PTEN induces chemosensitivity in PTEN-mutated prostate cancer cells by suppression of Bcl-2 expression. *J. Biol. Chem.* **276**, 38830–6.

26. Murillo, H., Huang, H., Schmidt, L.J., Smith, D.I. and Tindall, D.J. (2001) Role of PI3K signaling in survival and progression of LNCaP prostate cancer cells to the androgen refractory state. *Endocrinology* **142**, 4795–805.

27. Seki, M., Iwakawa, J., Cheng, H. and Cheng, P.W. (2002) p53 and PTEN/MMAC1/TEP1 gene therapy of human prostate PC-3 carcinoma xenograft, using transferrin-facilitated lipofection gene delivery strategy. *Hum. Gene. Ther.* **13**, 761–73.

28. Nakamura, N., Ramaswamy, S., Vazquez, F., Signoretti, S., Loda, M. and Sellers, W.R. (2000) Forkhead transcription factors are critical effectors of cell death and cell cycle arrest downstream of PTEN. *Mol. Cell. Biol.* **20**, 8969–82.

29. Graff, J.R., Konicek, B.W., McNulty, A.M. et al. (2000) Increased AKT activity contributes to prostate cancer progression by dramatically accelerating prostate tumor growth and diminishing p27Kip1 expression. *J. Biol. Chem.* **275**, 24500–5.

30. Mamillapalli, R., Gavrilova, N., Mihaylova, V.T. et al. (2001) PTEN regulates the ubiquitin-dependent degradation of the CDK inhibitor p27(KIP1) through the ubiquitin E3 ligase SCF(SKP2). *Curr. Biol.* **11**, 263–7.

31. Lynch, R.L., Konicek, B.W., McNulty, A.M. et al. (2005) The progression of LNCaP human prostate cancer cells to androgen independence involves decreased FOXO3a expression and reduced p27KIP1 promoter transactivation. *Mol. Cancer Res.* **3**, 163–9.
32. McDonnell, T.J., Troncoso, P., Brisbay, S.M. et al. (1992) Expression of the protooncogene bcl-2 in the prostate and its association with emergence of androgen-independent prostate cancer. *Cancer Res.* **52**, 6940–4.
33. Gleave, M., Tolcher, A., Miyake, H. et al. (1999) Progression to androgen independence is delayed by adjuvant treatment with antisense Bcl-2 oligodeoxynucleotides after castration in the LNCaP prostate tumor model. *Clin. Cancer Res.* **5**, 2891–8.
34. Huang, H., Zegarra-Moro, O.L., Benson, D. and Tindall, D.J. (2004) Androgens repress Bcl-2 expression via activation of the retinoblastoma (RB) protein in prostate cancer cells. *Oncogene* **23**, 2161–76.
35. Grossmann, M.E., Huang, H. and Tindall, D.J. (2001) Androgen receptor signaling in androgen-refractory prostate cancer. *J. Natl. Cancer Inst.* **93**, 1687–97.
36. Coffey, D.S. (1993) Prostate cancer. An overview of an increasing dilemma. *Cancer* **71**, 880–6.
37. Pienta, K.J. and Esper, P.S. (1993) Risk factors for prostate cancer. *Ann. Intern. Med.* **118**, 793–803.
38. Isaacs, J.T. (1984) Antagonistic effect of androgen on prostatic cell death. *Prostate* **5**, 545–57.
39. Berchem, G.J., Bosseler, M., Sugars, L.Y., Voeller, H.J., Zeitlin, S. and Gelmann, E.P. (1995) Androgens induce resistance to bcl-2-mediated apoptosis in LNCaP prostate cancer cells. *Cancer Res.* **55**, 735–8.
40. Kimura, K., Markowski, M., Bowen, C. and Gelmann, E.P. (2001) Androgen blocks apoptosis of hormone-dependent prostate cancer cells. *Cancer Res.* **61**, 5611–8.
41. Carson, J.P., Kulik, G. and Weber, M.J. (1999) Antiapoptotic signaling in LNCaP prostate cancer cells: a survival signaling pathway independent of phosphatidylinositol 3'-kinase and Akt/protein kinase B. *Cancer Res.* **59**, 1449–53.
42. Lin, J., Adam, R.M., Santiestevan, E. and Freeman, M.R. (1999) The phosphatidylinositol 3'-kinase pathway is a dominant growth factor-activated cell survival pathway in LNCaP human prostate carcinoma cells. *Cancer Res.* **59**, 2891–7.
43. Huang, H., Muddiman, D.C. and Tindall, D.J. (2004) Androgens negatively regulate forkhead transcription factor FKHR (FOXO1) through a proteolytic mechanism in prostate cancer cells. *J. Biol. Chem.* **279**, 13866–77.
44. Wang, S.I., Parsons, R. and Ittmann, M. (1998) Homozygous deletion of the PTEN tumor suppressor gene in a subset of prostate adenocarcinomas. *Clin. Cancer Res.* **4**, 811–5.
45. Suzuki, H., Freije, D., Nusskern, D.R. et al. (1998) Interfocal heterogeneity of PTEN/MMAC1 gene alterations in multiple metastatic prostate cancer tissues. *Cancer Res.* **58**, 204–9.
46. Cairns, P., Okami, K., Halachmi, S. et al. (1997) Frequent inactivation of PTEN/MMAC1 in primary prostate cancer. *Cancer Res.* **57**, 4997–5000.
47. Feilotter, H.E., Nagai, M.A., Boag, A.H., Eng, C. and Mulligan, L.M. (1998) Analysis of PTEN and the 10q23 region in primary prostate carcinomas. *Oncogene* **16**, 1743–8.
48. Vlietstra, R.J., van Alewijk, D.C., Hermans, K.G., van Steenbrugge, G.J. and Trapman, J. (1998) Frequent inactivation of PTEN in prostate cancer cell lines and xenografts. *Cancer Res.* **58**, 2720–3.
49. Gray, I.C., Stewart, L.M., Phillips, S.M., et al. (1998) Mutation and expression analysis of the putative prostate tumour-suppressor gene PTEN. *Br. J. Cancer* **78**, 1296–300.
50. Nemoto, S. and Finkel, T. 2002 Redox regulation of forkhead proteins through a p66shc-dependent signaling pathway. *Science* **295**, 2450–2.

51. Lee, S.R., Yang, K.S., Kwon, J., Lee, C., Jeong, W. and Rhee, S.G. (2002) Reversible inactivation of the tumor suppressor PTEN by H2O2. *J. Biol. Chem.* **277**, 20336–42.

52. Leslie, N.R. and Downes, C.P. (2002) PTEN: The down side of PI 3-kinase signalling. *Cell. Signal.* **14**, 285–95.

53. Vazquez, F., Ramaswamy, S., Nakamura, N. and Sellers, W.R. (2000) Phosphorylation of the PTEN tail regulates protein stability and function. *Mol. Cell. Biol.* **20**, 5010–8.

54. Torres, J. and Pulido, R. (2001) The tumor suppressor PTEN is phosphorylated by the protein kinase CK2 at its C terminus. Implications for PTEN stability to proteasome-mediated degradation. *J. Biol. Chem.* **276**, 993–8.

55. Cheng, J.Q., Godwin, A.K., Bellacosa, A. et al. (1992) AKT2, a putative oncogene encoding a member of a subfamily of protein-serine/threonine kinases, is amplified in human ovarian carcinomas. *Proc. Natl. Acad. Sci. USA* **89**, 9267–71.

56. Ruggeri, B.A., Huang, L., Wood, M., Cheng, J.Q. and Testa, J.R. (1998) Amplification and overexpression of the AKT2 oncogene in a subset of human pancreatic ductal adenocarcinomas. *Mol. Carcinog.* **21**, 81–6.

57. Liao, Y., Grobholz, R., Abel, U. et al. (2003) Increase of AKT/PKB expression correlates with gleason pattern in human prostate cancer. *Int. J. Cancer* **107**, 676–80.

58. Malik, S.N., Brattain, M., Ghosh, P.M. et al. (2002) Immunohistochemical demonstration of phospho-Akt in high Gleason grade prostate cancer. *Clin. Cancer Res.* **8**, 1168–71.

59. Kreisberg, J.I., Malik, S.N., Prihoda, T.J. et al. (2004) Phosphorylation of Akt (Ser473) is an excellent predictor of poor clinical outcome in prostate cancer. *Cancer Res.* **64**, 5232–6.

60. Mellon, K., Thompson, S., Charlton, R.G. et al. (1992) p53, c-erbB-2 and the epidermal growth factor receptor in the benign and malignant prostate. *J. Urol.* **147**, 496–9.

61. Chan, J.M., Stampfer, M.J., Giovannucci, E. et al. (1998) Plasma insulin-like growth factor-I and prostate cancer risk: a prospective study. *Science* **279**, 563–6.

62. Shayesteh, L., Lu, Y., Kuo, W.L. et al. (1999) PIK3CA is implicated as an oncogene in ovarian cancer. *Nat. Genet.* **21**, 99–102.

63. Samuels, Y. and Velculescu, V.E. (2004) Oncogenic mutations of PIK3CA in human cancers. *Cell Cycle* **3**, 1221–4.

64. Ogg, S., Paradis, S., Gottlieb, S. et al. (1997) The Fork head transcription factor DAF-16 transduces insulin-like metabolic and longevity signals in C. elegans. *Nature* **389**, 994–9.

65. Lin, K., Dorman, J.B., Rodan, A. and Kenyon, C. (1997) daf-16: An HNF-3/forkhead family member that can function to double the life-span of Caenorhabditis elegans. *Science* **278**, 1319–22.

66. Kops, G.J., de Ruiter, N.D., De Vries-Smits, A.M., Powell, D.R., Bos, J.L. and Burgering, B.M. (1999) Direct control of the Forkhead transcription factor AFX by protein kinase B. *Nature* **398**, 630–4.

67. Tang, E.D., Nunez, G., Barr, F.G. and Guan, K.L. (1999) Negative regulation of the forkhead transcription factor FKHR by Akt. *J. Biol. Chem.* **274**, 16741–6.

68. Biggs, W.H., 3rd, Meisenhelder, J., Hunter, T., Cavenee, W.K. and Arden, K.C. (1999) Protein kinase B/Akt-mediated phosphorylation promotes nuclear exclusion of the winged helix transcription factor FKHR1. *Proc. Natl. Acad. Sci. USA* **96**, 7421–6.

69. Medema, R.H., Kops, G.J., Bos, J.L. and Burgering, B.M. (2000) AFX-like Forkhead transcription factors mediate cell-cycle regulation by Ras and PKB through p27kip1. *Nature* **404**, 782–7.

70. Ramaswamy, S., Nakamura, N., Sansal, I., Bergeron, L. and Sellers, W.R. (2002) A novel mechanism of gene regulation and tumor suppression by the transcription factor FKHR. *Cancer Cell* **2**, 81–91.

71. Schmidt, M., Fernandez de Mattos, S., van der Horst, A. et al. (2002) Cell cycle inhibition by FOXO forkhead transcription factors involves downregulation of cyclin D. *Mol. Cell. Biol.* **22**, 7842–52.

72. Alvarez, B., Martinez, A.C., Burgering, B.M. and Carrera, A.C. (2001) Forkhead transcription factors contribute to execution of the mitotic programme in mammals. *Nature* **413**, 744–7.

73. Tran, H., Brunet, A., Grenier, J.M. et al. (2002) DNA repair pathway stimulated by the forkhead transcription factor FOXO3a through the Gadd45 protein. *Science* **296**, 530–4.

74. Furukawa-Hibi, Y., Yoshida-Araki, K., Ohta, T., Ikeda, K. and Motoyama, N. (2002) FOXO forkhead transcription factors induce G(2)-M checkpoint in response to oxidative stress. *J. Biol. Chem.* **277**, 26729–32.

75. Gilley, J., Coffer, P.J. and Ham, J. (2003) FOXO transcription factors directly activate bim gene expression and promote apoptosis in sympathetic neurons. *J. Cell Biol.* **162**, 613–22.

76. Modur, V., Nagarajan, R., Evers, B.M. and Milbrandt, J. (2002) FOXO proteins regulate tumor necrosis factor-related apoptosis inducing ligand expression. Implications for PTEN mutation in prostate cancer. *J. Biol. Chem.* **277**, 47928–37.

77. Accili, D. and Arden, K.C. (2004) FOXOs at the crossroads of cellular metabolism, differentiation, and transformation. *Cell* **117**, 421–6.

78. Hosaka, T., Biggs, W.H., 3rd, Tieu, D. et al. (2004) Disruption of forkhead transcription factor (FOXO) family members in mice reveals their functional diversification. *Proc. Natl. Acad. Sci. USA* **101**, 2975–80.

79. Furuyama, T., Kitayama, K., Shimoda, Y. et al. (2004) Abnormal angiogenesis in FOXO1 (Fkhr)-deficient mice. *J. Biol. Chem.* **279**, 34741–9.

80. Daly, C., Wong, V., Burova, E. et al. (2004) Angiopoietin-1 modulates endothelial cell function and gene expression via the transcription factor FKHR (FOXO1). *Genes Dev.* **18**, 1060–71.

81. Castrillon, D.H., Miao, L., Kollipara, R., Horner, J.W. and DePinho, R.A. (2003) Suppression of ovarian follicle activation in mice by the transcription factor FOXO3a. *Science* **301**, 215–8.

82. Nakae, J., Park, B.C. and Accili, D. (1999) Insulin stimulates phosphorylation of the forkhead transcription factor FKHR on serine 253 through a Wortmannin-sensitive pathway. *J. Biol. Chem.* **274**, 15982–5.

83. Rena, G., Guo, S., Cichy, S.C., Unterman, T.G. and Cohen, P. (1999) Phosphorylation of the transcription factor forkhead family member FKHR by protein kinase B. *J. Biol. Chem.* **274**, 17179–83.

84. Guo, S., Rena, G., Cichy, S., He, X., Cohen, P. and Unterman, T. (1999) Phosphorylation of serine 256 by protein kinase B disrupts transactivation by FKHR and mediates effects of insulin on insulin-like growth factor-binding protein-1 promoter activity through a conserved insulin response sequence. *J. Biol. Chem.* **274**, 17184–92.

85. Brunet, A., Sweeney, L.B., Sturgill, J.F. et al. (2004) Stress-dependent regulation of FOXO transcription factors by the SIRT1 deacetylase. *Science* **303**, 2011–5.

86. Sunayama, J., Tsuruta, F., Masuyama, N. and Gotoh, Y. (2005) JNK antagonizes Akt-mediated survival signals by phosphorylating 14-3-3. *J. Cell Biol.* **170**, 295–304.

87. Hu, M.C., Lee, D.F., Xia, W. et al. (2004) IkappaB kinase promotes tumorigenesis through inhibition of forkhead FOXO3a. *Cell* **117**, 225–37.

88. Plas, D.R. and Thompson, C.B. (2003) Akt activation promotes degradation of tuberin and FOXO3a via the proteasome. *J. Biol. Chem.* **278**, 12361–6.

89. Matsuzaki, H., Daitoku, H., Hatta, M., Tanaka, K. and Fukamizu, A. (2003) Insulin-induced phosphorylation of FKHR (FOXO1) targets to proteasomal degradation. *Proc. Natl. Acad. Sci. USA* **100**, 11285–90.

90. Aoki, M., Jiang, H. and Vogt, P.K. (2004) Proteasomal degradation of the FOXO1 transcriptional regulator in cells transformed by the P3k and Akt oncoproteins. *Proc. Natl. Acad. Sci. USA* **101**, 13613–7.

91. Huang, H., Regan, K.M., Wang, F. et al. (2005) Skp2 inhibits FOXO1 in tumor suppression through ubiquitin-mediated degradation. *Proc. Natl. Acad. Sci. USA* **102**, 1649–54.
92. Papandreou, C.N., Daliani, D.D., Nix, D. et al. (2004) Phase I trial of the proteasome inhibitor bortezomib in patients with advanced solid tumors with observations in androgen-independent prostate cancer. *J. Clin. Oncol.* **22**, 2108–21.
93. Yang, G., Ayala, G., De Marzo, A. et al. (2002) Elevated Skp2 protein expression in human prostate cancer: association with loss of the cyclin-dependent kinase inhibitor p27 and PTEN and with reduced recurrence-free survival. *Clin. Cancer Res.* **8**, 3419–26.
94. Ben-Izhak, O., Lahav-Baratz, S., Meretyk, S. et al. (2003) Inverse relationship between Skp2 ubiquitin ligase and the cyclin dependent kinase inhibitor p27Kip1 in prostate cancer. *J. Urol.* **170**, 241–5.
95. Latres, E., Chiarle, R., Schulman, B.A. et al. (2001) Role of the F-box protein Skp2 in lymphomagenesis. *Proc. Natl. Acad. Sci. USA* **98**, 2515–20.
96. Signoretti, S., Di Marcotullio, L., Richardson, A. et al. (2002) Oncogenic role of the ubiquitin ligase subunit Skp2 in human breast cancer. *J. Clin. Invest.* **110**, 633–41.
97. Nasrin, N., Ogg, S., Cahill, C.M. et al. (2000) DAF-16 recruits the CREB-binding protein coactivator complex to the insulin-like growth factor binding protein 1 promoter in HepG2 cells. *Proc. Natl. Acad. Sci. USA* **97**, 10412–7.
98. Fukuoka, M., Daitoku, H., Hatta, M., Matsuzaki, H., Umemura, S. and Fukamizu, A. (2003) Negative regulation of forkhead transcription factor AFX (FOXO4) by CBP-induced acetylation. *Int. J. Mol. Med.* **12**, 503–8.
99. Tissenbaum, H.A. and Guarente, L. (2001) Increased dosage of a sir-2 gene extends lifespan in Caenorhabditis elegans. *Nature* **410**, 227–30.
100. Motta, M.C., Divecha, N., Lemieux, M. et al. (2004) Mammalian SIRT1 represses forkhead transcription factors. *Cell* **116**, 551–63.
101. van der Horst, A., Tertoolen, L.G., de Vries-Smits, L.M., Frye, R.A., Medema, R.H. and Burgering, B.M. (2004) FOXO4 is acetylated upon peroxide stress and deacetylated by the longevity protein hSir2(SIRT1). *J. Biol. Chem.* **279**, 28873–9.
102. Daitoku, H., Hatta, M., Matsuzaki, H. et al. (2004) Silent information regulator 2 potentiates FOXO1-mediated transcription through its deacetylase activity. *Proc. Natl. Acad. Sci. USA* **101**, 10042–7.
103. Matsuzaki, H., Daitoku, H., Hatta, M., Aoyama, H., Yoshimochi, K. and Fukamizu, A. (2005) Acetylation of FOXO1 alters its DNA-binding ability and sensitivity to phosphorylation. *Proc. Natl. Acad. Sci. USA* **102**, 11278–83.
104. Debes, J.D., Sebo, T.J., Lohse, C.M., Murphy, L.M., Haugen de, A.L. and Tindall, D.J. (2003) p300 in prostate cancer proliferation and progression. *Cancer Res.* **63**, 7638–40.
105. Linja, M.J., Porkka, K.P., Kang, Z. et al. (2004) Expression of androgen receptor coregulators in prostate cancer. *Clin. Cancer Res.* **10**, 1032–40.
106. Comuzzi, B., Lambrinidis, L., Rogatsch, H. et al. (2003) The transcriptional co-activator cAMP response element-binding protein-binding protein is expressed in prostate cancer and enhances androgen- and anti-androgen-induced androgen receptor function. *Am. J. Pathol.* **162**, 233–41.
107. Di Cristofano, A., Pesce, B., Cordon-Cardo, C. and Pandolfi, P.P. (1998) Pten is essential for embryonic development and tumour suppression. *Nat. Genet.* **19**, 348–55.
108. Podsypanina, K., Ellenson, L.H., Nemes, A. et al. (1999) Mutation of Pten/Mmac1 in mice causes neoplasia in multiple organ systems. *Proc. Natl. Acad. Sci. USA* **96**, 1563–8.
109. Di Cristofano, A., De Acetis, M., Koff, A., Cordon-Cardo, C. and Pandolfi, P.P. (2001) Pten and p27KIP1 cooperate in prostate cancer tumor suppression in the mouse. *Nat. Genet.* **27**, 222–4.
110. Gao, H., Ouyang, X., Banach-Petrosky, W. et al. (2004) A critical role for p27kip1 gene dosage in a mouse model of prostate carcinogenesis. *Proc. Natl. Acad. Sci. USA* **101**, 17204–9.

111. You, M.J., Castrillon, D.H., Bastian, B.C. et al. (2002) Genetic analysis of Pten and Ink4a/Arf interactions in the suppression of tumorigenesis in mice. *Proc. Natl. Acad. Sci. USA* **99**, 1455–60.

112. Abate-Shen, C., Banach-Petrosky, W.A., Sun, X. et al. (2003) Nkx3.1; Pten mutant mice develop invasive prostate adenocarcinoma and lymph node metastases. *Cancer Res.* **63**, 3886–90.

113. Ma, L., Teruya-Feldstein, J., Behrendt, N. et al. (2005) Genetic analysis of Pten and Tsc2 functional interactions in the mouse reveals asymmetrical haploinsufficiency in tumor suppression. *Genes Dev.* **19**, 1779–86.

114. Lesche, R., Groszer, M., Gao, J. et al. (2002) Cre/loxP-mediated inactivation of the murine Pten tumor suppressor gene. *Genesis* **32**, 148–9.

115. Wu, X., Wu, J., Huang, J. et al. (2001) Generation of a prostate epithelial cell-specific Cre transgenic mouse model for tissue-specific gene ablation. *Mech. Dev.* **101**, 61–9.

116. Wang, S., Gao, J., Lei, Q. et al. (2003) Prostate-specific deletion of the murine Pten tumor suppressor gene leads to metastatic prostate cancer. *Cancer Cell* **4**, 209–21.

117. Trotman, L.C., Niki, M., Dotan, Z.A. et al. (2003) Pten dose dictates cancer progression in the prostate. *PLoS Biol.* **1**, E59.

118. Ma, X., Ziel-van der Made, A.C., Autar, B. et al. (2005) Targeted biallelic inactivation of Pten in the mouse prostate leads to prostate cancer accompanied by increased epithelial cell proliferation but not by reduced apoptosis. *Cancer Res.* **65**, 5730–9.

119. Backman, S.A., Ghazarian, D., So, K. et al. (2004) Early onset of neoplasia in the prostate and skin of mice with tissue-specific deletion of Pten. *Proc. Natl. Acad. Sci. USA* **101**, 1725–30.

120. Majumder, P.K., Febbo, P.G., Bikoff, R. et al. (2004) mTOR inhibition reverses Akt-dependent prostate intraepithelial neoplasia through regulation of apoptotic and HIF-1-dependent pathways. *Nat. Med.* **10**, 594–601.

121. Yoeli-Lerner, M., Yiu, G.K., Rabinovitz, I., Erhardt, P., Jauliac, S. and Toker, A. (2005) Akt blocks breast cancer cell motility and invasion through the transcription factor NFAT. *Mol. Cell.* **20**, 539–50.

14   New Perspectives in Prediction of Clinical Outcome of Prostate Cancer

TMPRSS2–ETS Gene Fusion in Prostate Cancer

Mark A. Rubin, MD

1. INTRODUCTION

Prostate cancer (PC) is a common and clinically heterogeneous disease with marked variability in progression. This chapter focuses on the recently identified novel *TMPRSS2–ETS* gene fusions in PC *(1–3)*. Hematological malignancies are often characterized by balanced, disease-specific chromosomal rearrangements (i.e., balanced translocations). The prototypic example

From: *Current Clinical Oncology: Prostate Cancer:*
Signaling Networks, Genetics, and New Treatment Strategies
Edited by: R. G. Pestell and M. T. Nevalainen © Humana Press, Totowa, NJ

is the malignant transformation of white blood cells to Chronic Myeloid Leukemia (CML) through a balanced translocation between chromosomes 9 and 22 resulting in the novel tyrosine kinase fusion protein BCR–ABL. Until recently, most solid tumors had been characterized only by non-specific chromosomal aberrations. By applying a new bioinformatics approach called Cancer Outlier Profile Analysis (COPA), Tomlins et al. identified a common translocation in PC, involving the tightly androgen-regulated gene *TMPRSS2* (21q22.3) and *ETS* transcription factor family members, either *ERG* (21q22.2) or *ETV1* (7p21.2) *(1)*. This translocation is detected in invasive PC and in approximately 20% of the precursor lesion, high-grade prostatic intraepithelial neoplasia (PIN) *(4)*, but not in bening prostate tissue or prostatic atrophy *(4)*. Significant associations with common morphological features and gene fusion have also been identified *(5)*. After logistic regression analysis five morphological features were independently associated with positive TMPRSS2-ERG fusion status: blue-tinged mucin, cribriform pattern, macronucleoli, intraductal tumor spread, and signet-ring cell-like features. The association between phenotype and TMPRSS2-ERG fusion suggest that there are molecular alterations associated with gene fusion PC. A previous role of *ETS* genes in PC progression has been suggested based on the over-expression of *ERG* and *ETV1* at the transcript level. *TMPRSS2* is one of the most highly androgen-regulated genes. The critical fusion of TMPRSS2 and ETS family members defines a novel and significant class of PC that, according to emerging data, may be at a higher risk of disease progression *(2,6,7)*. This chapter will provide an overview of this exciting development in our understanding of PC biology from the initial discovery to the potential applications in the clinical management setting.

2. DISCOVERY OF TMPRSS2–ETS GENE FUSION PC

The key to discovering the *TMPSS2–ETS* gene fusion was to develop a simple statistical approach to identifying oncogene profiles from expression array data sets and recognizing that this process also identified genes commonly associated with known genomic translocations. Tomlins et al., sought to identify to discover oncogenes as *using publicly available microarray data. The goal was to identify previously described oncogene such as her-2-neu* or *EGFR*—where overexpression is observed in only a subset of tumors from patients with breast or lung cancer, respectively—and importantly to discover novel oncogenes. The method called Cancer Outlier Profile Analysis (COPA) was developed based on the idea that evaluating variance in a data set using the median instead of the mean would maintain the peaks of outliers. COPA has three steps *(1,8)*. First, gene expression values are median centered, setting each gene's median expression value to zero. Second, the median absolute

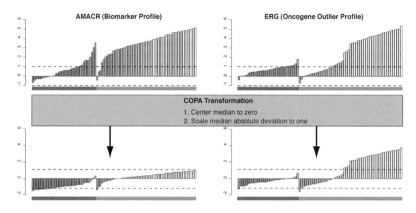

Fig. 1. Cancer Outlier Profile Analysis (COPA) evaluates variance in a data set using the median instead of the mean in order to maintain the peaks of outliers. The first to steps of COPA are presented here. First, gene expression values are median centered, setting each gene's median expression value to zero. Second, the median absolute deviation (MAD) is calculated and scaled to 1 by dividing each gene expression value by its MAD. This approach was used instead of centering data around the mean because it has less effect on the tails or outliers. In the third step (not shown), the 75th, 90th, and 95th percentiles of the transformed expression values are tabulated for each gene and then genes are rank-ordered by their percentile scores, leading to a prioritized list of outlier profiles.

deviation (MAD) is calculated and scaled to 1 by dividing each gene expression value by its MAD (*8*, Fig. 1). This approach was used instead of centering data around the mean because it has less effect on the tails or outliers. Third, the 75th, 90th, and 95th percentiles of the transformed expression values are tabulated for each gene and then genes are rank-ordered by their percentile scores, leading to a prioritized list of outlier profiles.

By applying COPA, 132 gene expression data sets representing 10,486 microarray experiments were interrogated for outlier genes (*1*). Examples of known genes that are overexpressed in a subset of a particular tumor type were identified, such as the oncogene *her-2-neu* and *E-Cadherin* (CDH1). Interestingly, genes such as RUNX1T1 (ETO) and PBX1 also scored high on COPA. These two genes are known to be associated with the *AML-ETO* and *E2A-PBX1* gene translocations in acute myeloid leukemia (AML) and acute lymphoblastic leukemia, respectively. Both of these translocations only occur in a subset of the cases (i.e., outlier cases). Two genes consistently scored high in PC microarray experiments, *ERG* and *ETV1*. Both of these genes are members of the ETS family of transcription factors. They were overexpressed in the majority (50–70%) of PCs and were mutually exclusive across several independent gene expression data sets, suggesting that they may be functionally

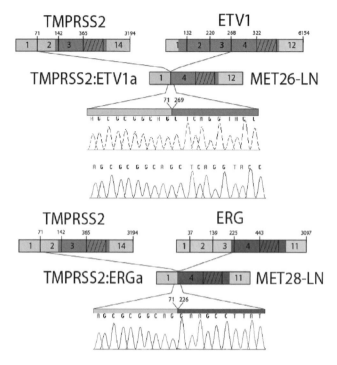

Fig. 2. Anatomy of the TMPRSS2 to ETS Family Gene Fusions Identified in Prostate Cancer. Adapted from ref. *1*.

redundant in PC development *(1)*. Because the ETS family of transcription factors has previously been seen in the genomic translocation of the Ewing's family tumors, AML and other rare tumors, the possibility that they were part of a translocation in PC was explored. When the ERG cDNA transcript was evaluated exon by exon, overexpression was seen at the distal (3′ end) but not the proximal portion (5′ end). By sequencing the cDNA transcripts, fusions of the 5′-untranslated region of *TMPRSS2* (21q22.3) with the *ETS* transcription factor family members, either *ERG* (21q22.2), *ETV1* (7p21.2) *(1)*, and more recently *ETV4 (3)*, were identified, suggesting a novel mechanism for overexpression of the ETS genes in PC (*1*, Fig. 2).

3. ETS FAMILY OF TRANSCRIPTION FACTORS AND PC

The ETS family of transcription factors consist of 30–40 genes that bind to DNA in a site-specific manner and mediate transcriptional activation and/or repression of other genes. The ETS family members are defined by an 87-amino-acid domain that is necessary and sufficient for this site-specific DNA binding, and this domain is flanked by protein–protein interaction domains that mediate transcriptional activation and/or repression. Some ETS genes have auto-inhibitory domains that block DNA binding in the absence of co-factors. Thus, ETS transcription factors appear to bind to promotor/enhancer elements of target genes leading to transcription activation and/or repression (for review of this topic, see refs. *(9,10)*). Although ETS target genes have been identified for many of the ETS family members, their exact role in transcription regulation is not well understood *in vivo (11)*. Only a few studies have specifically examined the role of ETS genes in prostate tissues. Gavrilov et al. described the protein expression of several ETS family members in prostate tissue samples using immunohistochemistry *(12)*. They reported nuclear expression of Elf-1 and Fli-1 in 16/25 and 20/25 of high-grade PCs, respectively. Interestingly, ERG expression was only observed in 44% (11/25) of PC samples tested, of which 7 were with Gleason score 7 or above. Recent work from Petrovics et al. identified *ERG* as the most frequently overexpressed oncogene in the transcriptome by using a combination of expression array analysis and quantitative real-time RT-PCR on 114 PC samples isolated with laser capture microdissection *(13)*. High *ERG* expression was identified as a predictor of higher prostate-specific antigen (PSA).

4. TMPRSS2 AND PC

The *TMPRSS2* gene located on chromosome 21 encodes a serine protease. *TMPRSS2* is a type II integral membrane protein similar to hepsin, which is also frequently upregulated in PC *(14)*. Lin et al. identified *TMPRSS2* as the most up upregulated gene in the LNCaP PC cell line after exposure to androgens *(15)*. TMPRSS2 expression was not affected by androgen stimulation in the androgen-unresponsive PC-3 and DU 145 cell lines. Human PC xenografts that were either androgen dependant or independent demonstrated TMPRSS2 expression. *In situ* hybridization demonstrated that TMPRSS2 RNA was expressed in the basal cell compartment of normal prostate epithelium and in PC. Vaarala et al. confirmed this observation independently demonstrating higher RNA levels in PC when compared with benign prostate tissue *(16)*. Vaarala et al. further demonstrated that TMPRSS2 is expressed at higher levels in high-grade PC and in hormonally treated PC. These data support the view that *TMPRSS2* may be associated with PC progression. Immuno-histochemical studies by Afar et al. identified TMPRSS2 protein expression in the cytoplasm of normal prostate epithelium and PC *(17)*. Interestingly, in mice lacking TMPRSS2 (*tmprss2$^{-/-}$*), there is no discernable phenotype suggesting that the dysregulation alone is insufficient to affect prostate pathophysiology *(18)*.

5. TMPRSS2–ETS FUSION IN PC

The identification of this fusion between the prostate-specific, strongly androgen-regulated gene *TMPRSS2 (21q22.3)* to *ERG* (21q22.2) or *ETV1* (7p21.2) was a surprising discovery. Using other methods to validate these findings [i.e., RT-PCR or fluorescence *in situ* hybridization (FISH)] in human PC samples, the TMPRSS2–ETS translocation has been seen in approximately 50–70% of all cases examined *(1–4,7,19,20)*. Because TMPRSS2 is regulated by androgens, a series of cell line experiments was performed to demonstrate that exposure to androgen would specifically regulate the fused ETS family member. In the PC cell line VCaP *(21)* with the TMPRSS2–ERG translocation, exposure to a dose of synthetic androgen specifically increased ERG expression. In the LNCaP PC cell lines without an ERG translocation, exposure to synthetic androgen did not alter ERG expression. Therefore, it appears that the TMPRSS2–ETS gene fusion is acting as a novel androgen responsive oncogene.

6. TMPRSS2–ERG GENE FUSION

The most common gene fusion is *TMPRSS2:ERG* found in approximately 50% of the surgical-based cohorts studied to date *(2,4,7,19,20)*. As both of these genes lie in close proximity on chromosome 21, 100K oligonucleotide Single Nucleotide Polymorphism (SNP) arrays could be used to characterize an intronic deletion observed by FISH analysis. By interrogating 30 PC samples, including cell lines, xenografts and hormone naïve and hormone refractory metastatic PC samples, genomic loss between *ERG* and *TMPRSS2* on chromosome 21q23 could be readily visualized (Fig. 2A–C from ref. *(2)*). The rearrangement status for *TMPRSS2:ERG* and *TMPRSS2:ETV1* was determined for these 30 PC by FISH and/or PCR (Fig. 2A, gray and light blue bar). None of the samples tested demonstrated a *TMPRSS2:ETV1* rearrangement. Discrete genomic loss was observed in *TMPRSS2:ERG* rearrangement positive samples involving an area between *TMPRSS2* and the *ERG* loci. The extent of these discrete deletions was heterogeneous. For a subset of samples 45% (5 of 11) the deletion occurs in proximity of *ERG* intron 3. For a majority of samples 64% (7 of 11) the deletion ends in proximity of the SNP located on *TMPRSS2* (the next SNP in the telomeric direction is about 100K bp distant). The VCaP cell line shows copy number gain along the entire chromosome 21. Interestingly, for *TMPRSS2:ERG* rearrangement-positive tumors, 71% (5 of 7) hormone refractory PC demonstrate a deletion between *TMPRSS2* and the *ERG* loci whereas deletion was only identified in 25% (1 of 4) hormone naïve metastatic PC samples (ULM LN 13). There is significant homogeneity for the deletion borders with two distinct sub-classes, distinguished by the start point of the deletion—either at 38.765 Mb or at 38.911 Mb. None of the

standard PC cell lines [PC-3, LNCaP, DU-145, or CWR22 (22Rv1)] demonstrated the *TMPRSS2:ERG* or *TMPRSS2:ETV1* fusion. Several of the LuCaP
xenografts demonstrate *TMPRSS2:ERG* fusion with deletion including LuCaP
49 (established from an omental mass) and LuCaP 93, both hormone-insensitive
[androgen receptor (AR)-negative] small-cell PCs. The VCaP cell line derived
from a hormone refractory PC demonstrated significant copy number gain on
chromosome 21 (*2*, Fig. 2A–C).

These findings are particularly intriguing when one considers the 9:22
translocation identified in CML. The *Bcr–Abl* translocation involving chromosomes 9 and 22 is believed to cause the malignant transformation to CML.
Evidence for this includes mouse models overproducing the fusion protein
(BCR–ABL), which lead to the development of a murine leukemia that
sometime resembles CML *(22)* and can be reversed by lowering the expression
of BCR–ABL *(23)*. The clinical prognosis of CML is heterogeneous with
significant differences in clinical features including the rate of progression to
blast crisis transformation.

The surprising development in CML was that in addition to the *Bcr–Abl*
translocation, a subset of cases harbor a deletion of the derivative chromosome
9 involved in the reciprocal translocation *(24–28)*. Multiple studies have now
confirmed that these deletions are large, show varying breakpoints, and occur at
the time of the translocation (i.e., an early event). The presence of the deletion is
associated with poor prognostic outcome *(24,25)*. Therefore, even if it appears
that CML with a common translocation should be a homogenous disease
with similar clinical course, the presence of this deletion may explain for the
considerable genetic heterogeneity. The mechanism is not understood *(29)*. The
deletion might be associated with a predisposition for genetic instability. The
deletion might alter the activity of the *Bcr–Abl* fusion protein. The deletion
might also be associated with loss of another gene leading to either entire loss
of activity ("two-hit" model) or decreased expression (haploinsufficiency).

The high percentage of *TMPRSS2:ERG* fusion PCs suggests that *ERG* is the
most common fusion partner. The hospital-based studies to date suggest that at
least 50% of PCs harbor the *TMPRSS:ERG* gene fusion. With the recent identification of a third molecular subtype (*TMPRSS2:ETV4*), one can anticipate
finding other translocation partners, such as *FLI1*, based on expression array
data. This would be similar to observation in the Ewing's family tumors, where
approximately 85% of tumors harbor a tumor-associated t(11;22)(q24;q12)
rearrangement resulting in the juxtaposition of the *EWS* gene (Ewing's Sarcoma
Gene) on chromosome 22 with the *FLI1* gene on chromosome 11. Four other
ETS family members have been identified as translocation partners of *EWS*.
The second most common *ETS* translocation partner is *ERG* seen in approximately 10% of cases *(30)*. Finally, the identification of the *TMPRSS2:ETS*
gene fusion in PC suggests that distinct molecular subtypes may further define
risk of disease progression.

7. TMPRSS2–ETS GENE FUSION PC AND DISEASE PROGRESSION

Currently, there is limited information regarding the association between gene fusion PC and disease progression. Because *TMPRSS2–ETV1* and *TMPRSS2–ETV4* fusion PCs are so infrequent, the following comments apply only to *TMPRSS2–ERG* fusion PC, which appear to behave in a

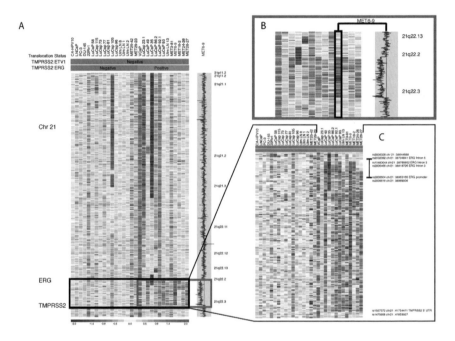

Fig. 3. A-C. Genomic deletions on chromosome 21 between ERG and TMPRSS2. Interrogating high density 100K SNP arrays (~110.000 loci on the genome) on a panel of 30 Prostate Cancer (PCA) samples, we observed a commonly deleted area on chromosome 21q22.2-22.3, spanning the region between ERG and *TMPRSS2*. A. Samples, including 6 cell lines, 13 xenografts and 11 metastatic PCA samples, were characterized for *TMPRSS2:ERG* and *TMPRSS2:ETV1* status (gray bars for negative and blue bar for positive status), by qPCR and/or by FISH. B. Magnification of the green framed box in A. Signal intensity on the right side is proportional to copy number intensity of a hormone refractory metastatic PCA sample (MET6-9). Interestingly, for *TMPRSS2:ERG* rearrangement positive tumors, the 71% (5 of 7) hormone refractory PCA demonstrate a deletion between *TMPRSS2* and the *ERG* loci whereas deletion was only identified in 1 of 4 hormone naïve metastatic PCA samples (ULM LN 13). C. Magnification of the black framed box in A. SNP data include 25 loci along *ERG*, distributed from the gene promoter to intron 5 and 1 SNP on the 3'UTR of *TMPRSS2*. There is significant homogeneity for the deletion borders with two sub-classes, distinguished by the start point of the deletion—either 38.765 Mb or 38.911 Mb).

more aggressive manner. Perner et al. explored for an association between rearrangement status and clinical and pathological parameters (2). Interestingly, *TMPRSS2–ERG* rearrangement with deletion was observed in a higher percentage of PC cases with advanced tumor stage (pT) ($p = 0.03$), and the presence of metastatic disease in regional pelvic lymph nodes (pN_0 versus pN_{1-2}) ($p = 0.02$). *TMPRSS2:ERG*-rearranged PC with deletions demonstrated a statistical trend for higher PSA biochemical recurrence when compared with translocation negative PC. Wang et al. also reporting on a radical prostatectomy series identified a significant association between gene fusion and higher tumor stage (i.e., seminal vesicle invasion) *(7)*.

Perhaps the strongest evidence to date to suggest an association between *TMPRSS2–ERG* fusion PC and more aggressive disease is from Demichelis et al. study on a population of men diagnosed with incidental PC and followed for up to 30 years on a watchful waiting protocol *(6)*. The Örebro Watchful Waiting cohort represents a treatment naïve population drawn from a strictly defined catchment area for 190,000 inhabitants living in Örebro *(31–35)*. The frequency of *TMPRSS2:ERG* gene fusion in this watchful waiting population is lower than the 55% (16/29) reported by Tomlins et al. *(1)* and 78% (14/18) reported by Soller et al. *(20)*. These differences might be explained by the lower percentage of high-grade cases in this watchful waiting cohort as compared to the other non-population based studies. In this report on a population-based cohort of men with localized PCs followed by expectant (watchful waiting) therapy, the *TMPRSS2:ERG* gene fusion was identified in 15% (17/111) of the tumors. There was a statistically significant association between *TMPRSS2:ERG* gene fusion and PC-specific death (cumulative incidence ratio = 2.7, $p < 0.01$, 95% confidence interval = 1.3–5.8) (Fig. 4A-C). Quantitative RT-PCR demonstrated high ERG expression to be associated with *TMPRSS2:ERG* gene fusion ($p < 0.005$) supporting the view that the transcription factor *ERG* is acting as an oncogene in this gene fusion. These data support the observation that *TMPRSS2:ERG* gene fusion PC have a more aggressive phenotype, possibly mediated through increased ERG expression.

Interestingly, high ERG expression has been associated with poor clinical outcome in AML *(36)* but was seen as protective in the only other PC study *(13)* examining ERG, where high levels of ERG transcription were associated with a lower incidence of PSA biochemical failure *(13)*. The findings by Petrovics need to be viewed with caution because PSA relapse or biochemical failure is a poor surrogate endpoint for clinically meaningful endpoints including clinical relapse and death as demonstrated by three recent studies. In a single institution study of men diagnosed with clinically localized PC in the pre-PSA screening era, Porter et al. observed 45.5% PSA biochemical failure in a radical prostatectomy series, but PC-specific death occurred in 18.5% of the population with a follow-up time of up to 25 years *(37)*. Carver et al. recently

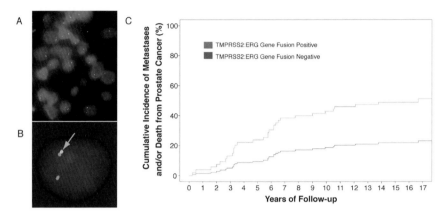

Fig. 4. A-C. The FISH assay detects the characteristic hybridization pattern associated with *TMPRSS2:ERG* gene fusion, which is associated with disease progression. **A-B).** For analyzing the *ERG* rearrangement on chromosome 21q22.2, a break apart probe assay was applied, consisting of the biotin-14-dCTP labeled BAC clone RP11-24A11 (eventually detected with cy3-avidin to produce a red signal) and the digoxigenin-dUTP labeled BAC clone RP11-137J13 (eventually detected with FITC-antidigoxygenin to produce a green signal), spanning the neighboring centromeric and telomeric region of the *ERG* locus, respectively. All BAC clones were obtained from the BACPAC Resource Center, Children's Hospital Oakland Research Institute (CHORI), Oakland, CA. Prior to tissue analysis, the integrity and purity of all probes were verified by hybridization to normal peripheral lymphocyte metaphase spreads. Tissue hybridization, washing, and fluorescence detection were performed as described previously (Garraway et al., 2005; Rubin et al., 2004). Using this multicolor FISH probe system, a nucleus without *ERG* rearrangement demonstrates two pairs of juxtaposed red and green signals. Juxtaposed red-green signals sometimes form a yellow fusion signal (**B, arrow**). A nucleus with an *ERG* rearrangement shows replacement of one juxtaposed red-green signal pair with a single red signal for the translocated allele (**B**). This hybridization pattern is consistent with a deletion of ~2.8 Mb of genomic DNA encompassing the telomeric portion of the ERG probe. A break apart assay is in general significantly easier to interpret as compared to a fusion assay and will detect rearrangement in all cases. Thus, a break apart assays is more specific and much faster to evaluate (Bridge et al., 2006). The samples were analyzed under a 100x oil immersion objective on an Olympus BX-51 fluorescence microscope equipped with appropriate filters, a CCD (charge-coupled device) camera and the CytoVision FISH imaging and capturing software (Applied Imaging, San Jose, CA). Evaluation of the tests was independently performed by two pathologists (SP and J-M M) both with experience in analyzing interphase FISH experiments. For each case, we attempted to score a minimum 100 nuclei per case. **C)** In a cumulative incidence regression model, we evaluated *TMPRSS2:ERG* as a determinant for the cumulative incidence or metastases or prostate cancer-specific death. We observed a significant difference in survival in favor of *TMPRSS2:ERG* gene fusion negative cases. The cumulative incidence ratio (CIR = 2.7, P < 0.01, 95% CI = 1.3 to 5.8) was performed using the weighted estimating equation as implemented in cmprsk R library (Team, 2004). Other statistical analysis was performed using SPSS 13.0 for Windows (SPSS Inc., Chicago, Il).

.

reported that in a population of high-risk men with T3 PC who underwent radical prostatectomy, 36% with PSA biochemical failure subsequently went on to have clinically relevant disease progression *(38)*. Ward et al. found that in a population of 3897 radical prostatectomy patients, only 8.3% of the men with PSA biochemical failure died of PC with a median follow-up time of 10 years *(39)*. PSA failure is thus associated with PC-specific death, but the majority of men with PSA biochemical failure will die of other causes. Therefore, the strongest data in initially untreated PCs is that there is a significant association between *TMPRSS2–ERG* gene fusion status and PC disease progression. Emerging work from multiple investigators is attempting to validate these findings.

8. SUMMARY

The *TMPRSS2:ERG* gene fusion is the most common of the three reported *TMPSS2–ETS* gene fusions. It is seen in approximately 50–70% of PC taken from hospital-based surgical cohorts. In a lower risk Watchful Waiting cohort, the incidence was approximately 15%. This supports other evidence that gene fusion is associated with higher tumor stage and a more aggressive disease progression. The gene fusion is seen only in PC and 20% of high-grade PIN but not benign prostate glands or atrophy suggesting that it is an early event in the development of invasive PC. In addition, this study supports the critical role of ERG as an oncogene in PC.

The discoverAy of the common *TMPRSS2:ETS* gene fusions in PC using COPA suggests that other translocations may be identified in common epithelial tumors. The combination of an organ specific promoter such as TMPRSS2 for PC fused to an oncogene may also be a common theme in carcinogenesis. Strategies for the development of diagnostic tests and targeted therapy are underway, perhaps leading to rational drug development similar to the development of imatinib (STI571, Gleevec) therapy for CML.

REFERENCES

1. Tomlins SA, Rhodes DR, Perner S, et al. Recurrent fusion of TMPRSS2 and ETS transcription factor genes in prostate cancer. *Science* 2005;310(5748):644–8.
2. Perner S, Demichelis F, Beroukhim R, et al. TMPRSS2:ERG fusion-associated deletions provide insight into the heterogeneity of prostate cancer. *Cancer Res* 2006;66(17):8337–41.
3. Tomlins SA, Mehra R, Rhodes DR, et al. TMPRSS2:ETV4 gene fusions define a third molecular subtype of prostate cancer. *Cancer Res* 2006;66(7):3396–400.
4. Perner S, Mosquera JM, Demichelis F, et al. TMPRSS2-ERG fusion prostate cancer: an early molecular event associated with invasion. *Am J Surg Pathol* 2007;31(6):882–8.
5. Mosquera JM, Perner S, Demichelis F, et al. Morphological features of TMPRSS2–ERG gene fusion prostate cancer. *J Pathol* 2007.
6. Demichelis F, Fall K, Perner S, et al. TMPRSS2:ERG gene fusion associated with lethal prostate cancer in a watchful waiting cohort. *Oncogene* 2007.

7. Wang J, Cai Y, Ren C, Ittmann M. Expression of variant TMPRSS2/ERG fusion messenger RNAs is associated with aggressive prostate cancer. *Cancer Res* 2006;66(17):8347–51.

8. Rubin MA, Chinnaiyan AM. Bioinformatics approach leads to the discovery of the TMPRSS2:ETS gene fusion in prostate cancer. *Lab Invest* 2006.

9. Sharrocks AD. The ETS-domain transcription factor family. *Nat Rev Mol Cell Biol* 2001;2(11):827–37.

10. Arvand A, Denny CT. Biology of EWS/ETS fusions in Ewing's family tumors. *Oncogene* 2001;20(40):5747–54.

11. Sementchenko VI, Watson DK. Ets target genes: past, present and future. *Oncogene* 2000;19(55):6533–48.

12. Gavrilov D, Kenzior O, Evans M, Calaluce R, Folk WR. Expression of urokinase plasminogen activator and receptor in conjunction with the ets family and AP-1 complex transcription factors in high grade prostate cancers. *Eur J Cancer* 2001;37(8):1033–40.

13. Petrovics G, Liu A, Shaheduzzaman S, et al. Frequent overexpression of ETS-related gene-1 (ERG1) in prostate cancer transcriptome. *Oncogene* 2005;24(23):3847–52.

14. Dhanasekaran SM, Barrette TR, Ghosh D, et al. Delineation of prognostic biomarkers in prostate cancer. *Nature* 2001;412(6849):822–6.

15. Lin B, Ferguson C, White JT, et al. Prostate-localized and androgen-regulated expression of the membrane-bound serine protease TMPRSS2. *Cancer Res* 1999;59(17):4180–4.

16. Vaarala MH, Porvari K, Kyllonen A, Lukkarinen O, Vihko P. The TMPRSS2 gene encoding transmembrane serine protease is overexpressed in a majority of prostate cancer patients: detection of mutated TMPRSS2 form in a case of aggressive disease. *Int J Cancer* 2001;94(5):705–10.

17. Afar DE, Vivanco I, Hubert RS, et al. Catalytic cleavage of the androgen-regulated TMPRSS2 protease results in its secretion by prostate and prostate cancer epithelia. *Cancer Res* 2001;61(4):1686–92.

18. Kim TS, Heinlein C, Hackman RC, Nelson PS. Phenotypic analysis of mice lacking the Tmprss2-encoded protease. *Mol Cell Biol* 2006;26(3):965–75.

19. Yoshimoto M, Joshua AM, Chilton-Macneill S, et al. Three-color FISH analysis of TMPRSS2/ERG fusions in prostate cancer indicates that genomic microdeletion of chromosome 21 is associated with rearrangement. *Neoplasia* 2006;8(6):465–9.

20. Soller MJ, Isaksson M, Elfving P, Soller W, Lundgren R, Panagopoulos I. Confirmation of the high frequency of the TMPRSS2/ERG fusion gene in prostate cancer. *Genes Chromosomes Cancer* 2006.

21. Korenchuk S, Lehr JE, L MC et al. VCaP, a cell-based model system of human prostate cancer. *In Vivo* 2001;15(2):163–8.

22. Pear WS, Miller JP, Xu L, et al. Efficient and rapid induction of a chronic myelogenous leukemia-like myeloproliferative disease in mice receiving P210 bcr/abl-transduced bone marrow. *Blood* 1998;92(10):3780–92.

23. Huettner CS, Zhang P, Van Etten RA, Tenen DG. Reversibility of acute B-cell leukaemia induced by BCR-ABL1. *Nat Genet* 2000;24(1):57–60.

24. Huntly BJ, Reid AG, Bench AJ, et al. Deletions of the derivative chromosome 9 occur at the time of the Philadelphia translocation and provide a powerful and independent prognostic indicator in chronic myeloid leukemia. *Blood* 2001;98(6):1732–8.

25. Sinclair PB, Nacheva EP, Leversha M, et al. Large deletions at the t(9;22) breakpoint are common and may identify a poor-prognosis subgroup of patients with chronic myeloid leukemia. *Blood* 2000;95(3):738–43.

26. Grand F, Kulkarni S, Chase A, Goldman JM, Gordon M, Cross NC. Frequent deletion of hSNF5/INI1, a component of the SWI/SNF complex, in chronic myeloid leukemia. *Cancer Res* 1999;59(16):3870–4.

27. Herens C, Tassin F, Lemaire V, et al. Deletion of the 5´-ABL region: a recurrent anomaly detected by fluorescence in situ hybridization in about 10% of Philadelphia-positive chronic myeloid leukaemia patients. *Br J Haematol* 2000;110(1):214–6.

28. Kolomietz E, Al-Maghrabi J, Brennan S, et al. Primary chromosomal rearrangements of leukemia are frequently accompanied by extensive submicroscopic deletions and may lead to altered prognosis. *Blood* 2001;97(11):3581–8.

29. Huntly BJ, Bench AJ, Delabesse E, et al. Derivative chromosome 9 deletions in chronic myeloid leukemia: poor prognosis is not associated with loss of ABL-BCR expression, elevated BCR-ABL levels, or karyotypic instability. *Blood* 2002;99(12):4547–53.

30. Delattre O, Zucman J, Melot T, et al. The Ewing family of tumors—a subgroup of small-round-cell tumors defined by specific chimeric transcripts. *N Engl J Med* 1994;331(5):294–9.

31. Johansson JE, Adami HO, Andersson SO, Bergstrom R, Holmberg L, Krusemo UB. High 10-year survival rate in patients with early, untreated prostatic cancer. *JAMA* 1992;267(16):2191–6.

32. Johansson JE, Holmberg L, Johansson S, Bergstrom R, Adami HO. Fifteen-year survival in prostate cancer. A prospective, population-based study in Sweden. *JAMA* 1997;277(6): 467–71.

33. Johansson JE, Adami HO, Andersson SO, Bergstrom R, Krusemo UB, Kraaz W. Natural history of localised prostatic cancer. A population-based study in 223 untreated patients. *Lancet* 1989;1(8642):799–803.

34. Johansson JE, Andren O, Andersson SO, et al. Natural history of early, localized prostate cancer. *JAMA* 2004;291(22):2713–9.

35. Andren O, Fall K, Franzen L, Andersson SO, Johansson JE, Rubin MA. How well does the Gleason score predict prostate cancer death? A 20-year followup of a population based cohort in Sweden. *J Urol* 2006;175(4):1337–40.

36. Marcucci G, Baldus CD, Ruppert AS, et al. Overexpression of the ETS-related gene, ERG, predicts a worse outcome in acute myeloid leukemia with normal karyotype: a Cancer and Leukemia Group B study. *J Clin Oncol* 2005;23(36):9234–42.

37. Porter CR, Kodama K, Gibbons RP, et al. 25-year prostate cancer control and survival outcomes: a 40-year radical prostatectomy single institution series. *J Urol* 2006;176(2): 569–74.

38. Carver BS, Bianco FJ, Jr., Scardino PT, Eastham JA. Long-term outcome following radical prostatectomy in men with clinical stage T3 prostate cancer. *J Urol* 2006;176(2):564–8.

39. Ward JF, Blute ML, Slezak J, Bergstralh EJ, Zincke H. The long-term clinical impact of biochemical recurrence of prostate cancer 5 or more years after radical prostatectomy. *J Urol* 2003;170(5):1872–6.

15 New Directions in Radiation Therapy of Prostate Cancer

Brachytherapy and Intensity-Modulated Radiation Therapy

Sean Collins, MD, PhD,
Donald McRae, PhD,
Gregory Gagnon, MD,
and Anatoly Dritschilo, MD

CONTENTS

From: *Current Clinical Oncology: Prostate Cancer:*
Signaling Networks, Genetics, and New Treatment Strategies
Edited by: R. G. Pestell and M. T. Nevalainen © Humana Press, Totowa, NJ

1. INTRODUCTION

Advances in the treatment of prostate cancer with radiation are the direct result of improved diagnostic accuracy, integration of imaging with treatment delivery, and enhanced precision in radiation dose distribution. The optimal approach is multidisciplinary, requiring the participation of urologists, radiation oncologists, and medical oncologists as patients are stratified on the basis of disease extent and prognosis, acceptance of risks and benefits, and consideration of changes in quality of life. Patients may select among several radical prostatectomy or radiation therapy options. Combinations of treatment modalities are frequently used. Radical prostatectomy may be followed by post-operative radiation therapy or androgen deprivation in conjunction with external radiation therapy for the treatment of patients at high risk for local disease recurrence. Furthermore, systemic treatments using neoadjuvant androgen deprivation therapy and chemotherapy are under investigation. Here, we will focus on the clinical considerations for application of currently available radiation therapy technology for patients with disease confined to the prostate and adjacent tissues.

With the advent of prostate-specific antigen (PSA) testing, prostate cancer is frequently diagnosed in men at early stages in which the tumor is organ confined and potentially curable by radical surgery, conformal external radiation, or brachytherapy *(1–4)*. Radical prostatectomy offers an excellent treatment for younger men (<65 years old) presenting with organ-confined disease and low to intermediate risk factors *(5)*. High-risk factors may include PSA > 30 ng/dl, Gleason's grade > 7, and clinical stage > II. Such patients may be more appropriately treated with radiation therapy. Furthermore, patients with pre-existing medical conditions such as obesity, diabetes mellitus, or cardiovascular diseases are generally not candidates for radical prostatectomy because of increased surgical morbidity. The accepted risks of surgery include higher incidences of incontinence and impotence, leading some men in the younger age group to select radiation therapy, based on quality of life considerations.

Although there are no suitable randomized clinical trials to directly compare surgical to radiation therapy options, data gleaned from selected retrospectively reviewed series support radical prostatectomy for long-term local tumor control (15 year follow-up), while radiation therapy provides comparable intermediate-term local control (10 year follow-up) with better quality of life *(6–8)*. Local tumor control impacts cancer cure, and failure to achieve local control in the treatment of prostate cancer has been correlated to the development of metastases *(9,10)*.

Patients who undergo radical prostatectomy also receive the benefit of pathological examination of the surgical specimen to determine fidelity of margins of resection and the absence or presence of extra-prostatic disease in surrounding tissues, the seminal vesicles, and lymph nodes. Patients with such findings are categorized with pathological stage III disease and may benefit from post-operative radiation therapy *(11)*.

Patients electing radiation therapy based on general health, stage, age, or quality of life criteria are offered conformal external beam radiation therapy, brachytherapy, or a combination of these modalities. In addition, patients presenting with Gleason's grade of 7 or greater, or a PSA greater than 30 ng/dl, are treated with androgen deprivation therapy in combination with radiation therapy and for an additional 1–3 years afterwards *(12,13)*.

Local tumor control with radiation is dose dependent, following a steep, sigmoid dose–response relationship (Fig. 1, curve a). A similar curve exists for normal tissue complications in response to radiation dose (Fig. 1, curve b), but this curve is shifted to the higher dose region. The therapeutic ratio is a useful concept defined by comparing the dose to achieve tumor control to the dose at which normal tissue complications are observed. Therefore, it becomes apparent that even small increases in the total dose in the prostate volume enhance local tumor control, offering a therapeutic benefit as long as normal tissue tolerances are respected.

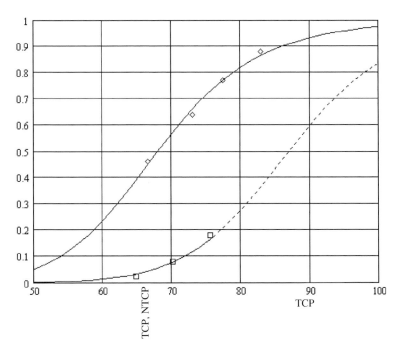

Fig. 1. Illustrative curve for dose–response relationship for tumor control probability (TCP) of prostate cancer (**a**) and normal tissue complication probability (NTCP) of rectum (**b**). Sigmoid curves were fit to reported data *(54,55)*.

Current techniques to achieve dose escalation include conformal external beam radiation therapy (3-D or IMRT), prostate brachytherapy, and particle (proton beam) irradiation. The latter is available at several specialized centers and offers theoretical advantages in dose distribution that have yet to meet clinical expectations. However, proton "boost" has been used to escalate prostate radiation dose to 79.2 Gy in a reported randomized series showing superiority as compared to 70.2 Gy using conventional external radiation *(14)*. It is unclear why the higher doses achieved by this approach offer an advantage over comparable doses achieved by linear accelerators or brachytherapy technology.

In this chapter, we review the evolution of technical advances in imaging, brachytherapy, and linear accelerator-based radiation delivery leading to improvements in dose distribution to the prostate tumor volume while sparing rectal and bladder tissues. Comments on evidence-based factors impacting the clinical decision process are offered, along with research direction for improving local-regional treatment of prostate cancer.

2. EVOLUTION OF EXTERNAL RADIATION TECHNIQUES

Although X-rays have been used to treat prostate cancer since the early 1900s, development of the clinical linear accelerator in the 1950s has marked the technological break-through that permitted Stanford University radiation oncologists to advance this treatment as an option to radical prostatectomy. Patients were treated to doses of 70 Gy, and 10-year disease-free survivals were reported (95% stage A1, 65–80% Stage A2, and 40–50% Stage B) *(15)*. These results were associated with observed high incidences of proctitis (10%) and cystitis (11%). Attempts to improve local tumor control by escalating the dose to the prostate beyond 70 Gy using conventional radiation therapy techniques led to increased late rectal toxicities *(16)*. Furthermore, reported late local recurrences were still seen in 30–40% of patients (Table 1).

3. THREE-DIMENSIONAL CONFORMAL RADIATION THERAPY

External beam radiation therapy is generally delivered using a medium- to high-energy linear accelerator (6 MeV or greater) and CT-based treatment planning. Such treatment has evolved from multiple fixed fields (4-field "box" technique), or small field rotational techniques (360° or 120° lateral arcs), to 6-field 3-D conformal techniques. The latter offers the ability to shape the radiation dose distribution to the tumor volume in three dimensions to assure that specified normal tissue tolerances are respected. In principle, this allows modest dose escalation in the tumor volume, dose reduction in normal tissues, or both.

Table 1
Results of Conventional EBRT for Patients with Organ-Confined Prostate Cancer

Study	PTS no.	Clinical stage	Total dose (Gy)	Local failure (10–15 years) (%)	DFS (years) (%)			OS (%)
					5	10	15	10
Bagshaw (29)	526	T_{1-2}	70–75	20–30	55–65	65–70	50–65	55–65
	385	T_3	70–75	38	38	50	35	38
Perez (30)	300	T_{1-2}	70	20–24	65–70	56–60		65–70
	412		70	40	42	38		42
Shipley (31)	307	T_{1-2}	67–70	16–18	66	70		66
Del Regato (32)	372	T_3	70–75		30	63	50	30
Hanks (38)	531	T_{1-2}	60–70	15–35	46–63	37–52		46–63
	296	T_3	60–70	30–42	32	20	17	32
Zagars (34)	551	T_3	60–70		47	40	40	47

DFS, disease-free survival; Gy, Gray; OS, overall survival.

Using strict biochemical failure criteria following conventional radiation doses, more than 50% of patients with clinically localized prostate cancer have been reported to experience disease progression with follow-up beyond 5 years (17,18). Technical advances associated with 3-D conformal radiation therapy have led to dose escalation, with a resultant decrease in local recurrence. However, dose escalation using 3-D conformal technology is still associated with increased rectal toxicity, supporting further research into technical improvements in dose distributions.

4. INTENSITY-MODULATED RADIATION THERAPY

To further control the dose distribution within the treated volume and to normal tissues, the concept of intensity modulation of the radiation beam has been applied. The edges of the field are shaped to conform to the cross-sectional silhouette of the tumor volume, and the radiation dose distribution within the treated volume is modulated. This technique has been reported to permit dose escalation within the prostate in excess of 80 Gy to limit dose to the rectal

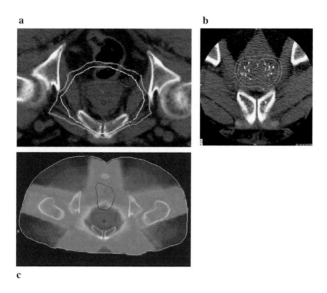

Fig. 2. This is an example of a radiation dose distribution for an IMRT plan prescribing 72 Gy to the prostate in panel **a**. Panel **b** shows the isodose distribution for a Pd-103 brachytherapy procedure, prescribing 125 Gy. Color map in panel **c** demonstrates the relationship of the high-dose irradiated region (red) to the lower dose regions irradiated in transit.

wall and to yield acceptable risks of radiation related morbidity *(19)*. Panels a and c in Fig. 2 illustrate the types of isodose plans that are readily achieved using IMRT technology, panel b illustrates a tight conformal dose achieved by brachytherapy.

5. IMAGE-GUIDED RADIATION THERAPY

Prostate and patient motion limit the precision achievable with IMRT. Therefore, image guidance has been employed to assure reproducibility in matching the delivered dose distribution to the tumor volume as it is positioned in the patient. Technologies to achieve this precision include cone beam tomography (reconstruction of a tomographic image on the treatment linear accelerator) or intermittent orthogonal diagnostic imaging to assure proper localization of fiducials or anatomic landmarks.

At Georgetown University Hospital, we have advanced the application of robotic stereotactic radiosurgical techniques for precise radiation delivery to partial volumes of the prostate (illustrated in Fig. 3) in conjunction with IMRT technology. This technology offers dose escalation capability beyond

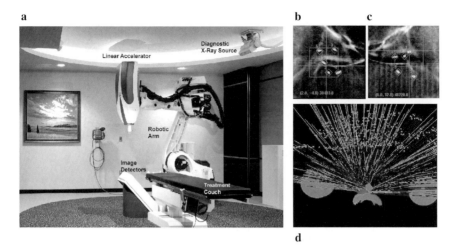

Fig. 3. The CyberKnife is shown in panel (**a**), five gold seed fiducials used for tracking the prostate are shown in orthogonal views in panels (**b**) and (**c**), and a representation of the 164 beams used to deliver the plan are shown in panel (**d**).

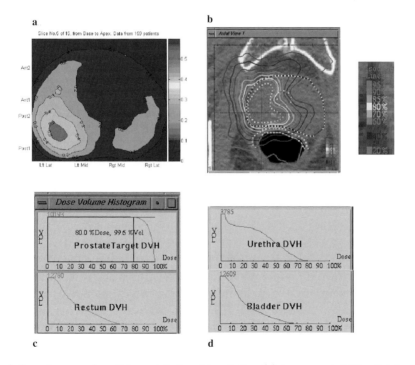

Fig. 4. Sample case of partial prostate boost irradiation (**a**) tumor probability distribution; (**b**) equivalent CT slice with isodose distribution lines from the treatment plan, and (**c** and **d**) dose-volume histograms (DVHs) for prostate, urethra, rectum, and bladder, respectively.

that achieved by conventional IMRT by focusing radiation delivery to partial prostate volumes. In Fig. 4, a tumor probability density map was developed (panel a) based on surgical findings in 159 patients treated with radical prostatectomy *(20)*. The resultant isodose plan for Cyberknife "boost" radiation is shown in panel b, and the respective dose volume histograms are presented in panels c and d, confirming dose escalation to a partial prostate volume.

6. USE OF ANDROGEN DEPRIVATION WITH RADIATION THERAPY OF PROSTATE CANCER

Androgen deprivation therapy has been appreciated as a therapeutic modality for the treatment of advanced, metastatic prostate cancer. However, the use of androgen deprivation therapy also offers advantages for selected patients receiving external radiation therapy or brachytherapy. Several studies report improved local control and disease-free survival in patients undergoing androgen deprivation before, during, and after radiation therapy *(12,13)*. Furthermore, prostate downsizing as a result of androgen deprivation limits obstructive urinary symptoms and permits brachytherapy in patients otherwise not suitable for technical reasons. The side effects of androgen deprivation include hot flashes, fatigue, impotence, and feminization. Current studies focus on determining the optimal duration for androgen deprivation as an integral part of treatment of unfavorable risk-localized prostate cancer.

7. BRACHYTHERAPY OF PROSTATE CANCER

Brachytherapy offers a technique for the delivery of radiation therapy by placing radioactive sources near or within the tumor volume. Prostate brachytherapy is achieved by interstitial placement of permanent (Pd-103 or I-125) or temporary (Ir-192) sources into the prostate through thin, hollow surgical needles. The technique is performed under trans-rectal ultra-sound (TRUS) guidance for accurate placement of the needle applicators although CT or MRI guidance may also be used *(21)*. If permanent radioactive sources are placed into the prostate, they are distributed to achieve a pre-planned dose distribution, and the sources are left in place. Pd-103 has a half-life of 17 days (60 days for I-125); therefore, the sources decay to background in a matter of months and remain in the patient as tiny inert metal cylinders. This treatment is referred to as low-dose rate (LDR) brachytherapy. Permanent TRUS-guided prostate brachytherapy is performed as a same day surgery procedure, with a short recovery time, leading to minimal disruption of the patient's schedule.

Temporary brachytherapy is generally delivered using high-dose rate (HDR) remote afterloading technology *(22)*. These procedures generally require patient hospitalization, with the patient confined to bed, remaining supine for 48 h to accommodate applicator positioning. The intensity of the radioactive Ir-192 source is quite high (~10 Ci) to allow for sufficient radiation exposure in a short time at pre-programmed positions in the interstitial applicators. The total radiation dose achieved within the prostate is the summation of the individual doses. The principal benefits of HDR brachytherapy lie in the greater flexibility of shaping the dose distributions within the prostate and that the patient does not retain radioactive sources after the procedure. This procedure is generally performed as a boost dose, in combination with 5–6 weeks of external radiation therapy.

Patients presenting with PSA ≤ 10ng/dl, Gleason's score < 7, and clinical stage < T_{2C} are considered favorable risk and are suitable candidates for LDR brachytherapy alone. Those with one adverse prognostic feature are considered intermediate risk and with two factors, high risk *(23,24)*. In general, patients falling into the favorable risk category can be effectively treated with LDR

Table 2
Low-Dose Rate Brachytherapy for Patients with Organ-Confined Prostate Cancer

Study	No. Pts.	Isotope and dose (Gy)	Risk stratification	PSA relapse-free survival	
				5 Years (%)	12 Years (%)
Beyer *(33)*	695	I-125 (160 Gy)	Favorable	88	
			Intermediate	79	
			Unfavorable	65	
Blasko *(35)*	230	Pd-103 (115 Gy)	Favorable	94	
			Intermediate	82	
			Unfavorable	65	
Zelefsky *(36)*	248	I-125 (160 Gy)	Favorable	88	
			Intermediate	77	
			Unfavorable	38	
Stock *(37)*	258	I-125/ Pd-103	Favorable	88	
			Unfavorable	60	
Potters *(38)*	1449	I-125/ Pd-103	Favorable		89
			Intermediate		78
			Unfavorable		63

Table 3
Low-Dose Rate Brachytherapy Plus EBRT for Patients with Organ-Confined Prostate Cancer

Study	No. Pts.	Isotope and dose (Gy)	EBRT (Gy)	Risk stratification	PSA relapse free survival (years) (%)
Ragde (39)	147	I-125 (160)		Favorable	66 (12)
	82	I-125 (120)	45	Unfavorable	79 (12)
Critz (40)	1,469	I-125 (115)	45	Favorable	93 (10)
				Intermediate	80 (10)
				Unfavorable	61 (10)
Dattoli (41)	258	Pd-103 (80)	41	Unfavorable	79 (10)
Sylvester (42)	232	I-125/Pd-103	45	Favorable	85 (10)
				Intermediate	77 (10)
				Unfavorable	45 (10)

Table 4
High-Dose Rate Brachytherapy in Patients with Organ-Confined Prostate Cancer

Study	No. Pts.	EBRT dose (Gy)	Fraction size (Gy)	Fraction no.	PSA relapse-free survival (years) (%)	Complications GU/GI (%)
Mate (22)	104	50.4	3–4	4	84 (5)	6.7/0
Martinez (43)	207	46	5.5–11.5	2–3	74 (5)	8/1
Galalae (44)	144	40–50	9.0	2	68–77 (5)	2.3/4.1
Demanes (45)	110	36	6.0	4	85 (3)	4/1
Syed (46)	200	45	5.5–6.5	4	89 (2.5)	2/1.5
Deger (47)	442	40–50.4	9–10	2	60 (5)	11/1

brachytherapy with the expectation of PSA relapse-free survival rates in excess of 80% at 5 years (Table 2).

Patients presenting with high-risk features are more appropriately treated with external radiation techniques or combinations of external radiation with brachytherapy boost (Tables 3 and 4). Factors underlying treatment failure following brachytherapy in high-risk patients include insufficient radiation dose to larger volumes of cancer and inadequate coverage of periprostatic tissues in patients at increased risk for extra prostatic disease. The combination of external radiation to a dose sufficient for treatment of microscopic disease followed by brachytherapy boost to areas of bulk disease offers the opportunity to dose escalate within the prostate and to limit the dose to the rectal wall. Both LDR and HDR brachytherapy may be employed for such treatments.

8. ADJUVANT RADIATION THERAPY AFTER RADICAL PROSTATECTOMY

Although radical prostatectomy provides excellent local control for patients whose cancers are confined to the prostate, the risk of local recurrence is increased if pathologic analysis of the prostatectomy specimen reveals that the tumor extends beyond the prostate, involves a surgical margin, or invades a seminal vesicle. Single institutional retrospective series have suggested that moderate doses (60–68 Gy) of external beam radiation therapy to the prostatic fossa effectively eradicate residual microscopic disease and reduce the risk of local failure *(25)*. Until recently, the timing of post-operative treatment was not well defined; radiation therapy could be offered immediately post-operatively or deferred until PSA failure was documented *(26,27)*. Salvage radiation therapy at the time of PSA progression offers potentially curative treatment and improves progression-free survival while avoiding radiation risks in some patients. However, supported by a large randomized trial, immediate post-operative external beam radiation therapy has been shown to provide superior biochemical progression-free survival and local control as compared to observation and deferred treatment in patients with high-risk features *(28)*. Immediate post-operative radiation therapy is currently advanced as the preferred treatment approach, generally in conjunction with androgen deprivation therapy.

9. FUTURE DIRECTIONS

The remaining challenges in advancing prostate cancer therapy center on optimization of techniques and patient selection for local treatment of disease, further development of salvage therapy for patients with recurrent disease, and the integration of cytotoxic and biologic therapy for systemic disease. In this review, we have provided an institutional point of view, based on current, albeit admittedly incomplete data.

We recognize that local control of prostate cancer by radiation therapy is dependent on diagnostic accuracy for definition of tumor extent and precision of radiation delivery. Clinical trials have established a dose–response relationship that falls within achievable therapeutic parameters. Furthermore, currently available technologies offer the capability to effectively treat localized prostate cancers.

We also recognize that biologic heterogeneity of prostate cancers offers further challenges to improvements in therapy. The goal of integrating knowledge of the molecular biology underlying prostate cancer with local and systemic treatment strategies is on the horizon, but beyond the scope of this review. However, current research efforts include the incorporation of biology with technology for prostate cancer treatment. Many of these may use currently available biologic assays or those on the horizon, such as gene expression or proteomic analyses.

We complete this review with several questions relevant to future improvements in prostate cancer therapy. Can prognostic information be determined from gene expression patterns to differentiate aggressive cancers with metastatic potential from those that are slowly growing and remain localized (48,49)? Can response to treatment and associated risks be predicted by polymorphisms in key candidate genes encoding DNA repair-related proteins (50)? Are there effective systemic agents suitable for use in patients with aggressive cancer (51)? Will there be an effective gene therapy approach to treat prostate cancer in conjunction with radiation therapy or as an independent approach (52,53)? Such research areas are currently under investigation using tools that can be readily integrated into clinical trials.

ACKNOWLEDGMENTS

This work was supported by PO1 CA74175.

REFERENCES

1. Papsidero, L.D., Wang, M.C., Valenzuela, L.A., Murphy, G.P., Chu, T.M. (1980) A prostate antigen in sera of prostatic cancer patients. *Cancer Res.* 40, 2428–32.
2. Holmberg, L., Bill-Axelson, A., Helgesen, F., Salo, J.O., Folmerz, P., Haggman, M., Andersson, S.O., Spangberg, A., Busch, C., Nordling, S., Palmgren, J., Adami, H.O., Johansson, J.E., Norlen, B.J. (2002) Scandinavian Prostatic Cancer Group Study Number 4. A randomized trial comparing radical prostatectomy with watchful waiting in early prostate cancer. *N Engl J Med.* 347, 781–9.

3. Zelefsky, M.J., Leibel, S.A., Gaudin, P.B., Kutcher, G.J., Fleshner, N.E., Venkatramen, E.S., Reuter, V.E., Fair, W.R., Ling, C.C., Fuks, Z. (1998) Dose escalation with three-dimensional conformal radiation therapy affects the outcome in prostate cancer. *Int J Radiat Oncol Biol Phys.* 41, 491–500.

4. Ragde, H., Elgamal, A.A., Snow, P.B., Brandt, J., Bartolucci, A.A., Nadir, B.S., Korb, L.J. (1998) Ten-year disease free survival after transperineal sonography-guided iodine-125 brachytherapy with or without 45-gray external beam irradiation in the treatment of patients with clinically localized, low to high Gleason grade prostate carcinoma. *Cancer.* 83, 989–1001. Review.

5. Walsh, P.C., Lepor, H. (1987) The role of radical prostatectomy in the management of prostatic cancer. *Cancer.* 60, 526–37.

6. Scardino, P.T., Hanks, G.E. (1997) A comparison of prostate cancer treatments: are therapeutic implications justified? *Cancer J Sci Am.* 3, 70–2.

7. Walsh, P.C. (1996) Re: potency-sparing radical retropubic prostatectomy: a simplified anatomical approach. *J Urol.* 155, 294.

8. Hanks, G.E., Hanlon, A.L., Pinover, W.H., Horwitz, E.M., Schultheiss, T.E. (1999) Survival advantage for prostate cancer patients treated with high-dose three-dimensional conformal radiotherapy. *Cancer J Sci Am.* 5, 152–8.

9. Kuban, D.A., el-Mahdi, A.M., Schellhammer, P.F. (1987) Effect of local tumor control on distant metastasis and survival in prostatic adenocarcinoma. *Urology.* 30, 420–6.

10. Fuks, Z., Leibel, S.A., Wallner, K.E., Begg, C.B., Fair, W.R., Anderson, L.L., Hilaris BS, Whitmore, W.F. (1991) The effect of local control on metastatic dissemination in carcinoma of the prostate: long-term results in patients treated with 125-I implantation. *Int J Radiat Oncol Biol Phys.* 21, 537–47.

11. Bolla, M., van Poppel, H., Collette, L., van Cangh, P., Vekemans, K., Da Pozzo, L., de Reijke, T.M., Verbaeys, A., Bosset, J.F., van Velthoven, R., Marechal, J.M., Scalliet, P., Haustermans, K., Pierart, M. (2005) European Organization for Research and Treatment of Cancer. Postoperative radiotherapy after radical prostatectomy: a randomised controlled trial (EORTC trial 22911). *Lancet.* 366, 572–8.

12. Bolla, M., Gonzalez, D., Warde, P., Dubois, J.B., Mirimanoff, R.O., Storme, G., Bernier, J., Kuten, A., Sternberg, C., Gil, T., Collette, L., Pierart, M. (1997) Improved survival in patients with locally advanced prostate cancer treated with radiotherapy and goserelin. *N Engl J Med.* 337, 295–300.

13. Pilepich, M.V., Winter, K., John, M.J., Mesic, J.B., Sause, W., Rubin, P., Lawton, C., Machtay, M., Grignon, D. (2001) Phase III radiation therapy oncology group (RTOG) trial 86–10 of androgen deprivation adjuvant to definitive radiotherapy in locally advanced carcinoma of the prostate. *Int J Radiat Oncol Biol Phys.* 50, 1243–52.

14. Zietman, A.L., DeSilvio, M.L., Slater, J.D., Rossi, C.J., Jr., Miller, D.W., Adams, J.A., Shipley, W.U. (2005) Comparison of conventional-dose vs high-dose conformal radiation therapy in clinically localized adenocarcinoma of the prostate: a randomized controlled trial. *JAMA.* 294, 1274–6.

15. Ray, G.R., Cassady, J.R., Bagshaw, M.A. (1973) Definitive radiation therapy of carcinoma of the prostate. A report on 15 years of experience. *Radiology.* 106, 407–18.

16. Smit, W.G., Helle, P.A., van Putten, W.L., Wijnmaalen, A.J., Seldenrath, J.J., van der Werf-Messing, B.H. (1990) Late radiation damage in prostate cancer patients treated by high dose external radiotherapy in relation to rectal dose. *Int J Radiat Oncol Biol Phys.* 18, 23–9.

17. Zietman, A.L., Chung, C.S., Coen, J.J., Shipley, W.U. (2004) 10-year outcome for men with localized prostate cancer treated with external radiation therapy: results of a cohort study. *J Urol.* 171, 210–4.

18. Zagars, G.K., Pollack, A., Smith, L.G. (1999) Conventional external-beam radiation therapy alone or with androgen ablation for clinical stage III (T3, NX/N0, M0) adenocarcinoma of the prostate. *Int J Radiat Oncol Biol Phys.* 44, 809–19.

19. Zelefsky, M.J., Fuks, Z., Happersett, L., Lee, H.J., Ling, C.C., Burman, C.M., Hunt, M., Wolfe, T., Venkatraman, E.S., Jackson, A., Skwarchuk, M., Leibel, S.A. (2000) Clinical experience with intensity modulated radiation therapy (IMRT) in prostate cancer. *Radiother Oncol.* 55, 241–9.

20. Opell, M.B., Zeng, J., Bauer, J.J., Connelly, R.R., Zhang, W., Sesterhenn, I.A., Mun, S.K., Moul, J.W., Lynch, J.H. (2002) Investigating the distribution of prostate cancer using three-dimensional computer simulation. *Prostate Cancer Prostatic Dis.* 5, 204–8.

21. Holm, H.H., Juul, N., Pedersen, J.F., Hansen, H., Stroyer, I. (1983) Transperineal 125-iodine seed implantation in prostatic cancer guided by transrectal ultrasonography. *J Urol.* 130, 283–6.

22. Mate, T.P., Gottesman, J.E., Hatton, J., Gribble, M., Van Hollebeke, L. (1998) High dose-rate afterloading 192-Iridium prostate brachytherapy: feasibility report. *Int J Radiat Oncol Biol Phys.* 41, 525–33.

23. Sylvester, J.E., Blasko, J.C., Grimm, P.D., Meier, R., Malmgren, J.A. (2003) Ten-year biochemical relapse-free survival after external beam radiation and brachytherapy for localized prostate cancer: the Seattle experience. *Int J Radiat Oncol Biol Phys.* 57, 944–52.

24. D'Amico, A.V., Whittington, R., Malkowicz, S.B., Schultz, D., Blank, K., Broderick G.A., Tomaszewski, J.E., Renshaw, A.A., Kaplan, I., Beard, C.J., Wein, A. (1998) Biochemical outcome after radical prostatectomy, external beam radiation therapy, or interstitial radiation therapy for clinically localized prostate cancer. *JAMA.* 280, 969–74.

25. Valicenti, R.K., Gomella, L.G., Perez, C.A. (2003) Radiation therapy after radical prostatectomy: a review of the issues and options. *Semin Radiat Oncol.* 13, 130–40. Review.

26. Stephenson, A.J., Shariat, S.F., Zelefsky, M.J., Kattan, M.W., Butler, E.B., Teh, B.S., Klein, E.A., Kupelian, P.A., Roehrborn, C.G., Pistenmaa, D.A., Pacholke, H.D., Liauw, S.L., Katz, M.S., Leibel, S.A., Scardino, P.T., Slawin, K.M. (2004) Salvage radiotherapy for recurrent prostate cancer after radical prostatectomy. *JAMA.* 291, 1325–32.

27. Schild, S.E., Buskirk, S.J., Wong, W.W., Halyard, M.Y., Swanson, S.K., Novicki, D.E., Ferrigni, R.G. (1996) The use of radiotherapy for patients with isolated elevation of serum prostate specific antigen following radical prostatectomy. *J Urol.* 156, 1725–9.

28. Bolla, M., van Poppel, H., Collette, L., van Cangh, P., Vekemans, K., Da Pozzo, L., de Reijke, T.M., Verbaeys, A., Bosset, J.F., van Velthoven, R., Marechal, J.M., Scalliet, P., Haustermans, K., Pierart, M. (2005) European Organization for Research and Treatment of Cancer. Postoperative radiotherapy after radical prostatectomy: a randomised controlled trial (EORTC trial 22911). *Lancet.* 366, 572–8.

29. Bagshaw, M.A., Cox, R.S., Ramback, J.E. (1990) Radiation therapy for localized prostate cancer. Justification by long-term follow-up. *Urol Clin North Am.* 17, 787–802.

30. Perez, C.A., Lee, H.K., Georgiou, A., Logsdon, M.D., Lai, P.P., Lockett, M.A. (1993) Technical and tumor-related factors affecting outcome of definitive irradiation for localized carcinoma of the prostate. *Int J Radiat Oncol Biol Phys.* 26, 581–91.

31. Shipley, W.U., Thames, H.D., Sandler, H.M., Hanks, G.E., Zietman, A.L., Perez, C.A., Kuban, D.A., Hancock, S.L., Smith, C.D. (1999) Radiation therapy for clinically localized prostate cancer: a multi-institutional pooled analysis. *JAMA.* 281, 1598–604.

32. del Regato, J.A., Trailins, A.H., Pittman, D.D. (1993) Twenty years follow-up of patients with inoperable cancer of the prostate (stage C) treated by radiotherapy: report of a national cooperative study. *Int J Radiat Oncol Biol Phys.* 26, 197–201.

33. Beyer, D.C. (2001) The evolving role of prostate brachytherapy. *Cancer Control.* 8, 163–70.

34. Zagars, G.K., von Eschenbach, A.C., Johnson, D.E., Oswald, M.J. (1987) Stage C adeno-carcinoma of the prostate. An analysis of 551 patients treated with external beam radiation. *Cancer.* 60, 489–99.

35. Blasko, J.C., Grimm, P.D., Sylvester, J.E., Badiozamani, K.R., Hoak, D., Cavanagh, W. (2000) Palladium-103 brachytherapy for prostate carcinoma. *Int J Radiat Oncol Biol Phys.* 46, 839–50.

36. Zelefsky, M.J., Hollister, T., Raben, A., Matthews, S., Wallner, K.E. (2000) Five-year biochemical outcome and toxicity with transperineal CT-planned permanent I-125 prostate implantation for patients with localized prostate cancer. *Int J Radiat Oncol Biol Phys.* 47, 1261–6.

37. Stock, R.G., Stone, N.N. (1997) The effect of prognostic factors on therapeutic outcome following transperineal prostate brachytherapy. *Semin Surg Oncol.* 3, 454–60.

38. Potters, L., Morgenstern, C., Calugaru, E., Fearn, P., Jassal, A., Presser, J., Mullen, E. (2005) 12-year outcomes following permanent prostate brachytherapy in patients with clinically localized prostate cancer. *J Urol.* 173, 1562–6.

39. Ragde, H., Korb, L.J., Elgamal, A.A., Grado, G.L., Nadir, B.S. (1997) Modern prostate brachytherapy. Prostate specific antigen results in 219 patients with up to 12 years of observed follow-up. *Semin Surg Oncol.* 13, 454–60.

40. Critz, F.A., Levinson, K. (2004) 10-year disease-free survival rates after simultaneous irradiation for prostate cancer with a focus on calculation methodology. *J Urol.* 172, 2232–8.

41. Dattoli, M., Wallner, K., True, L., Cash, J., Sorace, R. (2003) Long-term outcomes after treatment with external beam radiation therapy and Palladium 103 for patients with higher risk prostate carcinoma: influence of prostatic acid phosphatase. *Cancer.* 97, 979–83.

42. Sylvester, J.E., Blasko, J.C., Grimm, P.D., Meier, R., Malmgren, J.A. (2003) Ten-year biochemical relapse-free survival after external beam radiation and brachytherapy for localized prostate cancer: the Seattle experience. *Int J Radiat Oncol Biol Phys.* 57, 944–52.

43. Martinez, A.A., Gustafson, G., Gonzalez, J., Armour, E., Mitchell, C., Edmundson, G., Spencer, W., Stromberg, J., Huang, R., Vicini, F. (2002) Dose escalation using conformal high-dose-rate brachytherapy improves outcome in unfavorable prostate cancer. *Int J Radiat Oncol Biol Phys.* 53, 316–27.

44. Galalae, R.M., Kovacs, G., Schultze, J., Loch, T., Rzehak, P., Wilhelm, R., Bertermann, H., Buschbeck, B., Kohr, P., Kimmig, B. (2002) Long-term outcome after elective irradiation of the pelvic lymphatics and local dose escalation using high-dose-rate brachytherapy for locally advanced prostate cancer. *Int J Radiat Oncol Biol Phys.* 52, 81–90.

45. Demanes, D.J., Rodriguez, R.R., Altieri, G.A. (2000) High dose rate prostate brachytherapy: the California Endocurietherapy (CET) method. *Radiother Oncol.* 57, 289–96.

46. Syed, A.M., Puthawala, A., Sharma, A., Gamie, S., Londrc, A., Cherlow, J.M., Damore, S.J., Nazmy, N., Sheikh, K.M., Ko, S.J. (2001) High-dose-rate brachytherapy in the treatment of carcinoma of the prostate. *Cancer Control.* 8, 511–21.

47. Deger, S., Boehmer, D., Roigas, J., Schink, T., Wernecke, K.D., Wiegel, T., Hinkelbein, W., Budach, V., Loening, S.A.. (2005) High dose rate (HDR) brachytherapy with conformal radiation therapy for localized prostate cancer. *Eur Urol.* 47, 441–8.

48. Yu, Y.P., Landsittel, D., Jing, L., Nelson, J., Ren, B., Liu, L., McDonald, C., Thomas, R., Dhir, R., Finkelstein, S., Michalopoulos, G., Becich, M., Luo, J.H. (2004) Gene expression alterations in prostate cancer predicting tumor aggression and preceding development of malignancy. *J Clin Oncol.* 22, 2790–9.

49. Dhanasekaran, S.M., Barrette, T.R., Ghosh, D., Shah, R., Varambally, S., Kurachi, K., Pienta, K.J., Rubin, M.A., Chinnaiyan, A.M. (2001) Delineation of prognostic biomarkers in prostate cancer. *Nature.* 412, 822–6.

50. Nam, R.K., Zhang, W.W., Jewett, M.A., Trachtenberg, J., Klotz, L.H., Emami, M., Sugar, L., Sweet, J., Toi, A., Narod, S.A. (2005) The use of genetic markers to determine risk for prostate cancer at prostate biopsy. *Clin Cancer Res.* 11, 8391–7.

51. Armstrong, A.J., Carducci, M.A. (2006) New drugs in prostate cancer. *Curr Opin Urol.* 16, 138–45.

52. The, B.S., Ayala, G., Aguilar, L., Mai, W.Y., Timme, T.L., Vlachaki, M.T., Miles, B., Kadmon, D., Wheeler, T., Caillouet, J., Davis, M., Carpenter, L.S., Lu, H.H., Chiu, J.K., Woo, S.Y., Thompson, T., Aguilar-Cordova, E., Butler, E.B. (2004) Phase I-II trial evaluating combined intensity-modulated radiotherapy and in situ gene therapy with or without hormonal therapy in treatment of prostate cancer-interim report on PSA response and biopsy data. *Int J Radiat Oncol Biol Phys.* 58, 1520–9.

53. MacRae, E.J., Giannoudis, A., Ryan, R., Brown, N.J., Hamdy, F.C., Maitland, N., Lewis, C.E. (2006) Gene therapy for prostate cancer: current strategies and new cell-based approaches. *Prostate.* 66, 470–94.

54. Zelefsky, M.J., Leibel, S.A., Gaudin, P.B., Kutcher, G.J., Fleshner, N.E., Venkatramen, E.S., Reuter, V.E., Fair, W.R., Ling, C.C., Fuks, Z. (1998) Dose escalation with three-dimensional conformal radiation therapy affects the outcome in prostate cancer. *Int J Radiat Oncol Biol Phys.* 41, 489–90.

55. Zelefsky, M.J., Fuks, Z., Hunt, M., Yamada, Y., Marion, C., Ling, C.C., Amols, H., Venkatraman, E.S., Leibel, S.A. (2002) High-dose intensity modulated radiation therapy for prostate cancer: early toxicity and biochemical outcome in 772 patients. *Int J Radiat Oncol Biol Phys.* 53, 1111–6.

16 The Current Knowledge of Hormonal Therapy in the Treatment of Prostate Cancer

Mario A. Eisenberger, MD

CONTENTS

INTRODUCTION
THE BIOLOGY OF ANDROGEN DEPRIVATION
TREATMENT APPROACHES
CONTROVERSIES
REFERENCES

1. INTRODUCTION

The endocrine control of the growth and differentiation of prostate cancer has been major focus of study for more than half a century. The initial demonstration by Huggins and Hodges *(1)* that surgical castration and the administration of pharmacological doses of estrogens resulted in major objectives and subjective benefits to patients with advanced disease provides one of the best examples of the hormonal control of solid tumors in man. The biological events associated with androgen deprivation in this disease is an area of extensive study, and the clinical application of therapeutic modalities targeting androgen signaling pathways extend to virtually all stages of the disease including prevention approaches. Despite the uncontested efficacy of androgen deprivation as a treatment modality for prostate cancer, a number of critical questions regarding the best application of the various approaches continue to be the focus of major debates and controversies. Elimination of androgen production in the gonads or interference of its signaling steps

From: *Current Clinical Oncology: Prostate Cancer:*
Signaling Networks, Genetics, and New Treatment Strategies
Edited by: R. G. Pestell and M. T. Nevalainen © Humana Press, Totowa, NJ

induce a cascade of events that clinically reflects one of the most effective systemic palliative treatments known in solid tumors. In this chapter, we will review the treatment modalities currently available with this approach and outline some of the unresolved controversies. Specific issues related to the appropriate clinical applications of androgen deprivation treatment will be placed in what we consider is a proper perspective.

2. THE BIOLOGY OF ANDROGEN DEPRIVATION

The endocrine control of prostate growth is determined by the hypothalamic, pituitary, and gonadal axis. The synthesis and release of luteinizing hormone (LH) by the anterior pituitary gland is controlled by the hypothalamic luteinizing hormone-releasing hormone (LHRH) that is secreted in a pulsatile fashion. Pituitary release of LH induces the release of testosterone (T) by the Leydig cells in the testes *(2)*. Testicular testosterone represents more than 90% of the total pool of circulating testosterone. In the prostate, T is converted to dihydrotestosterone (DHT) by the enzyme 5-alpha reductase. DHT, the most potent androgen, binds to the intracellular androgen receptor, induces the nuclear activation of various target genes including the PSA gene, and induces the growth, differentiation, and proliferation of epithelial cells into a secretory state. The pituitary gland also secretes adrenocorticotropic hormone (ACTH) that in turn induces the synthesis and release of adrenal androgens that comprise about 10% of circulating androgens.

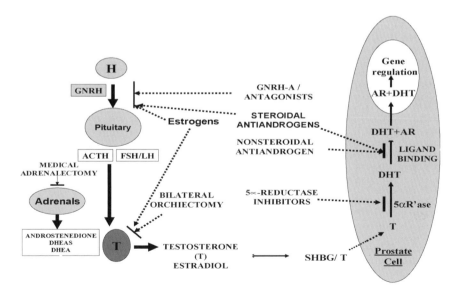

Fig. 1. Hypothalamic–pituitary–gonadal–adrenal axis: endocrine manipulations.

Over the past several years, attention has been devoted to the function of the androgen receptor in the control of prostate cancer growth. Studies in human prostate cancer xenograft models demonstrate that progression to androgen independence appears to be associated with increased expression of the androgen receptor and increased sensitivity to low levels of androgens *(2)*. The androgen receptor gene *(AR)* is the target of somatic genome alterations during progression to androgen independence *(2–19)*. Furthermore, *AR* mutations may result in altered ligand specificity, which appears to be the most plausible explanation for aberrant responses to various hormonal manipulations *(20)*. Similarly, pre-clinical studies have shown that prostate cancer cells of hormone-independent models containing the wild-type androgen receptor remain capable of androgen receptor signaling in response to different growth factor signaling pathways *(2)*. A diagrammatic representation of the events summarized above and the specific sites targeted by the various therapeutic maneuvers available today is shown in Fig. 1.

3. TREATMENT APPROACHES

Surgical removal of the testis has been considered for many years the gold standard for patients with metastatic disease. This relatively simple procedure, performed under local anesthesia, results in a rapid decrease of serum T and consequently early therapeutic benefits. Bilateral orchiectomy has been shown to be associated with dramatic symptomatic benefits in symptomatic patients in addition to frequent objective improvements in metastatic sites. It is still considered by many as the best approach for severely ill and symptomatic patients because of the rapidity of its therapeutic effects. The therapeutic effects of bilateral orchiectomy are directly related to the swift decline of T levels to the castrate range accomplished with the procedure. Serious local complications of bilateral orchiectomy are infrequent, and morbidity is primarily associated with the loss of T function, namely loss of libido and sexual impotence, fatigue, osteopenia and osteoporosis, anemia, decreased muscle mass, metabolic abnormalities, and weight gain. In addition, long-term effects of androgen deprivation include cognitive changes and significant psychological abnormalities.

The identification and isolation of the natural occurring hypothalamic gonadotropin-releasing hormone (GnRH) and definition of its physiology by Schally et al. *(21)* resulted in the synthesis of potent GnRH analogs (also known as LHRH analogs). GnRH analogs are able to down-regulate pituitary receptors and consequently, the synthesis and release of LH by the anterior pituitary gland and gonadal testosterone by the Leydig cells in the testes *(21)*. GnRH analogs are synthetic agonist compounds several times more potent than the naturally occurring hormone, which is accomplished by amino acid substitutions in the natural decapeptide molecule. These superagonist analogs initially induce a short rise in serum T (peak at about 48 h), which is known as an endocrine flare and can be associated with a clinical flare of the disease in about 10% of patients. This clinical flare of the disease is primarily characterized by the development or worsening of pain and possibly worsening of urinary obstructive symptomology and less frequently, neurological complications associated with spinal cord and nerve root compression. This initial rise in serum T is subsequently followed by a progressive decrease to castrate levels by the fourth week of treatment. Synthetic GnRH antagonists are still in development. The principal potential advantages of GnRH antagonist compounds relate to a more rapid decline in serum testosterone compared to agonists and the lack of a clinical flare of the disease associated with the initial rise in serum T. Abarelix, which is the only antagonist approved by the FDA, has shown dose-dependent hypersensitivity reactions, and its approval has been limited to very restricted situations.

Antiandrogens (AA) are compounds that competitively bind to the androgen receptor and hence prevent its activation by DHT and T. AAs are generally grouped in steroidal and non-steroidal compounds. Among steroidal AAs are megestrol acetate and cyproterone acetate *(21,22)*. These compounds are both progestational agents and thus also exert some of their effects on other sites of the pituitary–hypothalamic–gonadal axis, including inhibition of pituitary gonadotropins and gonadal androgen production. The AR/non-steroidal AA binding has been considered agonistic and, therefore, not extensively used in the USA in the initial treatment of prostate cancer. Cyproterone acetate has been shown in the laboratory and clinically to neutralize some of the effects of the initial stimulatory phase of GnRH analogs. However, it is not approved in the USA for clinical use.

Non-steroidal AAs are compounds that compete with DHT and T for AR binding. Current steroidal AAs approved in the USA are flutamide, bicalutamide, and nilutamide. AAs employed as single agents have been shown to be inferior to conventional forms of gonadal suppression in prospective randomized studies in patients with metastatic disease *(23,24)*, whereas the data in patients with M0 disease (high-risk adjuvant, biochemical relapses, and local-regional disease) at this time remains inconclusive. AAs are approved for use combined with GnRH analogs (see Subheading 4.2.). Bicalutamide is the most commonly used AA, and again, it is indicated for use in combination with GnRH analogs in patients with metastatic disease. Bicalutamide is metabolized in its S and R isomers, and most of its activity is due to the R-enantiomer. Bicalutamide displays prolonged absorption following a single dose of 50 mg with peak plasma concentrations observed at 16.3 ± 5.8 h. Bicalutamide is eliminated slowly from plasma with a T-half of 6.28 ± 0.50 days. With daily administration for 85 days, the accumulation ratio was 11.1 ± 0.7 after 50 mg ($n = 62$). Mean half-life was 7.6 ± 0.3 days. The daily recommended dose is 50 mg orally. Preclinical studies suggested a dose–response relationship in hormone-dependent prostate cancer models. However, this was not confirmed in subsequent clinical trials.

Two single-agent studies in patients with metastatic disease compared bicalutamide 50 mg and 150 mg/day to gonadal ablation alone *(23–28)*. A significant survival advantage for patients treated with gonadal ablation was observed in both studies. Therefore, it is generally agreed that AAs as single agents are not as effective as approaches aiming at a suppression of gonadal testosterone. Toxicity of bicalutamide is generally mild when used in combination with GnRH analogs. However, as a single agent, bicalutamide causes gynecomastia (frequently painful), usually modest gastrointestinal side effects (including nausea, vomiting, and diarrhea), hot flashes, and changes in libido (although not as common as with gonadal suppressive compounds). The two other AAs available in the USA are flutamide and nilutamide.

As mentioned above, T, the predominant circulating androgen in man, is reduced to DHT by the enzyme 5-α-reductase *(2)*, which induces responses in various androgen-sensitive target tissues. DHT is a potential etiological factor in benign prostate hypertrophy (BPH) and a potent stimulator of prostate cancer growth. Inhibition of DHT formation without affecting T levels may, theoretically, avoid some of the immediate and chronic side effects of conventional androgen deprivation therapy (ADT). However, unlike T suppression, the therapeutic role of selective inhibition of DHT in prostate cancer has not been clinically established. Inhibitors of the enzyme 5-α-reductase have not shown to have significant activity against prostate cancer *(29–33)*. However, their use in combination with non-stroidal AAs has been evaluated in clinical trials. The results of uncontrolled studies with the combination flutamide and finasteride (a type II 5-α-reductase inhibitor commercially available) do not suggest an additive effect with this combination. Dutasteride, an inhibitor of type I and II 5-α-reductase activity, has not been tested in prostate cancer at this time.

A broader appreciation of the biology of prostate cancer progression to androgen independence has renewed interest in various endocrine manipulations as a "second-line treatment." Second-line hormonal manipulations are employed in patients who demonstrate evidence of disease progression despite castrate levels of serum testosterone (<50 ng/dl). Approximately, 15–20% of men treated with AAs in addition to gonadal androgen suppression exhibit an "antiandrogen-withdrawal" syndrome, characterized by an improvement after stopping the AA but maintaining androgen deprivation *(34–37)*. Although not completely elucidated, it is suggested that *AR* mutations, encoding androgen receptors with altered ligand binding properties, may result in AAs to function as receptor agonists *(34–37)*. Several agents have been reported to provide beneficial responses, such as a drop in the serum PSA or an improvement in cancer symptoms, in patients progressing despite ADT. These agents include bicalutamide (20–24%), megestrol acetate (8–13%), DES (26–66%), ketoconazole with hydrocortisone (27–63%), glucocorticoids alone (18–22%) *(38,39)*, and recently more selective inhibitors of adrenal steroidogenesis (abiraterone). In general, responses to second-line endocrine therapies are brief, with median durations ranging between 3 and 4 months. Second-line hormonal therapies are frequently employed in patients with slowly progressing and asymptomatic metastatic disease before chemotherapy is offered. The long-term benefits on survival and quality of life with sequential endocrine approaches remain unestablished at this point. This is because there are no prospectively randomized comparisons between a sequential endocrine approach versus immediate docetaxel-based chemotherapy in patients with progressing castrate androgen-independent disease *(40)*.

4. CONTROVERSIES

4.1. Optimal Timing

The optimal time for initiation of hormonal therapy represents one of the most important unresolved issues regarding this treatment modality. Much of the controversy was triggered from the observations derived from the series of prospective randomized trials conducted by the Veterans Administration Cooperative Urological Research Group (VACURG) conducted from 1960 to 1975 *(41–44)*. On the VACURG study 1, patients with stages III and IV (C and D) were randomly allocated to receive either a placebo initially, daily oral 5.0 mg of diethylstilbestrol (DES), orchiectomy plus placebo, or orchiectomy plus 5.0 mg of DES. The main objective was to determine whether combined treatment with orchiectomy plus DES was superior to either treatment alone. Patients randomized to receive placebo initially were subsequently crossed over at the time of progression to one of the other three arms, and the choice of treatment was left at the discretion of the investigators. The main endpoint for study was survival. Also as part of study 1, patients with stages I and II (A and B) were randomly allocated to prostatectomy and placebo or prostatectomy + 5 mg of DES daily.

The most important observation in both studies was that 5.0 mg of DES was associated with an increased risk of death from cardiovascular disease. Among the complications associated with DES treatment were deep vein thrombosis/thrombophlebitis, angina pectoris, acute myocardial infarction, congestive heart failure, pulmonary embolus, and cerebral-vascular accidents. The final results on both segments of study 1 indicated no survival differences in favor of any treatment, thus suggesting that (i) for patients with early disease (stages I and II), the addition of 5.0 mg of DES to local surgical treatment does not provide additional benefits and in fact may adversely affect long-term prognosis because of a high incidence of severe, life-threatening, cardiovascular complications; (ii) the addition of 5.0 mg of DES to surgical castration in patients with stages III and IV disease was not superior to either modality alone, and this could at least partly be because of an increase in potentially lethal toxicity. The impact of treatment-related morbidity and mortality was more apparent on the stage III group. (iii) Patients randomized to the placebo arm and subsequently treated on one of the other treatment arms at the time of their symptomatic or objective progression had the same survival as those receiving these treatments at the time of randomization. Thus, it was interpreted that early treatment for asymptomatic prostatic cancer patients provides no survival advantage over treatment at the symptomatic stage only. This assessment of VACURG Study 1 has been the subject of significant criticism, particularly with regard to the conclusions derived with regard to the optimal timing for initiation of endocrine treatment. Is it appropriate to assume, based on the data from this unplanned analysis, that early treatment offers no advantages over delayed initiation of treatment? It is important to recognize that VACURG Study 1 was not designed to test the concept of early versus delayed treatment, and the appropriate design of a study to test this important question requires very specific definitions of endpoints to be assessed prospectively and not retrospectively as done in VACURG Study 1. As the decision of "progression" was not uniformly defined "a priori," it remains possible, and likely, that the crossover to an alternate treatment in actuality occurred at different biological times. Similarly, the choice of crossover treatment was left at the discretion of the treating physicians, thus resulting in different therapeutic approaches following progression. This incorporates additional complexity for final analysis. Furthermore, a substantial proportion of the patients included on study 1 died of either treatment-related toxicity or other co-morbidities, thus preventing reliable assessments of survival in relation to prostatic cancer.

In VACURG Study 2, patients with stages III and IV disease were randomized to receive placebo, 0.2 mg of DES, 1.0 mg of DES, and 5.0 mg of DES. This study was stopped early, once the increased risk of cardiovascular complications (particularly in stage III patients) with 5.0 mg of DES emerged. Both 5 and 1 mg/day of DES had a lower incidence of deaths due to prostate cancer compared to a dose of 0.2mg/day of DES or placebo. Unlike VACURG study 1, several patients on the placebo arm never received any treatment at the time of progression. Thus, study 2 is not perceived as an adequate test of the immediate versus deferred treatment question.

A number of important issues should be considered regarding these studies. First, the patient population in the VACURG studies is most likely distinct from contemporary series because of the stage migration phenomenon observed over the past two decades. Furthermore, estrogens and surgical castration have been replaced by medical castration with GnRH analogs and other compounds discussed above. As indicated, pharmacological doses of estrogens were associated with significant and potentially lethal toxicity, which contributed significantly to the survival observations on all VACURG studies. Because of these factors, extrapolation of the VACURG studies observations regarding early versus deferred treatment across other stages of prostate cancer and involving a different approach of ADT should most likely be avoided.

Several years later, the Medical Research Council (MRC) evaluated the survival of patients receiving immediate ADT compared to deferred ADT in patients with (M+) and without (M0) clinically evident metastatic disease *(45, 46)*. In the 434 patients with M+ disease, there was no difference in survival. However, the incidence of pathologic fractures, epidural cord compression, and renal failure was significantly lower in the group receiving immediate ADT. The M0 patients appeared to have a better disease-specific survival initially (lower number of prostate cancer deaths). However, a follow-up analysis failed to show a statistical significance. Furthermore, a significant number of patients randomized to the deferred arm never received treatment. The same comments regarding crossover criteria made on the VACURG studies apply to the MRC studies as well.

Despite all the methodological shortcomings at this point, the standard of care for patients with newly diagnosed metastatic prostate cancer should be immediate ADT. This is because it improves quality of life and reduces the incidence of disease-related morbidity. The best approach for patients with no metastasis, such as in the adjuvant setting or biochemically relapsed disease, remains undefined at this time. An extensive overview analysis reported by Aronson et al. *(47)*, discusses these issues in the context of benefits relative to risks and costs. Although it points out the lack of a clear-cut level 1 evidence to support either approach (early versus deferred), it remains clear that differences, if detected, are likely to be small. Consequently, definitive trials would require large numbers of patients and long follow-up time to provide the definitive results.

4.2. Complete Androgen Blockade

The hypothesis focusing on a combined approach of gonadal androgen suppression and a blockade of AR was promoted by Labrie et al. *(48,49)* over 20 years ago. These authors suggested that following gonadal ablation, prostate cancer cells continued a clinically significant hormone-dependent tumor growth. This was primarily due to the effects of androgens of adrenal origin. To neutralize the effects of adrenal androgens, a combined use of surgical or medical castration with a non-steroidal AA was proposed, and this approach was promoted as complete androgen blockade (CAB) *(49)*. A total of 27 prospectively randomized clinical trials involving more than 8000 patients were conducted to compare the efficacy of surgical or medical castration alone (monotherapy) to almost every possible combination of castration and AAs. The first published large scale prospectively randomized clinical trial was the NCI-sponsored INT-0036, published by Crawford et al. in 1989 *(50)*. Six hundred seventeen patients with stage D2 disease were randomly assigned to receive daily subcutaneous injections (1 mg/day) of leuprolide acetate plus flutamide versus leuprolide and placebo. The median overall survival with CAB and monotherapy was 35 and 29 months, respectively (two-sided $p = 0.03$). One prevalent argument was that the difference in outcome could have been a result of the neutralizing effects of the AA on the flare phenomenon associated with leuprolide treatment alone *(50,51)*. Indeed, on NCI INT-0036, it was evident during the first 12 weeks of treatment that patients randomized to the CAB arm had a more favorable trend in the directions of pain control, performance status, and serum acid phosphatase. The second explanation related to possible compliance problems with the daily injections that could result in inadequate testicular suppression and consequently favor those receiving leuprolide with flutamide. In view of these two unresolved issues, a confirmatory trial employing surgical castration as the underlying method of gonadal ablation was subsequently conducted under the auspices of the National Cancer Institute (NCI INT-0105). NCI INT-0105 was a prospectively randomized, double-blinded, placebo controlled trial comparing bilateral orchiectomy with and without flutamide in 1387 patients with stage D2 prostate cancer. With a median follow-up time of approximately 50 months and with 70% deaths occurring at the time of the final analysis, INT 0105 failed to confirm the initial findings of INT-0036. The median survival of patients on the CAB arm was 33 months compared to 30 months on the orchiectomy arm, which was not statistically significant (two-sided stratified $p = 0.14$, hazard ratio = 0.91, 90% CI = 0.81–1.01) *(52)*. The Australian multi-center trial reported by Zalcberg et al. *(53)* compared bilateral orchiectomy plus flutamide versus bilateral orchiectomy and placebo. This trial accrued 222 patients and was reported with a relatively short follow-up time. Interestingly, the Kaplan–Meier estimates of median survival favored the orchiectomy arm (31 and 23 months, respectively) although the difference was not statistically significant ($p = 0.21$).

The second positive trial was reported by Dijkman et al. *(54)* on a multinational prospectively randomized placebo-controlled study comparing orchiectomy plus nilutamide to orchiectomy alone. The results of this trial demonstrated a small but significant difference in median survival (27.3 vs. 23.6 months, $p = 0.032$), observed after 8.5 years follow-up, in favor of the CAB regimen. Crawford et al. subsequently reported the results of a prospective trial comparing the combination of leuprolide acetate plus nilutamide or leuprolide alone that demonstrated no difference in survival *(55)*. The third positive trial was conducted by the European Organization for Research and Treatment of Cancer (EORTC study 30853) comparing goserelin plus flutamide to bilateral orchiectomy in 327 patients mostly with stage D2 disease (M1 disease). The final analysis of EORTC 30853 demonstrated a 7-month difference in median overall survival ($p = 0.04$) in favor of the CAB arm *(56,57)*. The Danish Prostatic Cancer Group (DAPROCA) conducted a virtually identical study, with the same treatment arms and approximately the same number of patients. This study was completed around the same time as EORTC 30853 and showed a longer overall survival in favor of the monotherapy arm although the difference was not statistically significant *(58)*. An evaluation of DAPROCA and EORTC-30853 trials indicated comparable populations and study parameters. A combined analysis of both studies performed to enhance the power of comparisons did not show a significant survival difference *(59)*.

In 1995, the Prostate Cancer Trialists' Collaborative Group (PCTCG) reported on the first meta-analysis that was conducted as a measure to increase the statistical power of the observations of individual trials. Their report included data from 22 randomized trials comparing CAB to gonadal ablation alone in 5710 patients *(60)*. To achieve an intention-to-treat analysis, complete individual data was requested from the investigators on each randomized patient. Hazard ratio was calculated separately for every trial based on the raw data and then combined to all other trials using logrank statistics. The meta-analysis showed a 2.1% difference in mortality in favor of CAB treatment (6.4% reduction in annual odd of death), which is not statistically significant. The results were not influenced by the type AAs (flutamide, nilutamide or CPA) or the method of gonadal ablation.

The Agency for Health Care Policy and Research (AHCPR) published on the web (http://www.ahcpr.gov/clinic/index.html#evidence – AHCPR report No.99-E012) the result of a comprehensive meta-analysis based on all published CAB studies. This evaluation included all published studies in CAB that theoretically minimized a selection bias factor. This meta-analysis found no difference in 2-year survival rates (hazard ratio = 0.970; 95% CI = 0.866–1.087). Only the 10 of 27 trials reported both 2- and 5-year survival figures, and on these ten trials, the preliminary results suggested a minimal 5-year survival difference in favor of CAB that was considered of questionable clinical significance (hazard ratio = 0.871; 95% CI = 0.805–0.9887). QOL was prospectively evaluated in patients undergoing CAB treatment on NCI INT-0105 in a companion study reported by Moinpour et al. *(61)*. This randomized, double blind, placebo-controlled trial employed an evaluation of the SWOG QOL questionnaire during the initial 6 months of NCI INT-0105. Improvement in QOL over the baseline parameters was seen on both arms. However, this improvement was more pronounced on the placebo group (monotherapy). Patients on the CAB arm reported of a higher frequency of diarrhea and worsening of emotional functioning *(61)*. It was concluded that the QOL benefit resulting from orchiectomy in metastatic prostate cancer patients appeared to be offset by the addition of flutamide primarily because of an increased incidence of adverse effects.

The most compelling explanation for the predominantly negative results is that the overall CAB treatment effect size is indeed small and of questionable clinical significance (62). The role of AAs or other endocrine approaches (see below in second-line hormonal treatments) as second-line treatment options or sequentially applied for patients who demonstrate progression after gonadal ablation alone deserves further evaluation. The extensive clinical and laboratory investigation evolving from the initial reports on CAB have changed our approach to basic concepts involving hormone resistance in prostate cancer. Emphasis on AR biology research has challenged the traditional thought "that progression after initial castration represents evidence of categorical hormone resistance." Focus on sequential endocrine approaches and treatments targeting the AR are important avenues of research, which hopefully will advance new concepts of more significant relevance in clinical practice.

4.3. Intermittent Androgen Deprivation Therapy

Animal model studies have suggested that intermittent ADT might offer an advantage over conventional continuous T suppression in delaying prostate cancer (63). In the animal studies, mice were treated with bilateral orchiectomy and then subjected to tumor harvest after the tumors had regressed. Upon re-growth, the tumor was transplanted into intact mice treated again with bilateral orchiectomy. This intermittent treatment cycle was compared with continuous ADT until there was evidence of androgen-independence. Androgen independence was observed 51 days after initiation of continuous ADT compared to 147 days with intermittent androgen deprivation. The mechanism to explain the superiority of intermittent androgen deprivation has not been fully elucidated, and these initial observations were not confirmed by others employing a somewhat distinct approach. Rats carrying a transplantable androgen-dependent prostate cancer were treated with immediate bilateral orchiectomy or continuous DES, or intermittent DES. The animals treated continuously survived up to 50% longer than those treated intermittently (64). Clinical intermittent ADT is most often accomplished through careful monitoring of the serum PSA. The results of large randomized clinical trials of intermittent versus continuous ADT will test whether either approach is associated with a benefit in prostate cancer survival and/or in overall survival. Preliminary data thus far suggest that the intermittent approach is safe and possibly better tolerated than conventional continuous ADT. However, more information is needed from long-term treatment disease and treatment-related morbidity. Recently reported large prospectively randomized study conducted by Calais et al. (65) in patients with non-metastatic disease suggest that the intermittent approach is associated with a long disease-free survival comparable to the figures observed with continuous treatment. Clinical evidence of androgen independence (rising serum PSA levels) was low overall and comparable between arms. Longer follow-up time is necessary until more solid conclusions can be drawn.

REFERENCES

1. Huggins C, Hodges CV. Studies in prostatic cancer. I. The effect of castration, estrogens and androgen injections on serum phosphatases in metastic carcinoma of the prostate. *Cancer Res* 1:293, 1941.
2. Feldman BJ, Feldman D. The development of androgen-independent prostate cancer. *Nat Rev Cancer* 1:34–45, 2001.
3. Amler LC, Agus DB, LeDuc C, Sapinoso ML, Fox WD, Kern S, Lee D, Wang V, Leysens M, Higgins B, et al. Dysregulated expression of androgen-responsive and nonresponsive genes in the androgen-independent prostate cancer xenograft model CWR22-R1. *Cancer Res* 60:6134–41, 2000.
4. Mousses S, Wagner U, Chen Y, Kim JW, Bubendorf L, Bittner M, Pretlow T, Elkahloun AG, Trepel JB, Kallioniemi OP. Failure of hormone therapy in prostate cancer involves systematic restoration of androgen responsive genes and activation of rapamycin sensitive signaling. *Oncogene* 20:6718–23, 2001.
5. van der Kwast TH, Schalken J, Ruizeveld de Winter JA, van Vroonhoven CC, Mulder E, Boersma W, Trapman J. Androgen receptors in endocrine-therapy-resistant human prostate cancer. *Int J Cancer* 48:189–93, 1991.
6. Koivisto P, Kononen J, Palmberg C, Tammela T, Hyytinen E, Isola J, Trapman J, Cleutjens K, Noordzij A, Visakorpi T, et al. Androgen receptor gene amplification: a possible molecular mechanism for androgen deprivation therapy failure in prostate cancer. *Cancer Res* 57:314–9, 1997.
7. Visakorpi T, Hyytinen E, Koivisto P, Tanner M, Keinanen R, Palmberg C, Palotie A, Tammela T, Isola J, Kallioniemi OP. In vivo amplification of the androgen receptor gene and progression of human prostate cancer. *Nat Genet* 9:401–6, 1995.
8. Haapala K, Hyytinen ER, Roiha M, Laurila M, Rantala I, Helin HJ, Koivisto PA. Androgen receptor alterations in prostate cancer relapsed during a combined androgen blockade by orchiectomy and bicalutamide. *Lab Invest* 81:1647–51, 2001.
9. Marcelli M, Ittmann M, Mariani S, Sutherland R, Nigam R, Murthy L, Zhao Y, DiConcini D, Puxeddu E, Esen A, et al. Androgen receptor mutations in prostate cancer. *Cancer Res* 60:944–9, 2000.
10. Taplin ME, Bubley GJ, Shuster TD, Frantz ME, Spooner AE, Ogata GK, Keer HN, Balk SP. Mutation of the androgen-receptor gene in metastatic androgen-independent prostate cancer. *N Engl J Med* 332:1393–8, 1995.
11. Taplin ME, Bubley GJ, Ko YJ, Small EJ, Upton M, Rajeshkumar B, Balk SP. Selection for androgen receptor mutations in prostate cancers treated with androgen antagonist. *Cancer Res* 59:2511–5, 1999.
12. Tilley WD, Buchanan G, Hickey TE, Bentel JM. Mutations in the androgen receptor gene are associated with progression of human prostate cancer to androgen independence. *Clin Cancer Res* 2:277–85, 1996.
13. Veldscholte J, Ris-Stalpers C, Kuiper GG, Jenster G, Berrevoets C, Claassen E, van Rooij HC, Trapman J, Brinkmann AO, Mulder E. A mutation in the ligand binding domain of the androgen receptor of human LNCaP cells affects steroid binding characteristics and response to anti-androgens. *Biochem Biophys Res Commun* 173:34–40, 1990.

14. Schoenberg MP, Hakimi JM, Wang S, Bova GS, Epstein JI, Fischbeck KH, Isaacs WB, Walsh PC, Barrack ER. Microsatellite mutation (CAG24–>18) in the androgen receptor gene in human prostate cancer. *Biochem Biophys Res Commun* 198:74–80, 1994.
15. Suzuki H, Akakura K, Komiya A, Aida S, Akimoto S, Shimazaki J. Codon 877 mutation in the androgen receptor gene in advanced prostate cancer: relation to antiandrogen withdrawal syndrome. *Prostate* 29: 153–8, 1996.
16. Suzuki H, Sato N, Watabe Y, Masai M, Seino S, Shimazaki J. Androgen receptor gene mutations in human prostate cancer. *J Steroid Biochem Mol Biol* 46: 759–65, 1993.
17. Newmark JR, Hardy DO, Tonb DC, Carter BS, Epstein JI, Isaacs WB, Brown TR, Barrack ER. Androgen receptor gene mutations in human prostate cancer. *Proc Natl Acad Sci USA* 89:6319–23, 1992.
18. Gaddipati JP, McLeod DG, Heidenberg HB, Sesterhenn IA, Finger MJ, Moul JW, Srivastava S. Frequent detection of codon 877 mutation in the androgen receptor gene in advanced prostate cancers. *Cancer Res.* 54:2861–4, 1994.
19. Evans BA, Harper ME, Daniells CE, Watts CE, Matenhelia S, Green J, Griffiths K. Low incidence of androgen receptor gene mutations in human prostatic tumors using single strand conformation polymorphism analysis. *Prostate* 28:162–71, 1996.
20. Culig Z, Hobisch A, Cronauer MV, Cato AC, Hittmair A, Radmayr C, Eberle J, Bartsch G, Klocker H. Mutant androgen receptor detected in an advanced-stage prostatic carcinoma is activated by adrenal androgens and progesterone. *Mol Endocrinol* 7:1541–50, 1993.
21. Eisenberger MA, O'Dwyer PJ, Friedman MA. Gonadotropin hormone releasing hormone analogues: a new approach for prostate cancer. *J Clin Oncol* 4:414–24, 1986.
22. Laufer M, Denmeade SR, Sinibaldi VJ, Carducci MA, Eisenberger MA. Complete androgen blockade for prostate cancer: what went wrong? *J Urol* 164:3–9, 2000.
23. Schellhammer P, Sharifi R, Block N, Soloway MS, Venner PM, Patterson AL, Sarosdy MF, Vogelzang NJ, Schellenger JJ, Kolvenbag GJ. Clinical benefits of bicalutamide compared with flutamide in combined androgen blockade for patients with advanced prostatic carcinoma: final report of a double-blind randomized multicenter trial. *Urology* 50:330–6, 1997.
24. Schellhammer P, Sharifi R, Block N, Soloway MS, Venner PM, Patterson AL. Sarosdy MF Vogelzang NJ, Jones J, Kolvenbag GJ. A controlled trial of bicalutamide versus flutamide each in combination with lutenizing hormone-releasing hormone analogue therapy in patients with advanced prostate cancer. *Urology* 45:745–52, 1995.
25. Iversen P, Tveter K, Varenhorst E. Randomised study of Casodex 50 MG monotherapy vs. orchidectomy in the treatment of metastatic prostate cancer. The Scandinavian Casodex Cooperative Group. *Scan J Urol Nephrol* 30:93–8, 1996.
26. Tyrrell CJ, Kaisary AV, Iversen P, Anderson JB, Baert L, Tammela T, Chamberlain M, Webster A, Blackledge G. A randomized comparison of casodex 150-mg monotherapy versus castration in the treatment of metastatic and locally advanced prostate cancer. *Eur Urol* 33:447–56, 1998.
27. Iversen P, Tyrrell CJ, Kaisary AV, Anderson JB, Baert L, Tammela T, Chamberlain M, Carroll K, Gotting-Smith K, Blackledge GR. Casodex (bicalutamide) 150 mg monotherapy compared with castration in patients with previously untreated nonmetastatic prostate cancer: results from two multicenter randomized trials at a median follow-up of 4 years. *Urology* 51:389, 1998.
28. Iversen P, Tyrrell CJ, Anderson JB, et al. Comparison of "Casodex" (bicalutamide) 150 mg monotherapy with castration in previously untreated nonmetastatic prostate cancer: mature survival results. *J Urol* 163 (suppl):158, abstract 704, 2000.
29. Bruchovsky N, Wilson JD. The conversion of testosterone to 5-alpha-androstan-17-beta-ol-3-one by rat prostate in vivo and in vitro. *J Biol Chem* 243:2012–21, 1968.

30. Bruchovsky N, Wilson JD. Discovery of the role of dihydrotestosterone in androgen action. *Steroids* 64:753–9, 1999.

31. Cunha GR. Role of mesenchymal-epithelial interactions in normal and abnormal development of the prostate. *Cancer* 74:1030–44, 1994.

32. Bartsch G, Rittmaster RS, Klocker H. Dihydrotestosterone and the concept of 5 alpha reductase inhibition in human benign prostatic hyperplasia. *Eur Urol* 37: 367–80, 2000.

33. Andersson S, Bishop RW, Russell DW. Expression cloning and regulation of steroid 5 alpha-reductase, an enzyme essential for male sexual differentiation. *J Biol Chem* 1989; 264:16249–55.

34. Kelly WK, Scher HI. Prostate specific antigen decline after antiandrogen withdrawal: the flutamide withdrawal syndrome. *J Urol* 149:607–9, 1993.

35. Scher HI, Zhang ZF, Nanus D, Kelly WK. Hormone and antihormone withdrawal: implications for the management of androgen-independent prostate cancer. *Urology* 47:61–9, 1996.

36. Schellhammer PF, Venner P, Haas GP, Small EJ, Nieh PT, Seabaugh DR, Patterson AL, Klein E, Wajsman Z, Furr B, et al. Prostate specific antigen decreases after withdrawal of antiandrogen therapy with bicalutamide or flutamide in patients receiving combined androgen blockade. *J Urol* 157:1731–5, 1997.

37. Shi XB, Ma AH, Xia L, Kung HJ, de Vere White RW. Functional analysis of 44 mutant androgen receptors from human prostate cancer. *Cancer Res* 62:1496–502, 2002.

38. Oh WK. Secondary hormonal therapies in the treatment of prostate cancer. *Urology* 60: 87–92; discussion 93, 2002.

39. Small EJ, Halabi S, Dawson NA, Stadler WM, Rini BI, Picus J, Gable P, Torti FM, Kaplan E, Vogelzang NJ. Antiandrogen withdrawal alone or in combination with ketoconazole in androgen-independent prostate cancer patients: a phase III trial (CALGB 9583). *J Clin Oncol.* 22:1025–33, 2004.

40. Ryan C, Eisenberger M. Optimal timing for chemotherapy in HRPC. *J Clin Oncol.* 23(32):8242–8246, 2005.

41. Byar DP. The Veterans Administration Cooperative Urological Research Group's studies of cancer of the prostate. *Cancer* 32:1126–30, 1973.

42. Veterans Administration Cooperative Urological Research Group. Factors in the prognosis of carcinoma of the prostate: a cooperative study. *J Urol* 100:59–65, 1968.

43. Byar DP. Review of the Veteran's Administration studies of cancer of the prostate and new results concerning treatment of stage I and II tumors. In Pavone-Malacuso M, Smith PM, Edsmyr F (eds), *Bladder Tumors and Other Topics in Urological Oncology.* New York, Plenum Publishing, 471–92, 1980.

44. Byar DP, Corle DK. Hormone therapy for prostate cancer: results of the Veterans Administration Cooperative Urological Research Group studies. *NCI Monogr* 7; 165–70, 1988.

45. Prostate Cancer Working Party Investigators Group. Immediate versus deferred treatment for advanced prostatic cancer: initial results of the Medical Research Council. *Br J Urol* 79:235, 1997.

46. Kirk D. Immediate vs. deferred hormone treatment for prostate cancer: how safe is androgen deprivation? Medical Research Council Prostate Cancer Working Party Investigators Group. *Br J Urol* 86 (suppl):220, 2000.

47. Aronson N, Seidenfeld J, Samson DJ, et al. Relative *Effectiveness and Cost-Effectiveness of Methods of Androgen Suppression in the Treatment of Advanced Prostate Cancer.* Agency for Health Care Policy and Research publication No. 99-E012. Bethesda: United States Public Health Service, 1999.

48. Labrie F, Veillux R, Fournier A. Low androgen levels induce the development of androgen-hypersensitive cell clones shionogi mouse mammary carcinoma cells in culture. *J Natl Cancer Inst* 80:1138–47, 1988.

49. Labrie F, Dupont A, Belanger A, Giguere M, Lacoursiere Y, Emond J, Monfette G, Bergeron V. Combination therapy with flutamide and castration (LHRH agonists or orchiectomy) in advanced prostatic cancer: a marked improvement in response and survival. *J Steroid Biochem* 23:833–41, 1985.

50. Crawford ED, Eisenberger MA, Mcleod DG, Spaulding JT, Benson R, Dorr FA, Blumenstein BA, Davis MA, Goodman PJ. A controlled trial of leuprolide with and without flutamide in prostatic carcinoma. N Eng J Med 1989; 321:419–24.

51. Blumenstein BA. Some statistical considerations for the interpretations of trials of combined androgen therapy. Cancer 1993; 72:3834–40.

52. Eisenberger MA, Blumenstein BA, Crawford ED, Miller G, McLeod DG, Loehrer PJ, Wilding G, Sears K, Culkin DJ, Thompson IM Jr, Bueschen AJ, Lowe BA. A randomized and double-blind comparison of bilateral orchiectomy with or without flutamide for the treatment of patients with stage D2 prostate cancer: results of NCI Intergroup Study 0105. *N Eng J Med* 339:1036–42, 1998.

53. Zalcberg JR, Raghhaven D, Marshall V, Thompson PJ. Bilateral orchidectomy and flutamide versus orchidectomy alone in newly diagnosed patients with metastatic carcinoma of the prostate: an Australian multicentre trial. *Br J Urol* 77:865–9, 1996.

54. Dijkman GA, Janknegt RA, De Reijke TM, Debruyne FM. Long-term efficacy and safety of nilutamide plus castration in advanced prostate cancer and the significance of early prostate specific antigen normalization. *J Urol* 158:160–3, 1997.

55. Crawford ED, Kasimis BS, Gandara D, Smith JA, Soloway MS, Lange PH, Lynch DF, Al-Juburi A, Bracken RB, Wise HA, et al. A randomized controlled clinical trial of leuprolide and anandron vs. leuprolide and placebo for advanced prostate cancer. *Proc Annu Meet Am Soc Clin Oncol* 9:A523, 1990.

56. Denis LJ, Keuppens F, Smith PH, Whelan P, de Moura JL, Newling D, Bono A, Sylvester R. Maximal androgen blockade: final analysis of EORTC phase III trial 30853. *Eur Urol* 33:144–51, 1998.

57. Kuhn J-M, Billebaud T, Navratil H, Moulonguet A, Fiet J, Grise P, Louis JF, Costa P, Husson JM, Dahan R. Prevention of the transient adverse effects of a gonadotropin-releasing hormone analogue (buserelin) in metastatic prostatic carcinoma by administration of an antiandrogen (nilutamide). *N Engl J Med* 321:413–8, 1989.

58. Denis LJ, Carniero de Moura JL, Bono A, Sylvester R, Whelan P, Newling D, Depauw M. Goserelin acetate and flutamide versus bilateral orchiectomy: a phase III EORTC trial (30853). *Urology* 42:119–30, 1993.

59. Iversen P, Ramussen F, Klarskov P, Christensen IJ. Long-term results of Danish Prostatic Cancer Group Trial 86: goserelin acetate plus flutamide versus orchiectomy in advanced prostate cancer. *Cancer* 72:3851–4, 1993.

60. Prostate Cancer Clinical Trialists group. *The Lancet* 346:265–269, 1995.

61. Moinpour CM, Savage MJ, Troxel A, Lovato LC, Eisenberger MA, Veith RW, Higgins B, Skeel R, Yee M, Blumenstein BA, Crawford ED, Meyskens FL Jr. Quality of life in advanced prostate cancer: results of a randomized therapeutic trial. *J Natl Cancer Inst* 90:1537–44, 1998.

62. Laufer M, Denmeade SR, Sinibaldi VJ, Carducci MA, Eisenberger MA. Complete androgen blockade for prostate cancer: what went wrong? *J Urol* 164:3–9, 2000.

63. Akakura K, Bruchovsky N, Goldenberg SL, Rennie PS, Buckley AR, Sullivan LD. Effects of intermittent androgen suppression on androgen-dependent tumors. Apoptosis and serum prostate-specific antigen. *Cancer* 71:2782–90, 1993.

64. Russo P, Liguori G, Heston WD, Huryk R, Yang CR, Fair WR, Whitmore WF, Herr HW. Effects of intermittent diethylstilbestrol diphosphate administration on the R3327 rat prostatic carcinoma. *Cancer Res* 47:5967–70, 1987.

65. Calais da Silva FM, Goncalves F, et al. Phase III Study of intemmittent monotherapy vs. continuous androgen suppression in prostate cancer. *Proceedings Am Soc Chin Oncol* Part 1, vol 25, no 185(Supplement), 2007:S125.

66. Sadar MD, Gleave ME. Ligand-independent activation of the androgen receptor by the differentiation agent butyrate in human prostate cancer cells. *Cancer Res* 60:5825–31, 2000.

67. Craft N, Shostak Y, Carey M, Sawyers CL. A mechanism for hormone-independent prostate cancer through modulation of androgen receptor signaling by the HER-2/neu tyrosine kinase. *Nat Med* 5:280–5, 1999.

68. Hobisch A, Eder IE, Putz T, Horninger W, Bartsch G, Klocker H, Culig Z. Interleukin-6 regulates prostate-specific protein expression in prostate carcinoma cells by activation of the androgen receptor. *Cancer Res* 58:4640–5, 1998.

69. Nazareth LV, Weigel NL. Activation of the human androgen receptor through a protein kinase A signaling pathway. *J Biol Chem* 271:19900–7, 1996.

70. Tan J, Sharief Y, Hamil KG, Gregory CW, Zang DY, Sar M, Gumerlock PH, deVere White RW, Pretlow TG, Harris SE, et al. Dehydroepiandrosterone activates mutant androgen receptors expressed in the androgen-dependent human prostate cancer xenograft CWR22 and LNCaP cells. *Mol Endocrinol* 11:450–9, 1997.

71. Veldscholte J, Voorhorst-Ogink MM, Bolt-de Vries J, van Rooij HC, Trapman J, Mulder E. Unusual specificity of the androgen receptor in the human prostate tumor cell line LNCaP: high affinity for progestagenic and estrogenic steroids. *Biochim Biophys Acta* 1052:187–94, 1990.

17 Advances in Surgical Intervention of Prostate Cancer

Comparison of the Benefits and Pitfalls of Retropubic, Perineal, and Laparoscopic Radical Prostatectomy

Jay B. Basillote, MD, Thomas E. Ahlering, MD, FACS, and Douglas W. Skarecky, BS

CONTENTS

1. INTRODUCTION

Prostate cancer is the most common non-cutaneous malignancy in men and is the leading cause of death from cancer in men above 60 years of age *(1)*. Radical prostatectomy is one of the treatment options available for organ-confined disease. This procedure has undergone several technical modifications since Proust *(2)* and Young *(3)* introduced the perineal approach to radical

From: *Current Clinical Oncology: Prostate Cancer:*
Signaling Networks, Genetics, and New Treatment Strategies
Edited by: R. G. Pestell and M. T. Nevalainen © Humana Press, Totowa, NJ

prostatectomy. In 1945, Millin *(4)* described the retropubic approach, which was associated with significant blood loss and often resulted with negative post-operative sequelae such as impotence and urinary incontinence. In the early 1980s, a significant improvement in the technique was introduced by Walsh *(5,6)*, which has reduced the incidence of post-operative incontinence and impotence. Today, most open radical prostatectomies are performed using the principles outlined by Walsh.

Recently, with the desire to reduce post-operative morbidity and improve convalescence as the impetus, there has been increasing interest in the application of laparoscopy in the field of urology. Indeed, the feasibility and effectiveness of this technique has been shown in other urologic indications such as in nephrectomy *(7)* and adrenalectomy *(8)* and is emerging as the standard of care for these procedures. With the advantages clearly provided by laparoscopy in some urologic and other surgical procedures, several have now applied this technique to radical prostatectomy. Although, it was once thought to not offer advantages over the traditional open radical prostatectomy, recent advancements in instrumentation technology and accumulation of experience have significantly reduced the long operative times associated with this technically challenging procedure. Recent series have also shown promising results in the incontinence and impotence rates. In this review, we will discuss the history and development of this technique, the morbidities associated, and compare outcomes and convalescence with standard open prostatectomy.

2. HISTORY

Laparoscopic radical prostatectomy was first reported in 1992 *(9)* by Schuessler, Clayman, and Kavoussi; their experience after nine cases was reported in 1997 *(10)*. These authors described a transperitoneal approach. This report showed that laparoscopic radical prostatectomy was feasible but also technically challenging as these experienced surgeons reported an average operating time of 9.4 h with more than half of the operating time devoted to the vesicourethral anastomosis. These authors concluded that laparoscopy did not offer a clear advantage over open radical prostatectomy in terms of length of hospitalization, convalescence, continence, or potency and therefore was not, at that time, an efficacious alternative to open prostatectomy. However, they were optimistic that with innovations in instrumentation and other technical advances, it could develop into a procedure that would eventually be shorter, simpler, and possibly superior to open radical retropubic prostatectomy. In 1997 and 1998, Raboy et al. *(11,12)* described the extraperitoneal approach to laparoscopic radical prostatectomy; they performed this procedure in two patients. This technique mimicked the open approach described by Walsh *(5)*. Their incorporation of other refinements such as the use of the harmonic scalpel and use of tie clips instead of intracorporeal knot tying appeared to facilitate the procedure; in their two cases, they reported an average operative time of only 4.9 h. However, despite this initial sanguine report, there were no subsequent confirmatory publications from these surgeons.

It was not until 1999, when two groups of urologists in Paris adopted laparoscopic prostatectomy, that the technique entered its initial phase of acceptance (13,14). Guillonneau et al. (15), using a transperitoneal approach, reported impressive reductions in operative time to an average of less than 4 h. Similarly, Abbou et al. (16) reported median operating times of 4.3 h without and 5.1 h with pelvic lymphadenectomy in 33 patients after 10 were performed to standardize their technique. The key to the success of both groups appeared to be due to two factors: (i) greater individual surgeon experience with laparoscopy and in particular with laparoscopic suturing and (ii) a commitment to perform a sufficient number of cases prior to deciding on the feasibility of the procedure. The latter was the key to success as the "learning curve" was found to be in excess of four times the original series reported by Schuessler. Subsequently, both groups have reported results in a very large number of patients (17,18). It is a credit to their perseverance that laparoscopic prostatectomy has become an established procedure at many institutions worldwide.

Despite significant advances in technique and instrumentation, standard laparoscopy has inherent limitations. Because of the rigidity of the instrument shaft and the fixed position of the trocar on the abdominal wall, there is limitation in the degrees of freedom of movement to four (19). Another drawback is the two-dimensional view of the camera system that causes difficulty determining spatial distance. In addition, all movements are "counterintuitive," and control of the endoscope remains in the hands of an assistant or a "voice-controlled" robotic arm. Finally, there is minimal tactile feedback from the instrumentation. Accordingly, laparoscopic prostatectomy has been viewed as one of the most advanced laparoscopic procedures requiring the greatest of skill in both laparoscopic dissection and suturing.

With the advent of "robotic" laparoscopic surgery using a master–slave system, some of the drawbacks of standard laparoscopy can now be overcome. Originally developed for performing battlefield open trauma surgery with the surgeon controlling the surgical manipulators from a safe distance (i.e., telepresence surgery) (20), such systems have now been adapted for medical use in the civilian world and have already been proven to be effective in cardiac surgery as well as other procedures (21–23).

Several have now reported the application and feasibility of robotic systems (da Vinci, Intuitive Surgical, Sunnyvale, CA) to laparoscopic radical prostatectomy (24–27). To date, experience with the robotic prostatectomy has expanded to dozens of hospitals, but the high cost of the system has largely prevented its widespread application to radical prostatectomy. An increasing number of centers, however, have adopted this technique as their primary approach to radical prostatectomy with the largest series, comprising over 1700 cases reported by Menon *(28)*. Current reports, however, have been encouraging, and there is evidence that the learning curve may be shorter for laparoscopic radical prostatectomy performed with robotic surgical systems than standard laparoscopy *(27)*.

Laparoscopic radical prostatectomy whether performed using standard laparoscopic instruments or using robotic surgical systems is still being evaluated against open radical prostatectomy. Indeed, current results of expert urologic oncologists with open radical prostatectomy have set a high standard in terms of oncologic and functional outcomes *(29)*. To judge laparoscopic prostatectomy in light of present day radical prostatectomy, validated question-naires and analog assessment scales to determine the functional outcome will need to be combined with careful follow-up of oncologic outcome. Prospective studies of this nature, albeit non-randomized, are beginning to be reported by some centers.

3. TECHNIQUES

The technique of laparoscopic radical prostatectomy varies from center to center. Removal of the prostate gland can be performed in an antegrade (from apex to base) or retrograde fashion (from base to apex). The majority of cases reported have employed an intraperitoneal approach [e.g., Institut Mutualiste Montsouris (Paris, France), Klinikum Heilbronn (Heilbronn, Germany), Vattikutti Institute (Detroit, MI), and University of California, Irvine (Orange, CA)]. Guillonneau and Vallancien (Montsouris technique) (Fig. 1) standardized the intraperitoneal approach using an antegrade dissection of the prostate *(30)*. Rassweiler et al. *(31,32)* (Heilbronn technique) reported significant modifications of the Montsouris technique using a transperitoneal approach but performing the prostatic dissection retrogradely, mimicking open radical retropubic prostatectomy.

Following Raboy et al. *(11,12)*, several centers have also described the use of an extraperitoneal approach [e.g., Erasme Hospital (Brussels, Belgium) and University of Leipzig (Leipzig, Germany)]. This approach duplicates the conditions that urologists are accustomed to when performing open retropubic prostatectomy *(33–35)*.

Proponents of the extraperitoneal approach believe that it offers several advantages: less operative time and hospitalization *(36)*, no bowel injuries or peritonitis, and less ileus *(35,37)*. Others, however, could not find significant differences between both techniques and state this as a "false debate" *(38,39)*.

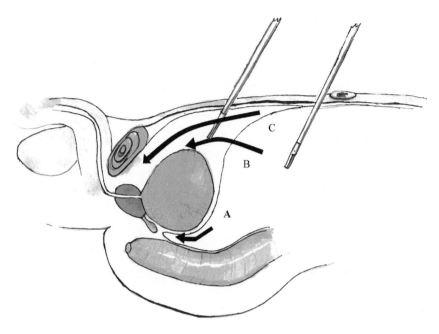

Fig. 1. Initial surgical approaches to laparoscopic radical prostatectomy: (**A**) Montsouris technique; (**B**) Heilbronn, VIP, and UCI technique (note: Heilbronn uses an ascending prostatic dissection while VIP and UCI uses a descending dissection); and (**C**) extraperitoneal technique. [Figure reprinted from *Surgical Endoscopy* (2004);18*(12)*: 1694–711, with kind permission of Springer Science and Business Media.]

Whether a transperitoneal or extraperitoneal approach should be used will just probably depend on surgeon preference.

3.1. Robotic Assisted Prostatectomy

The surgeons who have performed the most robotic assisted prostate-ctomy use the intraperitoneal descending technique described by the group in Montsouris with the modification of doing the retrovesical portion at the end, rather than the beginning, of the procedure *(40,41)*. Some surgeons have reported using an extraperitoneal technique; however, reported series thus far have been very small *(42,43)*.

Five or six ports are used, three of which are dedicated to the robot (i.e., da Vinci) (Fig. 2). Two of these ports are dedicated for the robotic arm instrumentation (R), and an umbilical port is dedicated for the robotic camera lens. The remaining ports are utilized for insertion of instruments by the laparoscopically astute assistant for retracting, suctioning, and passing sutures.

The robotic system that is FDA approved for laparoscopic prostatectomy is the da Vinci system (Fig. 3). This system provides the surgeon with a true

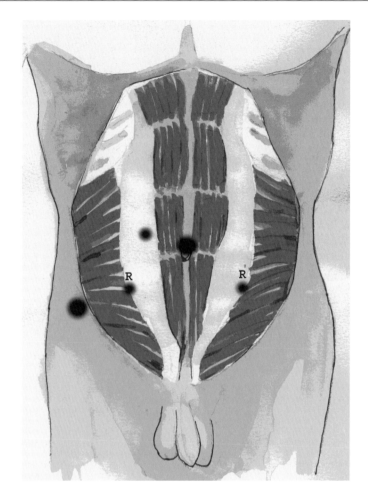

Fig. 2. Port-placements for UCI robotic technique. [Figure reprinted from *Surgical Endoscopy* (2004);18*(12)*: 1694–711, with kind permission of Springer Science and Business Media.]

three-dimensional image providing magnification up to 12×. This robot also incorporates the patented *Endowrist* technology that provides the dexterity of the surgeon's forearm and wrist at the operative site, thus providing six degrees of freedom (i.e., addition of pitch and yaw) as opposed to only four degrees of freedom allowed by standard laparoscopic instruments. In addition, this system has tremor control and 1:5 motion scaling. The entire system is operated from an ergonomic console at which the surgeon sits with their arms and head supported. The viewing system is one of "total immersion" and the control of the camera is by a foot pedal, which is both rapid and accurate.

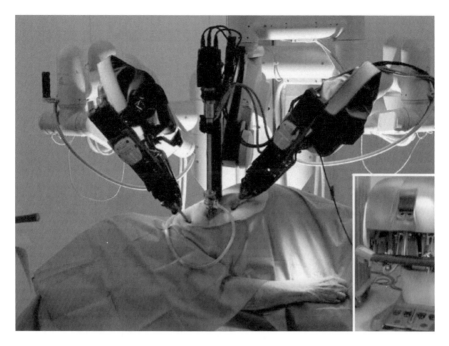

Fig. 3. The da Vinci Surgical System (Intuitive Surgical Inc., Sunnyvale, CA).

4. OPERATIVE RESULTS

4.1. Efficiency

As in any new procedure or technique, there is a learning curve for laparoscopic radical prostatectomy. The operative time for the first reported laparoscopic radical prostatectomy series averaged 9.4 h; however, centers with extensive experience have now reported operative times of 2.6–4.4 h (Table 1a). The learning curve for standard laparoscopic prostatectomy is estimated to be 40 cases based on operative time. Guillonneau et al. *(15)*, in their analysis of their first 120 patients, reported an average operative time of 4.7 h in their initial 40 patients, improving to 4.1 and 3.8 h in subsequent groups of 40 patients with an average of 3.95 h in the last 80 patients. Of note, with the robotic assisted laparoscopic prostatectomy in the hands of experienced open surgeons, the learning curve appears to be significantly truncated. Menon et al. *(27)* showed that robotic assistance enabled an experienced "open" surgeon to achieve operative times similar to an accomplished standard laparoscopist within only 18 cases. At UCI, the learning curve has been even faster despite the lack of laparoscopic training of the involved urologic oncologist; indeed, within 10 cases, the operative time had fallen to under 5 h (Fig. 4).

Table 1a
Pathologic Outcomes of Reported Laparoscopic Radical Prostatectomy Series

Authors	n	Age (years)	Method	OR time (h)	EBL (ml)	Tx. rate (%)	Anastomosis	Open conv. rate (%)
Schuessler (10)	9	65.6	Intraperitoneal descending	9.4	580	NA	Interrupted	0
Raboy (12)	2	60.5	Extraperitoneal ascending	4.9	500	50	Interrupted	0
van Velthoven (79)	22	NA	Intraperitoneal ascending	6.7	490	31	Running	23
Sundaram (55)	12	62.1	Intraperitoneal descending	9.6	327	8	Interrupted	8
Zippe (54)	50	64.9	Intraperitoneal descending	5.4	225	2	Interrupted	2
Roumeguere (80)	85	62.5	Extraperitoneal descending	4.8[a]	522	NA	Interrupted	2.4
Rassweiler (31)	180	64	Intraperitoneal ascending	4.5[c]	1230	31	Interrupted	4.4
Rassweiler (32)	450	65	Intraperitoneal ascending	4.4[d]	NA	24.8	Interrupted	2.0
Rassweiler (64)	500	64	Intraperitoneal ascending	Na	NA	NA	Interrupted	1.8
Hoznek (45)	134	64.8	Intraperitoneal descending	4.7[e]	NA	3	Running	0
Guillonneau (15)	120	64	Intraperitoneal descending	4.0[f]	402	10	Interrupted	5.8
Guillonneau (58)	350	64	Intraperitoneal descending	3.6[g]	354	5.7	Interrupted	2
Guillonneau (48)	567	63.5	Intraperitoneal descending	3.4[h]	380	4.9	Interrupted	1.2
Turk (77)	308	62.3	Intraperitoneal descending	3.9[i]	193	1.6	Interrupted	0
Stolzenburg (35)	70	63.4	Extraperitoneal descending	2.6[d]	350	1.4	Interrupted	0
Hara (82)	26	70	Intraperitoneal descending	7.5	850	100	Interrupted	0
Dahl (83)	70	60.8	Intraperitoneal descending	4.6	449	5.75	Interrupted	1.4
Salomon (47)	235	63.5	Intraperitoneal descending	4.4	NA	2	Running	0
Eden (84)	100	62.2	Intraperitoneal descending	4.0	313	3.0	Interrupted	1
Guillonneau (85)	1000	63	Intraperitoneal descending	NA	NA	NA	Interrupted	NA
Rozet (60)	600	62	Extraperitoneal descending	2.9[k]	380	1.2	Interrupted	0.2
Rassweiler (86)	33	68	Robotic intraperitoneal ascending	7.5[j]	NA	NA	Interrupted	0
Pasticier (26)	5	58	Robotic intraperitoneal descending	3.7	800	NA	Interrupted	0
Menon (27)	40	60.7	Robotic intraperitoneal descending	4.6	391	0	Interrupted	2.5
Tewari (87)	100	NA	Robotic intraperitoneal descending	Na	NA	NA	Interrupted	NA
Menon (88)	100	60	Robotic intraperitoneal descending	3.3	149	0	Running	NA
Bhandari (89)	300	60.3	Robotic intraperitoneal descending	2.96	109	0	Running	0
Patel (90)	200	59.5	Robotic intraperitoneal descending	2.35	75	0	Running	0
Ahlering (91)	60	62.9	Robotic intraperitoneal descending	3.9[m]	103	0	Running	0

OR time, operative time; RRP, radical retropubic prostatectomy; Tx rate, transfusion rate.

[a] Includes pelvic lymph node dissection in 17/85 patients.
[b] PSA<0.2 ng/ml.
[c] Includes pelvic lymph node dissection in 163/180 patients.
[d] Includes pelvic lymph node dissection.
[e] Includes pelvic lymph node dissection in 56/134 patients.
[f] Includes pelvic lymph node dissection in 36/120 patients.
[g] Includes pelvic lymph node dissection in 75/350 patients.
[h] Includes pelvic lymph node dissection in 110/567 patients.
[i] Includes pelvic lymph node dissection in 232/308 patients.
[j] Progression-free survival rate at 3 years.
[k] Includes pelvic lymph node dissection in 107/600 patients.
[l] Includes pelvic lymph node dissection in 27/33 patients.
[m] Excludes robotic setup.

margin positive rate							Gleason <7 or ≥7	Mean F/up (months)	PSA Progression Free Rate (<0.1 ng/ml)	cost
overall	pT2a	pT2b	pT2c	pT3a	pT3b	pT4				
11.1	NA	NA	NA	NA	NA	NA	14/0	26	NA	NA
50	NA	NA	NA	NA	NA	NA	50/0	NA	NA	NA
23	NA	NA	NA	NA	NA	NA	NA	NA	NA	NA
NA	NA	NA	NA	NA	NA	NA	NA	9.8	100	NA
20	NA	NA	NA	NA	NA	NA	NA	6.4	100	NA
25.8	7.8		NA		NA	NA	91.4[a]	NA		
16	2.3	NA	NA	15	34	100	NA	12	95	NA
18.6	NA	NA	NA	NA	NA	NA	NA	20.8	89[a]	NA
19	2.1	9.9	NA	25.2	42.0	89.4	NA	40	89[a]	NA
25	16.8		48.5	NA	NA	12	89.6	NA		
15	11	16	NA	0	50	NA	NA	2.2	89	$1237 less than RRP
15.4	3.6	14	NA	33	43.5	NA	NA	At least 12 m	92[a]	NA
NA	NA	NA	NA	NA	NA	NA	NA	NA	NA	NA
21.1	NA	NA	NA	NA	NA	NA	NA	NA	NA	NA
21.4	6.2		29.4	100	NA	At least 1 m	NA	NA		
NA	NA	NA	NA	NA	NA	NA	NA	4	100	NA
11.4	6.7	8.2	NA	60	NA	NA	NA	3 m	NA	NA
20.6	15.8		41.4	NA	NA	36	86.2[a]	NA		
16	NA	NA	NA	NA	NA	NA	NA	At least 3 m	100	NA
19.2	6.9	18	NA	30	32	NA	18.2/30	12	90.5[j]	NA
17.7	14.9	6.7	16.5	26.9	22.6	NA	NA	12	NA	NA
18	NA	NA	NA	NA	NA	NA	NA	NA	NA	NA
20	0	33	NA	NA	NA	NA	NA	NA	NA	NA
17.5	NA	NA	NA	NA	NA	NA	NA	6.5	NA	NA
NA	NA	NA	NA	NA	NA	NA	NA	NA	NA	NA
15	10.6		40	NA	14/16	5.5	NA	NA		
NA	NA	NA	NA	NA	NA	NA	NA	NA	NA	NA
10.5	5.7		28.5	20	33			9.7	95	
16.7	4.5		47	100	NA	NA	NA	NA		

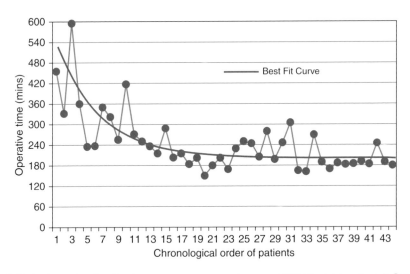

Fig. 4. Robotic-assisted radical prostatectomy experience at UCI (first 45 patients). [Figure reprinted from *Surgical Endoscopy* (2004);18(*12*): 1694–711, with kind permission of Springer Science and Business Media.]

The learning curve to 4-h proficiency was 12 cases with subsequent mean operating time of 3.45 h, thereafter *(44)*.

4.2. Equanimity

4.2.1. COMPLICATIONS

Complications of laparoscopic radical prostatectomy probably are related to surgeon's experience and occur generally during the learning curve. Indeed, Rassweiler et al. *(31)* reported that the complication rate for the first 120 cases was 22.5%, but it decreased significantly to 11.7% in their next group of 60 patients. Similarly, Hoznek et al. *(45)* reported a drop in their complication rate from 22.5% in their first 40 of 134 patients to 3.2% in the next 94 patients. Complications include bowel injuries, ureteric injuries, bladder injury, urine extravasation, urinoma formation, intra-abdominal hemorrhage, lymphorrhea, phlebitis, and port-site hernias *(31,45,46)*. The complication rates of reported series with at least 100 patients range from 8.9 to 25% (Table 1b). To date there has been only one death associated with laparoscopic radical prostatectomy *(47)*. The open conversion rates range from 0 to 23%, dependent upon the experience of the surgeon. Guillonneau et al. *(46)* reported a conversion rate of 10% in their initial 40 of 550 patients; in their next 40 patients, this decreased to 7.5%, and since then, they did not have any conversions in their next 470 patients.

Table 1b

Functional Outcomes of Laparoscopic Radical Prostatectomy Series in Table 1a

Authors	n	Age (years)	Method	No. hospital days	No. days with catheter	Time of assessment from surgery, months	Continence rate	Criteria	Erectile function (bns/uns)	Complication rate total, major/minor
Schuessler (10)	9	65.6	Intraperitoneal descending	7.3	NA	26	66%	0 pads	50%	33,11.1/22.2%
Raboy (12)	2	60.5	Extraperitoneal ascending	2.5	14	NA	NA	NA	NA	NA
van Velthoven (79)	22	NA	Intraperitoneal ascending	NA	NA	NA	NA	NA	NA	NA
Sundaram (55)	12	62.1	Intraperitoneal descending	2.4	NA	9.8	82%	0 pads	0%	64%
Zippe (54)	50	64.9	Intraperitoneal descending	1.6	NA	6	76%	0 pads	4%	20.6/14%
Rouneguere (80)	85	62.5	Extraperitoneal descending	NA	6	12	80.7%	0 pads	53.8% / NA	NA
Rassweiler (31)	180	64	Intraperitoneal descending	10	7	12	97%[a]	0 pads	NA	18.9,9/15%
Rassweiler (32)	450	65	Intraperitoneal ascending	NA	7	18	95%	0 pads	NA	14.7%
Hoznek (45)	134	64.8	Intraperitoneal descending	6.1	4.8	12	86.2%[b]	0 pads	46% / 28%[b,c]	8.9%
Guillonneau (15)	120	64	Intraperitoneal descending	6.6	6.6	6	73.3%[a]	0 pads	5%[a]/NA[a]	25.2/22.5%
Guillonneau (58)	350	64	Intraperitoneal descending	6	5.8	12	85.5%[a]	0 pads	32%[b,d] / NA	13.4%
Guillonneau (48)	567	63.5	Intraperitoneal descending	6.2	6.1	NA	NA	NA	NA	17.1,3.7/14.6%
Turk (81)	308	62.3	Intraperitoneal descending	7.5	7.8	12	92%	NA	38.5%[e]	13.6%
Stolzenburg (35)	70	60.8	Extraperitoneal descending	NA	8.2	6	90%[a]	0 or pad use for protection only	0%[f]	NA,1.4%/NA
Hara (82)	26	70	Intraperitoneal descending	NA	9	6	100%	0 pads	14%[g]	15.4%
Dahl (83)	70	60.8	Intraperitoneal descending	2.5	7	AT LEAST 3M	70.6%	0 pads	NA	20.7,1/12.9%
Salomon (47)	235	63.5	Intraperitoneal descending	6.8	5.7	NA	NA	NA	NA	15%
Eden (80)	100	62.2	Intraperitoneal descending	4.2	NA	12	90%	0 pads	62% / NA	8.6/2%
Guillomneau (85)	1000	63	Intraperitoneal descending	NA	NA	NA	NA	NA	NA	NA
Rozet (60)	600	62	Extraperitoneal descending	6.3	7.6	12	84%[a]	0 pads	64%[a,h]	11.2,2/9.2%

(Continued)

Table 1b
(Continued)

Authors	n	Age (years)	Method	No. hospital days	No. days with catheter	Time of assessment from surgery, months	Continence rate	Criteria	Erectile function (bns/ans)	Complication rate total, major/minor
Rassweiler (86)	33	68	Robotic intraperitoneal ascending	NA	6.8	NA	NA	NA	NA	NA
Pasticier (26)	5	58	Robotic intraperitoneal descending	NA	5.5	NA	NA	NA	NA	0%
Menon (27)	40	60.7	Robotic intraperitoneal descending	NA	NA	1.5	NA	NA	29% / NA	15.0/15%
Tewari (87)	100	NA	Robotic intraperitoneal descending	NA	NA	6M	NA	NA	58%[a]	NA
Menon (88)	100	60	Robotic intraperitoneal descending	0.96I	4.2	6M	92%	0 or pad use for protection only	NA	8%
Bhandari (89)	300	60.3	Robotic intraperitoneal descending	1.2	6.9	NA	NA	NA	NA	5.7%
Patel (90)	200	59.5	Robotic intraperitoneal descending	1.1	7.9	12	98%	0 pads	NA	NA
Ahlering (91)	60	62.9	Robotic intraperitoneal descending	1.1	7	3M	75%[a]	0 pads	NA	6.7%

[a] Assessed with validated patient questionnaire.
[b] Assessed with non-validated prospective questionnaire.
[c] Results for 1-month post-operative period.
[d] Results for 6-month post-operative period.
[e] With at least one neurovascular bundle preserved.
[f] 33% had erection with medical or mechanical assistance with at least one neurovascular bundle preserved; result for 2-month post-operative period.
[g] Result for 3-month post-operative period.
[h] In 89 patients who had bilateral neurovascular preservation and who were potent preoperatively.
[i] Length of hospital stay for 95/100 patients.

Reported blood loss for laparoscopic radical prostatectomy ranges from 103 to 1230 ml with an average of 431 ml. The wide range may be a result of variability in the technique. Rassweiler et al. *(31)* have reported the highest mean blood loss with this technique: 1230 ml and a transfusion rate of 30% among 180 patients. They attributed this to their technique that advocates early dissection and ligation of the dorsal venous complex followed by immediate division of the urethra. Indeed, Guillonneau and Vallancien with their approach of delayed dorsal venous complex transaction found an average blood loss of 380 cc and a transfusion rate of only 4.9% among 567 patients *(48)*. Similarly, Menon using a robotic assisted approach and delayed transection of the dorsal venous complex had a similar estimated blood loss of only 391 cc and a transfusion rate of 0% in their first 40 patients *(27)* and only 150 cc after 100 patients *(49)*. With further experience, however, Rassweiler et al. *(50)* have shown that blood loss decreased from 1550 cc in their initial group of patients to 800 cc in their most recent group. Accordingly, Guillonneau et al. *(15)* reported transfusion rates of 10% in their first 120 cases but decreased to 5% *(48)* after performing a total of 567 operations.

4.2.2. CONVALESCENCE

Studies assessing post-operative recovery and convalescence in laparoscopic radical prostatectomy are beginning to accumulate. Guillonneau and Vallancien assessed post-operative pain after laparoscopic radical prostatectomy and reported that only 9% of patients requested narcotic analgesics on post-operative day 1 and only 2% on day 2 *(15)*. Based on their own positive experience, Bollens et al. *(33)* have started giving intramuscular morphine on an as needed basis only during the first post-operative day to their patients after laparoscopic radical prostatectomy. Bhayani et al. *(51)* compared 24 patients who underwent laparoscopic radical prostatectomy and 36 patients who underwent open radical prostatectomy, and found that there was no difference in the in-hospital morphine-equivalent requirement between the two groups. Pain medication use at home was also less in the laparoscopic group as was the time to complete convalescence. Similarly, a prospective comparison of radical open retropubic prostatectomy and laparoscopic robotic assisted prostatectomy by Menon et al. *(52)* has shown significantly lower pain scores in the laparoscopic robotic assisted prostatectomy patients.

The length of hospital stay is often used as an instrument to measure patient recovery as it generally correlates with the time to return to basic activities. The length of hospital stay of reported laparoscopic radical prostatectomy series has ranged from 0.96 to 9.3 days (Table 1b), whereas the average hospitalization after open retropubic radical prostatectomy is between 2 and 3 days *(53)*. However, as the most reported and largest series of laparoscopic radical prostatectomy were performed in European centers, the variation can be explained by differences in health care systems in Europe and USA. Indeed, smaller series performed in the USA reported hospital stays of 0.96–2.4 days *(49,54,55)*.

4.2.3. PRESERVATION OF CONTINENCE

Evaluation of continence rates between reported series can be difficult due to subjectivity of the surgeon and patient, the use of different or non-validated questionnaires by surgeons, and the variability in the definition of continence by surgeons and patients. With at least 12 months follow-up, the urinary continence rates in reported laparoscopic series range from 66 to 97% (Table 1b). However, to truly assess this parameter, the use of a validated continence questionnaire is essential; data accumulated by any other instruments are "suspect." Salomon et al. *(56)*, using a questionnaire derived from the International Continence Society (ICS)-male self-administered questionnaire *(57)*, reported that of 100 consecutive patients who underwent laparoscopic radical prostatectomy, 45 and 90% had complete diurnal continence (no pads) 1 month and 12 months after surgery, respectively. Similarly, using the ICS-male questionnaire, Guillonneau et al. *(58)* reported that 86% had complete diurnal continence (no pads) 12 months post-operatively.

4.2.4. PRESERVATION OF SEXUAL FUNCTION

Preservation of sexual function has a significant impact on quality of life in men undergoing radical prostatectomy. As in the evaluation of continence rates, there is also difficulty in the evaluation of sexual function after laparoscopic radical prostatectomy due to variability in the use of validated questionnaires (e.g., IIEF). Again, in this regard, the use of validated questionnaires is essential to the discernment of accurate data that can then be used to compare erectile function with different operative techniques. In addition, rates of preservation of sexual function can be affected by patient's age, whether one or both neurovascular bundles were preserved and whether the patient was sexually active prior to surgery. In addition, it is important to differentiate between those patients who are normally potent and those who after the procedure are potent only with medications.

In reported series, men who were reportedly potent preoperatively had erectile function adequate for intercourse in 38.5–64% of patients after 1-year follow-up (Table 1b). There was no uniformity, however, on whether one or both neurovascular bundles were preserved on these patients. Also, some of these reports did not use a validated questionnaire. Guillonneau et al. *(58)*, using a non-validated questionnaire, reported that among 22 patients who were sexually active preoperatively, 13 (59%) had erections 6 months post-operatively, of which six patients had erections considered to be insufficient for intercourse and seven patients had non-medically-induced erections adequate for sexual intercourse. Similarly, Anastasiadis et al. *(59)*, using a non-validated questionnaire, reported that after bilateral neurovascular nerve preservation, 53% had erections suitable for intercourse without routine administration of sildenafil and without the use vacuum erection devices or pharmacologic injection/suppository therapy 12 months after surgery. At a median follow-up of 6 months, Rozet et al. *(60)*, using a validated questionnaire, reported the erection and intercourse rate was 64 and 43%, respectively, in men who were potent preoperatively while using 10 mg of tadalafil taken every 2 days post-operatively.

4.3. Economy

Guillonneau et al. *(15)* reported that standard laparoscopy is $1237 less expensive than open radical prostatectomy. In this study, however, the mean hospital stay was 8 days for the open radical prostatectomy group. This is in sharp contrast to the USA, where the typical hospital stay is 1–3 days; also the cost of operating room time is far less in Europe than in the USA. In a comparison between laparoscopic and open radical prostatectomy performed at a single US academic medical center, Anderson et al. *(61)* found that the cost for laparoscopy was significantly greater than the cost for open radical prostatectomy ($6760 vs. $5253). Most of this cost difference was due to the higher surgical supply costs and operating room costs in the laparoscopic group.

With respect to robot-assisted radical prostatectomy, cost is a significant concern as well. Indeed, the widespread use of this device has been prevented mostly by its price, which for the da Vinci unit runs $980,000 plus a $100,000/year service contract. While these expenses are determined by how many surgical disciplines are using the robot in a given year, there is a fixed cost of using the robot in that a $300–500 charge is incurred with its use to pay for the reposable robotic instruments used during a given procedure. Indeed, Lotan et al. *(62)* blamed the amortized cost of the robot purchase and maintenance and the cost of robotic instruments as the reason for the higher cost of robotic prostatectomy compared to both standard laparoscopic and open radical retropubic prostatectomy. The cost premium for robotic prostatectomy may be overcome, however, in higher volume centers *(63)*.

4.4. Effectiveness

The primary goal of laparoscopic radical prostatectomy is to attain a "surgical cure." The efficacy of this procedure in achieving this goal can be measured by margin status of the specimen and absence of chemical or clinical recurrence. The overall margin-positive rate in reported series of laparoscopic radical prostatectomy with at least 20 patients ranges from

10.5–25.8% (Table 1a). However, in pathologically organ-confined disease (pT2), the margin-positive rate is lower, ranging from 2.1–14.9%.

Assessment of biological progression is difficult to accomplish for laparoscopic radical prostatectomy considering the relatively short follow-up of reported series thus far. Rozet et al. *(60)* reported that after a mean follow-up of 12 months, 95% had a serum level of Prostate Specific Antigen (PSA) lower than 0.2 ng/ml while Rassweiler et al. reported that after a median follow-up of 12 months, 95% of patients were PSA progression free. With a longer mean follow-up of 20.8 months, these authors reported a PSA progression-free rate of 89% *(64)*. The PSA recurrence was 3.2, 6.5, 15.9, and 23.9% in stages pT2a, pT2b, pT3a, and pT3b/4, respectively.

5. COMPARISON TO OPEN RADICAL PROSTATECTOMY

A summary of radical retropubic and perineal prostatectomy series are illustrated in Table 2. Indeed, with their extensive experience and excellent results, open surgeons have provided a high standard for the laparoscopic surgeons to emulate.

On average, operative times for the laparoscopic approach are significantly longer than the open approach. Longer operative times are inherent to the laparoscopic approach because of the decreased tactile feedback and the two-dimensional nature of laparoscopic video imaging. With experience, however, several urologists have reported significant reductions in operative times. A 7–8 h operation is common for the first laparoscopic radical prostatectomies, but, after about 50 cases, less than 3.5–4 h is required for an experienced team *(58)*. Perhaps, with further experience, operative times will decrease further to operative times reported in large open radical prostatectomy series where the average operative time in experienced hands is 3 h *(15,65)*. Indeed, Menon now reports operative times for robotic laparoscopic prostatectomy is between 1.2 and 2.7 h. In fact at their hospital, the robotic prostatectomy takes less time to perform than open radical prostatectomy or standard laparoscopic prostatectomy *(66)*.

As in any other procedure that is in its early stage of development, the complication rate of laparoscopic radical prostatectomy is correlated with the surgeon's experience. Rassweiler et al. *(31)* reported a complication rate of 23.3% in their initial series. With experience, this, however, decreased to 11.7% in their last group of patients that is comparable to the experience of open surgeons.

Despite a short experience with the procedure, one area in which laparoscopic radical prostatectomy is better than open prostatectomy, in most hands, is in the aspect of blood loss. Reported blood loss for laparoscopic radical prostatectomy ranges from 103 to 1230 ml with an average of 430 ml; this is

Table 2
Reported Series of Open Radical Prostatectomy

Institution	n	Age (years)	Method	Oper. time (hrs)	EBL (ml)	Transfusion rate	# hospital days	Margin pos rate	Follow-up (months)	Continence Rate	Criteria	Erectile function (bns/uns)	PSA progression-free rate (<0.2 ng/ml)	Complication rate total, major/minor
Johns Hopkins (68)	64	57	RRP	NA	NA	NA	NA	1.5%	18	93%[a]	0 pads	86%[a,b]	NA	NA
Johns Hopkins (88)	2404	59	RRP	NA	NA	NA	NA	NA	75	NA	NA	NA	74%	NA
NYU (29)	1000	60.3	RRP	NA	819	9.7%	2.3	19.9%	NA	NA	NA	NA	NA	7%
NYU (89)	621	58.7	RRP	NA	NA	NA	NA	NA	24	82.4%[a]	0 pads	NA	NA	NA
Washington Univ (90)	1342	NA	RRP	NA	NA	11.5%	NA	NA	NA	NA	NA	NA	NA	7.4, 2.4/5.0%
Washington Univ (69)	1870	63	RRP	NA	NA	NA	NA	NA	18	92%	0 pads	68%/47%	NA	10%
Washington Univ (92)	3477	61	RRP	NA	NA	NA	NA	NA	18	93%	0 Pads	NA	NA	9%
Mayo Clinic (93)	3170	66	RRP	NA	600–1030	5–31%	NA	24%	60	NA	NA	NA	52% at 10years	NA
Baylor (94)	1000	62.9	RRP	NA	NA	NA	NA	12.8%	53.2	NA	NA	NA	75%(<0.4 ng/ml)	NA
Baylor (95)	472	62.2	RRP	3.0	NA	28.6%	6.2	NA	NA	NA	NA	NA	NA	27.8, 9.8/18.0%
Baylor (96)	314	60.5	RRP	NA	NA	NA	NA	NA	25.4	NA	NA	70/26%[c]	NA	NA
Baylor (97)	581	63	RRP	NA	800	NA	NA	NA	24	91%	0 pads	NA	NA	NA
Columbia Univ. (98)	480	62.6	RRP	NA	NA	NA	NA	NA	39.6	91.8%[d]	0 pads	NA	NA	NA
Mayo Clinic, Jacksonville (99)	325	66	RRP	NA	750	22%	4.2	18%	50	93%	NA	47%[e]	NA	NA
Henri Mondor (65)	147	65.1	RRP	3.3	NA	38%	15.2	31.7%	54	NA	0 leak	NA	NA	NA
Armed Forces, multi-institutional (100)	190	63.7	RRP	NA	1575	NA	NA	39.5%	41.4	69.9%	0 leak	8.9%[e]	77.1%	NA
Armed Forces, multi-institutional (100)	190	62.2	RPP	NA	802	NA	NA	43.1%	41.1	74.8%	0 leak	8.2%[e]	72.4%	NA

(Continued)

Table 2
(Continued)

Institution	n	Age (years)	Method	Oper. time (hrs)	EBL (ml)	Transfusion rate	# hospital days	Margin pos rate	Follow-up (months)	Continence Rate	Criteria	Erectile function (bns/uns)	PSA progression-free rate (<0.2 ng/ml)	Complication rate total, major/minor
Johannes Gutenberg-Univ Mainz (101)	630	65	RPP	NA	NA	5.9%	NA	15%	8	NA	NA	NA	NA	22.1,2.4/19.7%
Henri Mondor (65)	119	64.6	RPP	3.0	NA	19%	8.5	18.5%	46.8	NA	NA	NA	NA	NA
UCSF (102)	220	67	RPP	NA	600	NA	NA	NA	18	95%	0 pads	73/68%	NA	2%/NA
Duke Univ. (103)	1242	65.2	RPP	NA	NA	NA	NA	23%	48	NA	NA	NA	NA	NA
Mayo Clinic, Jacksonville (99)	500	66	RPP	NA	270	4%	1.5	16%	50	94%	NA	34%[e]	NA	NA
Tulane Univ. (104)	250	63	RPP	NA	NA	NA	2.7	NA	30	93%[a]	0–1 pads	41%[a]	NA	NA
Tulane Univ. (72)	25	NA	RPP	1.8	350	NA	0.7	NA	12	94.7%	NA	41%	NA	NA
Northern Inst. of Urology (105)	508	65.8	RPP	NA	NA	1%	NA	18%	48	96%	0 pads	83/74%	85.8%	NA

bns, bilateral nerve sparing; RPP, radical perineal prostatectomy; RRP, radical retropubic prostatectomy; uns, unilateral nerve sparing.
[a] Assessed with validated patient questionnaire.
[b] 89% of men underwent bilateral nerve-sparing procedure.
[c] Men age 60 years or younger; for 60.1–65 years = 49/15%; for age >65 years = 43/13%.
[d] Assessed with non-validated retrospective questionnaire.
[e] Nerve-sparing status unknown.

quite low when compared to other reported series of open radical prostatectomy in which the average blood loss is in the 800–1600 cc range (Table 2). As a result of decreased blood loss, the transfusion rates are also less in laparoscopic radical prostatectomy than open radical prostatectomy. Transfusion rates also tend to decrease as experience is attained. Guillonneau et al. *(15)* reported transfusion rates of 10% in their first 120 cases but decreased to 5% *(48)* after performing a total of 567 operations. In Menon's robotic series, none of over 1100 patients required intraoperative transfusion.

Functional outcomes such as urinary continence and potency are important; however, as previously noted, the lack of validated questionnaires in most series make the reported data largely anecdotal and thus unreliable. Indeed, it has been shown that self-administered questionnaires report poorer results than data obtained by the clinical interview that most series report *(67)*. This makes assessment and comparison of the laparoscopic technique to the open technique very difficult. Fortunately, at this time, many urologic oncologists as well as laparoscopic surgeons are now using mailed validated questionnaires to assess their continence, potency, and quality of life outcomes.

As previously noted, Guillonneau et al. *(58)*, using a validated question-naire, reported a urinary continence rate of 86% after 12 months follow-up while Rassweiler et al. *(31)* reported diurnal continence in 97% of patients after 12 months follow-up. These reported results are comparable to several reported series of open radical prostatectomy *(68,69)*. Indeed, Walsh et al. *(68)*, using a validated questionnaire, reported continence rates of 54, 80, and 93% at 3, 6, and 12 months after radical retropubic prostatectomy, respectively. Similarly, Rassweiler et al. *(31)* reported continence rates of 54, 74, and 97% for the same time points after laparoscopic radical prostatectomy. In a prospective comparative study between laparoscopic radical prostatectomy and open retropubic prostatectomy performed at their institution, Anastasiadis et al. *(59)* also showed that the two approaches showed similar continence rates using a validated self-administered questionnaire.

Comparison of potency rates between radical prostatectomy series is difficult for all of the aforestated reasons: variability in the procedure (none, one, or two bundles preserved), variability in assessing erectile function (i.e., failure to use validated sexual questionnaires), and failure to distinguish among potent males with regard to requirements for medical assistance (i.e., sildenafil). Given these variables, it is understandable that the range for potency after radical prostatectomy even in large series has been cited at 5–46%. These figures are similar to what has been the experience with laparoscopic radical prostatectomy to date. Obviously, further reporting with proper data collection instruments is needed to compare potency rates across various surgical techniques and to relate these potency rates to the patients' preoperative state.

The cornerstone oncologic principle of radical prostatectomy is complete elimination of all prostate cancer cells. Cancer control can be assessed by margin status of the surgical specimen and presence of biological recurrence. Margin-positive rates of reported laparoscopic series with more than 100 cases have been between 15 and 25%. These results are comparable to reported open radical prostatectomy series. Caution is advised, however, when comparing these results as adequate comparison can only be performed after adjustment of relevant covariates. It has been shown that surgical margin status is affected by the clinical stage, serum PSA, and biopsy Gleason score *(70)*. When assessing only patients with PSA < 10 ng/ml, the surgeons at Henri Mondor *(47)* reported a surgical margin-positive rate of 20.6%, which is less than their previous report of 25% when not restricting for PSA level *(45)*. Similarly, when only patients with favorable characteristics are considered (i.e., PSA < 10 ng/ml, cT1c disease, and Gleason grade of ≤7), Ahlering noted a decrease in the surgical margin-positive rate from 36 to 0% *(44)*.

Because of the relatively short follow-up thus far, assessment of biological recurrence is difficult to perform for laparoscopic radical prostatectomy. With mean 12 months follow-up, series with over 100 cases have reported overall PSA progression-free rates of 89–95%. Because of variations in the various covariates such as preoperative PSA, clinical stage, Gleason score, and surgical margin status, comparison with reported open radical prostatectomy series is difficult. Nevertheless, reported PSA progression-free rates of 89–95% at 1 year seem to be on a par with the overall actuarial 2-year progression-free rate of 82% in a recent cohort of patients with prostate cancer treated with open radical prostatectomy between 1997 and 2000 *(71)*.

Authors from institutions who have experience in both open and laparoscopic radical prostatectomy have also reported their results on both techniques. Rassweiler et al. *(50)* reported that the laparoscopic technique is superior to radical retropubic prostatectomy in terms of blood loss, transfusion rate, complication rate, post-operative analgesic use, and length of time of catheter use. Continence rates at 12 months were similar in both groups. Despite a stage shift (preoperative PSA and pathologic stage) in favor of the laparoscopic group, there was no difference in surgical margin status and biochemical recurrence between the two groups. Salomon et al. *(65)*, in comparing their historical series on radical retropubic and perineal prostatectomy with the laparoscopic approach, showed that there is no difference in the surgical margin-positive rates between the techniques despite changes in selection criteria over time resulting in stage migration in favor of the most recent cohort, the laparoscopic group. Lastly, in comparing 25 consecutive patients undergoing radical retropubic, laparoscopic, and perineal prostatectomy, Puri et al. *(72)* showed that the perineal approach is superior when it comes to operative time (106 min), post-operative pain, hospital stay, time to return to normal activity, and blood loss. These authors concluded that the laparoscopic approach has a steep learning curve and may not be readily acceptable in mainstream practice. To be sure, we are still in a phase in which laparoscopy is "new," additional studies are needed to determine whether it is actually "better" than what we already have.

6. IMPACT OF ROBOTICS

Robotic laparoscopic surgery offers several advantages over standard laparoscopy, such as tremor filtering, intuitive motion, true 3D vision, six degrees of freedom, plus motion scaling. The master unit allows full range of movement that is transmitted accurately to the surgical arms; indeed, the "wrist" on the robot is superphysiologic given its ability to rotate 540°.

Because of various technological advantages offered by robotic systems, Menon et al. *(27)* have shown that robotic assistance can decrease the learning curve for laparoscopic surgery when performed by a skilled "open" surgeon. Results at the University of California, Irvine (UCI), corroborate the results from the Vattikuti Urological Institute in Detroit. Indeed, we have reported our experience in which one of our urologic oncologists, untrained in laparoscopy, was able to successfully complete a laparoscopic prostatectomy following a 4-day course in robotic laparoscopy *(41)*. Figure 4 illustrates the operative time in the first 45 patients for robot-assisted radical prostatectomy at UCI all performed by the same laparoscopically naïve urologic oncologist. Operative time decreased with increased experience [Spearman's rank correlation (Rho) –0.74, 95% confidence interval = –0.86 to –0.55, $p < 0.001$) and operative times in his last five cases have averaged 3.2 h. The operative times started to stabilize approximately by the 11th case. This may suggest that the learning curve for robotic-assisted prostatectomy is significantly shorter than standard laparoscopy even when performed by a non-laparoscopic, though experienced open surgeon.

The one major, ongoing disadvantage of both pure laparoscopic and even more so with robotic systems is the lack of tactile feedback. The operating surgeon has to compensate for this by taking visual cues using the endoscopic image. Indeed, for handling tissues and for suturing, it is necessary for the surgeon to "feel" with his eyes; the tension on tissues or sutures is visually estimated rather than felt.

The initial results of potency outcomes have ushered in questions of iatrogenic damage to the neurovascular bundles (NVB) by electrocautery during laparoscopic prostatectomy. In an animal study evaluating function of the Neurovascular Bundle (NVB) following dissection with three different forms of thermal energy, Ong and associates *(73)* demonstrated the caustic effect of thermal energy near the NVB. In 2005, Eichel et al. *(74)* described for robotic prostatectomy and Gill and associates *(75)* for standard lap prostate-ctomy, a method for cautery-free dissection of the NVB. Initial short-term outcomes by Ahlering et al. *(76)* demonstrate a significant improvement in 3-month potency rates of the cautery-free technique (43%) as compared to their previous (thermal) technique (8.3%), but long-term follow-up is immature.

With the added dexterity provided by robotic systems, it is possible that more and more surgeons will take advantage of this technology. A patient-driven component seeking robotic systems may also be in motion. Its high cost, however, must be weighed carefully against its efficacy and efficiency especially as results with standard laparoscopy continue to improve with increasing experience. However, the fact remains that the skilled laparoscopist is the result of years of laparoscopic training on top of years of training as an open surgeon in contrast, the skilled robotic surgeon appears to be "ready-made" on the basis of open surgical training alone plus a minimal amount of training in basic laparoscopy (i.e., obtaining a pneumoperitoneum and port placement). The one major caveat is that for those surgeons transferring open skills to a robotic platform, the tables assistant needs to be a skilled laparoscopic surgeon. The assistant in these procedures is responsible for obtaining the pneumoperitoneum, proper port placement, and invaluable assistance during the procedure (i.e., retraction, suction, and introduction of suture material).

7. FUTURE DIRECTIONS

With the advantages provided by laparoscopy in other procedures, it was not surprising that this technique would eventually be applied to radical prosta-tectomy. With increasing interest in this technique, new task-oriented instru-ments/devices are already being developed to facilitate the procedure. Indeed, use of suturing devices for completing the vesicourethral anastomosis has already been described during open radical prostatectomy *(77)*.

Even more exciting is the possibility of developing a sutureless urethrovesical anastomosis. Grummet et al. *(78)* evaluated laser welding as an alternative method of forming the vesicourethral anastomosis. In a canine model, these authors reported that their technique was feasible and effective in producing a water-tight anastomosis.

8. SUMMARY

Despite initial pessimism in the procedure, laparoscopic radical prosta-tectomy has developed into a technique that has become the standardized procedure of choice at many medical centers worldwide. However, while this has occurred, it is sobering to note that laparoscopic radical prostatectomy provides clear advantages over the open retropubic technique only in the realms of blood loss, transfusion rates, and post-operative analgesic use. Even in these areas, it has not been shown to be superior to a radical perineal prostatectomy. In almost all other areas, it appears to be comparable to both open approaches: continence, potency, cancer cure, and hospital stay. In one area, at least in the USA, it remains presumably problematic: cost. However, definitive information in this regard has yet to be published.

While the performance of a prospective randomized study comparing laparoscopic and open prostatectomy would be the scientific ideal, the reality is that this type of study will likely never be accomplished. As such, the determination of the true place of laparoscopic radical prostatectomy in the surgical pantheon will be based on the completion of appropriate single-technique studies in which validated questionnaires for continence and potency are complemented by sufficient follow-up of similarly clinically staged patients. Hopefully, these data will result from co-operative intra-institutional or inter-institutional efforts in which open and laparoscopic patients proceed through a similar protocol of pre-operative and post-operative questionnaires and follow-up laboratory studies. Until such time, patients are unfortunately being subjected to "new" techniques without the knowledge of whether the approach is actually beneficial, equal, or worse than what already exists. This state of affairs is the result of a health care system in which market share rather than science all too often rules the day; evidence-based medicine has yet to forcefully and properly enter the realm of surgery. Hopefully, at some point in the future, all patients subjected to a "new" procedure will be part of a properly designed multi-institutional study prepared to determine the efficacy of that given procedure, such that the good can be adopted and the bad discarded with the least amount of harm to all concerned. If, as Einstein noted, "God doesn't play dice with the universe," neither should surgeons with their patients.

REFERENCES

1. Greenlee RT, Murray T, Bolden S, Wingo PA. Cancer Statistics 2000. *CA Cancer J Clin* 2000;**50**:7–33.
2. Proust R. Technique de la prostatectomie perineale. *Ass Franc Urol* 1901;**5**:361.
3. Young HH. The early diagnosis and radical cure of carcinoma of the prostate: being a study of 40 cases and presentations of radical operation which was carried out in 4 cases. *Johns Hopkins Hosp Bull* 1905;**16**:315.
4. Millin T. Retropubic prostatectomy: a new extravesicle technique. *Lancet* 1945;**2**:693.
5. Walsh PC, Donker PJ. Impotence following radical prostatectomy: insight into etiology and prevention. *J Urol* 1982;**128**(3):492–7.
6. Walsh PC, Lepor H, Eggleston JC. Radical prostatectomy with preservation of sexual function: anatomical and pathological considerations. *Prostate* 1983;**4**(5):473–85.
7. Clayman RV, Kavoussi LR, Soper NJ, et al. Laparoscopic nephrectomy: initial case report. *J Urol* 1991;**146**:278–82.

8. Winfield HN, Hamilton BD, Bravo EL. Laparoscopic adrenalectomy: the preferred choice? *J Urol* 1998;**160**:325–9.

9. Schuessler WW, Schulam PG, Clayman RV, Vancaille TH. Laparoscopic radical prostatectomy: initial case report. *J Urol* 1992;**147**:246 (Abstract 130).

10. Schuessler WW, Schulam PG, Clayman RV, Kavoussi LR. Laparoscopic radical prostatectomy: initial short-term experience. *Urology* 1997;**50**(6):854–7.

11. Raboy A, Ferzli G, Albert P. Initial experience with extraperitoneal endoscopic radical retropubic prostatectomy. *Urology* 1997;**50**(6):849–53.

12. Raboy A, Albert P, Ferzli G. Early experience with extraperitoneal endoscopic radical retropubic prostatectomy. *Surg Endosc* 1998;**12**(10):1264–7.

13. Abbou C, Antiphon P, Salomon L, et al. Laparoscopic radical prostatectomy: preliminary results. *J Endourol* 1999;**13**:A45.

14. Guillonneau B, Vallancien G. Laparoscopic radical prostatectomy: initial experience and preliminary assessment after 65 operations. *Prostate* 1999;**39**(1):71–5.

15. Guillonneau B, Vallancien G. Laparoscopic radical prostatectomy: the Montsouris experience. *J Urol* 2000;**163**(2):418–22.

16. Abbou CC, Salomon L, Hoznek A, et al. Laparoscopic radical prostatectomy: preliminary results. *Urology* 2000;**55**(5):630–4.

17. Guillonneau B, El-Fettouh H, Baumert H, et al. Laparoscopic radical prostatectomy: Oncological evaluation after 1,000 cases at Montsouris Institute. *J Urol* 2003;**169**:1261–6.

18. Ruiz L, Salomon L, Hoznek A, et al. Comparison of early oncologic results of laparoscopic radical prostatectomy by extraperitoneal versus transperitoneal approach. *Eur Urol* 2004;**46**(1):50–4; discussion 54–6.

19. Breedveld P, Stassen HG, Meijer DW, Stassen LPS. Theoretical background and conceptual solution for depth perception and eye-hand coordination problems in laparoscopic surgery. *Minim Invas Ther Allied Technol* 1999;**8**:227–34.

20. Green PE, Piantanida TA, Hill JW, et al. Telepresence: dexterious procedures in a virtual operating field. *Am Surg* 1991;**57**:192 (Abstract).

21. Shennib H, Bastawisy A, McLoughlin J, Moll F. Robotic computer-assisted telemanipulation enhances coronary artery bypass. *J Thorac Cardiovasc Surg* 1999;**117**:310–3.

22. Cardiere GB, Himpens J, Vertruyen M, Favretti F. The world's first obesity surgery performed by a surgeon at a distance. *Obes Surg* 1999;**9**:206–9.

23. Young JA, Chapman WHR, Kim VB, et al. Robotic-assisted adrenalectomy for adrenal incidentaloma: case and review of the technique. *Surg Laparosc Endosc Percutan Tech* 2002;**12**(2):126–30.

24. Binder J, Kramer W. Robotically-assisted laparoscopic radical prostatectomy. *BJU Int* 2001;**87**(4):408–10.

25. Abbou CC, Hoznek A, Salomon L, et al. Laparoscopic radical prostatectomy with a remote controlled robot. *J Urol* 2001;**165**(6 Pt 1):1964–6.

26. Pasticier G, Rietbergen JB, Guillonneau B, Fromont G, Menon M, Vallancien G. Robotically assisted laparoscopic radical prostatectomy: feasibility study in men. *Eur Urol* 2001;**40**(1):70–4.

27. Menon M, Shrivastava A, Tewari A, et al. Laparoscopic and robot assisted radical prostatectomy: establishment of a structured program and preliminary analysis of outcomes. *J Urol* 2002;**168**(3):945–9.

28. Menon M, Shrivastava A, Tewari A. Laparoscopic radical prostatectomy: conventional and robotic. *Urology* 2005;**66**(5 Suppl):101–4.

29. Lepor H, Nieder AM, Ferrandino MN. Intraoperative and postoperative complications of radical retropubic prostatectomy in a consecutive series of 1,000 cases. *J Urol* 2001;**166**(5):1729–33.

30. Guillonneau B, Vallancien G. Laparoscopic radical prostatectomy: the Montsouris technique. *J Urol* 2000;**163**(6):1643–9.
31. Rassweiler J, Sentker L, Seemann O, Hatzinger M, Rumpelt HJ. Laparoscopic radical prostatectomy with the Heilbronn technique: an analysis of the first 180 cases. *J Urol* 2001;**166**(6):2101–8.
32. Rassweiler J, Seemann O, Hatzinger M, Schulze M, Frede T. Technical evolution of laparoscopic radical prostatectomy after 450 cases. *J Endourol* 2003;**17**(3):143–54.
33. Bollens R, Vanden Bossche M, Roumeguere T, et al. Extraperitoneal laparoscopic radical prostatectomy. Results after 50 cases. *Eur Urol* 2001;**40**(1):65–9.
34. Stolzenburg JU, Do M, Pfeiffer H, Konig F, Aedtner B, Dorschner W. The endoscopic extraperitoneal radical prostatectomy (EERPE): technique and initial experience. *World J Urol* 2002;**20**(1):48–55.
35. Stolzenburg JU, Do M, Rabenalt R, et al. Endoscopic extraperitoneal radical prostatectomy: initial experience after 70 procedures. *J Urol* 2003;**169**(6):2066–71.
36. Eden CG, King D, Kooiman GG, Adams TH, Sullivan ME, Vass JA. Transperitoneal or extraperitoneal laparoscopic radical prostatectomy: does the approach matter? *J Urol* 2004;**172**(6 Pt 1):2218–23.
37. Hoznek A, Antiphon P, Borkowski T, et al. Assessment of surgical technique and perioperative morbidity associated with extraperitoneal versus transperitoneal laparoscopic radical prostatectomy. *Urology* 2003;**61**(3):617–22.
38. Cathelineau X, Cahill D, Widmer H, Rozet F, Baumert H, Vallancien G. Transperitoneal or extraperitoneal approach for laparoscopic radical prostatectomy: a false debate over a real challenge. *J Urol* 2004;**171**(2 Pt 1):714–6.
39. Erdogru T, Teber D, Frede T, et al. Comparison of transperitoneal and extraperitoneal laparoscopic radical prostatectomy using match-pair analysis. *Eur Urol* 2004;**46**(3):312–9; discussion 320.
40. Tewari A, Peabody J, Sarle R, et al. Technique of da vinci robot-assisted anatomic radical prostatectomy. *Urology* 2002;**60**(4):569.
41. Perer E, Lee DI, Ahlering TE, Clayman RV. Robotic revelation: laparoscopic radical prostatectomy by a non-laparoscopic surgeon. *J Am Coll Surg* 2003; **197**(4):693–696.
42. Antiphon P, Hoznek A, Gettman M, et al. Extraperitoneal laparoscopic robot assisted radical prostatectomy. *J Urol* 2003;**169**(4):Abstract V965.
43. Dakwar G, Ahmed M, Sawczuk I, Rosen J, Lanteri V, Esposito M. Extraperitoneal robotic prostatectomy: comparison of technique and results at one institution. *J Urol* 2003;**169**(4):Abstract 1660.
44. Ahlering TE, Skarecky DW, Lee DI, Clayman RV. Successful transfer of open surgical skills to a laparoscopic environment using a robotic interface: initial experience with laparoscopic radical prostatectomy. *J Urol* 2003;**169**(4):Abstract 1649.
45. Hoznek A, Salomon L, Olsson LE, et al. Laparoscopic radical prostatectomy. The Creteil experience. *Eur Urol* 2001;**40**(1):38–45.
46. Guillonneau B, Cathelineau X, Doublet J, Baumert H, Vallancien G. Laparoscopic radical prostatectomy: assessment after 550 procedures. *Crit Rev Oncol Hematol* 2002;**43**(2):123.
47. Salomon L, Levrel O, Anastasiadis A, et al. Outcome and complications of radical prostatectomy in patients with PSA < 10 ng/ml: comparison between the retropubic, perineal and laparoscopic approach. *Prostate Cancer Prostatic Dis* 2002;**5**:285–90.
48. Guillonneau B, Rozet F, Cathelineau X, et al. Perioperative complications of laparoscopic radical prostatectomy: the Montsouris 3-year experience. *J Urol* 2002;**167**(1):51–6.
49. Menon M, Tewari A, Peabody J. Vattikuti Institute prostatectomy: technique. *J Urol* 2003;**169**(6):2289–92.
50. Rassweiler J, Seemann O, Schulze M, Teber D, Hatzinger M, Frede T. Laparoscopic versus open radical prostatectomy: a comparative study at a single institution. *J Urol* 2003;**169**(5):1689–93.

51. Bhayani SB, Pavlovich CP, Hsu TS, Sullivan W, Su LM. Prospective comparison of short-term convalescence: laparoscopic radical prostatectomy versus open radical retropubic prostatectomy. *Urology* 2003;**61**(3):612–6.
52. Menon M, Tewari A, Baize B, Guillonneau B, Vallancien G. Prospective comparison of radical retropubic prostatectomy and robot-assisted anatomic prostatectomy: the Vattikuti Urology Institute experience. *Urology* 2002;**60**(5):864–8.
53. Schulam PG, Link RE. Laparoscopic radical prostatectomy. *World J Urol* 2000;**18**(4): 278–82.
54. Zippe C, Meraney AM, Sung GT, Gill I. Laparoscopic radical prostatectomy in the USA. Cleveland Clinic series of 50 patients. *J Urol* 2001;**165**:1341A.
55. Sundaram CP, Landman J, Rehman J, Clayman RV. Complications and technical difficulties during initial experience with laparoscopic radical retropubic prostatectomy. *J Urol* 2001;**165**:1445A.
56. Salomon L, Anastasiadis A, Katz R, et al. Urinary continence and erectile function: a prospective evaluation of functional results after radical laparoscopic prostatectomy. *Eur Urol* 2002;**42**(4):338.
57. Babaian RJ, Troncoso P, Bhadkamkar VA, Johnston DA. Analysis of clinicopathologic factors predicting outcome after radical prostatectomy. *Cancer* 2001;**91**(8):1414–22.
58. Guillonneau B, Cathelineau X, Doublet JD, Vallancien G. Laparoscopic radical prostatectomy: the lessons learned. *J Endourol* 2001;**15**(4):441–5; discussion 447–8.
59. Anastasiadis AG, Salomon L, Katz R, Hoznek A, Chopin D, Abbou CC. Radical retropubic versus laparoscopic prostatectomy: a prospective comparison of functional outcome. *Urology* 2003;**62**(2):292–7.
60. Rozet F, Galiano M, Cathelineau X, Barret E, Cathala N, Vallancien G. Extraperitoneal laparoscopic radical prostatectomy: a prospective evaluation of 600 cases. *J Urol* 2005;**174**(3):908–11.
61. Anderson JK, Murdock A, Cadeddu JA, Lotan Y. Cost comparison of laparoscopic versus radical retropubic prostatectomy. *Urology* 2005;**66**(3):557–60.
62. Lotan Y, Cadeddu JA, Gettman MT. The new economics of radical prostatectomy: cost comparison of open, laparoscopic and robot assisted techniques. *J Urol* 2004;**172**(4 Pt 1):1431–5.
63. Scales CD, Jr., Jones PJ, Eisenstein EL, Preminger GM, Albala DM. Local cost structures and the economics of robot assisted radical prostatectomy. *J Urol* 2005;**174**(6):2323–9.
64. Rassweiler J, Schulze M, Teber D, et al. Laparoscopic radical prostatectomy with the Heilbronn technique: oncological results in the first 500 patients. *J Urol* 2005;**173**(3): 761–4.
65. Salomon L, Levrel O, de la Taille A, et al. Radical prostatectomy by the retropubic, perineal and laparoscopic approach: 12 years of experience in one center. *Eur Urol* 2002;**42**(2):104.
66. Menon M, Tewari A, Peabody JO, et al. Vattikuti Institute prostatectomy, a technique of robotic radical prostatectomy for management of localized carcinoma of the prostate: experience of over 1100 cases. *Urol Clin North Am* 2004;**31**(4):701–17.
67. Ojdeby G, Claezon A, Brekkan E, Haggman M, Norlen BJ. Urinary incontinence and sexual impotence after radical prostatectomy. *Scand J Urol Nephrol* 1996;**30**(6):473–7.
68. Walsh PC, Marschke P, Ricker D, Burnett AL. Patient-reported urinary continence and sexual function after anatomic radical prostatectomy. *Urology* 2000;**55**(1):58–61.
69. Catalona W, Carvalhal G, Mager DE, Smith DS. Potency, continence and complication rates in 1,870 consecutive radical retropubic prostatectomies. *J Urol* 1999;**162**:433–8.
70. Partin AW, Pound CR, Clemens JQ, Epstein JI, Walsh PC. Serum PSA after anatomic radical prostatectomy. The Johns Hopkins experience after 10 years. *Urol Clin North Am* 1993;**20**(4):713–25.

71. Ung J, Richie J, Chen M, Renshaw A, D'Amico A. Evolution of the presentation and pathologic and biochemical outcomes after radical prostatectomy for patients with clinically localized prostate cancer diagnosed during the PSA era. *Urology* 2002;**60**(3):458.

72. Puri K, Ruiz-Deya G, Thomas R, Davis R. Radical prostatectomy: retropubic versus laparoscopic versus perineal. *J Urol* 2003; **169**(4):Abstract 1658.

73. Ong AM, Su LM, Varkarakis I, et al. Nerve sparing radical prostatectomy: effects of hemostatic energy sources on the recovery of cavernous nerve function in a canine model. *J Urol* 2004;**172**:1318–22.

74. Eichel L, Chou D, Skarecky DW, Ahlering TE. Feasibility study for laparoscopic radical prostatectomy cautery free neurovascular bundle preservation. *Urology* May 2005;**65**(5):994–7.

75. Gill IS, Osamu U. Mauricio R, et al. Lateral pedicle control during laparoscopic radical prostatectomy: refined technique. *Urology* 2005;**65**:23–27.

76. Ahlering TE, Eichel L, Skarecky DW. Early potency with cautery free neurovascular bundle preservation study in robotic laparoscopic radical prostatectomy. *J. Endourology* 2005 **19**(6):715–8.

77. Yamada Y, Honda N, Nakamura K, et al. New semiautomatic suturing device (Maniceps) for precise vesicourethral anastomosis during radical retropubic prostatectomy. *Int J Urol* 2002;**9**(1):71–2.

78. Grummet JP, Costello AJ, Swanson DA, Stephens LC, Cromeens DM. Laser welded vesicourethral anastomosis in an in vivo canine model: a pilot study. *J Urol* 2002;**168**:281–4.

79. van Velthoven R, Peltier A, Hawaux E, Vandewalle J. Transperitoneal laparoscopic anatomic radical prostatectomy, preliminary results. *J Urol* 2000;**163**(4):141 (Abstract 621).

80. Roumeguere T, Bollens R, Vanden Bossche M, et al. Radical prostatectomy: a prospective comparison of oncological and functional results between open and laparoscopic approaches. *World J Urol* 2003;**20**(6):360–6.

81. Tuerk IA, Fabrizio MD, Deger S, et al. Laparoscopic radical prostatectomy – the combined experience from Berlin and Norfolk with 308 patients. *J Urol* 2002;**164**(4):341 (Abstract).

82. Hara I, Kawabata G, Miyake H, et al. Feasibility and usefulness of laparoscopic radical prostatectomy: Kobe University experience. *Int J Urol* 2002;**9**:635–40.

83. Dahl DM, L'Esperance JO, Trainer AF, et al. Laparoscopic radical prostatectomy: initial 70 cases at a U.S. university medical center. *Urology* 2002;**60**(5):859–63.

84. Eden CG, Cahill D, Vass JA, Adams TH, Dauleh MI. Laparoscopic radical prostatectomy: the initial UK series. *BJU Int* 2002;**90**:876–82.

85. Guillonneau B, El-Fettouh H, Baumert H, et al. Laparoscopic radical prostatectomy: oncological evaluation after 1,000 cases at Montsouris Institute. *J Urol* 2003;**169**:1261–66.

86. Rassweiler J, Binder J, Frede T. Robotic and telesurgery: will they change our future? *Curr Opin Urol* 2001;**11**:309–20.

87. Tewari A, Dasari R, Hasan M, et al. Sexual function following robotic prostatectomy: a study using expanded prostate cancer index composite (EPIC) quality of life instrument. *J Urol* 2003;**169**(4):Abstract 1615.

88. Menon M, Shrivastava A, Sarle R, Hemal A, Tewari A. Vattikuti Institute prostatectomy: a single-team experience of 100 cases. *J Endourol* 2003;**17**:785–90.

89. Bhandari A, McIntire L, Kaul SA, Hemal AK, Peabody JO, Menon M. Perioperative complications of robotic radical prostatectomy after the learning curve. *J Urol* 2005;**174**(3):915–8.

90. Patel VR, Tully AS, Holmes R, Lindsay J. Robotic radical prostatectomy in the community setting–the learning curve and beyond: initial 200 cases. *J Urol* 2005;**174**(1):269–72.

91. Ahlering TE, Woo D, Eichel L, Lee DI, Edwards R, Skarecky DW. Robot-assisted versus open prostatectomy: a comparison of one surgeon's outcomes. *Urology* 2004; **63**(5):819–822.

92. Kandu SD, Roehl K, Eggener SE, Antenor JA, Han M, Catalona W. Potency, continence and complications in 3,477 consecutive radical retropubic prostatectomies. *J Urol* 2004;**172**:2227–31.

93. Zincke H, Oesterling JE, Blute ML, Bergstralh EJ, Myers RP, Barrett DM. Long-term (15 years) results after radical prostatectomy for clinically localized (stage T2c or lower) prostate cancer. *J Urol* 1994;**152**(5 Pt 2):1850–7.

94. Hull GW, Rabbani F, Abbas F, Wheeler TM, Kattan MW, Scardino PT. Cancer control with radical prostatectomy alone in 1,000 consecutive patients. *J Urol* 2002;**167**(2 Pt 1):528–34.

95. Dillioglugil O, Leibman BD, Leibman NS, Kattan MW, Rosas AL, Scardino PT. Risk factors for complications and morbidity after radical retropubic prostatectomy. *J Urol* 1997;**157**(5):1760–7.

96. Rabbani F, Stapleton AM, Kattan MW, Wheeler TM, Scardino PT. Factors predicting recovery of erections after radical prostatectomy. *J Urol* 2000;**164**(6):1929–34.

97. Eastham JA, Kattan MW, Rogers E, et al. Risk factors for urinary incontinence after radical prostatectomy. *J Urol* 1996;**156**(5):1707–13.

98. Goluboff ET, Saidi JA, Mazer S, et al. Urinary continence after radical prostatectomy: the Columbia experience. *J Urol* 1998;**159**(4):1276–80.

99. Parra RO. Analysis of an experience with 500 radical perineal prostatectomies in localized prostate cancer. *J Urol* 2000;**163**(4):Abstract 1265.

100. Lance RS, Freidrichs PA, Kane C, et al. A comparison of radical retropubic with perineal prostatectomy for localized prostate cancer within the Uniformed Services Urology Research Group. *BJU Int* 2001;**87**(1):61–5.

101. Gillitzer R, Melchior SW, Hampel C, Wiesner C, Fichtner J, ThUroff JW. Specific complications of radical perineal prostatectomy: a single institution study of more than 600 cases. *J Urol* 2004;**172**(1):124–8.

102. Weldon VE, Tavel FR, Neuwirth H. Continence, potency and morbidity after radical perineal prostatectomy. *J Urol* 1997;**158**(4):1470–5.

103. Iselin CE, Robertson JE, Paulson DF. Radical perineal prostatectomy: oncological outcome during a 20-year period. *J Urol* 1999;**161**(1):163–8.

104. Ruiz-Deya G, Davis R, Srivastav SK, M Wise A, Thomas R. Outpatient radical prostatectomy: impact of standard perineal approach on patient outcome. *J Urol* 2001;**166**(2):581–6.

105. Harris MJ. Radical perineal prostatectomy: cost efficient, outcome effective, minimally invasive prostate cancer management. *Eur Urol* 2003;**44**(3):303–8; discussion 308.

18 Targeted Therapy Trials for Prostate Cancer

Elisabeth I. Heath, MD,
and Michael A. Carducci, MD

CONTENTS

1. INTRODUCTION

There are over 200 novel agents currently under evaluation in clinical trials for the treatment of prostate cancer. Drug development in prostate cancer continues to improve as a better understanding of prostate cancer biology is achieved, providing further insight into the complex extracellular and intracellular-signaling networks as potential targets. As current treatment options for men with advanced disease are not curative, there is a growing need for additional agents to be developed for men with progressive prostate cancer.

From: *Current Clinical Oncology: Prostate Cancer:*
Signaling Networks, Genetics, and New Treatment Strategies
Edited by: R. G. Pestell and M. T. Nevalainen © Humana Press, Totowa, NJ

This chapter will discuss current clinical trials of targeted agents in prostate cancer with specific focus on agents that modulate growth factors and their receptors, signal transduction pathways, cell survival pathways, and angiogenesis pathways. Several novel agents evaluated in combination clinical trials with cytotoxic agents briefly discussed in this chapter will be addressed elsewhere in more detail.

2. GROWTH FACTOR AND GROWTH FACTOR RECEPTOR INHIBITORS

Multiple growth factors and growth factor receptors have been identified as critical proteins in the prostate cancer-signaling network. Novel agents currently in clinical trials are designed to target specific protein families such as the epidermal growth factor receptor (EGFR) family, platelet-derived growth factor (PDGF) receptor family, and endothelin (ET) receptor family. Although the agents discussed are identified as specific growth factor inhibitors, it is well recognized that many of the agents are quite promiscuous in their target selection. The "one inhibitory agent-one receptor family" concept is not applicable for several targeted agents as there is a significant amount of cross-talk between signaling pathways, especially between the EGFR family and the vascular endothelial growth factor (VEGF) receptor family. In this chapter, novel agents that modulate the VEGF receptor family will be discussed with other agents that target angiogenesis.

2.1. Epidermal Growth Factor Receptor Inhibitors

The EGFR family has been recognized as a very important family of proteins, one which frequently impacts the cellular network within many different solid tumors including lung cancer, breast cancer, colon cancer, and prostate cancer *(1)*. Novel agents, including intravenous monoclonal antibodies, targeting the extracellular domain of the EGFR and oral, small molecules targeting the intracellular tyrosine kinase are in various stages of clinical trials. Although there are four growth receptor proteins in the EGFR family, most clinically available agents primarily target EGFR (HER-1 and c-erbB-1) and HER-2 (c-erbB-2) but not HER-3 (c-erbB-3) and HER-4 (c-erbB-4). Targeting EGFR and HER-2 is reasonable in prostate cancer as 40–90% of prostate cancer cells overexpress EGFR and up to 50% of prostate cancer cells overexpress HER-2 *(2,3)*.

2.1.1. MONOCLONAL ANTIBODIES

Monoclonal antibodies that target the extracellular domain of the EGFR may directly modulate the receptor itself or behave as competitors of the natural cellular ligands, such as the epidermal growth factor, transforming growth factor-α, and amphiregulin. Modulation of the extracellular domain of the EGFR ultimately results in disruption of intracellular signaling by affecting the Ras–Raf mitogen-activated protein kinase pathway and the phosphatidyl inositol 3′-kinase and Akt pathway (4,5).

Erbitux (Imclone), a monoclonal antibody initially FDA approved for colorectal cancer in 2004, directly affects the extracellular domain of the EGFR. Single-agent Erbitux (Imclone) has not been specifically studied in prostate cancer patients, but the combination of Erbitux (Imclone) and doxorubicin has been evaluated in a phase I/II trial in men with hormone-insensitive prostate cancer (6). One patient experienced a greater than 50% decline in serum PSA.

ABX-EGF (Abgenix) is a fully humanized monoclonal antibody that is slightly different from Erbitux (Imclone) in that it blocks ligand binding to EGFR (7). With promising preclinical observations of anti-tumor activity in prostate cancer cell lines and xenografts, a phase II trial of ABX-EGF (Abgenix) is under way in patients with hormone-insensitive prostate cancer (8,9). Another monoclonal antibody, EMD 72000 (EMD Pharmaceuticals), also targets the extracellular portion of the EGFR. To date, no prostate cancer-specific trials have been reported or published with this agent.

Agents targeting the HER-2 receptor have made rapid clinical advancements in other solid tumors. Herceptin (Genentech), FDA approved for metastatic breast cancer since 1998, has been evaluated in men with metastatic hormone-insensitive prostate cancer in a single-agent phase II trial (10). There were no partial or complete responders in eighteen patients. Two patients experienced stable disease on maintenance doses of 4 mg/kg i.v. weekly.

Combination studies with Herceptin (Genentech) and cytotoxic chemotherapy have been challenging to design and conduct. Morris et al. planned on conducting an elegantly designed four-arm trial in which patients with androgen-sensitive, androgen-insensitive, HER-2-positive, or HER-2-negative disease were all treated initially with Herceptin (Genentech) until progression, at which time paclitaxel a dose of 100 mg/m^2 intravenously was added (11). Of the 130 patients screened for the HER-2 status, only six eligible patients had HER-2-positive disease. Of the 15 prostate cancer patients who received Herceptin (Genentech) in combination with paclitaxel, three patients experienced a greater than 50% decline in their PSA levels. Similar difficulties were encountered with a phase II trial of Herceptin (Genentech) and docetaxel (12,13). HER-2 overexpression was found in only 20% of prostate cancer patients screened for the trial, and the trial eventually closed because of nonfeasibility.

Treatment of hormone-insensitive prostate cancer with single-agent Herceptin (Genentech) did not show any significant activity toward lowering of the serum PSA (10). The potential for synergy by adding Herceptin (Genentech) to a known effective cytotoxic regimen was determined in a phase I trial of Herceptin (Genentech), docetaxel, and estramustine in men with hormone-insensitive prostate cancer *(14)*. Indeed, greater than 69% of patients experienced a greater than 50% PSA decline. However, it is unclear as to how much benefit Herceptin (Genentech) truly added to this already active regimen.

Neither targeting EGFR nor HER-2 with novel agents either alone or in combination with cytotoxic chemotherapy has translated into significant reductions of the serum PSA levels in men with advanced prostate cancer. Concerns for lack of "activity" led to the development of a more broad-spectrum-targeting agent. Omnitarg (Genentech) is a humanized monoclonal antibody that inhibits ligand-associated heterodimerization of HER-2 with other EGFR family members *(15)*. A phase II study of Omnitarg (Genentech) is currently under way in patients with advanced prostate cancer utilizing the dose determined from the phase I trial conducted in 21 patients with advanced solid tumors *(16)*.

2.1.2. SMALL MOLECULE INHIBITORS

Clinical development of small molecule inhibitors of the EGFR family in patients with prostate cancer has been equally challenging as the clinical development of monoclonal antibody inhibitors. Small molecule inhibitors are oral agents that inhibit the binding of ATP to the tyrosine kinase domain of the receptor. Gefitinib (Astra Zeneca, Wilmington, DE, USA) and Erlotinib (Genentech) are two agents that obtained FDA approval for lung cancer in 2003 and 2004, respectively. Current status of Gefitinib (Astra Zeneca) is tenuous due to a largely negative confirmatory phase III trial in lung cancer. Erlotinib (Genentech) remains a viable treatment option for patients with advanced lung cancer.

Defining the clinical activity of Gefitinib (Astra Zeneca) in patients with prostate cancer has been as challenging as the Herceptin (Genentech) clinical trial experience. Two randomized phase II clinical trials with Gefitinib (Astra Zeneca) administered to patients with metastatic hormone-insensitive prostate cancer at two different dose levels did not result in any PSA response or any disease response *(17,18)*. Based on encouraging preclinical data suggesting the potential for synergistic activity, a pilot phase I trial of Gefitinib (Astra Zeneca) with docetaxel and estramustine was conducted in 30 patients with metastatic hormone-insensitive prostate cancer *(19)*. Indeed, 9 of 30 patients experienced a PSA response, but again, the contribution of Gefitinib (Astra Zeneca) to the already active regimen of docetaxel and estramustine was unclear. To date, larger confirmatory trials of this combination have not been initiated.

There are additional oral small molecule inhibitors in development. Erlotinib (Genentech), currently FDA approved in 2004 for lung cancer, has recently been studied in patients with advanced prostate cancer. Gravis et al. reported the results of an ongoing phase II study of Erlotinib (Genentech,

San Francisco, CA, USA) which was administered at a starting dose of 150 mg/day for 3 weeks with a dose escalation to 200 mg/day at week 4 in patients with metastatic hormone-insensitive prostate cancer *(20)*. One patient experienced disease stabilization with maintenance dose of 200 mg/day, but no PSA responses were noted.

Additional agents in development include PKI 166 (Novartis), GW572016 (GlaxoWelcome), EKB569 (Wyeth), BIBW 2992 (Boehringer Ingelheim), and CI-1033 (Pfizer, NY, NY, USA) *(21–25)*. GW572016 (GlaxoWelcome) is currently being evaluated in a phase II clinical trial in prostate cancer patients with rising PSA after local therapy (E5803).

The current potential for monoclonal antibodies and small molecule inhibitors that impact the EGFR family remains unclear. Few trials reported any activity toward prostate cancer, and when they did, it was not clear whether the activity was due to the effects of cytotoxic chemotherapy or the combination of a targeted agent with cytotoxic chemotherapy. This challenge will become a familiar theme in this chapter as the role of targeted agents as single agents or in combination with cytotoxic chemotherapy continues to evolve.

2.2. Platelet-Derived Growth Factor Inhibitors

The role of the PDGF family in prostate cancer has been actively investigated in preclinical studies. High levels of PDGF overexpression in primary and metastatic prostate cancers and its role in tumor growth and bone metastases have been reported *(26,27)*. Two novel agents specifically targeting the PDGF receptor include Gleevec (Novartis) and SU101 (Pfizer). Gleevec (Novartis) was FDA approved in 2001 for treatment of patients with refractory chronic myeloid leukemia. A phase II trial of single-agent Gleevec (Novartis) in patients with hormone-sensitive, PSA relapse prostate cancer resulted in minimal drug activity but considerable drug toxicity *(28)*. Of the 16 patients treated on this trial, no patient experienced a greater than 50% decrease in PSA. Another phase II trial of Gleevec (Novartis) was conducted in patients with metastatic, hormone-insensitive prostate cancer with evidence of biochemical relapse *(29)*. Unfortunately, none of the patients experienced a greater than 50% decline in PSA level. In addition, significant grade 3/4 toxicities including fatigue, rash, arthralgias, and peripheral edema resulted in dose delays and/or discontinuation in 50% of patients.

Hoping to achieve synergistic activity, Gleevec (Novartis) in combination with additional agents was evaluated *(30)*. Unfortunately, a combination trial of Gleevec (Novartis) and zoledronic acid resulted in no PSA response *(31)*. However, a phase I trial of Gleevec (Novartis) in combination with docetaxel did result in some patients experiencing PSA decreases, and this experience eventually led to an ongoing Phase II trial of Gleevec (Novartis) and docetaxel *(32)*.

Single-agent SU101 (Pfizer) did not result in any meaningful reduction in the serum PSA of 39 patients with metastatic hormone-insensitive prostate cancer *(33)*. The lack of meaningful clinical results for both PDGF inhibitors as single agents may not necessarily be surprising but is certainly disappointing. Further preclinical studies to improve our understanding and help define a more appropriate role of PDGF inhibitors in the treatment of prostate cancer may be necessary.

2.3. Endothelin Receptor Inhibitors

The family of ET receptors (consisting of three ET ligands and two receptors) is important in autocrine and/or paracrine growth signaling, signal transduction, nociception, and osteogenesis. Oral ET receptor inhibitors include Atrasentan (Abbott Park, IL, USA), Bosentan (already FDA approved for pulmonary hypertension), and ZD4054 (Astra Zeneca) *(34–36)*. Atrasentan (Abbott), a highly potent and selective antagonist to the ET_A receptor has moved forward into phase III clinical trials based on encouraging results from phase II studies in men with prostate cancer *(37,38)*. One such encouraging phase II trial that evaluated Atrasentan (Abbott) versus placebo in the treatment of patients with asymptomatic metastatic hormone-insensitive prostate cancer reported an increase in the time to PSA progression of 155 days with Atrasentan (Abbott) versus 71 days with placebo was reported $(p = 0.0002)$ *(37)*. However, this result was not achieved in the intent-to-treat phase III study of Atrasentan (Abbott) versus placebo in patients with metastatic hormone-insensitive prostate cancer (M00-211). Although no significant difference in time to progression between Atrasentan (Abbott) versus placebo in the M00-211 trial was reported, Atrasentan (Abbott) did provide a significant delay in time to onset of metastatic pain events *(38)*. Additional trial results of Atrasentan (Abbott) in patients with non-metastatic, hormone-insensitive prostate cancer (M00-244) are still pending.

A phase I/II study of Atrasentan (Abbott) and docetaxel in men with metastatic hormone-insensitive prostate cancer is currently ongoing *(39)*. A combination phase III trial of Atrasentan (Abbott) with docetaxel and prednisone versus docetaxel and prednisone alone is underway in the Southwest Oncology Group (SWOG 0421). Although Atrasentan (Abbott) as a single agent was not able to show a difference in time to progression compared to placebo in a large, phase III trial (M00-211), this may yet be achieved with Atrasentan (Abbott) administered in combination with cytotoxic chemotherapy.

2.4. Insulin-Like Growth Factor Receptor Inhibitors

The insulin-like growth factor receptor (IGFR) family of receptors is emerging as a potentially important pathway in prostate tumorigenesis *(40)*. Two ligands, IGF-1 and IGF-2, have been shown to be overexpressed in higher grade prostate tumors compared to lower grade tumors *(40)*. Downstream-signaling effects of the mitogen-activated protein kinase pathway and the phosphatidyl inositol 3´-kinase and Akt pathway appears to be dependant on the stage of prostate cancer progression (benign versus malignant) *(41)*. There are also preclinical studies to suggest that successful modulation of the IGFR axis may not depend on androgen receptor status *(42,43)*. High serum levels of IGF-binding protein-3 may also be associated with an increased risk of prostate cancer *(44)*.

Novel agents targeting the IGFR axis are in active development. IMC-A12 (Imclone), a monoclonal antibody targeting the IGF-1R, is currently in phase I clinical trials in patients with advanced solid tumors. Another IGFR inhibitor undergoing preclinical testing includes NVP-ADW742 (Novartis) *(45)*. Silibinin, a naturally occurring flavonoid which targets IGFR axis along with EGFR axis, is currently in early clinical trials in men with prostate cancer *(46)*. As our knowledge base grows with respect to the IGFR pathway, additional novel agents targeting the IGFR axis will most likely enter clinical trials, perhaps for consideration of both treatment and prevention of prostate cancer.

Additional promising signaling pathways emerging in prostate cancer not discussed in this chapter include transforming growth factor-beta receptor family, keratinocyte growth factor family, and fibroblast growth factor family *(47,48)*.

3. SIGNAL TRANSDUCTION PATHWAY INHIBITORS

The signal transduction pathways within a prostate cancer cell are extremely complex. As we gain a better understanding of the various pathways and their relationships to one another, the more challenging it becomes to rationally develop new agents. Inhibition of one particular point within a signal transduction pathway may indeed affect multiple other parallel or downstream pathways. Novel agents targeting the farnesyl protein transferase, raf kinase, rapamycin kinase, mammalian target of rapamycin, and heat shock proteins have been developed and are undergoing current evaluation in clinical trials. Development of relevant biomarkers that truly assess drug activity is very difficult with this group of targeted agents as there are many different proteins in the signaling pathways that may be important in advancing our understanding of the drug's mechanism of action.

Moreover, the development of novel biomarkers other than PSA to determine efficacy is necessary to advance drug development in prostate cancer. A novel agent that does not modulate the serum PSA level will have difficulty in moving forward in clinical trials as both patients and physicians will not be excited in pursuing the treatment.

3.1. Farnesyl Protein Transferase Inhibitors

The enzyme farnesyl protein transferase catalyzes a step in the post-translational addition of an isoprenoid side chain at the carboxyl terminus of a number of important proteins, including Ras protein. Thus, inhibitors of farnesyl protein transferase become inhibitors of the Ras protein. Current farnesyl protein transferase inhibitors (FTI) include L-778,123 (Merck, Whitehouse Station, NJ, USA), Zarnestra (Johnson and Johnson, New Brunswick, NJ, USA), SCH-66336 (Schering Plough, Kenilworth, NJ, USA), and BMS-214662 (Bristol Myers Squibb, NY, NY, USA) *(49)*. To date, development of L-778,123 (Merck) was discontinued because of prolonged QT toxicity. Zarnestra (Johnson and Johnson) has been evaluated in a phase II trial in men with metastatic hormone-insensitive prostate cancer *(50)*. Unfortunately, none of the patients achieved a PSA response. Combination therapy trials with cytotoxic agents are under way in other solid tumors but not specifically in prostate cancer. Similar to the other single-agent data discussed with the growth factor receptor inhibitors, single-agent treatment of a farnesyl transferase inhibitor did not result in PSA activity.

3.2. Raf Kinase Inhibitor

BAY 43-9006 (Bayer) is an oral, small molecular tyrosine kinase inhibitor of c-Raf-1 and B-raf amongst other proteins. The Ras-Raf kinase cascade plays an integral part in signal transduction. BAY 43-9006, recently FDA approved in 2005 for renal cell carcinoma, has been evaluated in clinical trials in patients with prostate cancer. Encouraging preclinical data suggested a reasonable rationale for evaluating BAY 43-9006 in prostate cancer. Steinbild et al. recently reported the results of a phase II study of BAY 43-9006 in patients with metastatic hormone-insensitive prostate cancer who are chemonaive *(51)*. Of 55 patients, two patients achieved a PSA response.

Another phase II trial of BAY 43-9006 in patients with metastatic hormone-insensitive prostate cancer with 60% of patients receiving prior chemotherapy treatment reported no PSA response *(52)*. Of interest, two patients in this study experienced a dramatic improvement in their bone scans but were considered to have progressed on treatment based on their rising PSA levels. Again, similar to other single-agent trials, BAY 43-9006 administered as a single-targeted agent did not effectively modulate serum PSA.

3.3. Rapamycin Kinase Inhibitors (MTOR Pathway)

The phosphatidyl inositol 3′-kinase and Akt pathway is another critical pathway in signal transduction. PTEN (phosphatase and tensin homolog deleted on chromosome 10) further regulates the signaling pathway as inactivation of PTEN has been associated with prostate cancer tumorigenesis. The mammalian target of Rapamycin (Wyeth), MTOR, that is further downstream is another critical protein in the pathway *(53)*. Recently, an inhibitor of MTOR has been shown to be effective in renal cell carcinoma when compared to interferon *(54)*. Four MTOR inhibitors currently available in clinical trials include Rapamycin (Wyeth), CCI779 (Wyeth), RAD001 (Novartis), and AP-23573 (Ariad Pharmaceuticals, Cambridge, MA, USA) with only Rapamycin (Wyeth) having achieved FDA approval in 1999 for prophylaxis of organ rejection in renal transplant patients *(55)*. Rapamycin (Wyeth) has been investigated in patients with hormone-insensitive prostate cancer with or without evidence of metastases *(56)*. Of 11 patients enrolled, two patients experienced a change in their PSA velocity.

Preclinical data with CCI779 (Wyeth) has shown evidence of activity in prostate cancer *(53,57)*. A clinical trial of CCI779 (Wyeth) administered in a neoadjuvant setting in a prostate cancer trial has been completed. Prior to undergoing prostatectomy, patients were assigned to one of three different dose levels of CCI779 (Wyeth) administered for 4 weeks *(58)*. Inhibition of MTOR pathway was demonstrated in prostate cancer tissue in a dose-dependent manner. Additional prostate cancer clinical trials with this class of drugs will most likely be completed within the next year.

3.4. Multiple Targeted Pathways: Heat Shock Protein 90 Inhibitor

Heat shock protein 90 (Hsp90) is one of the most abundant and versatile cellular chaperone proteins. One of the many proteins in the heat shock protein family, Hsp90, acts as a molecular chaperone for various other client proteins including those critical in signal transduction *(59)*. 17-allylamino-17- demethoxygeldanamycin (17-AAG) (Kosan) and 17-dimethyl aminoethylamino-17-demethoxygeldanamycin (17DMAG) (Kosan) are novel intravenous agents that bind to Hsp90, thereby disrupting Hsp90's function as a cellular chaperone to multiple other cellular proteins *(60)*. A phase II trial of 17-AAG was conducted in patients with metastatic hormone-insensitive prostate cancer who have been previously treated with at least one prior chemotherapy *(61)*. The trial has been completed and is undergoing data analysis. DMAG has been evaluated in a phase I trial and may be pursued in prostate cancer.

4. CELL SURVIVAL PATHWAYS

4.1. Proteasome Inhibitor

Cellular proteasome, a large protein unit that degrades ubiquinated proteins, has a very important role in cell survival, cell-cycle regulation, and apoptosis. In prostate cancer xenografts, administration of a proteasome inhibitor resulted in modulation of apoptosis and angiogenesis *(62)*. Velcade (Millenium, Cambridge, MA, USA) is a proteasome inhibitor currently FDA approved as of 2003 for the treatment of patients with refractory multiple myeloma. Velcade

(Millenium) was evaluated in a phase I trial of mostly patients with metastatic hormone-insensitive prostate cancer *(63)*. Two of the 24 patients with prostate cancer achieved a greater than 50% decline in serum PSA. A combination phase I trial of Velcade (Millenium) in combination with docetaxel was conducted *(64)*. Of the two patients with prostate cancer, one achieved stable disease by PSA. Additional trials evaluating the potential combination of Velcade (Millenium) and docetaxel will need to be conducted to determine true efficacy or synergy with cytotoxic chemotherapy.

4.2. Histone Deacetylase Inhibitor

Histone acetylation is a normal posttranslational modification of the core nucleosomal histones in the DNA. The histones are key proteins of chromatin that ultimately impact gene regulation. Histones are deacetylated by histone deacetylase (HDAC). Inhibitors of HDAC have the potential to reverse epigenetic transcriptional silencing that may or may not have occurred as a result of DNA methylation.

HDAC inhibitors currently being evaluated in clinical trials include phenylbutyrate (Medicis, Scottsdale, AZ, USA), suberoylanilide hydroxamic acid (SAHA) (Merck), MGCD0103 (MethylGene), PXD101 (CuraGen and TopoTarget), FK228 (Gloucester Pharmaceuticals), and MS275 (Schering AG). Phenylbutyrate (Medicis) and SAHA (Merck) have been evaluated in phase I trials of patients who were primarily metastatic hormone-insensitive prostate cancer patients. No PSA responses were reported *(65,66)*. Similar with the other single-agent trials of targeted agents in prostate cancer, there is an underwhelming response with regards to effect on serum PSA despite encouraging preclinical studies *(67)*. SAHA (Merck), PXD101 (CuraGen and TopoTarget), and MS275 (Schering AG) are currently in phase II trials or combination phase I trials.

FK228 (Gloucester Pharmaceuticals) was evaluated in a phase II trial in patients with hormone-insensitive prostate cancer *(68)*. Two of 18 patients have had a 50% decrease in PSA and 1 of 18 patients had a 40% decrease in PSA. Accrual into the study is still ongoing.

5. ANTI-VASCULAR TARGETING AGENTS

5.1. Angiogenesis Inhibitors

Most novel angiogenesis inhibitors modulate the VEGFs and their signaling pathways. Inhibitors of angiogenesis stop new blood vessel formation from existing vasculature by potentially halting supply of blood and nutrients to the cancer. There is a vast group of novel agents and pathways not discussed in this chapter including matrix metalloproteinase inhibitors, cyclooxygenase-2 inhibitors as well as many other growth factor families discussed earlier in this chapter that have either direct or indirect effects on angiogenesis.

5.1.1. VASCULAR ENDOTHELIAL GROWTH FACTOR INHIBITORS

Six VEGF ligands and three VEGF receptors play critical roles in tumorigenesis and angiogenesis. Binding of the ligands to their receptors results in initiation of the intracellular-signaling pathways, both in parallel and downstream. It is the parallel, cross-talk signaling that occurs with VEGF and other nearby proteins that make the process of angiogenesis complex.

VEGF has been determined to be overexpressed in serum and tissue in patients with prostate cancer, and high levels of serum VEGF correlated with decreased survival in men with metastatic hormone-insensitive prostate cancer (69,70). Preclinical studies support the inhibition of the VEGF axis with a single agent or in combination with cytotoxic chemotherapy (71).

Bevacuzimab (Genentech), a humanized murine monoclonal antibody against the VEGF receptor, evaluated in combination with docetaxel and estramustine resulted in 81% of metastatic, hormone-insensitive prostate cancer patients experiencing a greater than PSA decrease of 50% (72). There were significant thrombotic events attributed to estramustine in this study, and a subsequent phase III study of docetaxel/prednisone plus bevacuzimab (Genentech) or placebo (CALGB90401) is underway without the estramustine. Accrual to this study was initiated in April 2005.

Additional monoclonal antibody in development includes 2C3 (Peregrine) and VEGF-Trap (Bristol Meyers Squibb). Prostate cancer-specific trials with these two agents have not yet been reported.

There are many oral small molecule compounds that target the tyrosine kinase domain of VEGFR being developed and are currently in clinical trials. Sunitinib (SU011248) (Pfizer), AZD2171 (Astra Zeneca), ZD6474 (Astra Zeneca), PTK787/ZK 222584 (Novartis), and CP-547,632 (OSI Pharmaceuticals) are among the many agents in clinical trials in solid tumors. AZD2171 (Astra Zeneca) was evaluated in a phase I trial in patients with metastatic hormone-insensitive prostate cancer (73). No prostate cancer-specific phase II trials have yet been reported. Combination studies with cytotoxic chemotherapy in phase I trials are also underway.

5.1.2. THALIDOMIDE

Thalidomide, initially developed in the early 1950s to treat morning sickness in pregnant women, resulted in severe teratogenic effects and was subsequently taken off the market. Thalidomide has resurfaced as treatment for erythema nodosum leprosum, HIV disease, graft-versus-host disease, and multiple myeloma. Improved understanding and appropriate precautions have enabled thalidomide to be used as a potential anti-angiogenic agent in prostate cancer.

Thalidomide as a single agent was evaluated in a randomized, phase II trial in men with metastatic hormone-insensitive prostate cancer (74). Eighteen percent of the men treated with a lower dose of 200 mg orally daily experienced a greater than 50% reduction in the PSA response rate. The combination of thalidomide with docetaxel also resulted in 53% of men experiencing a greater than 50% reduction in the PSA response rate compared to 37% of men treated with docetaxel alone, albeit these results were not statistically significant *(75)*. However, drug toxicity in these trials was significant including venous thrombosis, transient ischemic attack, and stroke.

Neoadjuvant thalidomide along with GM-CSF is currently being investigated in high-risk men undergoing radical prostatectomy *(76)*. Administration of thalidomide for a maximum of 8 weeks prior to surgery resulted in induction of PTEN, Ki-67, CD3, and CD68. Patient accrual and additional biomarker studies are ongoing.

Newer generation thalidomide analogs known as immunomodulatory drugs (IMiDs) including lenalidomide (Revlimid) (Celgene), CC-5013 (Celgene), and CC-4047 (Celgene) are currently being evaluated in phase I and II trials in men with prostate cancer either as single agents or in combination with docetaxel chemotherapy *(77–79)*. Lenalidomide (Revlimid) (Celgene) was approved in June 2006 for the treatment of multiple myeloma.

5.1.3. Vascular Targeting Agents

The two types of vascular targeting agents (VTAs) currently available include ligand-directed agents effecting tumor endothelium and small molecule agents indirectly effecting tumor endothelium. Agents evaluated in clinical trials include combretastatin (Oxigene), ZD6126 (Astra Zeneca), DMXAA (Antisoma PLC), and AVE8062 (Aventis) *(80–82)*. Results are eagerly awaited for DMXAA (Antisoma PLC) as it is now in a phase II prostate cancer clinical trial in combination with docetaxel.

6. SUMMARY

There are many targeted agents being evaluated in prostate cancer, all at different phases of clinical trials. Additional promising targeted agents include radiopharmaceuticals, bisphosphonates, nutriceuticals, and immunotherapy. Most of the novel targeted agents discussed in this chapter had promising preclinical data that supported the initial clinical study rationale. The global lack of PSA response achieved in men treated with single-agent-targeted agent is disappointing. Moreover, the contribution of the targeted agent to an already active chemotherapy regimen in the combination trials is difficult to ascertain. There is an urgent need for additional biomarkers other than PSA to help guide drug development of these novel targeted agents.

This chapter did not describe treatment-related toxicities of each novel agent in great detail. Fortunately, most of the single-agent-targeted treatment trials did not result in traditional toxicities as frequently seen in with cytotoxic chemotherapy, but the toxicities that were reported, such as rash and fatigue, were significant and should not be underestimated. The toxicities associated with angiogenesis inhibitors including hypertension, thrombosis, bleeding, proteinuria, and bowel perforation are not common, but when they occur, they may be devastating.

Drug development of targeted novel agents is a rapidly evolving and complex process. Appropriate trial design, identification of new and relevant biomarkers to better assess drug activity and disease response, and close monitoring of toxicities will efficiently advance the discovery of exciting and new treatments for patients with prostate cancer.

ACKNOWLEDGMENTS

Funding for some of the studies described in this chapter was provided by Abbott Laboratories or Sanofi-Aventis. Dr. Michael A. Carducci is a paid consultant to Abbott Laboratories, MethylGene, Cougar Biotech, Dendreon, Glaxo-Smith Kline, and Centocor and on the speaker's bureau for Sanofi-Aventis and Abbott Laboratories. The terms of this arrangement are being managed by the Johns Hopkins University in accordance with its conflict of interest policies.

CONFLICT OF INTEREST STATEMENT

Dr. Elisabeth I. Heath has no financial disclosures to declare.

REFERENCES

1. Blackledge G. Growth factor receptor tyrosine kinase inhibitors; clinical development and potential for prostate cancer therapy. *J Urol.* 2003;170(6 Pt 2):S77–83;discussion S83.
2. Di Lorenzo G, Tortora G, D'Armiento FP, et al. Expression of epidermal growth factor receptor correlates with disease relapse and progression to androgen-independence in human prostate cancer. *Clin Cancer Res.* 2002;8(11):3438–44.
3. Di Lorenzo G, Autorino R, De Laurentiis M, et al. HER-2/neu receptor in prostate cancer development and progression to androgen independence. *Tumori.* 2004;90(2):163–70.
4. Le Page C, Koumakpayi IH, Lessard L, Mes-Masson AM, Saad F. EGFR and Her-2 regulate the constitutive activation of NF-kappaB in PC-3 prostate cancer cells. *Prostate.* 2005;65(2):130–40.
5. El Sheikh SS, Domin J, Abel P, Stamp G, Lalaniel N. Phosphorylation of both EGFR and ErbB2 is a reliable predictor of prostate cancer cell proliferation in response to EGF. *Neoplasia.* 2004;6(6):846–53.
6. Slovin SF, Kelly WK, Cohen R, et al. Epidermal growth factor receptor (EGFr) monoclonal antibody (MoAb) C225 and doxorubicin (DOC) in androgen independent (AI) prostate cancer (PC): results of a phase Ib/IIa study. *Proc Am Soc Clin Oncol.* 1997:1108.

7. Foon KA, Yang XD, Weiner LM, et al. Preclinical and clinical evaluations of ABX-EGF, a fully human anti-epidermal growth factor receptor antibody. *Int J Radiat Oncol Biol Phys.* 2004;58(3):984–90.
8. Yang X, Wang P, Fredlin P, et al. ABX-EGF, a fully human anti-EGF receptor monoclonal antibody: inhibition of prostate cancer in vitro and in vivo. *Proc Am Soc Clin Oncol.* 2002:2454.
9. Figlin R, Belldegrun AS, Crawford J, et al. ABX-EGF, a fully human antiepidermal growth factor receptor (EGFR) monoclonal antibody (mAb) in patients with advanced cancer: phase I clinical results. *Proc Am Soc Clin Oncol.* 2002:35.
10. Ziada A, Barqawi A, Glode LM, et al. The use of trastuzumab in the treatment of hormone refractory prostate cancer: phase II trial. *Prostate.* 2004;60(4):332–7.
11. Morris MJ, Reuter VE, Kelly WK, et al. HER-2 profiling and targeting in prostate carcinoma. *Cancer.* 2002;94(4):980–6.
12. Lara PN, Jr., Chee KG, Longmate J, et al. Trastuzumab plus docetaxel in HER-2/neu-positive prostate carcinoma: final results from the California Cancer Consortium Screening and Phase II Trial. *Cancer.* 2004;100(10):2125–31.
13. Lara PN, Jr., Meyers FJ, Gray CR, et al. HER-2/neu is overexpressed infrequently in patients with prostate carcinoma. Results from the California Cancer Consortium Screening Trial. *Cancer.* 2002;94(10):2584–9.
14. Small EJ, Bok R, Reese DM, Sudilovsky D, Frohlich M. Docetaxel, estramustine, plus trastuzumab in patients with metastatic androgen-independent prostate cancer. *Semin Oncol.* 2001;28(4 Suppl 15):71–6.
15. Franklin MC, Carey KD, Vajdos FF, Leahy DJ, de Vos AM, Sliwkowski MX. Insights into ErbB signaling from the structure of the ErbB2-pertuzumab complex. *Cancer Cell.* 2004;5(4):317–28.
16. Agus DB, Gordon MS, Taylor C, et al. Phase I clinical study of pertuzumab, a novel HER dimerization inhibitor, in patients with advanced cancer. *J Clin Oncol.* 2005;23(11):2534–43.
17. Canil CM, Moore MJ, Winquist E, et al. Randomized phase II study of two doses of gefitinib in hormone-refractory prostate cancer: a trial of the National Cancer Institute of Canada-Clinical Trials Group. *J Clin Oncol.* 2005;23(3):455–60.
18. Schroeder FH, Wildhagen MF, et al. ZD1839 (gefitinib) and hormone resistant (HR) prostate cancer-final results of a double blind randomized placebo-controlled phase II study. *Proc Am Soc Clin Oncol.* 2004:4698.
19. Wilding G, Soulie P, Trump D, Das-Gupta A. Small E. Results from a pilot phase I trial of gefitinib combined with docetaxel and estramustine in patients with hormone-refractory prostate cancer. *Cancer.* 2006;106(9):1917–24.
20. Gravis G, Goncalves A, Bladou F, Salem N, Esterni B, Bagattini S, Viens P. A phase II study of erlotinib in advanced prostate cancer. *2006 Prostate Cancer Symposium.* 2006:245.
21. Hoekstra R, Dumez H, Eskens FA, et al. Phase I and pharmacologic study of PKI166, an epidermal growth factor receptor tyrosine kinase inhibitor, in patients with advanced solid malignancies. *Clin Cancer Res.* 2005;11(19 Pt 1):6908–15.
22. Burris HA, III, Hurwitz HI, Dees EC, et al. Phase I safety, pharmacokinetics, and clinical activity study of lapatinib (GW572016), a reversible dual inhibitor of epidermal growth factor receptor tyrosine kinases, in heavily pretreated patients with metastatic carcinomas. *J Clin Oncol.* 2005;23(23):5305–13.
23. Erlichman C, Hidalgo M, Boni JP, et al. Phase I study of EKB-569, an irreversible inhibitor of the epidermal growth factor receptor, in patients with advanced solid tumors. *J Clin Oncol.* 2006;24(15):2252–60.
24. Slichenmyer WJ, Elliott WL, Fry DW. CI-1033, a pan-erbB tyrosine kinase inhibitor. *Semin Oncol.* 2001;28(5 Suppl 16):80–5.

25. Lewis N, Marshall J, Amelsberg A, Cohen RB, Stopfer P, Hwang J, Malik S. A phase I dose escalation study of BIBW 2992, an irreversible dual EGFR/HER2 receptor tyrosine kinase inhibitor, in a 3 week on 1 week off schedule in patients with advanced solid tumors. *Proc Am Soc Clin Oncol.* 2006:3091.

26. Hofer MD, Fecko A, Shen R, et al. Expression of the platelet-derived growth factor receptor in prostate cancer and treatment implications with tyrosine kinase inhibitors. *Neoplasia.* 2004;6(5):503–12.

27. Uehara H, Kim SJ, Karashima T, et al. Effects of blocking platelet-derived growth factor-receptor signaling in a mouse model of experimental prostate cancer bone metastases. *J Natl Cancer Inst.* 2003;95(6):458–70.

28. Rao K, Goodin S, Levitt MJ, et al. A phase II trial of imatinib mesylate in patients with prostate specific antigen progression after local therapy for prostate cancer. *Prostate.* 2005;62(2):115–22.

29. Sinibaldi V, Carducci MA, Elza-Brown E, Rosenbaum E, Denmeade S, Pili R, Walczak J, Garrett-Mayer E, Moore-Cooper S, Eisenberger MA. A phase II evaluation of imitanib mesylate (G) in stage M0 prostate cancer (PC) patients (pts) on hormonal therapy (HT) with evidence of biochemical relapse. *Proc Am Soc Clin Oncol.* 2006:14612.

30. Kubler HR, van Randenborgh H, Treiber U, et al. In vitro cytotoxic effects of imatinib in combination with anticancer drugs in human prostate cancer cell lines. *Prostate.* 2005;63(4):385–94.

31. Tiffany NM, Wersinger EM, Garzotto M, Beer TM. Imatinib mesylate and zoledronic acid in androgen-independent prostate cancer. *Urology.* 2004;63(5):934–9.

32. Mathew P, Fidler IJ, Logothetis CJ. Combination docetaxel and platelet-derived growth factor receptor inhibition with imatinib mesylate in prostate cancer. *Semin Oncol.* 2004;31(2 Suppl 6):24–9.

33. Ko YJ, Small EJ, Kabbinavar F, et al. A multi-institutional phase II study of SU101, a platelet-derived growth factor receptor inhibitor, for patients with hormone-refractory prostate cancer. *Clin Cancer Res.* 2001;7(4):800–5.

34. Yuyama H, Noguchi Y, Fujimori A, et al. Superiority of YM598 over atrasentan as a selective endothelin ETA receptor antagonist. *Eur J Pharmacol.* 2004;498(1–3):171–7.

35. Dreicer R, Curns N, Morris C, et al. ZD4054 specifically inhibits endothelin A receptor-mediated effects, but not endothelin B receptor-mediated effects. *ASCO Prostate Cancer Symposium.* 2005:237.

36. Carducci MA, Nelson JB, Padley RJ, et al. The endothelin-1 receptor antagonist atrasentan (ABT-627) delays clinical progression in hormone refractory prostate cancer: a multinational, randomized, double-blind, placebo-controlled trial. *Proc Am Soc Clin Oncol.* 2001:694.

37. Carducci MA, Padley RJ, Breul J, et al. Effect of endothelin-A receptor blockade with atrasentan on tumor progression in men with hormone-refractory prostate cancer: a randomized, phase II, placebo-controlled trial. *J Clin Oncol.* 2003;21(4):679–89.

38. Carducci MA, Nelson JB, Saad F, et al. Effects of atrasentan on disease progression and biological markers in men with metastatic hormone-refractory prostate cancer: phase 3 study. *Proc Am Soc Clin Oncol;*2004:4508.

39. Moore C, Creel P, Petros W, Torain T, Yenser S, Gockerman J, Hurwitz H, Garcia Turner A, Sleep DJ, George DJ. Phase I/II study of docetaxel and atrasentan in men with metastatic hormone-refractory prostate cancer (HPRC). *Proc Am Soc Clin Oncol.* 2006:14504.

40. Liao Y, Abel U, Grobholz R, et al. Up-regulation of insulin-like growth factor axis components in human primary prostate cancer correlates with tumor grade. *Hum Pathol.* 2005;36(11):1186–96.

41. Cardillo MR, Monti S, Di Silverio F, Gentile V, Sciarra F, Toscano V. Insulin-like growth factor (IGF)-I, IGF-II and IGF type I receptor (IGFR-I) expression in prostatic cancer. *Anticancer Res.* 2003;23(5A):3825–35.

42. Wu JD, Odman A, Higgins LM, et al. In vivo effects of the human type I insulin-like growth factor receptor antibody A12 on androgen-dependent and androgen-independent xenograft human prostate tumors. *Clin Cancer Res.* 2005;11(8):3065–74.
43. Wu JD, Haugk K, Woodke L, Nelson P, Coleman I, Plymate SR. Interaction of IGF signaling and the androgen receptor in prostate cancer progression. *J Cell Biochem.* 2006;99(2): 392–401.
44. Severi G, Morris HA, MacInnis RJ, et al. Circulating insulin-like growth factor-I and binding protein-3 and risk of prostate cancer. *Cancer Epidemiol Biomarkers Prev.* 2006;15(6): 1137–41.
45. Warshamana-Greene GS, Litz J, Buchdunger E, Garcia-Echeverria C, Hofmann F, Krystal GW. The insulin-like growth factor-I receptor kinase inhibitor, NVP-ADW742, sensitizes small cell lung cancer cell lines to the effects of chemotherapy. *Clin Cancer Res.* 2005;11(4):1563–71.
46. Singh RP, Agarwal R. Prostate cancer chemoprevention by silibinin: bench to bedside. *Mol Carcinog.* 2006;45(6):436–42.
47. Planz B, Oltean H, Deix T, et al. Effect of keratinocyte growth factor and activin on cell growth in the human prostatic cancer cell line LNCaP. *World J Urol.* 2004;22(2):140–4.
48. Roznovanu SL, Amalinci C, Radulescu D. Molecular mechanisms in hormone-resistant prostate cancer. *Rev Med Chir Soc Med Nat Iasi.* 2005;109(3):577–83.
49. Mazieres J, Pradines A, Favre G. Perspectives on farnesyl transferase inhibitors in cancer therapy. *Cancer Lett.* 2004;206(2):159–67.
50. Haas N, Peereboom D, Ranganathan S, et al. Phase II trial of R115777, an inhibitor of farnesyltransferase, in patients with hormone refractory prostate cancer. *Proc Am Soc Clin Oncol.* 2002:721.
51. Steinbild S, Mross K, Morant D, Koberle D, Dittrich C, Strumberg D, Hochhaus A, Hanauske A, Burkholder I, Scheulen ME. Phase II study of sorafenib (BAY 43–9006) in hormone refractory patients with prostate cancer: a study of the Central European Society for Anticancer Drug Research-EWIV (CESAR). *Proc Am Soc Clin Oncol.* 2006:3094.
52. Dahut W, Scripture CD, Posadas EM, Wu S, Arlen PM, Gulley JL, Wright J, Chen CC, JOnes E, Figg WD. Bony metastatic disease responses to sorafenib (BAY 43–9006) independent of PSA in patients with metastatic androgen independent prostate cancer. *Proc Am Soc Clin Oncol.* 2006:4506.
53. Wu L, Birle DC, Tannock IF. Effects of the mammalian target of rapamycin inhibitor CCI-779 used alone or with chemotherapy on human prostate cancer cells and xenografts. *Cancer Res.* 2005;65(7):2825–31.
54. Hudes G, Carducci M, Tomczak P, Dutcher J, Figlin R, Kapoor A, Staroslawska E, O'Toole T, Park Y, Moore L. A phase 3, randomized, 3-arm study of temsirolimus (TEMSR) or interferon-alpha (IFN) or the combination of TEMSR + IFN in the treatment of first-line, poor-risk patients with advanced renal cell carcinoma (adv RCC). *Proc Am Soc Clin Oncol.* 2006:LBA4.
55. Dancey JE. Inhibitors of the mammalian target of rapamycin. *Expert Opin Investig Drugs.* 2005;14(3):313–28.
56. Jac J, Sharef S, Khan M, Amato RJ. Rapamycin for androgen-independent prostate cancer (AIPC). *2006 Prostate Cancer Symposium*;2006:257.
57. Tolcher AW. Novel therapeutic molecular targets for prostate cancer: the mTOR signaling pathway and epidermal growth factor receptor. *J Urol.* 2004;171(2 Pt 2):S41–3; discussion S44.
58. Thomas G, Speicher L, Reiter R, Ranganathan S, Hudes G, Strahs A, Pisters L, Greenberg R, Ryan J, Logothetis C, Sawyers C. *Demonstration that Temsirolimus Preferentially Inhibits the mTOR Pathway in the Tumors of Prostate Cancer Patients with PTEN Deficiences.* AACR-NCI-EORTC International Conference. Philadelphia, PA;2005:C131.

59. Solit DB, Scher HI, Rosen N. Hsp90 as a therapeutic target in prostate cancer. *Semin Oncol.* 2003;30(5):709–16.
60. Neckers L. Heat shock protein 90 inhibition by 17-allylamino-17- demethoxygel-danamycin: a novel therapeutic approach for treating hormone-refractory prostate cancer. *Clin Cancer Res.* 2002;8(5):962–6.
61. Heath EI, Gaskins M, Pitot HC, et al. A phase II trial of 17-allylamino-17- demethoxygel-danamycin in patients with hormone-refractory metastatic prostate cancer. *Clin Prostate Cancer.* 2005;4(2):138–41.
62. Williams S, Pettaway C, Song R, Papandreou C, Logothetis C, McConkey DJ. Differential effects of the proteasome inhibitor bortezomib on apoptosis and angiogenesis in human prostate tumor xenografts. *Mol Cancer Ther.* 2003;2(9):835–43.
63. Papandreou CN, Daliani DD, Nix D, et al. Phase I trial of the proteasome inhibitor borte-zomib in patients with advanced solid tumors with observations in androgen-independent prostate cancer. *J Clin Oncol.* 2004;22(11):2108–21.
64. Messersmith WA, Baker SD, Lassiter L, et al. Phase I trial of bortezomib in combination with docetaxel in patients with advanced solid tumors. *Clin Cancer Res.* 2006;12(4):1270–5.
65. Carducci MA, Gilbert J, Bowling MK, et al. A Phase I clinical and pharmacological evaluation of sodium phenylbutyrate on an 120-h infusion schedule. *Clin Cancer Res.* 2001;7(10):3047–55.
66. Kelly WK, Richon VM, O'Connor O, et al. Phase I clinical trial of histone deacetylase inhibitor: suberoylanilide hydroxamic acid administered intravenously. *Clin Cancer Res.* 2003;9(10 Pt 1):3578–88.
67. Butler LM, Agus DB, Scher HI, et al. Suberoylanilide hydroxamic acid, an inhibitor of histone deacetylase, suppresses the growth of prostate cancer cells in vitro and in vivo. *Cancer Res.* 2000;60(18):5165–70.
68. Molife R, Patterson S, Riggs C, Higano C, Stadler WM, Dearmaley D, Parker C, McCulloch W, Shala Bono A. Phase II study of FK228 in patients with metastatic hormone refractory prostate cancer (HPRC). *Proc Am Soc Clin Oncol.* 2006:217.
69. Strohmeyer D, Strauss F, Rossing C, et al. Expression of bFGF, VEGF and c-met and their correlation with microvessel density and progression in prostate carcinoma. *Anticancer Res.* 2004;24(3a):1797–804.
70. Duque JL, Loughlin KR, Adam RM, Kantoff PW, Zurakowski D, Freeman MR. Plasma levels of vascular endothelial growth factor are increased in patients with metastatic prostate cancer. *Urology.* 1999;54(3):523–7.
71. Retter AS, Figg WD, Dahut WL. The combination of antiangiogenic and cytotoxic agents in the treatment of prostate cancer. *Clin Prostate Cancer.* 2003;2(3):153–9.
72. Picus J, Halabi S, Rini B, et al. The use of bevacizumab (B) with docetaxel (D) and estramustine (E) in hormone refractory prostate cancer (HPRC): initial results of CALGB 90006. *Proc Am Soc Clin Oncol*;2003:1578.
73. Ryan C, Stadler WM, Roth B, Puchalski T, Morris C, Small E. Phase I evaluation of AZD2171, a highly potent VEGFR tyrosine kinase inhibitor, in pateints with hormone refractory prostate cancer (HPRC). *2006 Prostate Cancer Symposium*;2006:261.
74. Figg WD, Dahut W, Duray P, et al. A randomized phase II trial of thalidomide, an angiogenesis inhibitor, in patients with androgen-independent prostate cancer. *Clin Cancer Res.* 2001;7(7):1888–93.
75. Dahut WL, Gulley JL, Arlen PM, et al. Randomized phase II trial of docetaxel plus thalidomide in androgen-independent prostate cancer. *J Clin Oncol.* 2004;22(13):2532–9.
76. Garcia J, Magi-Galluzzi C, Rothaermel J, Elson P, Zhou M, Klein E, Dreicer R. Neoadjuvant GM-CSF and thalidomide in men with high risk prostate carcinoma undergoing radical prostatectomy. *Proc Am Soc Clin Oncol*.2006:4564.

77. Moss R, Shelton G, Mella J, Mohile SG, Petrylak DP. A phase I open label, dose escalation study to determine the maximum tolerated dose and to evaluate the safety profile of lenalidomide with every three week docetaxel in subjects with androgen independent prostate cancer. *Proc Am Soc Clin Oncol.* 2006:14618.
78. Tohnya TM, Ng SS, Dahut WL, et al. A phase I study of oral CC-5013 (lenalidomide, Revlimid), a thalidomide derivative, in patients with refractory metastatic cancer. *Clin Prostate Cancer.* 2004;2(4):241–3.
79. Sison B, Bond T, Amato RJ, et al. Phase II study of CC-4047 in patients with metastatic hormone-refractory prostate cancer (HPRCa). *Proc Am Soc Clin Oncol* 2004:4701.
80. Rustin GJ, Galbraith SM, Anderson H, et al. Phase I clinical trial of weekly combretastatin A4 phosphate: clinical and pharmacokinetic results. *J Clin Oncol.* 2003;21(15):2815–22.
81. Evelhoch JL, LoRusso PM, He Z, et al. Magnetic resonance imaging measurements of the response of murine and human tumors to the vascular-targeting agent ZD6126. *Clin Cancer Res.* 2004;10(11):3650–7.
82. Thorpe PE. Vascular targeting agents as cancer therapeutics. *Clin Cancer Res.* 2004;10(2):415–27.

19 New Perspectives on Chemotherapy in Prostate Cancer

Julia H. Hayes, MD,
and Philip Kantoff, MD

CONTENTS

1. INTRODUCTION

Androgen deprivation therapy (ADT) has been the standard of care for patients with advanced prostate cancer for over 50 years. Most patients will respond to this treatment by demonstrating symptomatic improvement, regression of metastases, and almost always by a decline in prostate-specific antigen (PSA) levels. However, the median duration of response to ADT is only 18–24 months for patients with metastatic disease. About 30–40% of patients will respond to further hormonal manipulation, but the majority will become refractory to this treatment within months. Metastatic hormone refractory prostate cancer (HRCaP) is both morbid and rapidly progressive, with a median survival of 18–20 months.

Until recently, no therapy had been shown to prolong life in men with HRCaP. Chemotherapy was historically viewed as ineffective in part because for many years, the development and recognition of effective treatment for prostate cancer was hampered by difficulties in assessing patient response to therapy. Early

From: *Current Clinical Oncology: Prostate Cancer:*
Signaling Networks, Genetics, and New Treatment Strategies
Edited by: R. G. Pestell and M. T. Nevalainen © Humana Press, Totowa, NJ

clinical trials of chemotherapy in patients with advanced disease enrolled patients with symptomatic disease, high tumor burden, and poor performance status: in these trials, objective response rates were less than 10% *(1,2)*. However, the recognition of both PSA response as an assessment of activity and palliation of symptoms as clinically meaningful led to a reevaluation of the role of chemotherapy in treating HRCaP.

In the 1990s, PSA became widely available and was soon adopted as a marker of response in clinical trials. Evidence showed that post-therapy decline of PSA, and the dynamics of that decline, could be used to predict survival *(3–5)*. In 1999, the Prostate Specific Antigen Working Group established a definition of PSA response as a $\geq$50% decline in PSA, confirmed by a second PSA value at least 4 weeks later, without clinical or radiographic evidence of disease progression *(6)*, a definition now widely accepted by clinical investigators. During these years, investigators also focused on examining whether chemotherapy was of palliative benefit *(7)*. Pain scales and instruments to assess quality of life were developed and used as primary and secondary endpoints in clinical trials. The approval by the US Food and Drug Administration of the first chemotherapeutic agent for use in HRCaP, mitoxantrone, was based largely on its palliative benefit.

In 2004, two important phase III studies were published demonstrating for the first time that chemotherapy provided a survival benefit in patients with HRCaP. Both studies utilized docetaxel-based regimens, and docetaxel plus prednisone has become the standard first-line therapy for HRCaP. As a result, efforts to study this agent both in combination with other established agents and with novel chemotherapeutic and targeted therapies have been initiated. Following the model established in other solid tumors such as colon and breast, docetaxel is also being investigated in patients with early-stage disease at high risk of recurrence after definitive therapy. This chapter will discuss the evolution of chemotherapy's role in the treatment of HRCaP, the recent trials showing a survival benefit to docetaxel-based regimens in HRCaP, and the promising new regimens for both advanced and early stage disease now in development.

2. CHEMOTHERAPY FOR ANDROGEN-INDEPENDENT PROSTATE CANCER

2.1. Demonstrating Clinical Benefit to Chemotherapy in HRCaP: Mitoxantrone and Prednisone

Clinical trials of chemotherapy conducted in the 1970s and 1980s demonstrated limited benefit in patients with HRCaP. However, with the introduction of PSA response as a measure of drug activity, quality of life measures as clinical outcomes, better supportive therapy, and a more rational approach to the selection of chemotherapeutic agents, there were hints that chemotherapy might provide clinical efficacy. In the late 1990s, two randomized trials demonstrated for the first time that chemotherapy provided a clinical benefit to patients with HRCaP. In these trials, mitoxantrone plus corticosteroids were shown to provide significant improvements in quality of life and pain relief in these patients. Despite the fact that neither trial provided evidence of improvement in survival, this regimen became the standard against which future regimens were judged.

The first of these trials was a Canadian multi-institutional study published in 1996 that examined mitoxantrone plus prednisone versus prednisone alone in men with symptomatic HRCaP *(8)*. One hundred sixty-one men were randomized to receive mitoxantrone (12 mg/m^2 every three weeks) plus prednisone (10 mg daily) or prednisone alone. The primary endpoint was pain response, as assessed by a sustained two-point reduction in pain on a six-point validated pain questionnaire; secondary endpoints included degree and duration of response, quality of life, and overall survival. Patients who failed treatment with prednisone were permitted to crossover to the mitoxantrone arm. In men treated with combination therapy, a significant improvement in pain response (29 vs. 12%, $p = 0.01$) and duration of pain relief was seen (43 vs. 18 weeks, log-rank $p < 0.0001$). Toxicities included grade 3 or 4 neutropenia in 45%, though only 1% developed neutropenic fever, and possible cardiac toxicity associated with mitoxantrone in 4%. Of note, there was no statistically significant difference in PSA response rate, but men with a PSA response were significantly more likely to achieve a palliative response.

In a subsequent Cancer and Leukemia Group B (CALGB) trial, 242 men with HRCaP were randomized to mitoxantrone (14 mg/m^2 every 3 weeks) plus hydrocortisone (40 mg daily) or to hydrocortisone alone *(9)*. The primary endpoint was overall survival; secondary endpoints included time to disease progression, time to treatment failure, PSA and objective response rates, and quality of life. This trial included both symptomatic and asymptomatic men: more than 90% of patients had bone metastases, but 37% of patients had no analgesic requirement at baseline.

As in the Canadian study, no significant difference in survival was found (12.6 months for hydrocortisone vs. 12.3 months for combined therapy). However, small but statistically significant increases in median time to disease progression and to treatment failure were seen in the mitoxantrone arm (2.3 vs. 3.7 months, $p = 0.0254$ for treatment failure and $p = 0.0218$ for disease progression). A post hoc analysis revealed that a higher proportion of patients in the mitoxantrone arm achieved a PSA response (38 vs. 22, $p = 0.008$). In addition, patients who achieved $\geq 50\%$ decline in PSA from baseline had a significant improvement in survival (20.5 vs. 10.3 months; log-rank, $p < 0.001$). Although overall pain response was not significantly improved with the addition of mitoxantrone, statistically significant improvements in the frequency and severity of pain were reported. This difference between the two trials is most likely due to the number of patients reporting pain at enrollment: while pain was a prerequisite for enrollment in the Tannock study, only one-third of patients required analgesics at the time of study entry in the CALGB study.

As a result of these two studies, mitoxantrone plus prednisone emerged as the standard against which other chemotherapeutic regimens were measured.

2.2. Demonstrating a Survival Benefit in HRCAP: Docetaxel-Based Chemotherapy

In 2004, two important randomized controlled trials demonstrated for the first time that chemotherapy improved overall survival in patients with HRCaP. Both the TAX 327 and SWOG 9916 trials compared docetaxel-based regimens to the standard mitoxantrone plus prednisone. These trials provided confirmation of data from multiple phase II studies that had suggested that docetaxel-based regimens could prolong life in these patients.

2.2.1. SINGLE-AGENT DOCETAXEL

In the 1990s, preclinical data emerged suggesting that prostate cancer cells were particularly sensitive to mitotic spindle inhibitors such as vinca alkaloids and the taxanes (10). Both preclinical and phase II clinical trials identified docetaxel as particularly promising. Docetaxel is a semisynthetic taxane thought to exert its cytotoxic effect through at least two distinct mechanisms. It is known to bind to tubulin subunits, inhibiting microtubule disassembly and leading to apoptosis. It has also been shown to inactivate the antiapoptotic protein bcl-2, which is overexpressed in prostate cancer cells and has been associated with both chemotherapy and androgen resistance (11,12).

Phase II studies using docetaxel as a single agent evaluated both weekly and every 3-week schedules (Table 1). PSA response rates were similar between the two schedules, ranging from 38 to 64%, and up to 40% of the patients in these trials had an objective response. Myelosuppression was more pronounced in the every 3-week schedule, with neutropenia seen in up to 70% of patients. TAX 327 was a phase III trial designed to compare these two dosing regimens to each other and to the standard mitoxantrone plus prednisone.

2.2.2. TAX 327

The TAX 327 clinical trial compared two schedules of docetaxel plus prednisone to mitoxantrone plus prednisone (13). This phase III multi-center clinical trial enrolled 1006 men with hormone refractory disease to (i) mitoxantrone 12 mg/m^2 every 3 weeks, (ii) docetaxel 30 mg/m^2 weekly for 5 weeks of a 6-week cycle, or (iii) docetaxel 75 mg/m^2 every 3 weeks. All patients also received prednisone 10 mg daily. The primary endpoint of this study

Table 1
Phase II Trials of Single Agent Docetaxel in HRCaP

Trial	Docetaxel Dose	PSA response rate (%)	Meaurable response rate (%)	Overall survival
Picus and Schultz (58)	75 mg/m^2 Q3W	46	24	27 months
Friedland et al. (59)	75 mg/m^2 Q3W	38	29	67% at 15 months
Berry et al. (60)	36 mg/m^2 weekly for 6 of 8 weeks	41	33	9.4 months
Beer et al. (61)	36 mg/m^2 weekly for 6 of 8 weeks	46	40	10 months
Gravis et al. (62)	35 mg/m^2 weekly for 6 of 8 weeks	48	28 (stable disease)	20 months
Ferrero et al. (63)	40 mg/m^2 weekly for 6 of 8 weeks	64	17	58% at 1 year

Q3W, every three weeks.

was overall survival; secondary endpoints included PSA response, objective tumor response, pain response, and quality of life endpoints using the validated Functional Assessment of Cancer-Prostate (FACT-P) questionnaire. Ninety percent of patients had bone metastases at enrollment, and 45% of men reported pain at baseline.

After a median follow-up of 20.7 months, men treated with docetaxel on either schedule demonstrated a significant prolongation in overall survival as compared to mitoxantrone [hazard ratio (HR) 0.83, $p = 0.04$]. This improvement in median survival was significant for the every 3-week docetaxel arm as compared to mitoxantrone (18.9 vs. 16.5 months, HR = 0.76, $p = 0.0009$) but not for weekly docetaxel: median survival in the weekly docetaxel arm was 17.4 months ($p = 0.36$).

Docetaxel also demonstrated statistically significant improvements in secondary endpoints. Compared to mitoxantrone, every 3-week docetaxel also demonstrated statistically significant improvements in PSA response rate (45 vs. 32%, $p < 0.0005$) and pain response rate (35 vs. 22%, $p = 0.01$). Patients receiving docetaxel also had significant improvement in quality of life measures as compared to patients treated with mitoxantrone.

The treatment was well tolerated in general though adverse events were more common in the docetaxel arms. The highest rate of completion of all cycles of therapy was seen in the every 3-week docetaxel arm and the lowest in the mitoxantrone arm (although this rate may reflect the lower activity of mitoxantrone). Grade 3–4 neutropenia was seen in 32% of patients in the every 3-week docetaxel arm, as compared to 1.5% of men in the weekly docetaxel arm and 22% men in the mitoxantrone arm. However, febrile neutropenia was a rare occurrence, seen in less than 3% of men. Left ventricular ejection fraction impairment was more common in the mitoxantrone arm compared with the docetaxel arms (22% in the mitoxantrone arm, 10% in the every 3-week docetaxel arm, and 8% in weekly docetaxel, $p = 0.0015$ for each comparison). Adverse events leading to treatment discontinuation included musculoskeletal or nail changes, peripheral neuropathy, fatigue, infection in the docetaxel arms, and cardiac dysfunction.

2.2.3. DOCETAXEL PLUS ESTRAMUSTINE

The second study, SWOG 9916, compared docetaxel plus estramustine to mitoxantrone plus prednisone *(14)*. The inclusion of estramustine in this regimen was based on preclinical and early clinical data suggesting that estramustine acted synergistically with docetaxel and that this combination was an active and well-tolerated regimen.

2.2.3.1. Estramustine. Estramustine is a conjugate of estradiol and nornitrogen mustard initially synthesized as targeted therapy: the estrogen moiety was to target cancerous cells, and the nitrogen mustard would be cleaved in the cell, activating the alkylating agent. However, subsequent study has revealed that its antitumor activity is the result of binding to tubulin and microtubule-assembly proteins. Although estramustine has modest single-agent activity in prostate cancer, it has been shown to act synergistically with other microtubule-inhibiting agents such as the vinca alkaloids and the taxanes and with etoposide *(15–17)*. Phase II clinical trials of estramustine in combination with agents including paclitaxel, vinorelbine, vinblastine, and etoposide confirmed the activity of these regimens (Table 2).

2.2.3.2. Estramustine Plus Docetaxel. Beginning in the early 2000s, phase II clinical trials suggested that estramustine plus docetaxel was the most active of these combinations (Table 3). PSA responses ranging from 4*(5)* to 68% were reported, and median survival ranged from 13.5 to 20 months. Although this regimen also appeared superior to docetaxel plus prednisone in phase II data (Table 1), significant toxicity is seen with the addition of estramustine. Grade 3 nausea and vomiting are more frequently reported, and estramustine has been associated with a thrombosis rate of approximately 10%, most likely as a result of its estrogen moiety. The use of prophylactic anticoagulation with low-dose warfarin and/or aspirin has been suggested, but to date, no randomized controlled trial has demonstrated its utility in preventing thromboembolic complications. However, given its promising clinical performance, docetaxel plus estramustine was evaluated in a phase III clinical trial.

2.2.4. SWOG 9916

In the SWOG 9916 trial, 674 eligible patients with hormone-refractory prostate cancer were randomized to every 3-week cycles of docetaxel (60 mg/m^2) and estramustine (280 mg three times daily for days 1–5) or the standard dose of mitoxantrone plus continuous prednisone as described above *(14)*. Doses of docetaxel and mitoxantrone were escalated to 70 and 14 mg/m^2, respectively, for men who did not experience grades 3 and 4 toxicities after the first cycle. The primary endpoint was overall survival; secondary endpoints included progression-free survival, objective response rate, and degree of PSA decline. Eight-five percent of patients had bone metastases, but the majority of patients reported minimal pain.

After a follow-up of 32 months, median survival was 20% longer in patients receiving docetaxel plus estramustine (17.5 vs. 15.6 months, $p = 0.02$). Median time to progression was also significantly improved in the docetaxel arm (6.3 vs. 3.2 months, log-rank $p < 0.001$), as was the PSA response rate (50 vs. 27%, $p < 0.001$). The objective tumor response rate was also improved but did not achieve statistical significance (17 vs. 11%, $p = 0.15$). Self-reported pain relief was not significantly different between the two arms.

Adverse events were more common in the docetaxel–estramustine cohort. More patients discontinued treatment in the docetaxel–estramustine arm than in

Table 2
Phase II Trials of Non-Docetaxel-Based Combination Chemotherapy in HRCAP

Chemotherapy	Dose	Author	N	PSA response rate (%)	Measurable response rate (%)	Median survival (months)
Paclitaxel-Based						
Paclitaxel, EMP	Paclitaxel 120 mg/m² CI over 96 h Q3W; EMP 600 mg/m²/d	Hudes et al. (64)	34	53	44	17
Paclitaxel, EMP	Paclitaxel 90 mg/m² weekly; EMP 140 mg TID × 3 day;	Vaughn et al. (65)	66	42	15	16
Paclitaxel, EMP	Paclitaxel 150 mg/m² weekly; EMP 280 mg TID	Vaishampayan et al. (66)	28	62	38	13
Paclitaxel, EMP, etoposide	Paclitaxel 135 mg/m² Q3W; EMP 280 mg TID; etoposide 100 mg/day × 14 day	Smith et al. (67)	40	65	45	12.8

Regimen	Dose	Study				
Paclitaxel, EMP, etoposide	Paclitaxel 50 mg/m²/week; EMP 280 mg TID day 1–10; Etoposide 50 mg BID day 1–10	Meluch et al. (68)	40	35	24	9.5
Paclitaxel, EMP, carbo-platin(TEC)	Paclitaxel 60–100 mg/m²; EMP 10 mg/kg; carboplatin AUC 6	Kelly et al. (69)	56	67	45	19.9
Paclitaxel, EMP, carbo-platin(TEC)	Paclitaxel 100 mg/m² QW; EMP 10 mg/kg × 5 day:Carboplatin AUC 6 day1 of 28 day cycle	Urakami et al. (70)	32	100	61	24
EMP based EMP, etoposide	EMP 15 mg/kg/day; etoposide 50 mg/m²/day	Pienta et al (71)	42	58	50	N/R

(Continued)

Table 2
(Continued)

Chemotherapy	Author	N	PSA response rate (%)	Measurable response rate (%)	Median survival (months)
EMP, etoposide	Dimopoulos et al. (72)	56	58	45	13
	Dose: EMP 140 mg PO TID; etoposide 50 mg/m² × 21 day				
EMP, vinorelbine	Smith et al. (73)	25	24	0	14.1
	Dose: EMP 14 mg TID x 14 day; Vinorelbine 25 mg/m² day 1,8 of 21 day cycle				
EMP, vinorelbine	Sweeney et al. (74)	23	71	12.5	15.1
	Dose: EMP 280 mg TID × 3 day; vinorelbine 20 mg/m² weekly				
EMP, vinblastine	Seidman et al. (75)	25	54	40	N/R
	Dose: EMP 10 mg/kg/day; vinblastine 4 mg/m² weekly				

410

EMP, vinblastine	EMP 600 mg/m²/day; vinblastine 4 mg/m²	Hudes et al. (76)	36	61	14	>4 months
Other agents						
Doxorubicin, ketoconazole	Dox 20 mg/m² by CI; ketoconazole 1200 mg QD	Sella et al. (77)	39	55	58	15.5

EMP, Estramustine; CI, continuous infusion; QD, daily; Q3W, Every three weeks; N/R, Not Reported.

Table 3
Phase II Trials of Docetaxel-Based Combination Therapy in HRCaP

Trial	Docetaxel	Combinations	PSA response rate (%)	Measurable response rate (%)	Overall survival
EMP					
Petrylak et al. (79)	70 mg/m² Q3W	EMP 280 mg TID days 1–5	68	55	77% at 1 year
Sinibaldi et al. (80)	70 mg/m² Q3W	EMP 280 mg Q6h days 1–5	45	20	13.5 months
Savarese et al. (81)	70 mg/m² Q3W	EMP 10 mg/kg/day divided in 3 doses days 1–5; hydrocortisone 40 mg daily	68	50	20 months
Oudard et al. (82)	70 mg/m² day 2 OR 35 mg/m² day 2 and 9 of 21d cycle	EMP 280 mg TID d 1–5 and 8–12	67%	20	18.6 months
			63%	7	18.4 months

EMP, estramustine; TID, three times per day.

412

the mitoxantrone arm (16 vs. 10%). Grade 3–4 neutropenia rates were similar between the two arms, but the incidence of febrile neutropenia was significantly higher in the docetaxel–estramustine arm (5 vs. 2%, $p = 0.01$). Thromboembolic events were also more common in the docetaxel–estramustine group despite the addition of prophylactic warfarin and aspirin half-way through the trial.

Based on the evidence of clinical benefit of docetaxel-based chemotherapy provided by these trials, the US FDA approved every 3-week docetaxel (75 mg/m^2) plus prednisone (10 mg daily) as first-line chemotherapy in HRCaP in May 2004.

2.2.5. THE ROLE OF ESTRAMUSTINE

Whether estramustine confers an additional survival benefit to docetaxel therapy is not known. The preclinical and phase II data discussed above suggest that response rates of docetaxel were improved with the addition of estramustine. However, the overall survival seen in TAX 327 (18.9 months) was similar to that achieved with the addition of estramustine in SWOG 9916 (18 months). To date, no direct comparison of the two regimens has been made.

Indirect evidence of the potential contribution of estramustine stems from two clinical trials comparing single-agent therapy, vinblastine or paclitaxel, with or without estramustine (18,19). In both trials, PSA response and median survival were improved with the addition of estramustine. For example, Berry et al. randomized 163 men to weekly paclitaxel (100 mg/m^2) with or without estramustine (EMP) (280 mg for 3 days per week) for 3 of every 4 weeks (19). Median survival was significantly prolonged for patients receiving estramustine (16.1 vs. 13.1 months, $p = 0.049$). PSA response rate was also significantly greater in patients who received estramustine (47 vs. 27%, $p < 0.01$). However, thrombotic events were more frequent in patients who received estramustine.

Although EMP may improve response rate and possibly survival, the absence of direct evidence supporting the use of estramustine in combination with docetaxel, the toxicities of estramustine must also be considered: as discussed above, estramustine is associated with significant thromboembolic and gastrointestinal adverse events. As a result, most clinicians prescribe every 3-week docetaxel plus prednisone as first-line therapy for HRCaP.

2.3. Future Directions in Cytotoxic Chemotherapy in HRCaP

2.3.1. TAXANE-BASED CHEMOTHERAPY

Given the survival advantage seen with docetaxel-based chemotherapy, efforts to identify other active taxane-based regimens continue. Investigators have focused on combining docetaxel with other agents in both doublet and triplet regimens. For example, doublet regimens combining docetaxel with vinorelbine have shown promise in phase II studies, and a phase III trial comparing this regimen with docetaxel plus estramustine is ongoing *(20,21)*.

A second approach has been to add a third agent to the combination of estramustine and docetaxel or paclitaxel. Although early phase II trials combining paclitaxel, estramustine, and etoposide did not demonstrate additional benefit over doublet regimens, carboplatin has emerged as a promising addition to both paclitaxel and docetaxel (Table 2) *(22,23)*. For example, the CALGB conducted a phase II study of docetaxel (70 mg/m^2), carboplatin (AUC 5), and estramustine (280 mg three times daily for 5 days) on a 21-day cycle *(24)*. Of 34 evaluable patients, 68% demonstrated a PSA response; patients with measurable disease exhibited a 52% response rate (95% CI = 30–74). Median time to progression was 8.1 months and overall survival was 19 months. The addition of carboplatin significantly increases toxicity, however, and randomized trials will be needed to demonstrate additive benefit.

2.3.2. NOVEL CHEMOTHERAPEUTIC AGENTS

2.3.2.1. Satraplatin. Satraplatin is an oral third-generation platinum agent with *in vitro* efficacy against taxane-resistant prostate cancer cell lines. Five phase II or III clinical trials of satraplatin in prostate cancer have been initiated, but three were terminated prior to achieving target accrual. One phase III clinical trial, conducted by the European Organisation for Cancer Research (EORTC), of satraplatin plus prednisone versus prednisone alone as first-line therapy for HRCaP was recently terminated early by the sponsoring corporation *(25)*. However, preliminary results in the 50 men (out of a planned 380) enrolled, both progression-free survival (5.2 vs. 2.5 months, *p* = 0.023) and PSA response (33 vs. 9%, *p* = 0.046), were significantly improved. A phase III clinical trial of satraplatin plus prednisone versus prednisone alone as second-line therapy for men with HRCaP, the Satraplatin and Prednisone against Refractory Cancer (SPARC), has been completed.

2.3.2.2. Ixabepilone. The epothilones are a novel class of chemotherapeutic agents that act as microtubule inhibitors, but with a chemical structure distinct from the taxanes. The epothilone B analog ixabepilone (BMS-247550) has been shown to have preclinical and clinical activity in both taxane-sensitive and -insensitive HRCaP cells *(26)*. In a phase II clinical trial of single-agent ixabepilone in HRCaP, 39% (16/41) of patients had a PSA response and median progression was 6 months *(27)*. The most frequent adverse effects included hematologic and neurologic toxicity, with 17% reporting grade 3 neuropathy. A randomized phase II clinical trial of ixabepilone (35 mg/m^2) every 3 weeks with or without estramustine (280 mg PO three times a day for days 1–5) in 92 patients with chemotherapy-naïve HRCaP reported a PSA response rate of 69% (95% CI = 55–82) in patients in the combination therapy arm and 48% (95% CI = 33–64) of patients who received ixabepilone alone *(28)*. Time to

PSA progression was 5.2 months in the combination arm and 4.4 months in patients treated with ixabepilone alone. Major toxicities included neutropenia and neuropathy, and 9% patients receiving estramustine had a thrombotic event.

2.3.3. CHEMOTHERAPY COMBINED WITH AGENTS WITH NOVEL MECHANISMS OF ACTION

The overall survival benefit seen with docetaxel-based chemotherapeutic regimens has led to a concerted effort to combine docetaxel with other agents with novel mechanisms of action. In recent years, an explosion in our understanding of the molecular basis of prostate cancer has taken place. Among the many novel agents designed to exploit this knowledge are those that target the endothelin-A (ET-A) receptor, calcitriol, mTOR inhibitors, epidermal growth factor inhibitors, vascular endothelial growth factor-2 inhibitors, antisense Bcl-2 oligonucleotide, bortezomib, and inhibitors of histone deacetylase. Several of these classes of agents have been evaluated in combination with docetaxel.

2.3.3.1. Inhibitors of VEGF. VEGF may play a role in prostate cancer growth and metastasis. Plasma and urine levels of VEGF are increased in men with localized prostate cancer as compared to normal men and are even higher in men with metastatic prostate cancer *(29)*. Plasma VEGF levels at diagnosis predict clinical and biochemical progression, and the level at diagnosis of HRCaP may correlate with survival *(30)*.

Bevacizumab is a humanized monoclonal antibody targeting VEGF. A single-armed clinical trial conducted by the CALGB has combined bevacizumab, docetaxel, and estramustine in 79 chemotherapy-naïve hormone-refractory prostate cancer patients.

Initial results demonstrated that 53% (9/17) of patients had an objective tumor response, and 65% of evaluable patients demonstrated a PSA response *(31)*. The CALGB is currently conducting a phase III double-blinded, placebo-controlled clinical trial comparing docetaxel with or without bevacizumab. Clinical trials employing combinations of docetaxel and the multiple kinase inhibitors Sutent (SU11248) or Sorafinib (BAY 43-9006) are also underway.

Although thalidomide has pleiotropic effects, its mechanism of action is unknown. One randomized phase II clinical trial evaluated thalidomide in men with HRCaP, randomizing 75 men to treatment with weekly docetaxel (30 mg/m^2) with or without thalidomide (200 mg/day) *(32)*. After a median follow-up of 26.4 months, a trend toward increased PSA response in the combined therapy cohort was seen (53 vs. 37%). Median progression-free survival was also increased at 5.9 versus 3.7 months. Of note, 18% (nine patients) in the thalidomide arm had thromboembolic events, and prophylactic low molecular weight heparin (LMWH) was added part-way through the trial. Although the trial was not designed to test for a survival benefit, after a median follow-up of 46.7 months, the median overall survival was significantly increased in men receiving combination therapy for docetaxel plus thalidomide (25.9 vs. 14.7 months, $p = 0.04$) *(33)*.

The National Cancer Institute is conducting an ongoing phase II clinical trial of combination therapy with docetaxel, prednisone, thalidomide, and bevacizumab in men with chemotherapy-naïve progressive HRPC.

2.3.3.2. Endothelin-A Receptor Antagonists. ET-1 has been implicated in the progression of prostate cancer as well as in the development of bony metastases. ET-1 is a potent vasoconstrictor and has been shown to mediate osteoblast growth and function *(34)*. Levels of ET-1 rise with increasing tumor burden, with the highest levels found in men with HRPC *(35)*. The effects of ET-1 are mediated through the ET_A receptor subtype, overexpressed in prostate cancer *(36)*. Atrasentan is a small molecule selective ET_A receptor antagonist. In one randomized, placebo-controlled phase II clinical trial, atrasentan significantly delayed clinical progression in men with asymptomatic HRPC *(34)*. A phase III study failed to show clinical benefit in an intention to treat analysis although signals of activity were once again noted *(37)*.

Atrasentan in combination with chemotherapeutic agents is under investigation. Initial results of a phase I/II clinical trial revealed no clinically significant pharmacokinetic interactions between atrasentan and docetaxel although the MTD has not yet been reached *(38)*. There is also currently a randomized, placebo-controlled phase III clinical trial conducted by SWOG (S0421) underway evaluating docetaxel and prednisone with or without atrasentan.

2.3.3.3. Calcitriol and Related Agents. High-dose calcitriol has been demonstrated to have both antiproliferative and proapoptotic effects. Although the mechanisms of calcitriol's antitumor effects are incompletely understood, they are thought to include induction of cellular differentiation of prostate cancer cells, resulting in apoptosis *(39)*. Calcitriol has also been shown to potentiate the effects of chemotherapeutic agents in the preclinical setting *(40–42)*.

High-dose oral calcitriol in combination with weekly docetaxel demonstrated high-response rates in a phase II clinical trial in 37 patients. Eighty-one percent of men demonstrated PSA response and 22% had >75% reduction in PSA. Fifty-three percent of patients had an objective response. Median time to progression was 11.4 months; median overall survival was 19.5 months *(43)*. The regimen was well tolerated and similar to single-agent docetaxel.

The androgen-independent prostate cancer study of calcitriol enhancing taxotere (ASCENT) was a double-blinded, placebo-controlled clinical trial evaluating weekly docetaxel with or without DN-101 (a novel formulation of high-dose calcitriol) *(44)*. The combination showed a superior PSA response, the primary outcome, but this benefit did not achieve statistical significance (58 vs. 49%, $p = 0.16$). Skeletal morbidity-free survival was also insignificantly prolonged (HR = 0.78, $p = 0.13$). However, median survival was significantly increased (23.5 vs. 16.4 months, HR = 0.67, $p = 0.035$). DN-101 was well tolerated, with no increase in AE in the experimental arm. A larger, randomized phase III trial is underway.

2.3.4. SECOND-LINE CHEMOTHERAPY FOR HRCaP

Despite the encouraging survival benefit seen in men treated with docetaxel-based chemotherapeutic regimens, most men will eventually progress through first-line chemotherapy. Beekman et al. found that in a cohort of 108 men treated initially with antimicrotubule-based chemotherapy, 81% of men received second-line and 40% received third-line chemotherapy *(45)*. Median survival time from the start of therapy was 13 months for second-line and 12 months for third-line therapy. However, few prospective clinical trials have evaluated second-line chemotherapeutic regimens in HRPC.

2.3.4.1. Retreatment with Docetaxel. Most men who discontinue docetaxel as first-line therapy do so either because of progressive disease or unacceptable side effects. For patients who take a "chemotherapy holiday" because of minor side effects while still responding, retreatment with docetaxel at the time of progression is appropriate *(46)*. For patients who exhibited significant myelosuppression on every 3-week docetaxel, weekly docetaxel could be considered.

One small study has suggested that lower-dose weekly docetaxel may be effective in patients who progress on first-line docetaxel. Twenty-five patients with PSA progression after docetaxel therapy were treated with docetaxel (35 mg/m^2) for 3 weeks of a 4-week cycle. PSA response was seen in 72% (18/25) patients and lasted a mean of 5.8 months *(47)*.

2.3.4.2. Sequential Therapy with Mitoxantrone and Docetaxel. No prospective clinical trials have evaluated the use of mitoxantrone or docetaxel as second-line therapy. However, several investigators have examined patient response when treated with mitoxantrone followed by docetaxel or the reverse (Table 4).

One study that followed 68 patients who received mitoxantrone followed by docetaxel or the reverse found that the PSA response rate to second-line therapy was significantly higher in men receiving docetaxel as second-line therapy (44 vs. 15%, $p = 0.012$) *(48)*. Median survival was no different between the groups, but second-line chemotherapy was associated with a high incidence of adverse events: 64% of patients receiving second-line docetaxel and 46% of those receiving second-line mitoxantrone required dose modification or cessation of therapy.

Oh et al. evaluated 68 patients who had been treated with a taxane followed by mitoxantrone or the reverse *(49)*. Responses to first-line taxane therapy were seen in 69% of patients and to second-line taxane in 61%; however, only 12% of patients responded to first line, and 6% to second-line mitoxantrone. Progression-free survival was longer with taxane-based therapy, whether used first or second, but overall progression-free survival and overall survival were

Table 4
Second-Line Chemotherapy for HRCAP

Trial	Mitoxantrone → Taxane[a]	Taxane[a] → Mitoxantrone
Michels et al. (48)	N = 33, RR 44%	N = 35, RR 15%
Saad et al. (83)	N = 20, RR 85%	N/A
Oh et al. (49)	N = 33, RR 60%, PFS 4 months	N = 35, RR 6%, PFS 6 weeks
Joshua et al. (84)	N = 20, RR 45%, TTP 5 months	N/A

[a] All patients received docetaxel except for those enrolled in Oh et al. which included patients who had received paclitaxel.

Adapted from Berthold et al. (46)

equivalent in both groups. Two further phase II trials of patients receiving docetaxel after mitoxantrone are ongoing (Table 4).

2.3.4.3. Addition of Carboplatin. Although single-agent carboplatin has modest activity in HRPC, carboplatin has been postulated to exert a synergistic effect when combined with docetaxel-based chemotherapy in taxane-refractory patients. A case series of four consecutive taxane-refractory patients treated with docetaxel (60–70 mg/m^2) plus carboplatin (AUC 4-5) has been reported, demonstrating a 100% PSA response rate (4/4) and 75% improvement in symptoms (3/4) (50). Treatment was well tolerated, and survival ranged from 4.5 to 12 months. A clinical trial investigating this regimen is currently underway.

2.3.4.4. Taxane-Based Chemotherapy after Ixabepilone. Rosenberg et al. recently published a retrospective analysis of 49 patients who received taxane-based chemotherapy after receiving ixabepilone with or without estramustine as part of a randomized phase II clinical trial (51). Second-line PSA responses were seen in 51% of patients (95% CI = 33–66). Patients who achieved a PSA response with first-line therapy were more likely to respond to second-line therapy with taxanes than those who did not (61 vs. 33%, $p = 0.08$). In addition, fewer patients who discontinued ixabepilone therapy due to disease progression demonstrated a PSA response than patients who discontinued based on toxicity or patient preference (36 vs. 71%, $p = 0.01$).

3. CHEMOTHERAPY IN HIGH-RISK LOCALIZED PROSTATE CANCER

Patients with high-risk, localized prostate cancer at diagnosis have been shown to have a significantly reduced cancer-specific survival. Although men with low- to intermediate-risk disease have long-term survival rates exceeding

90%, men with high-risk disease (defined as clinical stage T3 or T4, a PSA of ≥20 ng/mL, or a Gleason score of 8–10) have a 45% risk of death at 10 years *(52)*.

Chemotherapeutic agents effective in advanced disease have been shown to be beneficial in earlier stage disease in studies of breast, colon, and lung cancer. Given the survival benefit seen with docetaxel-based chemotherapeutic regimens metastatic in prostate cancer, efforts are underway to assess the efficacy of such regimens in early stage prostate cancer.

3.1. Adjuvant Chemotherapy

In the 1980s, a clinical trial conducted by the National Prostate Cancer Group randomized 184 patients with localized, advanced prostate cancer to (i) 2 years of oral cyclophosphamide, (ii) 2 years of estramustine, or (iii) observation *(53)*. After 10 years of follow-up, the estramustine cohort demonstrated an improvement in relapse-free survival, but no overall survival benefit was seen. Currently, SWOG is conducting a randomized controlled trial comparing hormone therapy with or without 2 years of mitoxantrone plus prednisone in high-risk patients after radical prostatectomy.

With the demonstration of survival benefit with docetaxel in advanced disease, adjuvant therapy with docetaxel-based regimens is being investigated. A phase III study (TAX-3501) will randomize patients to observation, androgen deprivation therapy (ADT) for 18 months, or ADT plus docetaxel (75 mg/m^2 every three weeks) for six cycles. Patients in the observation arm who progress will be randomized to ADT, with or without docetaxel.

3.2. Neoadjuvant Chemotherapy

The theoretical advantage to neoadjuvant chemotherapy is to eradicate micrometastases and to decrease local tumor burden prior to definitive local therapy. Studies in breast cancer have shown that neoadjuvant chemotherapy with doxorubicin-based regimens can result in pathologic complete responses at surgery, and these responses have been shown to predict disease-specific survival *(54,55)*. Neoadjuvant ADT has been shown to decrease the positive surgical margin rate in prostate cancer, but no correlation with increased disease-free survival has yet been found *(56,57)*.

Many phase I and phase II studies have examined the use of neoadjuvant chemotherapy in high-risk prostate cancer (Table 5). These trials have shown that neoadjuvant chemotherapy is feasible and safe; in general, surgical complications were not significantly increased in these trials. However, the clinical significance of the clinical outcomes reported is unknown. Many trials included

Table 5

Phase I/II Trials of Neoadjuvant Chemotherapy Prior to RP in Locally Advanced and High-Risk Prostate Cancer

Trial	Number of patients	Inclusion criteria	Hormonal therapy	Chemotherapy regimen	Clinical outcome
Pettaway et al. (85)	33	cT1-2, GI≥8; cT2b-c, GI≥7; PSA > 10 ng/mL	Yes	KAVE* × 12 weeks	50% PSA NMA after treatment
Clark et al. (86)	18	cT2b-c or T3 with PSA ≥ 15 or GI ≥ 8	Yes	EMP/VP-16 × 12 weeks	50% PSA NMA after treatment
Konety et al. (87)	36	cT1-2 with PSA ≥ 20; cT3-4; GI ≥ 8	Yes	TEC × 12-16 weeks	Median PSA nadir 0.17 ng/mL
Ko et al. (88)	12	cT3; PSA ≥ 20; GI ≥ 8	Yes	EMP + Docetaxel	75% PSA NMA after treatment
Hussain et al. (89)	21	≥ cT2b; PSA ≥ 15; GI ≥ 8	Yes	EMP + Docetaxel	100% PSA >50% decline
Gleave et al. (90)	72	GI ≥ 8 or PSA ≥ 20 plus 2 positive cores; T3a or GI ≥ 7 plus PSA ≥ 10 plus ≥ 3 positive cores	Yes	Weekly docetaxel	2 CR, 14 microfoci pT2

Febbo et al. (91)	19	Gl $\geq$ 8; PSA> 20; cT3	No	Docetaxel weekly	58% PSA response; 25% tumor volume reduction in 68%
Beer et al. (92)	22	CT2c; cT3a; PSA $\geq$ 15; Gl $\geq$ 4+3	No	Docetaxel weekly + mitoxantrone	Median PSA decline 41%
Drecier et al. (93)	29	T2b-T3; PSA > 15; Gl $\geq$ 8	No	Docetaxel weekly	24% PSA response after chemotherapy

*KAVE, Table adapted from Gleave et al. (94)

hormone therapy or estramustine, which lowers testosterone and, not surprisingly, PSA responses were seen in these trials.

Studies of docetaxel-based regimens reported PSA response rates of 24 to 60%, supporting the effectiveness of docetaxel in hormone-naïve prostate cancer. A clinical trial currently underway at our institution randomizes patients to definitive radiation therapy (70 Gray) plus 6 months of ADT with or without weekly docetaxel. Patients randomized to the docetaxel arm receive 3 cycles of neoadjuvant docetaxel at 20 mg/m^2 for three of every 4 weeks prior to radiation therapy, followed by seven cycles of docetaxel following the same schedule during radiation therapy. The primary endpoint is a 5-year overall survival. The CALGB (90203) will soon open a clinical trial that will randomize patients to neoadjuvant ADT plus radical prostatectomy with or without up to eight cycles of neoadjuvant docetaxel, with a primary endpoint of 5 year progression-free survival.

At present, neoadjuvant chemotherapy is considered investigational and should be undertaken only in the context of a clinical trial.

4. SUMMARY

The publication of two phase III clinical trials demonstrating a modest but clinically significant survival advantage with docetaxel-based chemotherapy has firmly established this regimen as the standard of care in HRPC. However, despite these encouraging results, the prognosis of men with HRPC remains poor. The future of prostate cancer treatment may need to build upon the success of docetaxel in HRPC by combining docetaxel with both known and novel chemotherapeutic agents and, as our understanding of the molecular basis of prostate cancer grows, with novel "noncytotoxic" therapies. With the benefits of chemotherapy in HRPC now undisputed, the application of these established and novel therapies to earlier stage disease holds great promise. Research efforts over the last decades have significantly improved both the life expectancy and the quality of life experienced by men with advanced disease. However, further research is needed to make prostate cancer a less debilitating and less lethal disease.

REFERENCES

1. Eisenberger M, Simon R, O'Dwyer PJ, et al. (1985) A reevaluation of nonhormonal cytotoxic chemotherapy in the treatment of prostatic carcinoma. *J Clin Oncol.* **3**, 827–841.
2. Yagoda A, Petrylak D. (1993) Cytotoxic chemotherapy for advanced hormone-resistant prostate cancer. *Cancer.* **71**, 1098–1109.
3. Scher H, Kelly WM, Zhang ZF, et al. (1999) Post-therapy serum prostate-specific antigen level and survival in patients with androgen-independent prostate cancer. *J Natl Cancer Inst.* **91**, 244–251.

4. Vollmer R, Dawson NA, Vogelzang NJ. (1998) The dynamics of prostate specific antigen in hormone refractory prostate carcinoma: an analysis of cancer and leukemia group B study 9181 of megestrol acetate. *Cancer.* **83**, 1989–1994.
5. Crawford E, Pauler DK, Tangen CM, et al. (2004) Three-month change in PSA as a surrogate endpoint for mortality in advanced hormone-refractory prostate cancer: data from Southwest Oncology Group Study 9916. *Proc Am Soc Clin Oncol.* **23**, 382s.
6. Bubley G, Carducci M, Dahut W, et al. (1999) Eligibility and response guidelines for phase II clinical trials in androgen-independent prostate cancer: recommendations from the Prostate-Specific Antigen Working Group. *J Clin Oncol.* **17**, 3461–3467.
7. Moore M, Osoba D, Stockler MR, et al. (1994) Use of palliative end points to evaluate the effects of mitoxantrone and low-dose prednisone in patients with hormonally resistant prostate cancer. *J Clin Oncol.* **12**, 689–694.
8. Tannock IF, Osoba D, Stockler MR, et al. (1996) Chemotherapy with mitoxantrone plus prednisone or prednisone alone for symptomatic hormone-resistant prostate cancer: a Canadian randomized trial with palliative endpoints. *J Clin Oncol.* **14**, 1756–1764.
9. Kantoff P, Halabi S, Conaway M, et al. (1999) Hydrocortisone with or without mitoxantrone in men with hormone-refractory prostate cancer: results of the Cancer and Leukemia Group B 9182 Study. *J Clin Oncol.* **17**, 2506–2513.
10. Pienta K, Smith DC. (2005) Advances in prostate cancer chemotherapy: a new era begins. *CA Cancer J Clin.* **55**, 300–318.
11. Raffo A, Perlman H, Chen MW, et al. (1995) Overexpression of BCL-2–2 protects prostate cancer cells from apoptosis in vitro and confers resistance to androgen depletion in vivo. *Cancer Res.* **55**, 4438–4445.
12. Leung S, Miyake H, Zellweger T, et al. (2001) Synergistic chemosensitization and inhibition of progression to androgen independence by antisense BCL-2–2 oligodeoxynucleotide and paclitaxel in the LNCaP prostate tumor model. *Int J Cancer.* **91**, 846–850.
13. Tannock IF, de Wit MD, Berry WR, et al. (2004) Docetaxel plus prednisone or mitoxantrone plus prednisone for advanced prostate cancer. *NEJM.* **351**, 1502–1512.
14. Petrylak DP, Tangen CM, Hussain MH, et al. (2004) Docetaxel plus prednisone or mitoxantrone plus prednisone for advanced prostate cancer. *NEJM.* **351**, 1513–1520.
15. Hartley-Asp B. (1984) Estramustine-induced mitotic arrest in two human prostatic carcinoma DU 145 and PC-3. *Prostate.* **5**, 93–100.
16. Mareel M, Storme GA, Dragonetti CH, et al. (1988) Antiinvasive activity of estramustine on malignant M04 mouse cells and on DU-145 human prostate carcinoma cells in vitro. *Cancer Res.* **48**, 1842–1849.
17. Hartley-Asp B, Kruse E. (1986) Nuclear protein matrix as a target for estramustine-induced cell death. *Prostate.* **9**, 387–395.
18. Hudes G, Einhorn L, Ross E, et al. (1999) Vinblastine versus vinblastine plus oral estramustine phosphate for patients with hormone-refractory prostate cancer: a Hoosier Oncology Group and Fox Chase Network phase III trial. *J Clin Oncol.* **17**, 3160–3166.
19. Berry W, Hathorn JW, Dakhil SR, et al. (2004) Phase II randomized trial of weekly paclitaxel with or without estramustine phosphate in progressive, metastatic, hormone-refractory prostate cancer. *Clin Prostate Cancer.* **3**, 104–111.
20. Goodin S, Rao KV, Kane M, et al. (2005) A phase II trial of docetaxel and vinorelbine in patients with hormone-refractory prostate cancer. *Cancer Chemother Pharmacol.* **56**, 199–204.
21. Di Lorenzo G, Pizza C, Autorino R, et al. (2004) Weekly docetaxel and vinorelbine (VINDOX) as first line treatment in patients with hormone refractory prostate cancer. *Eur Urol.* **46**, 709–711.
22. Smith D, Esper P, Strawderman M, et al. (1999) Phase II trial of oral estramustine, oral etoposide, and intravenous paclitaxel in hormone-refractory prostate cancer. *J Clin Oncol.* **17**, 1664–1671.

23. Meluch A, Greco FA, Morrisey LH, et al. (2003) Weekly paclitaxel, estramustine phosphate, and oral etoposide in the treatment of hormone-refractory prostate carcinoma: results of a Minnie Pearl Cancer Research Network phase II trial. *Cancer.* **98**, 2192–2198.

24. Oh W, Halabi S, Kelly WK, et al. (2003) A phase II study of estramustine, docetaxel, and carboplatin with granulocyte-colony-stimulating factor support in patients with hormone refractory prostate carcinoma: Cancer and Leukemia Group B 99813. *Cancer.* **98**, 2592–2598.

25. Sternberg C, Whelan P, Hetherington J, et al. (2005) Phase III trial of satraplatin, an oral platinum plus prednisone vs. prednisone alone in patients with hormone-refractory prostate cancer. *Oncology.* **68**, 2–9.

26. Lee F, Borzilleri R, Fairchild CR, et al. (2001) BMS-247550: a novel epothilone analog with a mode of action similar to paclitaxel but possessing superior antitumor efficacy. *Clin Cancer Res.* **7**, 1429–1437.

27. Lee D. (2004) Activity of epothilone B analogues ixabepilone and patupilone in patients with hormone-refractory prostate cancer. *Clin Prostate Cancer.* **3**, 80–82.

28. Galsky M, Small EJ, Oh WK, et al. (2005) Multiinstitutional randomized phase II trail of the epothilone B analog ixabepilone (BMS-247550) with or without estramustine phosphate in patients with progressive metastatic prostate cancer following castration. *J Clin Oncol.* **23**, 1439–1446.

29. Shariat SF, Anwuri VA, Lamb DJ, et al. (2004) Association of preoperative plasma levels of vascular endothelial growth factor and soluble vascular cell adhesion molecule-1 with lymph node status and biochemical progression after radical prostatectomy. *J Clin Oncol.* **22**, 1655–1663.

30. George D, Gockerman JP, Petros W, et al. (2005) A phase I/II study of docetaxel and atrasentan in men with metastatic hormone-refractory prostate cancer (HRPC). *Proc Am Soc Clin Oncol.* **23**, 419s.

31. Picus J, Halabi S, Rini B, et al. (2003) The use of bevacizumab (B) with docetaxel (D) and estramustine (E) in hormone refractory prostate cancer (HRPC): initial results of CALGB 90006. *J Clin Oncol.* **22**, 393s.

32. Dahut W, Gulley JL, Arlen PM, et al. (2004) Randomized phase II trial of docetaxel plus thalidomide in androgen-independent prostate cancer. *J Clin Oncol.* **22**, 2532–2539.

33. Retter A, Ando Y, Price DK, et al. (2005) Follow-up analysis of a randomized phase II study of docetaxel (D) and thalidomide (T) in androgen-independent prostate cancer (AIPC): updated survival data and stratification by CYP2C19 mutation status. *Proc Am Soc Clin Oncol.* **23**, 443s.

34. Carducci MA, Padley RJ, Breul J, et al. (2003) Effect of endothelin-A receptor blockade with atrasentan on tumor progression in men with hormone-refractory prostate cancer: a randomized, phase II, placebo-controlled trial. *J Clin Oncol.* **21**, 679–689.

35. Nelson JB, Hedican SP, George DJ, et al. (1995) Identification of endothelin-1 in the pathophysiology of metastatic adenocarcinoma of the prostate. *Nat Med.* **1**, 944–949.

36. Gohji K, Kitazawa S, Tamada H, et al. (2001) Expression of endothelin receptor A associated with prostate cancer progression. *J Urol.* **165**, 1033–1036.

37. Carducci M, Nelson JB, Saad F, et al. (2004) Effects of atrasentan on disease progression and biological markers in men with metastatic hormone-refractory prostate cancer: phase 3 study. *J Clin Oncol* **22**, abstr 4508 (Post-Meeting Edition).

38. George D, Gockerman JP, Petros W, et al. (2005) A phase I/II study of docetaxel and atrasentan in men with metastatic hormone-refractory prostate cancer (HRPC). *J Clin Oncol.* **23**, 419s.

39. Beer T, Myrthue A. (2004) Calcitriol in cancer treatment: from the lab to the clinic. *Mol Cancer Ther.* **3**, 373–381.

40. Beer T, Hough KM, Garzotto M, et al. (2001) Weekly high-dose calcitriol and docetaxel in advanced prostate cancer. *Semin Oncol.* **28**, 49–55.
41. Hershberger P, Yu WD, Modzelewski RA, et al. (2001) Calcitriol (1,25-dihydroxycholecalciferol) enhances paclitaxel antitumor activity in vitro and in vivo and accelerates paclitaxel-induced apoptosis. *Clin Cancer Res.* **7**, 1043–1051.
42. Moffatt K, Johannes WU, Miller GJ. (1999) 1alpha,25hydroxyvitaminD3 and platinum drugs act synergistically to inhibit the growth of prostate cancer cell lines. *Clin Cancer Res.* **5**, 695–703.
43. Beer TM, Eilers KM, Garzotte M, et al. (2003) Weekly high-dose calcitriol and docetaxel in metastatic androgen-independent prostate cancer. *J Clin Oncol.* **21**, 123–128.
44. Beer TM, Ryan CW, Venner PM, et al. (2005) Interim results from ASCENT: a double-blind randomized study of DN-101 (high-dose calcitriol) plus docetaxel in androgen-independent prostate cancer (AIPC). *Proc Am Soc Clin Oncol.* **23**, 382s (oral presentation).
45. Beekman K, Fleming MT, Scher HI, et al. (2005) Second-line chemotherapy for prostate cancer: patient characteristics and survival. *Clin Prostate Cancer.* **4**, 86–90.
46. Berthold D, Sternberg CN, Tannock IF. (2005) Management of advanced prostate cancer after first-line chemotherapy. *J Clin Oncol.* **23**, 8247–8252.
47. Ohlmann C, Engelmann UH, Heidenreich A. (2005) Second-line chemotherapy with docetaxel for prostate-specific antigen (PSA) relapse in men with hormone-refractory prostate cancer (HRPC) previously treated with docetaxel-based chemotherapy. *Proc Am Soc Clin Oncol.* **23**, 423s.
48. Michels J, Montemurro T, Kollmannsberger C, et al. (2005) First- and second-line chemotherapy with docetaxel or mitoxantrone in patients with hormone-refractory prostate cancer (HRPC): does sequence matter? *Proc Am Soc Clin Oncol.* **23**, 405s.
49. Oh W, Manola J, Babcic V, et al. (2005) Response to second-line chemotherapy in patients with hormone refractory prostate cancer (HRPC) receiving two sequences of mitoxantrone (M) and taxanes (T). *Proc Am Soc Clin Oncol.* **23**, 406s.
50. Oh W, George DJ, Tay MH. (2005) Response to docetaxel/carboplatin in patients with hormone-refractory prostate cancer not responding to taxane-based chemotherapy. *Clin Prostate Cancer.* **4**, 61–64.
51. Rosenberg JE, Galsky MD, Rohs NC, et al. (2006) A retrospective evaluation of second-line chemotherapy response in hormone-refractory prostate carcinoma. *Cancer.* **106**, 58–62.
52. D'Amico AV, Cote K, Loffredo M, et al. (2002) Determinants of prostate-cancer specific survival after radiation therapy for patients with clinically localized prostate cancer. *J Clin Oncol.* **20**, 4567–4573.
53. Schmidt J, Gibbones R, Murphy G. (1996) Adjuvant therapy for clinical localized prostate cancer treated with surgery or irradiation. *Eur Urol.* **29**, 425–433.
54. Shannon C, Smith I. (2003) Is there still a role for neoadjuvant therapy in breast cancer? *Crit Rev Oncol Hemol.* **45**, 77–90.
55. Kuerer H, Newman LA, Smith TL, et al. (1999) Clinical course of breast cancer patients with complete pathologic primary tumor and axillary lymph node response to doxorubicin-based neoadjuvant chemotherapy. *J Clin Oncol.* **17**, 460–469.
56. Soloway MS, Pareek K, Sharifi R, et al. (2002) Neoadjuvant androgen ablation before radical prostatectomy in cT2bNxM0 prostate cancer: 5-year results. *J Urol.* **167**, 112–116.
57. Gleave ME, Goldenberg SL, Chin JL, et al. (2001) Randomized comparative study of 3 versus 8 month neoadjuvant hormonal therapy before radical prostatectomy: biochemical and pathologic effects. *J Urol.* **166**, 500–506.

20 Concluding Remarks
The Future of Prostate Cancer

Leonard G. Gomella, MD,
and Richard K. Valicenti, MD

1. INTRODUCTION

We have witnessed dramatic improvements in the management of localized prostate cancer over the last 20 years. Stage migration, most likely due to screening efforts, combined with improvements in treatment modalities have been largely responsible for these outcomes. Despite the improvements in early detection and the slowly declining death rates, we will face significant challenges in managing this disease in the near future.

Prostate cancer, the most common solid tumor in men and predominantly a disease of older males, will become an increasingly important health care issue. Because of declining death rates in the USA as a result of improvements in conditions such as cardiovascular and smoking-related diseases and

From: *Current Clinical Oncology: Prostate Cancer:*
Signaling Networks, Genetics, and New Treatment Strategies
Edited by: R. G. Pestell and M. T. Nevalainen © Humana Press, Totowa, NJ

the emergence of the "baby boom" generation, we will unfortunately see an increasing incidence of prostate cancer in the coming years. According to US population statistics, about 10% of the US population was over the age of 65 in 2000. By the year 2030, it is expected that these men will comprise approximately 19% of the US population *(1)*.

Over treatment of prostate cancer is already a real concern today. SEER (Surveillance, Epidemiology, and End Results—Program) data suggest that up to 15% of prostate-specific antigen (PSA)-detected prostate cancer in Caucasian men and up to 37% in African-American men might never have presented clinically within the patient's lifetime *(2)*. At the present time, recommendation is that "older" men with significant co-morbidities or a less than 10-year life expectancy be considered for active surveillance if they had a non-aggressive form of prostate cancer. A recent observational study has now questioned this approach, as it suggested a reduced risk of mortality associated with active treatment for low- and intermediate-risk prostate cancer in the elderly Medicare population *(2)*. As noted, this elderly cohort of men will grow dramatically. Our burden going forward will be to more accurately characterize this disease and define the optimum modality for a given patient, be it prevention, active surveillance, or active intervention.

Risk stratification, a new concept in the 1990s, has assisted in optimizing treatments in men, particularly those with high-grade disease *(3)*. The good news is that the use of PSA screening as a tool for early detection appears to have reduced the numbers of men diagnosed with aggressive disease *(4)*. This concept of the multimodality approach to patients with adverse features and at high risk for progression following definitive therapy is now generally considered the standard of care. This multimodality strategy, used extensively in the management of breast cancer, has lead to improved outcomes. Multiple studies have confirmed the benefit of this approach in prostate cancer causing combinations of hormonal therapy, chemotherapy, radiation, and surgery.

At present, standard clinical parameters such as clinical stage, Gleason score, and PSA provide our most reliable predictors of disease outcome. While good, none of these parameters are adequate to completely characterize the disease in an individual patient. Additional markers, be they biochemical, molecular, histologic, or imaging based are needed to further define the aggressiveness of this common disease.

Early detection using PSA, multiple biopsies, and improved treatments currently represent the most useful strategies to decrease prostate cancer (PC) mortality. These parameters, coupled with improved treatments, have been reliable markers of disease and have been key components that define our treatment approaches to prostate cancer. What does the future of prostate cancer diagnosis and treatment hold to further reduce the burden of prostate cancer?

2. PROSTATE CANCER SCREENING

While the temporal evidence strongly suggests that screening for prostate cancer has been beneficial, screening remains controversial. Many large organizations such as the American Cancer Society and the American Urologic Association endorse screening in men with a life expectancy greater than 10 years, but this is not uniformly accepted. The significant stage and grade migration seen over the last 10 years is most likely due to PSA-based screening. However until two large prospective screening trials are completed in the next few years, screening for prostate cancer will likely remain controversial *(5)*.

3. THE FUTURE OF PROSTATE CANCER IMAGING

Accurate staging of prostate cancer at initial diagnosis, as well as accurate staging and tumor localization with biochemical recurrence, is inaccurate with current imaging techniques. Newer modalities are being investigated to accurately identify patients with prostate cancer at different stages of disease that will replace standard bone scans, CT, and MRI in the future.

Enhanced transrectal ultrasonography modalities, including ultrasound contrast agents, color and power Doppler, and elastrography, have demonstrated incremental benefit when combined with standard gray-scale ultrasonography to accurately target and diagnose prostate cancer *(6)*. Endorectal MRI, with contrast enhancement and spectroscopic imaging, shows promise in the initial staging of prostate cancer and help with treatment planning with either surgical or radiotherapeutic approaches. The use of PET scan for prostate cancer remains to be defined but may help delineate the site of recurrence with biochemical failure after local therapy. Combining molecular imaging modalities and the increased use of image-guided therapies will be the hallmark of the future of prostate cancer imaging.

4. MOLECULAR CHARACTERIZATION AND NEW TREATMENTS

As many authors have described in this book, efforts are underway to identify and characterize molecular genetic alterations in all stages of PC and integrate these findings into clinical practice *(7)*. It will be important to understand the timing of these alterations to identify evolution of the tumor as well as the prognosis and potential of the specific targeted treatment therapies. The so-called molecular profiling studies are needed that provide broad-spectrum genomic and proteomic data that could prove useful for the discovery of these new drugs and biomarkers.

The molecular genetics of PC are complex, and various different factors interact in the development of the disease. As has been discussed in this book, prostate cancer at presentation has one of the most diverse potential natural histories of any solid tumor. As in other tumors, the development of PC is a multistep process through a series of morphologically distinct lesions initiated by genetic and epigenetic changes. The extraordinarily high autopsy incidence of prostate cancer in older men suggests that many complex factors interact from the initiation to the progression of the disease. Prostate cancer has been more difficult to study than other solid tumors due in part to the fact that in a single radical prostatectomy specimen, multiple and heterogeneous transformed clones may be present, of which only one may give rise to the clinically significant cancer *(8)*.

At present , it appears that the precursor to PC appears to be the high-grade prostatic intraepithelial neoplasia (PIN). Proliferative inflammatory atrophy (PIA) has been proposed as a precursor of PIN. Both these lesions are characterized with high proliferative activity, and some chromosomal and genetic changes that are present are also in invasive PC *(8–10)*. Chemoprevention of prostate cancer is of high priority and identification of pathways other than hormonally activated will be a future challenge. PIN as a possible marker lesion does provide an opportunity for prostate cancer chemoprevention.

Inheritable and somatic genetic alterations including deregulation of oncogenes, tumor suppressor genes, and metastasis suppressor genes have been proposed to be involved in the progression from localized androgen-dependent states into androgen-independent and metastatic forms. As the prostate is a hormonally sensitive organ, androgens, estrogens, and regulated growth factors are intimately involved in carcinogenesis *(10)*. The role of genes involved in stress response, detoxification, and fatty acid metabolism is being clarified and represent completely new areas of diagnostic, therapeutic, and prevention targets *(12–15)*.

The signaling and genetic concepts reviewed in this book will be rapidly translated into patient care. In other cancers such as renal cell carcinoma, similar basic science discoveries have been rapidly accepted. As just one example, targeted therapies using the tyrosine kinase inhibitors sorafenib and sunitinib have been approved in patients with advanced disease and are under study as an early adjuvant therapy in renal cell carcinoma *(16)*. The approvals of these and other pending new drugs that block specific molecular targets are just one example of the new direction in all future cancer therapies including prostate cancer.

5. FUTURE SURGICAL APPROACHES

In some respects, the future of prostate cancer care is already here and thriving. After the years of widespread acceptance of nerve-sparing radical prostatectomy, laparoscopic prostatectomy became an accepted procedure in the late 1990s. The latest in minimally invasive radical prostatectomy was rapidly eclipsed by robotic prostatectomy *(17)*. The early data suggests that the oncologic outcomes are equivalent for patient with low-risk prostate cancer, and there may be advantages in blood loss and overall recovery. However, the advent of robotic prostatectomy has come a major paradigm shift, one that is likely to repeat itself in the new millennium. While robotic prostatectomy is appealing to patients for its cutting-edge technology it has been largely driven by marketing and consumer demand before there was a consensus amongst experts in the field of prostate cancer. As surgeons gain experience with merging this technology with oncologic and patient outcome analysis and are able to critique the benefits and limitations, these more technologic approaches to prostate cancer treatment will likely become more commonplace.

High-intensity-focused ultrasound (HIFU) is another high-tech approach to prostate cancer ablation that has wide acceptance worldwide but is not yet approved in the USA *(18)*. There are several trials ongoing that will likely make this technique available in the USA in the next 2–3 years. Cryotherapy remains an area of continuing refinement with improvements in delivery and image guidance expected to continue *(19)*.

6. FUTURE OF RADIATION THERAPY

Over the last 20 years, several technological advances have enhanced the planning and delivery of radiation therapy (RT), thereby allowing for dose escalation to the whole prostate, with reduced toxicity and better disease-free survival. These benefits have now been demonstrated in large multi-institutional databases as well as multiple clinical trials and have resulted from a wide range of RT delivery systems including proton beam therapy *(20–24)*. The current "standard" dose of primary RT for prostate cancer is now between 75 to 80 Gy, which is 10–20% higher than what can be safely delivered with conventional RT techniques. To translate these therapeutic benefits into routine clinical imaging of the prostate gland, verification of its position is required during RT.

To accomplish this goal, image-guided radiation therapy (IGRT) has been developed, to be used in conjunction with external beam RT and prostate brachytherapy (PBT). Successful execution of IGRT depends on a detailed understanding of a multiplicity of imaging modalities, the guidelines for target definition, and the methods of real-time monitoring of prostate gland motion. A review of published data shows that, with accurate imaging, accurate target volume delineation, and target positioning, tumor control improves to 90% without any notable increase in acute or late toxicity *(25–31)*. Significant toxicities associated with the critical normal tissues such as the rectum, small intestine, bladder, urethra, and penile base limit the potential of dose escalation, particularly when concurrent hormonal therapy is given. Beyond simple target localization and verification, there are active investigations to use fiducial dosimeters to monitor on-line for the first time actual radiation dose delivery.

The development of IGRT created a new era of RT for prostate cancer. Real-time electronic portal imaging with intraprostatic fiducial markers has been shown to improve targeting accuracy in 25–50% of the cases, and the commercial ultrasound-based system–BAT (developed by the Nomos Corporation, Swickley, PA)—is used to track prostate movement by the acquisition of transverse and sagittal suprapubic ultrasound images. The next generation of target volume localization and verification involves megavoltage CT and cone beam CT. The implementation of IGRT has opened the door to achieve higher than 85% disease control by focused, hypofractionated, high biologically equivalent dose of RT with IMRT or PBT. Proton beam RT has a proven benefit for prostate cancer due to its physical characteristics, which lead to high and conformal dose distribution in the tumor while reducing the entrance dose and stopping the beam distal to the target. These novel approaches rely heavily on IGRT technology and were considered experimental just a couple of years ago. However, because of the potential significant improvement of clinical outcome and accumulating clinical data, they are quickly becoming standard treatment for prostate cancer at cancer centers and are moving into community practice. While proton beam therapy facilities are very costly, it is hoped that smaller and more cost-efficient units will allow the technology to be more widely applied.

7. PROSTATE CANCER PRIMARY PREVENTION

The mandate from the NCI is to expand opportunities for cancer prevention by 2015. As part of the NCI's 2015 Challenge Vision for the Nation's Cancer Program, the NCI will "accelerate the discovery, development, and delivery of cancer prevention interventions by investing in research focused on systems biology, behavior modifications, environmental and policy influences, medical and nutritional approaches, and training and education for research and health professionals" (32). Prostate cancer is a high governmental priority in this area. Prostate cancer provides a useful target for chemoprevention because of its high prevalence, long latency time, hormonal dependency, identifiable precursor lesions, and the presence of a serum marker, namely PSA. For these reasons, many NCI and industry-sponsored trials are underway to identify effective chemopreventive agents.

The National Cancer Institute has sponsored two large trials to investigate prostate cancer chemoprevention: the Prostate Cancer Prevention Trial (PCPT) and the Selenium and Vitamin E Cancer Prevention Trial (SELECT) (33). The latter evaluates treatment with selenium, vitamin E, and both supplements together. The PCPT used the 5-α-reductase inhibitor finasteride (Proscar, Merck, Rahway, NJ, USA) to prevent prostate cancer in a prospective randomized trial. The PCPT demonstrated a nearly 25% reduction in the prevalence of prostate cancer compared to placebo. The Reduction by Dutas-teride of Prostate Cancer Events (REDUCE) trial, launched in 2002, is a 4-year, placebo-controlled trial of the second-generation 5-α-reductase inhibitor dutasteride (Avodart, GSK Middlesex, United Kingdom), 0.5 mg/day, in decreasing prostate cancer in high-risk patients *(34)*. Other trials are investigating cyclooxygenase-2 (COX-2) inhibitors with the results of these and the SELECT and REDUCE trials expected in the next 2–3 years.

The first cancer-specific vaccine was made available for general use in 2006 *(35)*. This targets several members of the HPV family of viruses that are linked to cervical cancer. While no virus has been consistently identified in the pathogenesis of prostate cancer, other non-viral-based vaccine strategies are rapidly moving toward approval *(36)*. These immunotherapeutic approaches will be initially directed at advanced disease but will likely find a place for study in the primary prevention of prostate cancer.

8. QUALITY OF LIFE ISSUES IN PROSTATE CANCER

While development treatment modalities have traditionally taken center stage, issues surrounding the quality of life of these different treatments is becoming increasingly recognized as an important issue *(37)*. These patient-centric issues are particularly important to quantify in a disease such as prostate cancer where so many different treatment approaches are available from active surveillance to aggressive and potentially life-saving multimodality therapy. As noted, RT improvements have greatly enhanced the safety profile of this modality. There is active research in areas such as penile rehabilitation, tissue engineering, and nerve grafting following definitive therapy for prostate cancer as well as strategies to reduce or eliminate side effects related to systemic approaches such as hormonal therapy *(38,39)*.

9. SUMMARY

In 1991, a panel convened by the NCI stated "Because prostate cancer rates are increasing and the United States population is aging, the death rate will increase again by 50% in the next 15 years." *(40)*. The report also cited 28,000 prostate cancer deaths per year. Although the male population has aged, it is reassuring that these dire warnings have not been realized. Our 2007 estimate for prostate cancer deaths in the US is 27,720; approximately the same as the 28,000 number from 1991 *(41)*.

There was a gradual increase in prostate cancer deaths that peaked in the early 1990s, prompting concerns that PSA screening had lead to a prostate cancer epidemic and over treatment of many men. Despite the dramatic changes in the rates of prostate cancer diagnosis, reductions in the death rate from prostate cancer have been noted, with mortality from prostate cancer in the USA declining 2.6% a year from 1990 to 1998 *(41)*. We recognize that many men are currently over treated and rely upon investigations noted in this book and elsewhere to provide the markers of aggressiveness to guide therapy in the future. However, a key question is will the increase in the overall health and length of life of males in the future make prostate cancer the increasing clinical problem we fear today? Or, will we find the unexpected advances that changed our colleagues 1991 predictions change our future predictions as well?

The authors of this 1991 report had no idea of the impact that PSA screening would have on this disease. Since the mid 1980s, the average index cancer found at radical prostatectomy decreased by over 50%, from 5.3 to 2.4 cc due almost exclusively to PSA screening *(42)*. However, the golden age of the believers of PSA screening may be over. As these authors note, "today the majority of PSA measured in men with prostate cancer is due to the BPH component and not due to cancer." The recently completed PCPT trial also raised concern when many men with low PSA levels (<4.0 ng/ml) were found to have prostate cancer, with many of them noted to be aggressive *(43)*. PSA is and continues to be a key component in our ability to detect the disease, guide treatment decisions, and follow up for recurrence. The need for new markers is clearly at hand and a high priority for the coming years, not only for screening but for the identification of the clinically significant cancers considered by many to be the "holy grail" of prostate cancer research today.

These slow and steady collective advances in this disease have been beneficial although many questions and controversies remain. The "multidisciplinary" approach to prostate cancer in both scientific investigation and patient care represents the new approach to this disease. Our colleagues in diverse fields of epidemiology, sociology, nutrition, molecular biology, cellular biology, radiology, urologic surgery, radiation oncology, and medical oncology are to be commended and credited with this progress. Our patients and their families are now also considered to be a part of this multidisciplinary team.

The new understanding about prostate cancer by patients is due to the increased profile of prostate cancer in the media and unprecedented access to medical information. Since the first book specifically dedicated to prostate cancer was written in 1993, dozens more have followed as the public seeks more and more information on prostate cancer *(44)*. Today over 16 million web sites contain some type of information on prostate cancer, some, unfortunately, of questionable value. Prominent statesman, entertainers, and sports figures have been willing to discuss this disease publicly over the last 15 years, something never seen before. The leadership of individuals such as Michael Milken, a well-known public advocate for prostate cancer research and education, who penned the introduction to this book, has done much to bring a more organized approach to research in this field. We look forward to the next 15 years of developments and what the latest in scientific and technologic advances will bring to further eradicate the morbidity and mortality of this disease.

REFERENCES

1. http://www.census.gov/ipc/www/usinterimproj/natprojtab02a.pdf (accessed February 17, 2007).
2. Wong YN, Mitra N, Hudes G, Localio R, Schwartz JS, Wan F, Montagnet C, Armstrong K. Survival associated with treatment vs. observation of localized prostate cancer in elderly men. *JAMA* 2006;296(22):2683–93. Erratum in *JAMA* 2007;297(1):42.
3. Roach M III, Weinberg V, Nash M, Sandler HM, McLaughlin PW, Kattan MW. Defining high risk prostate cancer with risk groups and nomograms: implications for designing clinical trials. *J Urol* 2006;176(6 Pt 2):S16–20.
4. Cooperberg MR, Moul JW, Carroll PR. The changing face of prostate cancer. *J Clin Oncol* 2005;23(32):8146–51.
5. Ilic D, O'connor D, Green S, Wilt T. Screening for prostate cancer: a Cochrane systematic review. *Cancer Causes Control* 2007;18:279–85.
6. Trabulsi EJ, Merriam W, Gomella LG. New imaging techniques in prostate cancer. *Curr Urol Rep.* 2006;7(3):175–80.
7. Vecchione A, Gottardo F, Gomella LG, Wildemore B, Fassan M, Bragantini E, Pagano F, Baffa R. Molecular genetics of prostate cancer: clinical translational opportunities. *J Exp Clin Cancer Res* 2007;26(1):515–27.
8. Bostwick DG, Shan A, Qian J, Darson M, Maihle NJ, Jenkins RB, Cheng L. Independent origin of multiple foci of prostatic intraepithelial neoplasia: comparison with matched foci of prostate carcinoma. *Cancer* 1998;83:1995–2002.
9. De Marzo AM, Nelson WG, Isaacs WB, Epstein JI. Pathological and molecular aspects of prostate cancer. *Lancet* 361:955–64, 2003
10. Kopper L., Tìmàr J.: Genomics of prostate cancer: is there anything to "translate"? *Pathol Oncol Res* 2005;11:197–203.
11. Mimeault M, Batra SK. Recent advances on multiple tumorigenic cascades involved in prostatic cancer progression and targeting therapies. *Carcinogenesis* 2006;27:1–22,.
12. Harden SV, Guo Z, Epstein JI, Sidransky D. Quantitative GSTP1 methylation clearly distinguishes benign prostatic tissue and limited prostate adenocarcinoma. *J Urol* 2003;169:1138–42.
13. Bastian PJ, Palapattu GS, Lin X, et al. Preoperative serum DNAGSTP1 CpG island hypermethylation and the risk of early prostate-specific antigen recurrence following radical prostatectomy. *Cancer Res* 2005;65:4762–8.
14. Van de Sande T, Roskams T, Lerut E, et al. High-level expression of fatty acid synthase in human prostate cancer tissues is linked to activation and nuclear localization of Akt/PKB. *J Pathol* 2005;206:214–9.
15. Mubiru JN, Valente AJ, Troyer DA. A variant of the alphamethyl-acyl-CoA racemase gene created by a deletion in exon 5 and its expression in prostate cancer. *Prostate* 2005;65:117–23.
16. Motzer RJ, Bukowski RM. Targeted therapy for metastatic renal cell carcinoma. *J Clin Oncol* 2006;24(35):5601–8.
17. Patel VR, Chammas MF Jr, Shah S. Robotic assisted laparoscopic radical prostatectomy: a review of the current state of affairs. *Int J Clin Pract* 2007;61(2):309–14.

18. Rewcastle JC. High intensity focused ultrasound for prostate cancer: a review of the scientific foundation, technology and clinical outcomes. *Technol Cancer Res Treat* 2006;5(6):619–25.
19. Mouraviev V, Polascik TJ. Update on cryotherapy for prostate cancer in 2006. *Curr Opin Urol* 2006;16(3):152–6.
20. Zietman AL, DeSilvio ML, Slater JD, et al. Comparison of conventional-dose vs. high-dose conformal radiation therapy in clinically localized adenocarcinoma of the prostate. JAMA 2005;294:1233–9.
21. Zelefsky MJ, Kuks Z, Hunt M, et al. High-dose radiation delivered by intensity modulated conformal radiotherapy improves the outcome of localized prostate cancer. *J Urol* 2001;166:1132–6.
22. Shipley WU, Verhey U, Munzenrider JE, et al. Advanced prostate cancer:the results of a randomized comparative trial of high-dose irradiation boosting with conformal protons compared with conventional dose irradiation using photons alone. *Int J Radiat Oncol Biol Phys* 1995;32:3–12.
23. Pollack A, Zagars GK, Smith LG, et al. Preliminary results of a randomized radiotherapy dose-escalation study comparing 70 Gy with 78 Gy for prostate cancer. *J Clin Oncol* 2000;18:3904–11.
24. Ryu JK, Winter K, Michalski JM, et al. Interim report of toxicity from three-dimensional radiotherapy for prostate cancer with RTOG 99406 dose level III (79.2 Gy). *Int J Radiat Oncol Biol Phys* 2002;54:1036–46.
25. Litzenberg DW, Balter JM, Lam KL, et al. Retrospective analysis of prostate cancer patients with implanted gold markers using off-line and adaptive therapy protocols. *Int J Radiat Oncol Biol Phys* 2005;63:123–33.
26. Willoughby TR, Kupelian PA, Pouliot J, et al, Target localization and real-time tracking using the Calypso 4D localization system in patients with localized prostate cancer. *Int J Radiat Oncol Biol Phys* 2006;65:528–34.
27. Balter JM, Wright JN, Newell LJ, et al. Accuracy of a wireless localization system for radiotherapy. *Int J Radiat Oncol Biol Phys* 2005;61:933–7.
28. Langen KM, Jones DT. Organ motion and its management. *Int J Radiat Oncol Biol Phys* 2005;50:265–78.
29. Lattanzi J, McNeeley S, Hanlon A, et al. Ultrasound-based stereotactic guidance of precision conformal external beam radiation therapy in clinically localized prostate cancer. *Urology* 2000;55:73–8.
30. Bel A, van Herk M, Bartelink H, et al. A verification procedure to improve patient set-up accuracy using portal images. *Radiother Oncol* 1993;29:253–60.
31. Yan D, Lockman D, Brabbins D, et al. An off-line strategy for constructing a patient-specific planning target volume in adaptive treatment process for prostate cancer. *Int J Radiat Oncol Biol Phys* 2000;48:289–02.
32. http://www.cancer.gov/aboutnci/2015/(accessed Feb 15, 2007).
33. Brand TC, Canby-Hagino ED, Pratap Kumar A, Ghosh R, Leach RJ, Thompson IM. Chemoprevention of prostate cancer. *Hematol Oncol Clin North Am* 2006;20(4):831–43.
34. Gomella LG. Chemoprevention of prostate cancer using dutasteride: the REDUCE trial. Review. *Curr Opin Urol* 2005;15(1):29–32.
35. Leggatt GR, Frazer IH. HPV vaccines: the beginning of the end for cervical cancer. *Curr Opin Immunol* 2007;19:232–8.
36. Sowery RD, So AI, Gleave ME. Therapeutic options in advanced prostate cancer: present and future. *Curr Urol Rep* 2007;8(1):53–9.
37. Henderson A, Andreyev HJ, Stephens R, Dearnaley D. Patient and physician reporting of symptoms and health-related quality of life in trials of treatment for early prostate cancer: considerations for future studies. *Clin Oncol (R Coll Radiol)* 2006;18(10):735–43.

38. Dall'era JE, Mills JN, Koul HK, Meacham RB. Penile rehabilitation after radical prostatectomy: important therapy or wishful thinking? *Rev Urol* 2006;8(4):209–15.
39. Gomella LG. Contemporary use of hormonal therapy in prostate cancer: managing complications and addressing quality-of-life issues. *BJU Int* 2007;99 (Suppl 1):25–9.
40. Chiarodo A. National Cancer Institute Roundtable on Prostate Cancer: Future Research Directions Cancer Research. *Cancer Res*1991;51, 2498–505.
41. http://www.cancer.org/downloads/STT/CAFF2007PWSecured.pdf (accessed February 15, 2007).
42. Stamey TA, Caldwell M, McNeal JE, Nolley R, Hemenez M, Downs J. The prostate specific antigen era in the United States is over for prostate cancer: what happened in the last 20 years? *J Urol* 2004;172(4 Pt 1):1297–301.
43. Thompson IM, Pauler DK, Goodman PJ, et al. Prevalence of prostate cancer among men with a prostate-specific antigen level < or = 4.0 ng per milliliter. *N Engl J Med* 2004;350(22):2239–46. Erratum in: *N Engl J Med* 2004;351(14):1470.
44. Gomella LG, Fried J. *Recovering from Prostate Cancer*, Harper Collins, New York, 1993.

Index

Printed in the United States of America